NO POSTAGE
NECESSARY
IF MAILED
IN THE
UNITED STATES

BUSINESS REPLY MAIL
FIRST CLASS PERMIT NO. 358
PACIFIC GROVE, CALIFORNIA 93950-5098

POSTAGE WILL BE PAID BY:

Brooks/Cole Publishing Company
511 Forest Lodge Road
Pacific Grove, CA 93950-5098

Solutions Manual for

Organic Chemistry

Ralph J. Fessenden
Joan S. Fessenden
University of Montana

Brooks/Cole Publishing Company
Pacific Grove, California

ISBN 0-534-12254-X

Printed in the United States of America

10 9 8 7 6 5 4 3 2 1

To the Student

This *Solutions Manual* contains the answers to the end-of-chapter problems in Fessenden and Fessenden, *Organic Chemistry, 4th Edition*. (Answers to the problems within the chapters are at the end of the text.)

Many of the answers here contain explanations of *how* the problems are solved. Answers for other, more routine, problems are presented without explanation.

The chapter-end problems are organized into two main groups. The first group is presented in order of the topics in the text. The Additional Problems at the end of each problem set mix the subject material presented in the chapter and, in some cases, draw upon material from previous chapters. Both sets of problems contain both drill problems and thought problems.

Be sure to answer the problems yourself before checking the solutions in this manual or before referring to the text. If your answers are correct, you have a reasonable mastery of the material. If you have several incorrect answers, you need additional study.

The *Study Guide* that accompanies the text contains hints for studying organic chemistry. The *Study Guide* also contains sets of additional drill problems and their answers.

Ralph J. Fessenden
Joan S. Fessenden

Contents

Chapter 1

Atoms and Molecules - A Review

1.19 (a) $1s^2$ $2s^2$ $2p^2$ (b) $1s^2$ $2s^2$ $2p^6$ $3s^2$ $3p^2$

(c) $1s^2$ $2s^2$ $3p^6$ $3s^2$ $3p^3$ (d) $1s^2$ $2s^2$ $2p^6$ $3s^2$ $3p^4$

1.20 sum of the exponents = number of e^- = atomic number

(a) $1s^2$ $2s^2$ $2p^6$ $3s^1$ $2 + 2 + 6 + 1 = 11$, the atomic number of Na

(b) Cl (c) Ne

1.21 (b) and (c), because each has the same number of electrons in its outer shell.

1.22 90°, which is the angle between *p* orbitals

1.23

```
      H H H H              H H H                       H
      ¨ ¨ ¨ ¨              ¨ ¨ ¨                       ¨
(a) H:C:C:C:C:H      (b) H:C:C:C:H          (c)      H:C:H
      ¨ ¨ ¨ ¨              ¨   ¨                        ¨
      H H H H              H ¨ H                 H H  ¨  H
                                                 ¨ ¨     ¨
                           H:C:H               H:C:C:C:C:H
                             ¨                   ¨   ¨ ¨
                             H                   H ¨ H H
                                                  :Cl:
                             ¨                     ¨
                            :O:
                            .  .
      ¨ ¨
(d) H:O:O:H          (e) H:C  :  C:H
      ¨ ¨                  ¨     ¨
                           H     H
```

1.24 (a)

for C:	4 - 4 - 0 = 0	for S:	6 - 4 - 0 = +2
for each O:	6 - 1 - 6 = -1	for Cl:	7 - 1 - 6 = 0

(b) zero for each atom (c) each C, zero; O, zero; right-hand O, -1

(d) each C, zero; S, +1; O, -1

(e) for each C in a CH_3 group: 4 - 4 - 0 = 0
for the C in the center: 4 - 3 - 0 = +1

(f) C, zero; O, +1

1.25 (a)

```
       H   H   H   H
       |   |   |   |
   H — C — C — C — C — H
       |   |   |   |
       H   |  :Br: H
           |   ..
       H — C — H
           |
           H
```

(b)

```
                                H
                                |
       H   H   :O:          H — C — H
       |   |   ||               |            H
   H — C — C — C — N̈ — C — C = C /
       |   |       |        |       |     \
       H   |       H        H       H      H
           |
       H — C — H
           |
           H
```

(c)

```
                 H
                 |
             H — C — H
       H         |           H
       |         |           |
   H — C — C  =  C — C ≡ C — C — H
       |   |                 |
       H   |                 H
       H — C — H
           |
           H
```

1.26 (a) $CH_3\ddot{\underset{\cdot\cdot}{O}}H$ (b) $CH_3CH_2\ddot{N}H_2$ (c) $(CH_3)_2\ddot{N}H$ or $CH_3\ddot{N}HCH_3$

1.27 All are $C_4H_{10}O$.

1.28 (a)

```
       H     H     H     H
       |     |     |     |
   H — C  =  C  —  C  =  C  —  C ≡ N:
```

(b)

```
       H   :O:   H
       |    ||   |
   H — C —  C  — C — H
       |         |
       H         H
```

(c)

```
       H   :O:
       |    ||    ..
   H — C —  C  —  O — H
       |          ..
       H
```

(d)

```
      :O:   H   :O:
       ||   |    ||
   H — C  — C  — C — H
            |
            H
```

1.29 (a) $CH_3\ddot{N}H_2$ (b) $(CH_3)_3N{:}$ (c) none (d) $CH_3\ddot{\underset{\cdot\cdot}{O}}H$

(e) $(CH_3)_3C\ddot{\underset{\cdot\cdot}{O}}H$ (f) none (g) $H_2C{=}\ddot{\underset{\cdot\cdot}{O}}$

1.30 (a) 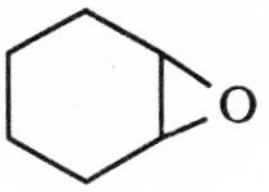 O (b) 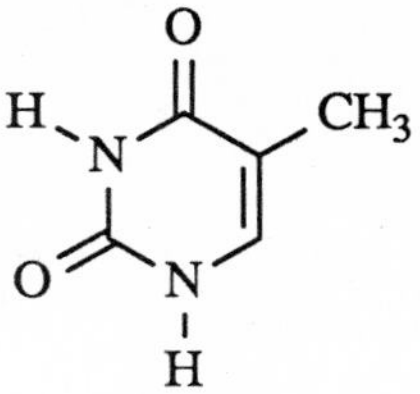 O, H, N, CH_3, O, N, H (c) 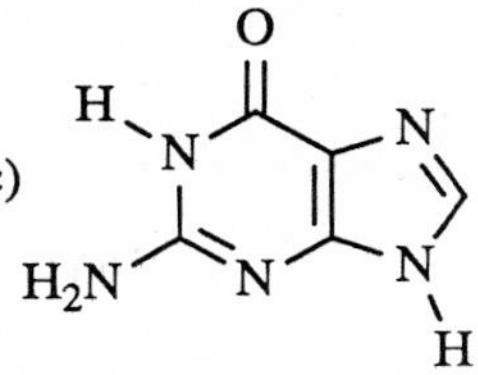 O, H, N, N, H_2N, N, N, H

(d)

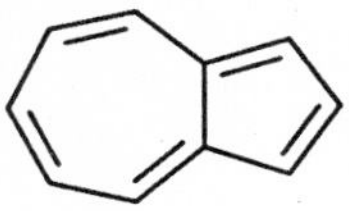

1.31 (a) 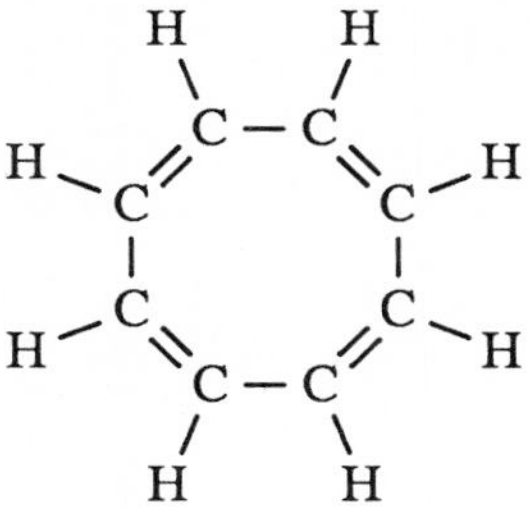 (b) 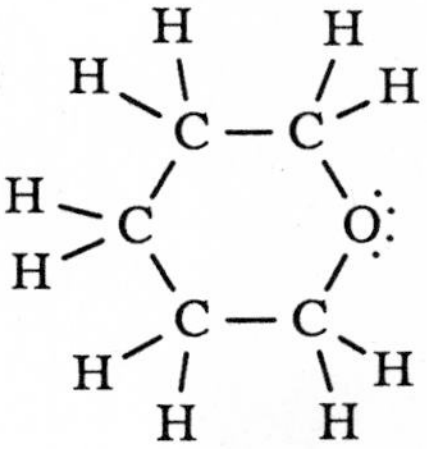 (c) 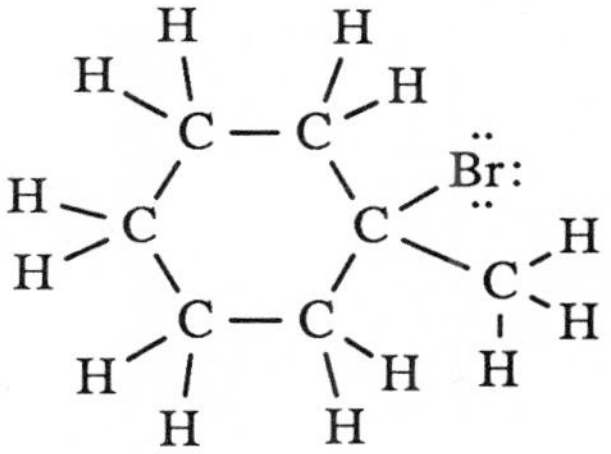 Br: (d)

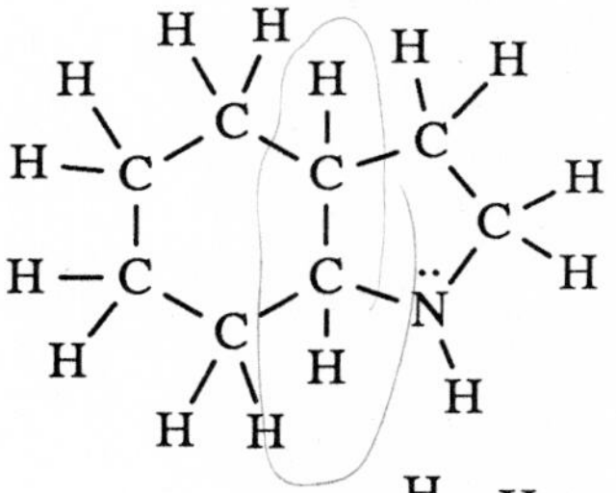

1.32 (a)  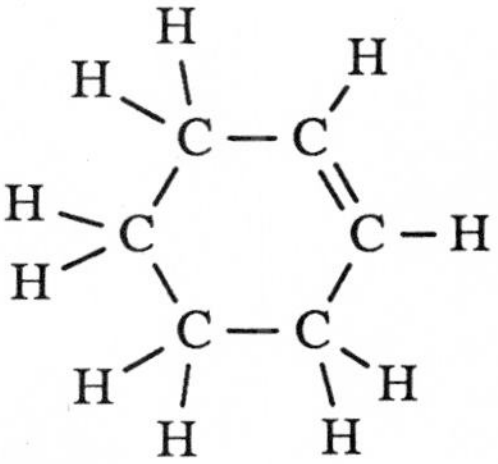 (b) 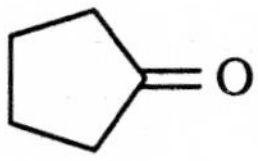 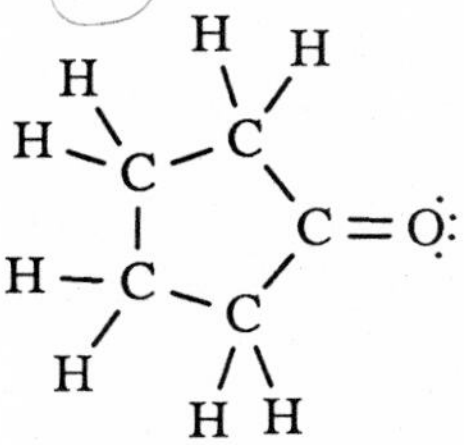 O, C=O:

A formula with the double bond or carbonyl group in a different position represents the same compound and is also correct. For example:

or

O

				ΔH	
1.33 (a)	F - F	→	2 F·	+ 37	the most exothermic--more exothermic than chlorination by 77.5 kcal/mole
	H_3C-H	→	CH_3· + H·	+104	
	CH_3· + F·	→	CH_3F	-108	
	H· + F·	→	HF	-135	
				-102 kcal/mol	
(b)	Cl-Cl	→	2 Cl·	+ 58	
	CH_3-H	→	CH_3· + H·	+104	
	CH_3· + Cl·	→	CH_3-Cl	- 83.5	
	H· + Cl·	→	HCl	-103	
				- 24.5 kcal/mol	
(c)	Br-Br	→	2 Br ·	+ 46	
	CH_3-H	→	CH_3· + H·	+104	
	CH_3· + Br ·	→	CH_3Br	- 70	
	H· + Br ·	→	HBr	- 87	
				- 7 kcal/mol	
(d)	I-I	→	2 I·	+ 36	
	CH_3-H	→	CH_3· + H·	+104	
	CH_3· + I·	→	CH_3I	- 56	
	H· + I·	→	HI	- 71	
				+ 13 kcal/mol	

1.34

	homolytic		heterolytic	
(a)	CH_3CH_2—$\ddot{C}l$: →	$CH_3\dot{C}H_2$ + :$\ddot{C}l$·	CH_3CH_2—$\ddot{C}l$: →	$CH_3\overset{+}{C}H_2$ + :$\ddot{C}l$:$^-$
(b)	H—$\ddot{O}$H →	H· + ·$\ddot{O}$H	H—$\ddot{O}$H →	H^+ + $^-$:$\ddot{O}$H
(c)	H—$\ddot{N}H_2$ →	H· + ·$\ddot{N}H_2$	H—$\ddot{N}H_2$ →	H^+ + $^-$:$\ddot{N}H_2$
(d)	CH_3—$\ddot{O}$H →	CH_3· + ·$\ddot{O}$H	CH_3—$\ddot{O}$H →	CH_3^+ + $^-$:$\ddot{O}$H
(e)	$CH_3\ddot{O}$—H →	$CH_3\ddot{O}$· + ·H	$CH_3\ddot{O}$—H →	$CH_3\ddot{O}$:$^-$ + H^+

In heterolytic cleavage, the bonding electrons always go with the more electronegative atom.

1.35 (a) $\overset{\delta-}{C}$—$\overset{\delta+}{Mg}$ (b) $\overset{\delta+}{C}$—$\overset{\delta-}{Br}$ (c) $\overset{\delta+}{C}$—$\overset{\delta-}{O}$ (d) $\overset{\delta+}{C}$—$\overset{\delta-}{Cl}$ (e) $\overset{\delta-}{C}$—$\overset{\delta+}{H}$ (f) $\overset{\delta-}{C}$—$\overset{\delta+}{B}$

The direction of the dipole is determined by comparison of the electronegativities of the bonded atoms.

1.36 (a) $\overset{\delta+}{CH_3}$—$\overset{\delta-}{\text{(O)}}$—$\overset{\delta+}{H}$ (b) $CH_3\overset{\delta-}{\overset{\text{(O)}}{\overset{\|}{\underset{\delta+}{C}}}}CH_3$ (c) $\overset{\delta-}{\text{(F)}}$—$\overset{\delta+}{CH_2}CH_2$—OH

(d) $(CH_3)_2NCH_2\overset{\delta+}{CH_2}$—$\overset{\delta-}{\text{(Ö)}}$—$\overset{\delta+}{H}$

In (a) and (d), the most electronegative atom is bonded to two other atoms; therefore, it is necessary to show the polarity of both its bonds. In (c), the O is also electronegative, but not as electronegative as F.

1.37 (a) $CH_3CH_2CH_3$, $CH_3CH_2CH_2NH_2$, $CH_3CH_2CH_2OH$

(b) $CH_3CH_2CH_2I$, $CH_3CH_2CH_2Br$, $CH_3CH_2CH_2Cl$

The O-H and C-Cl bonds have the greatest electronegativity difference in each series.

1.38 (a) $(CH_3)_2\underset{H}{N}$:······$H\ddot{N}(CH_3)_2$ (f) H—O⋮$CH_3\ddot{O}$—CH_2—CH_2 (intramolecular: H—O—CH_2—CH_2—$\ddot{O}CH_3$ with H···O) $CH_3\ddot{O}CH_2CH_2OH$ ⋮ $HOCH_2CH_2OCH_3$

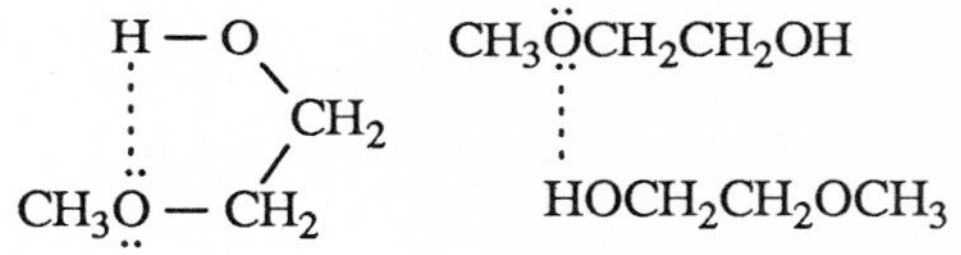

$CH_3OCH_2CH_2\underset{H}{\ddot{O}}$:······$HOCH_2CH_2OCH_3$

The compounds in (b) - (e) cannot undergo hydrogen bonding with other molecules of their own kind because they do not contain H bonded to O, N, or F.

1.39 (a) $(CH_3)_2\underset{H}{N}$:······$H\ddot{N}(CH_3)_2$ (b) $(CH_3)_2\ddot{N}H$······$:\ddot{O}H_2$

(c) $(CH_3)_2\overset{H}{N}$:······$\overset{H}{H\ddot{O}}$: (d) $H_2\ddot{O}$:······$\overset{H}{H\ddot{O}}$:

Nitrogen is less electronegative than oxygen; therefore, its unshared electrons are more loosely held and more available for hydrogen bonding. The OH bond is very polar (and the H is more positive) because of the high electronegativity of oxygen. For these two reasons, the N---HO hydrogen bond in (c) is the strongest hydrogen bond in the solution.

1.40 (a) CH_3I because an I atom has a greater mass than does a Cl atom.

(b) $CH_3CH_2CH_2OH$ because it can form hydrogen bonds while $CH_3CH_2CH_2Cl$ cannot.

(c) $CH_3CH_2CH_2CH_2OH$ because longer-chain molecules can form stronger van der Waals attractions. Also, hydrogen bonding is weaker in $(CH_3)_3COH$ because of the bulky groups near OH.

(d) $HOCH_2CH_2OH$ because its molecules can form twice as many hydrogen bonds.

1.41 (a) $CH_3OH + {}^-OH$ (b) $CH_3NH_3^+ + Cl^-$ (c) ${}^-O_2CCO_2^- + 2\ H_2O$

(d) cyclohexyl–$\overset{+}{N}H_2$ (e) CH_3CO_2H (f) $CH_3NH_3^+ + CH_3CO_2^-$

(g) $CH_3CO_2^- + CH_3OH$ (h) $CH_3NH_2 + CH_3OH$ (i) $CO_3^{2-} + 2\ CH_3OH$

1.42 (a) $(CH_3)_2\overset{+}{N}H_2$ (b) $CH_3CH_2CH_2CO_2^-$

1.43 (c) < (d) < (b) < (a). Remember, the conjugate base of a stronger acid is a weaker base.

1.44 (a) 4.76 (b) 10.00 (c) 4.28 (d) ~16 (e) ~50

(e) < (d) < (b) < (a) < (c)

The calculation of pK_a values is discussed in Section 1.10E of the text. These values may also be obtained directly from a calculator with a log function. Remember, however, to change the sign: pK_a = -log K_a. Also, note that a pK must be carried to the second decimal place to represent a K value such as 5.2×10^{-5}.

1.45 (a) 13.37 (b) 9.37 (c) 3.17 (a) < (b) < (c)

1.46 (a) $HBr + F^- \rightleftharpoons HF + Br^-$

HBr: stronger acid; HF: weaker acid

(b) $CH_3CH_2\ddot{N}H^- + CH_3\ddot{O}H \rightleftharpoons CH_3CH_2\ddot{N}H_2 + CH_3\ddot{O}:^-$

1.47 The 0.0100 mol of NaOH would neutralize one half the acetic acid yielding 0.0100 mol of sodium acetate. After neutralization, 0.0100 mol of acetic acid would remain in solution.

The pH of the acetic acid-acetate buffer solution is calculated from the following equation:

$$pH = pK_a + \log \frac{[\text{acetate ion}]}{[\text{acetic acid}]}$$

Because [acetate ion] equals [acetic acid], pH equals pK_a. The pK_a of acetic acid is 4.75 (from Table 1.8 in the text) and, thus, the pH is also 4.75.

1.48 $PbCl_4$ is a covalent compound, while $PbCl_2$ is ionic. Less energy is needed to melt a covalent compound because of the relatively weak intermolecular attractions. More energy is needed to melt an ionic compound because of the strong electrostatic attractions among the ions.

1.49 (a) ${}^-CN$, because carbon has a formal charge of -1 and N has a formal charge of 0.

(b) ${}^-C{\equiv}CH$, because the left-hand carbon has a formal charge of -1 and the right-hand carbon has a formal charge of 0.

(c) $CH_3OH_2^+$, because oxygen has a formal charge of +1 and the other atoms have formal charges of 0.

(d) CH_3O^-, because carbon has a formal charge of 0, and oxygen has a formal charge of -1.

(e) $^-NH_2$, because N has a formal charge of -1 and the H atoms have formal charges of 0.

(f) $CH_3C(=O)O^-$, because the right-hand oxygen has a formal charge of -1 and the other atoms have formal charges of 0.

1.50 (a) [cyclopropane] (b) [oxirane] (c) [tetrahydrofuran]

Four- and three-membered rings would also be correct for (c). For example,

[cyclobutyl]–OH or CH_3–[cyclopropyl]–OH or CH_3CH_2–[oxiranyl]

1.51 (a) $CH_3CH{=}CH_2$ (b) $CH_3CH(=O)$

Although $CH_2{=}CHOH$ must be considered a correct answer to this problem, it is unstable and is rapidly converted to $CH_3CH(=O)$ (see Section 10.8).

(c) $HC(=O)CH_2CH_2CH_3$, $CH_3C(=O)CH_2CH_3$, $CH_2{=}CHOCH_2CH_3$, $CH_2{=}CHCH_2OCH_3$, and $CH_2{=}CHCH_2CH_2OH$ are all correct answers. There are also other correct answers.

1.52 (a)

			ΔH
Cl_2	$\longrightarrow$	$2\ Cl\cdot$	+ 58
CH_3CH_2-H	$\longrightarrow$	$CH_3CH_2\cdot + H\cdot$	+ 98
$CH_3CH_2\cdot + Cl\cdot$	$\longrightarrow$	CH_3CH_2Cl	- 81.5
$H\cdot + Cl\cdot$	$\longrightarrow$	HCl	- 103
			- 28.5 kcal/mol

(b)

			ΔH
Br_2	$\longrightarrow$	$2\ Br\cdot$	+ 46
CH_3CH_2-H	$\longrightarrow$	$CH_3CH_2\cdot + H\cdot$	+ 98
$CH_3CH_2\cdot + Br\cdot$	$\longrightarrow$	CH_3CH_2Br	- 68
$H\cdot + Br\cdot$	$\longrightarrow$	HBr	- 87
			- 11 kcal/mol

The reaction in (a) liberates more energy.

1.53 (a) $(CH_3)_2CH\text{—}\ddot{Br}:$ → $(CH_3)_2\overset{+}{C}H$ + $:\ddot{Br}:^-$

(b) $CH_3CH_2\text{—}Li$ → $CH_3\ddot{C}H_2^-$ + Li^+

(Lithium is less electronegative than carbon.)

(c) $(CH_3)_2CH\text{—}\ddot{O}\text{—}CH(CH_3)_2$ → $(CH_3)_2CH\ddot{O}:^-$ + $\overset{+}{C}H(CH_3)_2$

(Cleavage of the other C-O bond yields the same ions.)

(d) CH_2CH_2 (ring with $\ddot{O}$) → $:\ddot{O}:^-$ on $CH_2\overset{+}{C}H_2$

(Because a ring bond is broken, the + and - charges remain in the same structure.)

1.54 a planar molecule: 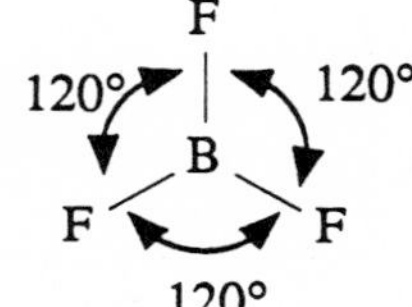 The three BF bond moments cancel in vector addition.

1.55 (b) < (c) < (a). The order is based upon hydrogen bonding with water. Compound (b) forms no hydrogen bonds. Compound (c) does form hydrogen bonds with water (but not with other ether molecules). Compound (a) has both hydrogen-bonding C=O and OH groups.

1.56 Compound A can form an intramolecular hydrogen bond and thus forms fewer hydrogen bonds between molecules. By contrast, compound B forms hydrogen bonds only with other molecules because it cannot form an intramolecular hydrogen bond. Therefore, there are more attractions among the molecules of compound B. (All possible hydrogen bonds are not shown in the following formulas. The squiggles indicate that the bond goes to the rest of the molecules not shown.)

A

B

1.57 The two compounds are equally soluble in water because they have the same molecular weights and both form hydrogen bonds with water. Pure 1-butanol can form hydrogen bonds in the pure state and therefore has a higher boiling point than diethyl ether, which does not have a partially positive hydrogen and thus cannot form hydrogen bonds in its pure state.

hydrogen bonding in water:

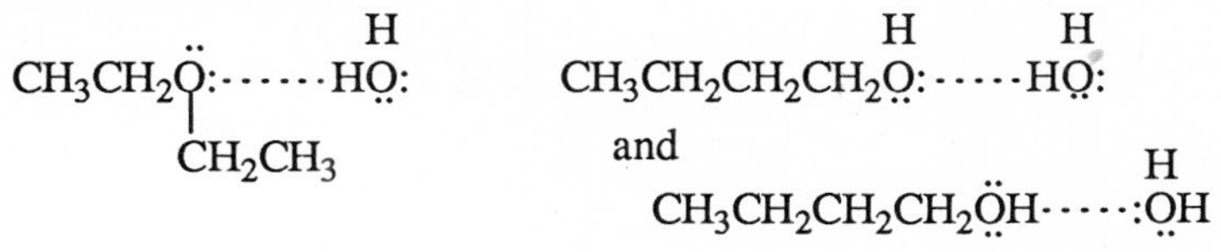

$CH_3CH_2\ddot{O}(CH_2CH_3){:}\cdots\cdots H\ddot{O}{:}H$ $\quad$ $CH_3CH_2CH_2CH_2\ddot{O}(H){:}\cdots\cdots H\ddot{O}{:}H$

and

$CH_3CH_2CH_2CH_2\ddot{O}H\cdots\cdots{:}\ddot{O}H(H)$

hydrogen bonding in 1-butanol:

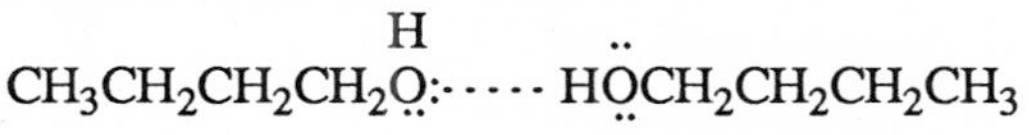

$CH_3CH_2CH_2CH_2\ddot{O}(H){:}\cdots\cdots H\ddot{O}CH_2CH_2CH_2CH_3$

1.58 (a) Lewis acid, $AlCl_3$; Lewis base, $(CH_3)_3CCl$

(b) Lewis acid, $(CH_3)_3C^+$; Lewis base, $CH_2{=}CH_2$

(c) Lewis acid, $(CH_3)_3C^+$; Lewis base, H_2O

(d) Lewis acid, Br_2; Lewis base, $CH_2{=}CH_2$

1.59 (a) piperazine (H–N(ring)N–H, arrows to each N lone pair) (b) $C_6H_5\ddot{O}{:}^-$ (c) $CH_3\ddot{O}H$ (d) $CH_3CH_2\ddot{Cl}{:}$

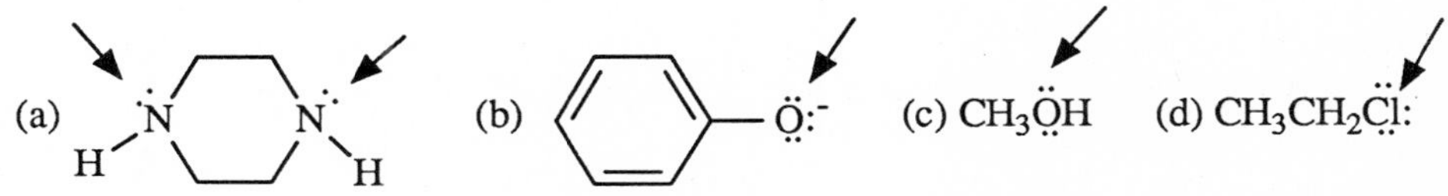

In each case the Lewis acid attacks an unshared pair of electrons on an electronegative atom.

1.60 (a) $CH_3\ddot{O}H + H{-}\ddot{O}S(=O)_2OH \rightleftharpoons CH_3\overset{+}{O}H_2 + {}^-{:}\ddot{O}S(=O)_2OH$

(b) $CH_3\ddot{O}{-}H + {}^-{:}\ddot{N}H_2 \rightleftharpoons CH_3\ddot{O}{:}^- + {:}NH_3$

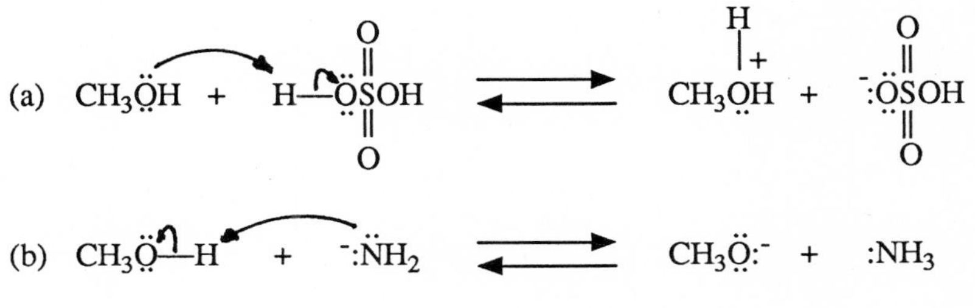

1.61 (a)

$$CH_3CH(OH){-}C(=O){-}\ddot{O}{-}H + CH_3\ddot{N}H_2 \rightleftharpoons CH_3CH(OH){-}C(=O){-}\ddot{O}:^- + CH_3\overset{+}{N}H_3$$

Both the —OH (hydroxyl group) and the —CO_2H (carboxyl group) contain H atoms bonded to O atoms. Of the two, the H of the —CO_2H group is lost preferentially to a base because the —CO_2H group is more acidic.

(b)

$$CaCO_3 + 2\ CH_3C(=O)\ddot{O}H \longrightarrow Ca^{2+} + 2\ CH_3C(=O)\ddot{O}:^- + \left[HOC(=O)OH\right]$$

$$\left[HOC(=O)OH\right] \longrightarrow H_2O + CO_2$$

Carbonic acid, H_2CO_3, is shown in brackets because it is unstable and decomposes to H_2O and CO_2.

(c)

$$CH_3C(CH_3)_2{-}\ddot{O}:^- + H{-}\ddot{O}H \rightleftharpoons CH_3C(CH_3)_2{-}OH + {}^-:\ddot{O}H$$

1.62 (a)

$$CH_3CH(OH){-}C(=O){-}O{-}H + HCO_3^- \rightleftharpoons CH_3CH(OH){-}C(=O){-}O^- + [H_2CO_3]$$

$$[H_2CO_3] \rightleftharpoons H_2O + CO_2$$

(b)

$$CH_3CH(OH){-}C(=O){-}O{-}H + HPO_4^{2-} \rightleftharpoons CH_3CH(OH){-}C(=O){-}O^- + H_2PO_4^-$$

1.63

$$\underset{\text{phenol}}{C_6H_5OH} \underset{}{\overset{H_2O}{\rightleftharpoons}} C_6H_5O^- + H^+$$

$$K_a = \frac{[C_6H_5O^-]\ [H^+]}{[C_6H_5OH]} \qquad pK_a = 10.00;\ \text{therefore, } K_a = 1.0 \times 10^{-10}$$

$$[H^+] = [C_6H_5O^-] = x \qquad [C_6H_5OH] = 0.010\ M - x \cong 0.010\ M$$

(Because the equilibrium constant is very small, the amount of phenol that ionizes is negligible compared to 0.010 *M*.)

Substituting,

$$\frac{x^2}{0.010} = 1.0 \times 10^{-10} \qquad x^2 = 1.0 \times 10^{-12} \qquad x = 1.0 \times 10^{-6} = [H^+]$$

$$pH = 6.0$$

1.64

$$\begin{aligned} \text{pH} &= \text{p}K_a + \log \frac{[\text{acetate ion}]}{[\text{acetic acid}]} \\ &= 4.75 + \log \frac{0.200}{0.100} \\ &= 4.75 - 0.30 \\ &= 4.45 \end{aligned}$$

Chapter 2

Orbitals and Their Role in Covalent Bonding

2.17 All are the same because C is tetrahedral.

2.18 (a) (b)

2.19 (a) All carbons are sp^3 hybridized.

(b) $H-C\equiv C-C(=CH_2)-CH_3$ (sp: the two triple-bonded carbons; sp^2: the $C=CH_2$ carbons; sp^3: the CH_3 carbon)

(c) All carbons are sp^2 hybridized.

2.20 In (a) and (c), the indicated carbons atoms are bonded to sp^2 carbons and therefore lie in the same plane as the sp^2 carbons.

In (b), the carbon atoms bonded directly to the sp^2 carbons lie in the plane of the sp^2 carbons. However, the indicated methyl group may or may not lie in this plane.

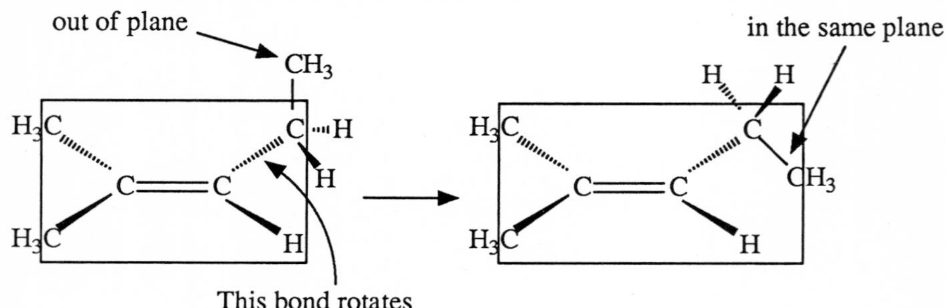

2.21 (a) 120° around sp^2 carbon.

(b) 180° around sp carbon.

(c) 120° around sp^2 carbon.

(d) 120° around sp^2 carbon.

(e) 109° around sp^3 carbon.

(f) 109° around sp^3 carbon.

The actual bond angles might differ slightly from those predicted because of dipole-dipole repulsions and attractions.

2.22 (a) H—O⁺—H with a lone pair (⇅) on O and a third H on O

H—$\overset{+}{O}$—H
 |
 H

(b) 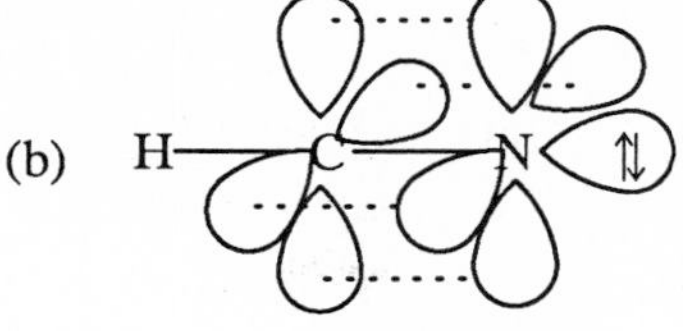

(c) $\overset{-}{O}$—H with three lone pairs (⇅) on O

(d) H—N—N—H
 | |
 H H
(one lone pair ⇅ on each N)

(e)

H
|
H — C — H
H | H
| | |
H — C —— N⁺ —— C — H
| | |
H | H
H — C — H
|
H

(f) 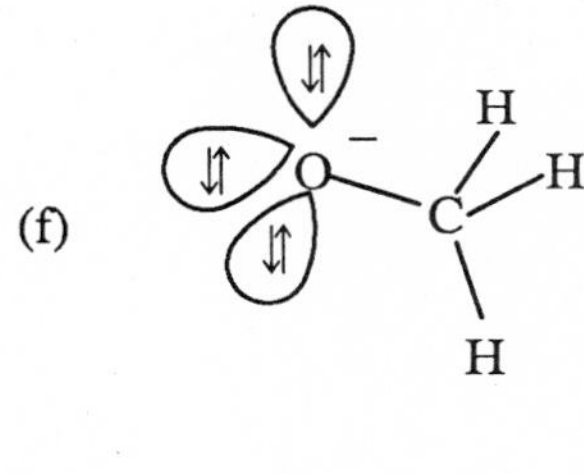

(g) 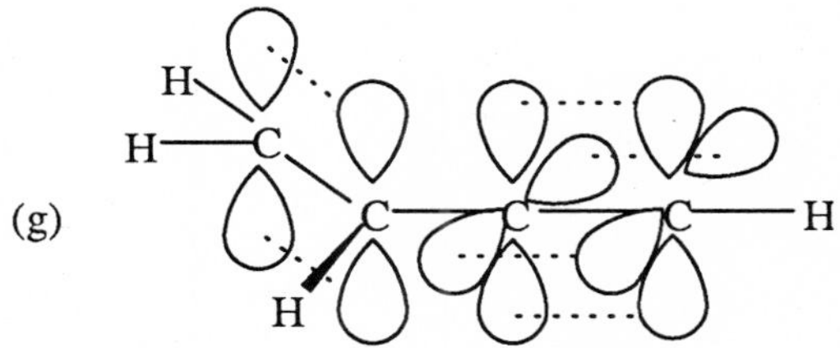

2.23

(a)		(b)	
σ^*	___	σ^*	___
π^*_2	___	π^*_2	↓
π_1	⇅	π_1	↑
σ	⇅	σ	⇅

In (a), the ground state, the electrons are paired and fill the two lowest-energy orbitals. Upon excitation, an electron is promoted from the highest occupied molecular orbital (π_1) to the lowest unoccupied molecular orbital (π^*_2). Any other transition would require a greater amount of energy.

2.24 (a) 2 (b) 2 (c) 2 (d) 2

The answers are based on the relative amounts of *s* character in the bonds. The greater the amount of *s* character, the shorter is the bond length.

2.25 (a) Each atom in the ring is sp^3.

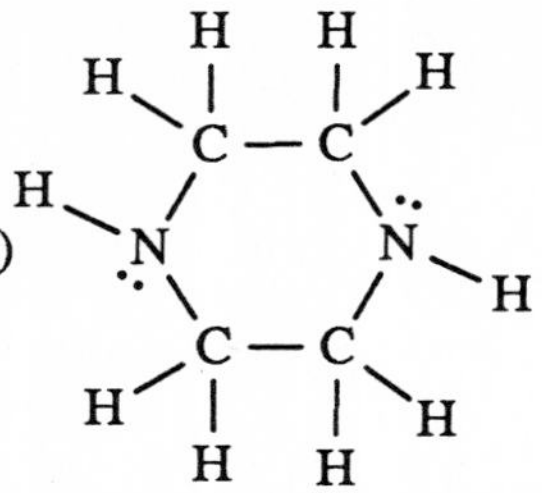

(b) Each atom in the ring is sp^2.

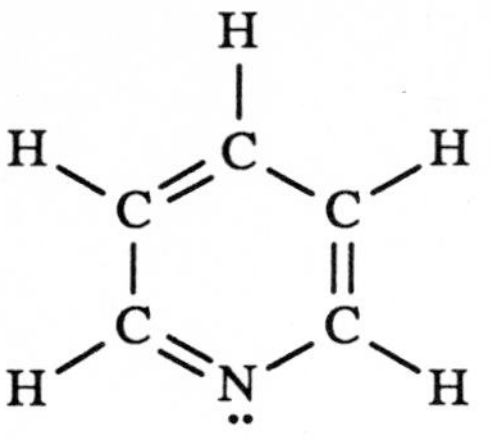

2.26

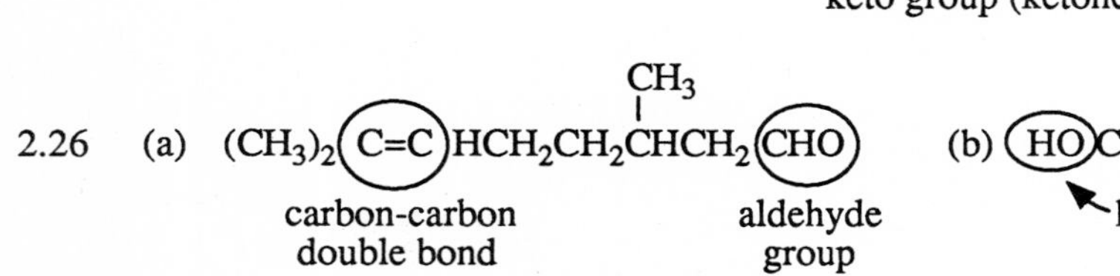

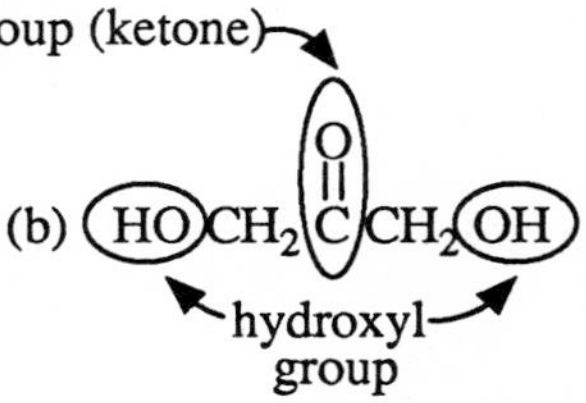

It is also correct to circle the carbon atoms bonded to the ketone carbonyl group.

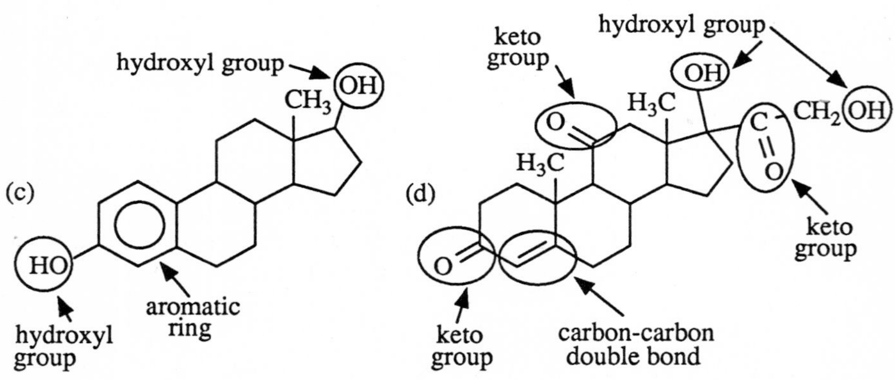

2.27 (a) $CH_3CH{=}CH_2$ (b) $CH_3C{\equiv}CH$ (c) $CH_3CH_2OCH_3$

(d) $CH_3CH_2CH_2OH$ or $(CH_3)_2CHOH$

(e) $CH_3CH_2CH_2NH_2$, $(CH_3)_2CHNH_2$, $(CH_3)_3N$, or $CH_3CH_2NHCH_3$

(f) $(CH_3)_2C{=}O$ (g) $CH_3CH_2\overset{O}{\overset{\|}{C}}H$ (h) $CH_3CH_2CO_2H$

2.28 (a) $CH_3CH_2CH_2\overset{O}{\overset{\|}{C}}OH$ (b) 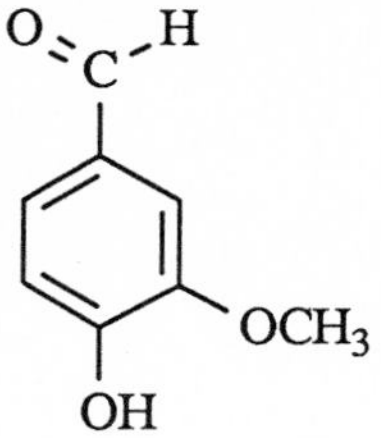 [benzene ring bearing CHO (O=C–H), OCH$_3$ and OH] (c) $CH_3\overset{O}{\overset{\|}{C}}OCH_2CH_2CH(CH_3)_2$

2.29 (a) $CH_3CH_2\overset{O}{\overset{\|}{C}}CH_3$ (b)  [cyclohexane ring bearing OH] (c) $CH_3CH_2CH_2CH{=}CH_2$ (d) CH_3CH_2CHO

2.30 (a) contains a pair of conjugated double bonds and thus contains delocalized pi electrons. In (b), the double bond of the carbonyl group is conjugated with one of the pi bonds in the triple bond and thus has delocalized pi electrons.

In (c), the carbon-carbon double bond is not conjugated with the carbonyl pi bond; therefore, these pi electrons are localized.

2.31 (a) $CH_2{=}CHCH{=}CHCH_3$ (b) $CH_2{=}CHCH_2CH{=}CH_2$

2.32 [cyclohexenone ring with =O; arrows labelled sp^2 point to the C=C and C=O carbons, arrows labelled sp^3 point to the remaining ring carbons]

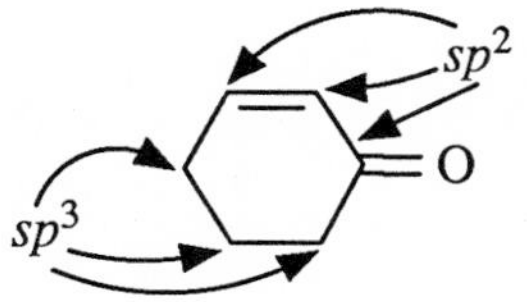

2.33 (a), (b), and (d) show resonance structures (structures that differ only in the positions of electrons), while (c) and (e) show structures in equilibrium (structures with different arrangements of atoms).

2.34 (a) 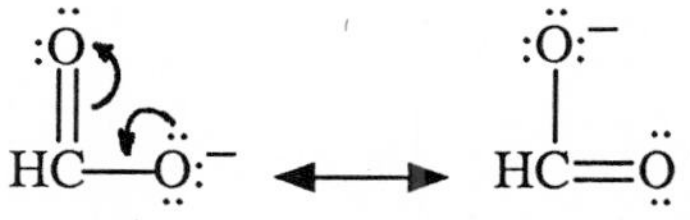 $H\overset{:\ddot{O}}{\overset{\|}{C}}{-}\ddot{\underset{..}{O}}:^-$ ⟷ $H\overset{:\ddot{O}:^-}{\overset{|}{C}}{=}\ddot{\underset{..}{O}}$ (b) $HO\overset{:\ddot{O}}{\overset{\|}{C}}{-}\ddot{\underset{..}{O}}:^-$ ⟷ $HO\overset{:\ddot{O}:^-}{\overset{|}{C}}{=}\ddot{\underset{..}{O}}$

(c) [benzene ring with curved arrows]–CH_3 ⟷ [benzene ring, alternate Kekulé structure]–CH_3

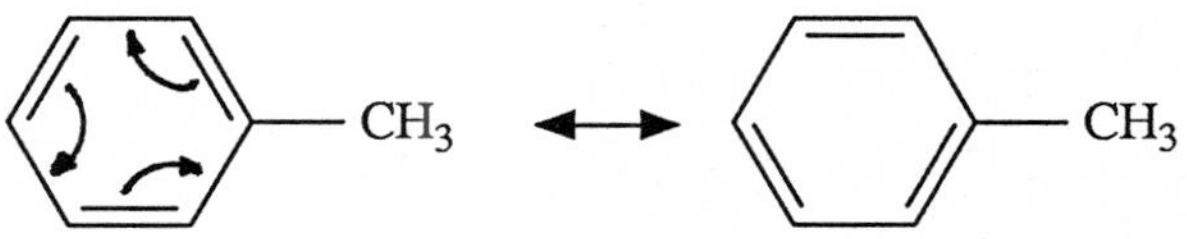

(d)  $CH_3\overset{+}{C}H{-}CH{=}CH_2$ ⟷ $CH_3CH{=}CH{-}\overset{+}{C}H_2$

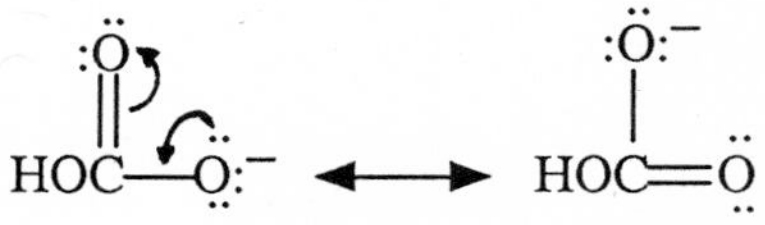

(e) 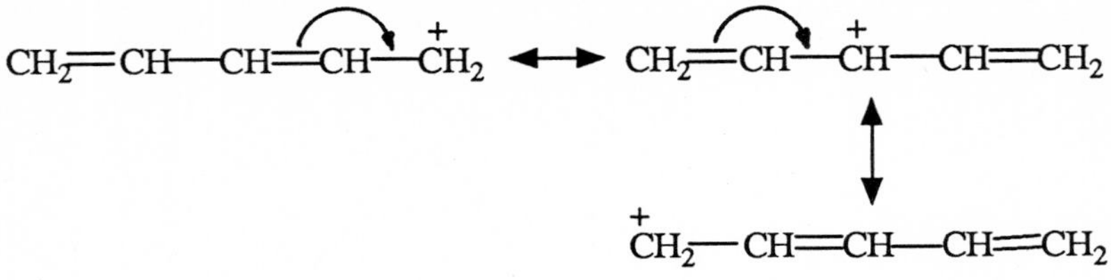 $CH_2{=}CH{-}CH{=}CH{-}\overset{+}{C}H_2 \longleftrightarrow CH_2{=}CH{-}\overset{+}{C}H{-}CH{=}CH_2 \longleftrightarrow \overset{+}{C}H_2{-}CH{=}CH{-}CH{=}CH_2$

(f) 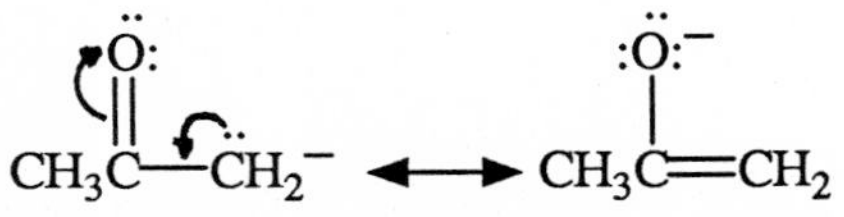 $CH_3C(=\ddot{O}:){-}\ddot{C}H_2^- \longleftrightarrow CH_3C(:\ddot{O}:^-){=}CH_2$

The following structure would be considered a less important contributor because of a greater charge separation and because one carbon does not have a complete octet of electrons.

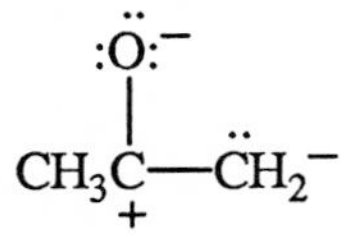

$CH_3\overset{+}{C}(:\ddot{O}:^-){-}\ddot{C}H_2^-$

(g) $:N{\equiv}C{-}\bar{\ddot{C}}H{-}C{\equiv}N: \longleftrightarrow :N{\equiv}C{-}CH{=}C{=}\ddot{N}:^- \longleftrightarrow {}^-:\ddot{N}{=}C{=}CH{-}C{\equiv}N:$

(h)

${}^-:\ddot{O}{-}C_6H_4{-}CH{=}\ddot{O}: \longleftrightarrow {}^-:\ddot{O}{-}C_6H_4{-}\overset{+}{C}H{-}:\ddot{O}:^- \longleftrightarrow {}^-:\ddot{O}{-}C_6H_4{-}CH{=}\ddot{O}:$

$\longleftrightarrow :\ddot{O}{=}C_6H_4^-{-}CH{=}\ddot{O}: \longleftrightarrow :\ddot{O}{=}C_6H_4^-{-}CH{=}\ddot{O}:$

$\longleftrightarrow :\ddot{O}{=}C_6H_4{=}CH{-}:\ddot{O}:^- \longleftrightarrow :\ddot{O}{=}C_6H_4^-{-}CH{=}\ddot{O}:$

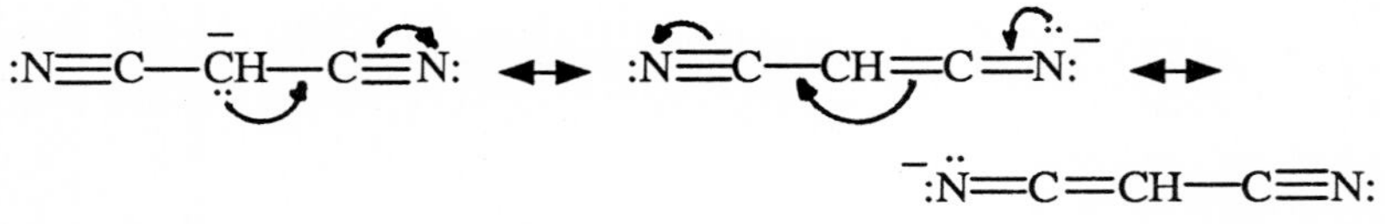

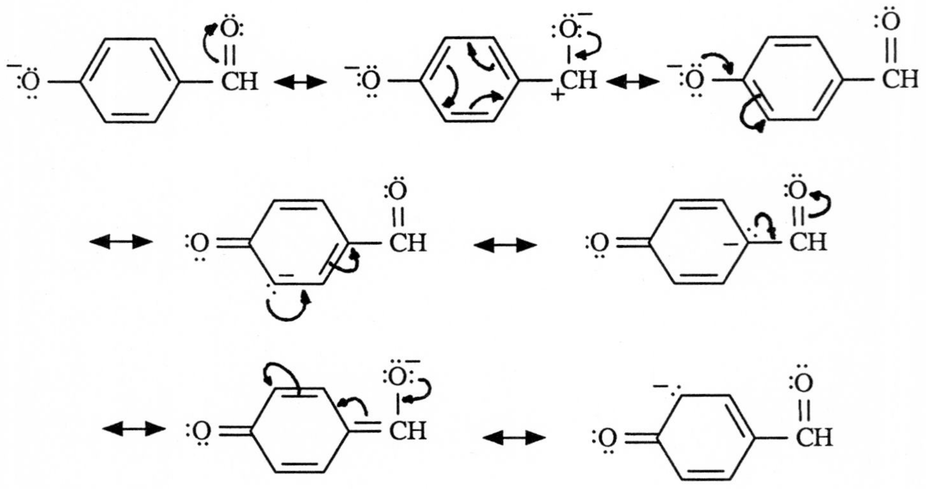

(i)

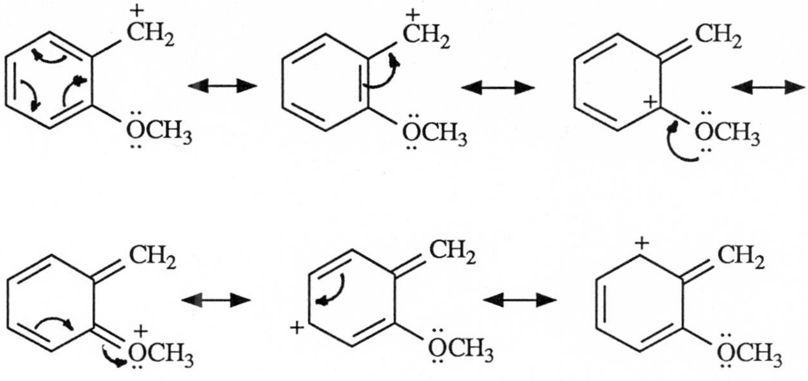

$$\overset{+}{C}H_2 \text{ / } \ddot{\underset{..}{O}}CH_3 \text{ resonance structures}$$

2.35 (a) $CH_3CH_2\overset{O}{\overset{||}{C}}NH_2$ because each C and N has an octet and there is no charge separation.

(b) $CH_3\overset{O}{\overset{||}{C}}Cl$ because each carbon, oxygen, and chlorine has an octet and there is no charge separation.

(c) 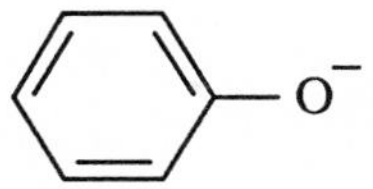 $C_6H_5-O^-$ because the electronegative oxygen carries the negative charge.

2.36 (a)

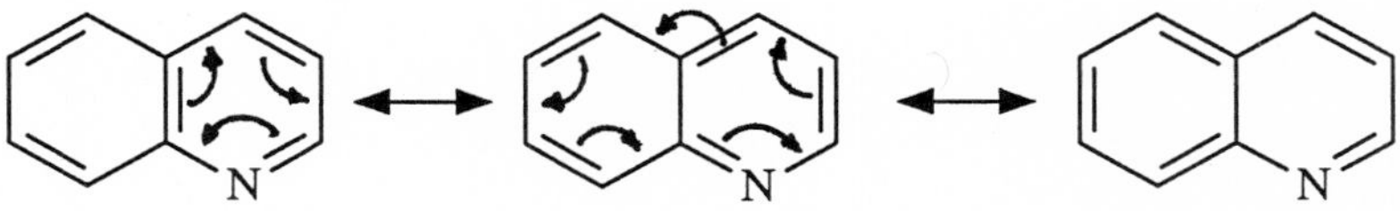

Note that the following two formulas represent the same structure.

(b)

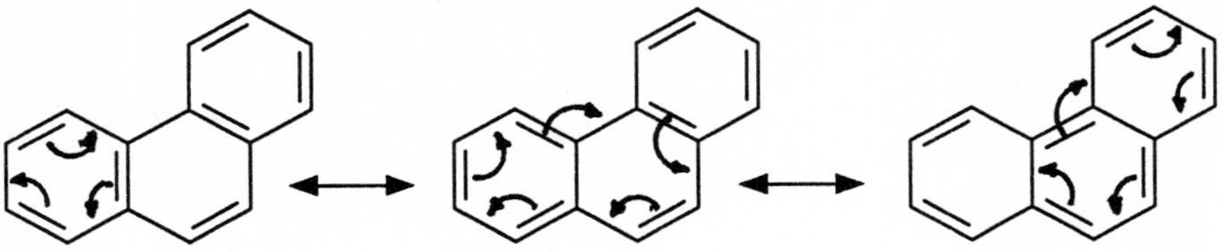

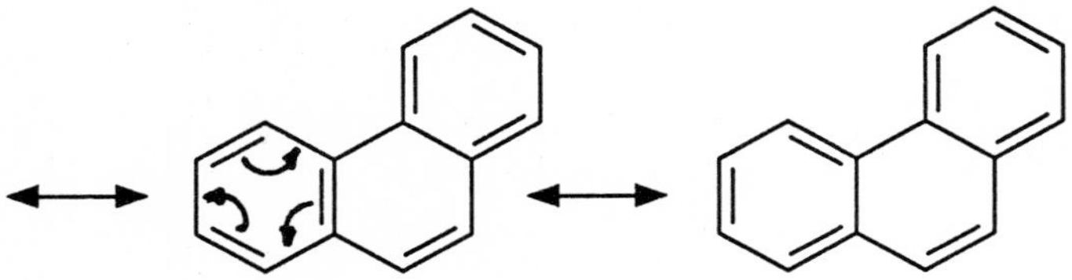

(c)

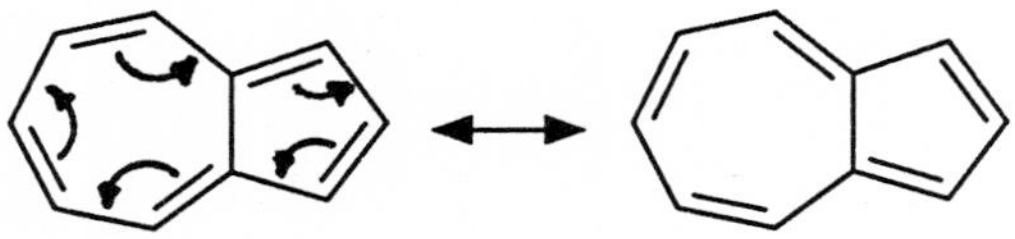

2.37 (a) The first cation is more stabilized, because its positive charge can be delocalized by the ring.

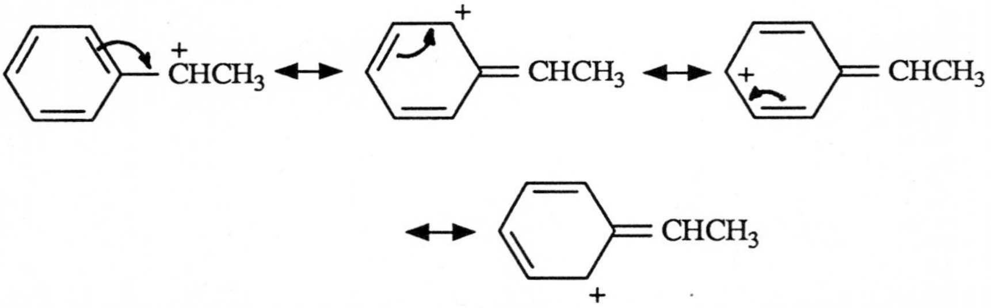

(b) The second cation because its positive charge can be delocalized by the double bond.

$CH_2{=}CH{-}\overset{+}{C}HCH_3 \longleftrightarrow \overset{+}{C}H_2{-}CH{=}CHCH_3$

(c) The second anion because the nitro group helps share the negative charge.

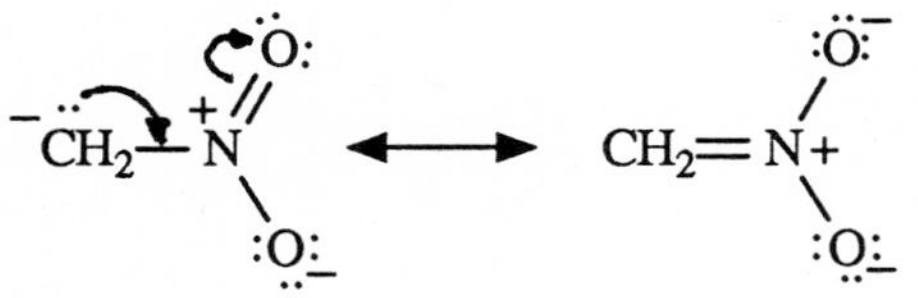

2.38 (a) $CH_2{=}C{=}CH_2$ (b)

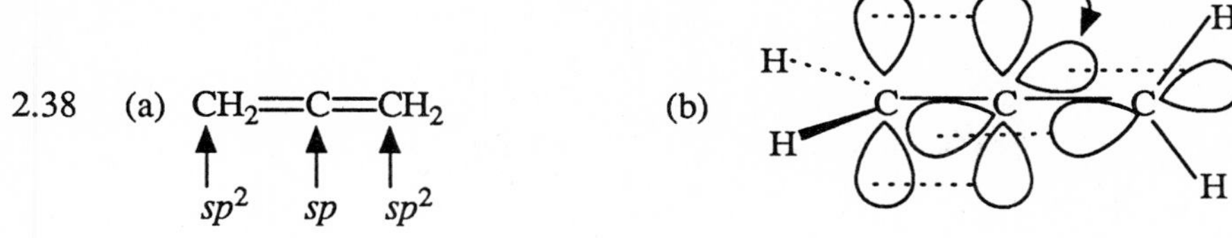

(c) No, they cannot overlap their sides.

(d) No, two pi bonds cannot delocalize electronic charge without overlap.

2.39 $\ddot{O}::C::\ddot{O}$

sp carbon

With an O–C–O bond angle of 180° and two pi bonds, the carbon atom must be *sp* hybridized.

2.40 H_2^+, or $H{\cdot}H^+$, σ^* ___ ; σ $\uparrow$ H_2^-, or $H{:}H^{\cdot -}$, σ^* $\uparrow$; σ $\uparrow\downarrow$

The H_2^+ ion contains one less electron than H_2 - that is, the group contains one electron. The H_2^- ion contains one more electron than H_2. Since the sigma bonding orbital in H_2^- is filled, the third electron must be in the antibonding orbital.

2.41 Carbon is in the *sp* state and has two *sp-s* bonds with the hydrogen atoms. The carbon atom has two *p* orbitals with one electron in each *p* orbital.

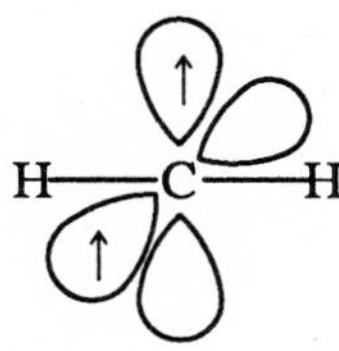

2.42 (a) Structures A and B are not resonance structures because they differ in the sequence of atoms. For example, structure A shows N bonded to two H atoms while B shows the N bonded to only one H atom.

(b) On paper, these structures are interconvertible by electron shifts.

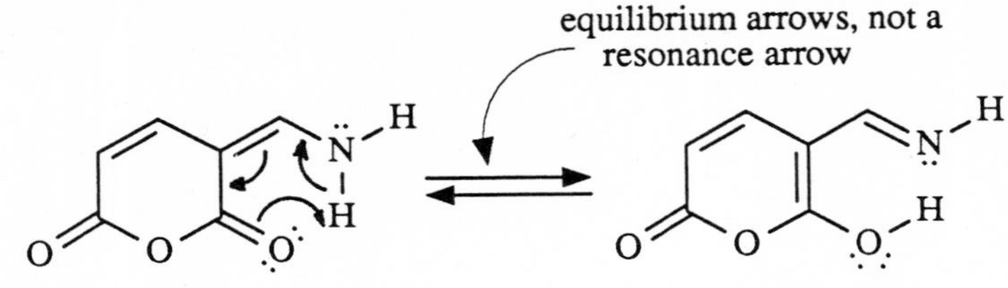

equilibrium arrows, not a resonance arrow

H bonded to N (closer to N than to O) $\rightleftharpoons$ H bonded to O (closer to O than to N)

2.43 (a) $CH_3\ddot{N}H—C(=\overset{+}{N}H_2)—\ddot{N}H_2 \longleftrightarrow CH_3\ddot{N}H—C(—\ddot{N}H_2)=\overset{+}{N}H_2 \longleftrightarrow CH_3\overset{+}{N}H=C(—\ddot{N}H_2)—\ddot{N}H_2$

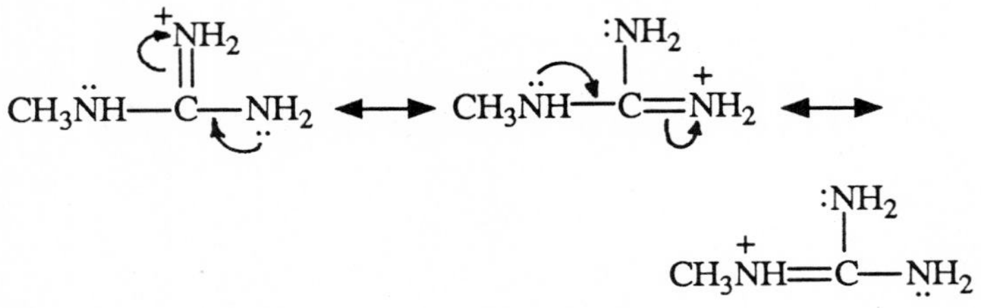

(b) $CH_3C(=\overset{+}{N}H_2)—\ddot{N}H_2 \longleftrightarrow CH_3C(—\ddot{N}H_2)=\overset{+}{N}H_2$

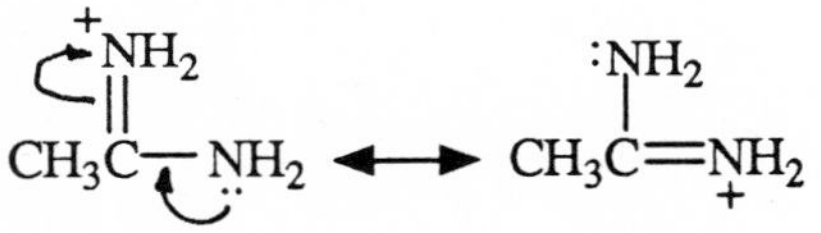

Chapter 3

Structural Isomerism, Nomenclature, and Alkanes

3.10 (a) is unsaturated because it contains a benzene ring. Its saturated counterpart would contain a cyclohexane ring.

saturated: 3-hydroxycyclohexyl–$\overset{OH}{|}$CHCH_2NHCH_3 (HO on the ring; OH on the side-chain carbon)

(b) is unsaturated because it contains both a carbon-carbon double bond and a carbon-nitrogen triple bond. Its saturated counterpart would be $CH_3CH_2CH_2NH_2$.

(c) is saturated because it contains no pi bond or aromatic pi cloud.

(d) is unsaturated because the carboxyl group contains a carbon-oxygen double bond.

3.11 (a) $CH_3CH{=}CH_2$, which is the same compound as $CH_2{=}CHCH_3$

(b) $CH_3OCH_2CH_2CH_3$, $CH_3OCH(CH_3)_2$, $CH_3CH_2CH_2CH_2OH$,

$(CH_3)_2CHCH_2OH$, $CH_3CH_2\underset{CH_3}{\underset{|}{C}}HOH$, $(CH_3)_3COH$

(c) $F-\overset{F}{\underset{Cl}{C}}-\overset{F}{\underset{H}{C}}-Br$, $F-\overset{H}{\underset{Cl}{C}}-\overset{F}{\underset{F}{C}}-Br$, $F-\overset{F}{\underset{H}{C}}-\overset{F}{\underset{Cl}{C}}-Br$

Note that redistributing H, Br, Cl, or F on the same carbon atom does not yield an isomer.

$F-\overset{F}{\underset{F}{C}}-\overset{Cl}{\underset{Br}{C}}-H$ $F-\overset{F}{\underset{F}{C}}-\overset{Br}{\underset{Cl}{C}}-H$

the same compound

3.12 (a) same compound. In each case, Cl is bonded to carbon 2 of a 5-carbon chain. To verify the identity, rotate the first formula 180° in the plane of the paper.

$CH_3\overset{Cl}{\overset{|}{C}}HCH_2CH_2CH_3 \longrightarrow CH_3CH_2CH_2\underset{Cl}{\underset{|}{C}}HCH_3$

rotate 180°

Alternatively, determine the IUPAC name for each formula. Both of these are 2-chloropentane.

(b) isomers. Cl is bonded to a C with three $-CH_3$ groups in one structure and to a C of a $-CH_2-$ group in the other. Expanding the formulas shows this relationship more clearly.

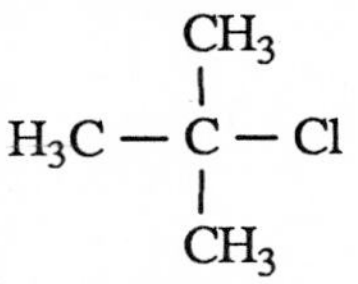

$$\begin{array}{c} CH_3 \\ | \\ H_3C - C - Cl \\ | \\ CH_3 \end{array} \quad \text{not the same as} \quad \begin{array}{c} CH_3 \\ | \\ H_3C - C - H \\ | \\ CH_2Cl \end{array}$$

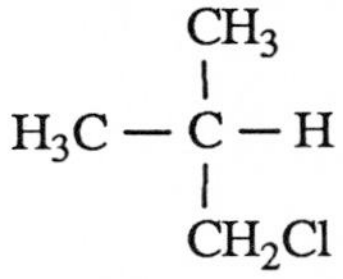

2-chloro-2-methylpropane — 1-chloro-2-methylpropane

(c) same compound (3-methyl-1-pentanol). Each contains a 5-carbon chain with $-OH$ at position 1 and $-CH_3$ at position 3.

(d) isomers. If the ring carbon atoms are numbered beginning with the $-OH$ carbon, the double bond begins at carbon 2 in the first formula and at carbon 3 in the second formula.

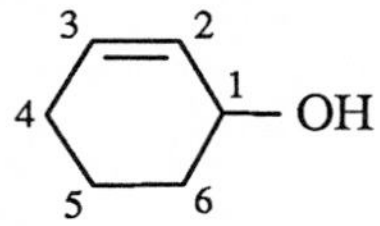

2-cyclohexenol 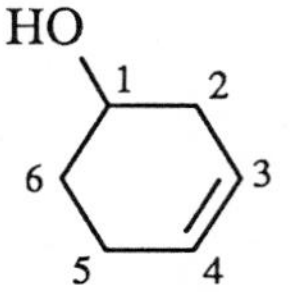 — 3-cyclohexenol

3.13 (a) $CH_3(CH_2)_4CH_3$, $(CH_3)_2CHCH_2CH_2CH_3$, $(CH_3)_3CCH_2CH_3$,

$(CH_3)_2CHCH(CH_3)_2$, $\begin{array}{c} CH_3 \\ | \\ CH_3CH_2CHCH_2CH_3 \end{array}$

(b) $CH_3CH_2CH_2CH_2OH$, $(CH_3)_2CHCH_2OH$, $\begin{array}{c} CH_3 \\ | \\ CH_3CH_2CHOH \end{array}$, $(CH_3)_3COH$

(c) $CH_3CH_2CH_2CH_2NH_2$, $(CH_3)_2CHCH_2NH_2$, $\begin{array}{c} CH_3 \\ | \\ CH_3CH_2CHNH_2 \end{array}$, $(CH_3)_3CNH_2$,

$CH_3CH_2CH_2NHCH_3$, $(CH_3)_2CHNHCH_3$, $(CH_3CH_2)_2NH$, $CH_3CH_2N(CH_3)_2$

(d) $CH_3CH_2CHBrCl$, $CH_3CHBrCH_2Cl$, $BrCH_2CH_2CH_2Cl$, $CH_3CHClCH_2Br$,

$(CH_3)_2CBrCl$

3.14 (a) C_8H_{14}, C_nH_{2n-2} (b) C_7H_{12}, C_nH_{2n-2}

3.15 (a) 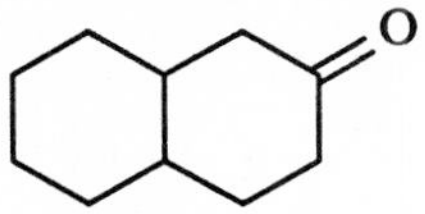 [decalone structure] (b) 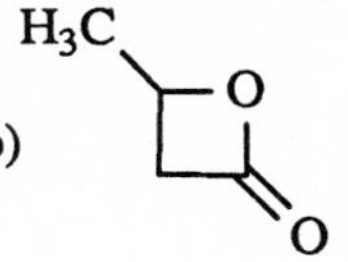 [H3C-substituted β-lactone structure] (c)  [1,4-cyclohexadiene structure]

(There are other correct answers)

3.16 (a) $CH_3CH_2C{\equiv}CCH_2CH_3$ (b) $CH_3\overset{O}{\overset{\|}{C}}OH$ (c) $CH_3\overset{O}{\overset{\|}{C}}CH_3$

(There may be other correct answers)

3.17 The side chain R must be a series of *n*-alkyl groups containing one, two, and three carbons; two, three, and four carbons; three, four, and five carbons; or four, five and six carbons.

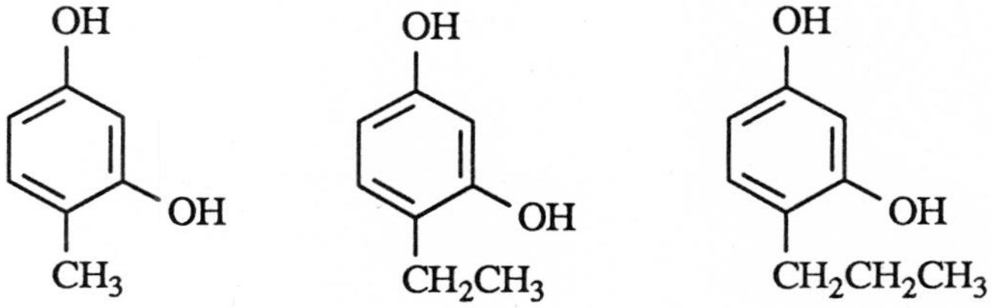

OH, OH, CH_3 OH, OH, CH_2CH_3 OH, OH, $CH_2CH_2CH_3$

or CH_2CH_3 $CH_2CH_2CH_3$ $CH_2(CH_2)_2CH_3$

or $CH_2CH_2CH_3$ $CH_2(CH_2)_2CH_3$ $CH_2(CH_2)_3CH_3$

or $CH_2(CH_2)_2CH_3$ $CH_2(CH_2)_3CH_3$ $CH_2(CH_2)_4CH_3$

3.18 $CH_3(CH_2)_4Br$, $CH_3(CH_2)_5Br$, $CH_3(CH_2)_6Br$, $CH_3(CH_2)_7Br$, $CH_3(CH_2)_8Br$, $CH_3(CH_2)_9Br$

3.19 (a) $(CH_3)_3C(CH_2)_5CH_3$

(b)
CH_2CH_3
$(CH_3CH_2)_2CHCHCH_2CH_2CH_3$

(c)
CH_3
$(CH_3)_2CHCH_2C(CH_2)_4CH_3$
CH_2CH_3

(d)
$CH(CH_3)_2$
$CH(CH_3)_2$

(e) CH_3CH_2CH-
CH_3

(f) $(CH_3)_3C-$

(g) $(CH_3)_2CHCH_2-$

(h) $CH_3(CH_2)_4-$
CH_3

(i) $(CH_3CH_2CH_2)_2CHCH(CH_3)_2$

3.20 (a) 3-ethylpentane

(b) 2,3-dimethylbutane

(c) 5-ethyl-1,2,3-trimethylcyclohexane

(d) *t*-butylcyclopentane

(e) 2,2,4,4-tetramethylheptane

(f) 2,2,3,4,4-pentamethylpentane

Be sure that the parent carbon chain is numbered from the end that will give the smaller numbers. In (d), no number is necessary for a monosubstituted cycloalkane.

3.21 Write out the formula from the incorrect name; then determine the correct name.

(a)
CH_3
$CH_3CH_2CH(CH_2)_4CH_3$ is correctly 3-methyloctane when the chain is numbered from the end closer to the branch.

(b)
CH_3
$CH_3C\text{-}CH_3$ is correctly 2,2-dimethylpropane because it contains a
CH_3
three-carbon continuous chain.

(c) $CH_3CH_2C(CH_3)(CH(CH_3)CH_3)—CH(CH_3)(CH_2)_5CH_3$

$$\begin{array}{c} \quad CH_3\ \ CH_3 \\ CH_3CH_2\overset{|}{\underset{|}{C}}—\overset{|}{C}H(CH_2)_5CH_3 \\ CH_3CHCH_3 \end{array}$$

start numbering here

This structure is correctly 3-ethyl-2,3,4-trimethyldecane because numbering should begin closer to the branches and the methyl groups should be grouped together.

hydroxyl

3.22 (a) $CH_3CH(OH)-CO_2H$ carboxyl (carboxylic acid)

(b) [structure: ring with H_3C, H_3C, CH_3, CH_3 substituents; side chain $CH=CH-C(CH_3)=O$]

keto (ketone)

carbon-carbon double bond

(c) [structure: cyclohexanone ring with =O and Br]

keto (ketone)

halo (bromo)

(d) $HC(=O)-CH_2CH_2-C(=O)-CH_3$

keto (ketone)

aldehyde

3.23 (a) $CH_3(CH_2)_4Cl$, $CH_3CHCl(CH_2)_2CH_3$, $CH_3CH_2CHClCH_2CH_3$

(b) [structure: cyclopentane ring with Cl]

(c) $ClCH_2C(CH_3)_2CH_2CH_3$, $(CH_3)_3CCHClCH_3$, $(CH_3)_3CCH_2CH_2Cl$

(d) $(CH_3)_3CCH_2Cl$

3.24 (a) 1,2,3,4,5,6-hexachlorocyclohexane

(b) 1,1,2-trichloroethane

(c) nitromethane

In (b), 1,2,2-trichloroethane would be incorrect.

3.25 (a) 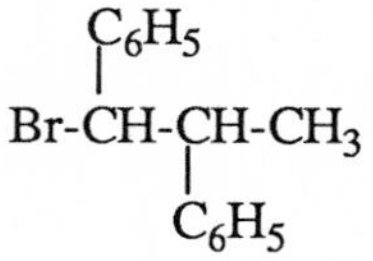 $Br\text{-}CH(C_6H_5)\text{-}CH(C_6H_5)\text{-}CH_3$

(b) CCl_3CCl_3

(c) $HOCH_2CHI(CH_2)_5CH_3$

(d) 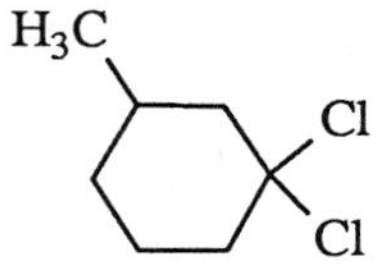 H_3C Cl Cl

3.26 (a) 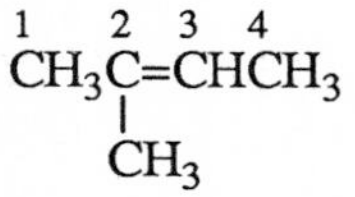 $\overset{1}{C}H_3\overset{2}{C}(CH_3)\text{=}\overset{3}{C}H\overset{4}{C}H_3$ 2-methyl-2-butene

Start numbering at the end closer to the branch.

(b) 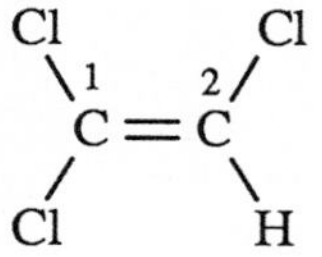 $Cl_2\overset{1}{C}\text{=}\overset{2}{C}(Cl)H$ trichloroethene (or 1,1,2-trichloroethene)

(c) 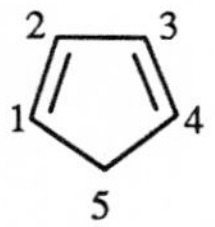 1 2 3 4 5 cyclopentadiene (or 1,3-cyclopentadiene)

Note that the ring is numbered so that consecutive numbers pass through the double bonds. This numbering is necessary because "1" in the prefix means that the double bond joins carbons 1 and 2, while "3" means that the double bond joins carbons 3 and 4.

Incorrect numbering: 5 4 1 3 2

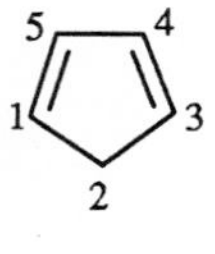

(d) 3-methyl-1,3,5-hexatriene

(e) 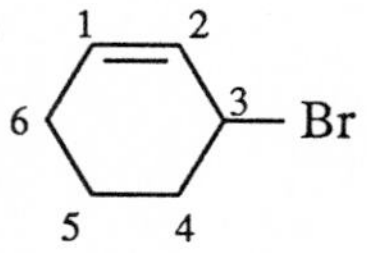 1 2 3 Br 6 5 4 3-bromocyclohexene (or 3-bromo-1-cyclohexene)

Number the ring to give the suffix functional group the smaller number.

(f) Expand the formula:

1 2 3 4 5 6
$C_6H_5-C\equiv C-C\equiv C-C\equiv C-C_6H_5$

1,6-diphenyl-1,3,5-hexatriyne

3.27 (a)

CH_3 O — 2-methylcyclopentanone

H_3C O — 3-methylcyclopentanone

You may have included the following stereoisomers, which will be discussed in Chapter 4.

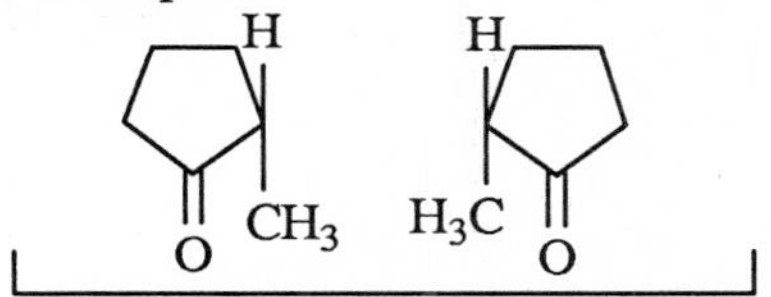

2-methylcyclopentanone

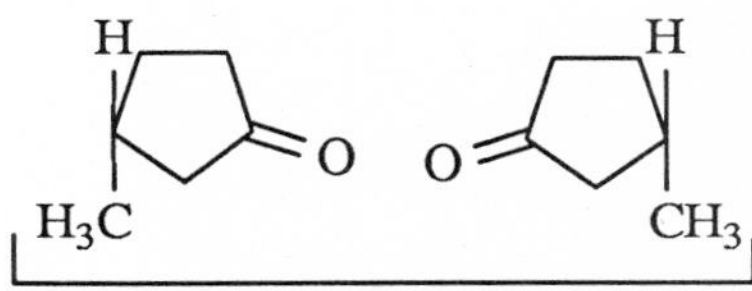

3-methylcyclopentanone

(b)

3 2 1 4 5 $CH(CH_3)_2$ — 1-isopropylcyclopentene

$CH(CH_3)_2$ — 3-isopropylcyclopentene

$(CH_3)_2CH$ — 4-isopropylcyclopentene

For 3-isopropylcyclopentene, you may have included the following stereoisomers, which will be discussed in Chapter 4.

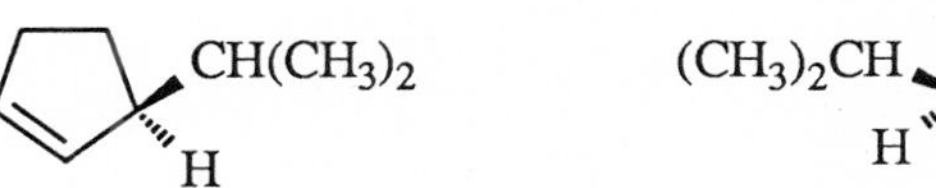

3.28 (a) dichloroethanoic acid (commonly called dichloroacetic acid). No prefix numbers are needed because there is only one dichloroethanoic acid.

(b) chloroethanal. No prefix number is needed.

(c) bromopropanone (commonly called bromoacetone). Again, no prefix number is needed.

3.29 Draw the carbon skeleton of the parent compound, number the chain or ring, add the substituents, and then fill in the hydrogen atoms.

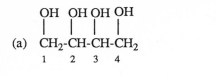

(c) $ClCH{=}CHC(OH)(CH_2CH_3)C{\equiv}CH$

(d) $CH_3C(CH_3){=}CHCH_2CH_2C(CH_3){=}CHCH{=}O$
8 7 6 5 4 3 2 1

(e)

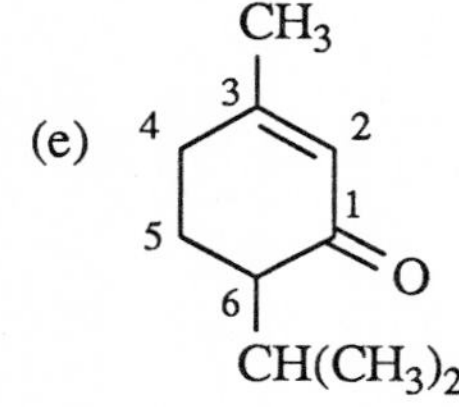

Note that the carbonyl carbon is position 1 and that numbering around the ring proceeds toward the carbon-carbon double bond.

3.30 (a) $-\Delta H$ for one $-CH_2-$ group can be calculated by subtracting $-\Delta H$ for $CH_3CH_2CH_3$ from that for $CH_3CH_2CH_2CH_3$:

$-\Delta H$ for $-CH_2-$ = 688 kcal/mol - 531 kcal/mol = 157 kcal/mol

$-\Delta H$ for $CH_3(CH_2)_6CH_3$ is ($-\Delta H$ for $CH_3CH_2CH_2CH_3$) + ($-\Delta H$ for <u>four</u> CH_2 groups)

$-\Delta H$ for octane = 688 kcal/mol + 4(157) kcal/mol

= approx. 1316 kcal/mol

(b) No, the heat of combustion of 2-methylheptane could not be predicted with any accuracy because this compound is branched, not continuous-chain.

3.31 (a) hexane (b) 2-butene (c) 1-pentanol

Branching decreases the boiling point by interfering with van der Waals attractions.

3.32 [cyclobutane]—OH; cyclopropane with OH and CH_3; cyclopropane with CH_2OH; cyclopropane with H_3C and OH

The general formula is $C_nH_{2n}O$; therefore, there must be either one site of unsaturation or one ring. Because the problem states that the compound is an alcohol and contains no carbon-carbon double bonds, the structure must contain either a three- or four-membered ring.

3.33 $\overset{O}{\overset{\|}{H}}CCH_2CH_2CO_2H$, $HC\overset{O}{\overset{\|}{}}CHCO_2H$ (with CH_3 on the second carbon) Both the aldehyde and the carboxyl groups must be at the ends of a chain. Thus, the other two carbon atoms must be positioned between these two functional groups. (If you allow other functional groups, such as C=C, there are other correct answers.)

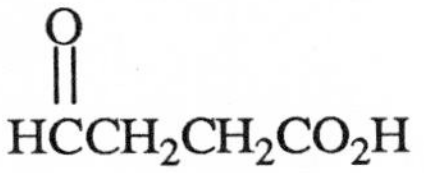

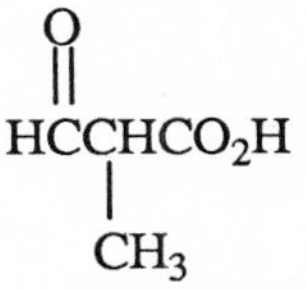

3.34 If $C_4H_6O_2$ ($C_nH_{2n-2}O_2$) contains no rings or carbon-carbon double bonds and is a ketone, it must contain two carbonyl groups. The only possible ketone that is not a keto aldehyde is 2,3-butanedione.

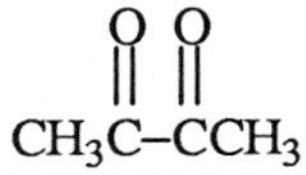

$$CH_3\overset{O}{\overset{\|}{C}}-\overset{O}{\overset{\|}{C}}CH_3$$

3.35 If C_2H_4O ($C_nH_{2n}O$) contains no double or triple bonds, it must contain one ring. A ring must contain at least three ring atoms; therefore, the structure must be as follows:

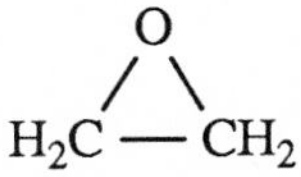

$$H_2C\overset{O}{\triangle}CH_2$$

3.36 (a) $CH_3CH_2CH_3 + Cl_2 \xrightarrow{h\nu} CH_3CH_2CH_2Cl + CH_3CHClCH_3 + HCl$

(b) $ClCH_2CH_2CH_2Cl$, $CH_3CHClCH_2Cl$, $CH_3CH_2CHCl_2$, $CH_3CCl_2CH_3$

(c) There are six CH_3 hydrogens and two CH_2 hydrogens in propane. Therefore, a completely random substitution would result in 6 parts $CH_3CH_2CH_2Cl$ and 2 parts $CH_3CHClCH_3$. The ratio of 1-chloroethane to 2-chloroethane would be 6:2, or 3:1.

3.37 $CH_3CH_2CH_2 \vdots CH_2CH_2CH_2CH_3 \xrightarrow{heat} CH_3CH{=}CH_2 + CH_3CH_2CH_2CH_3$

propene butane

Because propene contains three carbons from heptane, the alkane must contain the remaining four. Because an alkane is saturated, this product must be butane.

3.38 (a) $CH_3(CH_2)_4CH_3 \xrightarrow[\text{catalyst}]{\text{heat}}$ benzene $+ \ 4\,H_2$

(b) $CH_3(CH_2)_5CH_3 \xrightarrow[\text{catalyst}]{\text{heat}}$ C_6H_5–CH_3 $+ \ 4\,H_2$

(c) $CH_3(CH_2)_6CH_3 \xrightarrow[\text{catalyst}]{\text{heat}}$

- C_6H_5–CH_2CH_3 $+ \ 4\,H_2$
- C_6H_5–$CH{=}CH_2$ $+ \ 5\,H_2$
- C_6H_5–$C{\equiv}CH$ $+ \ 6\,H_2$

Chapter 4

Stereochemistry

4.20

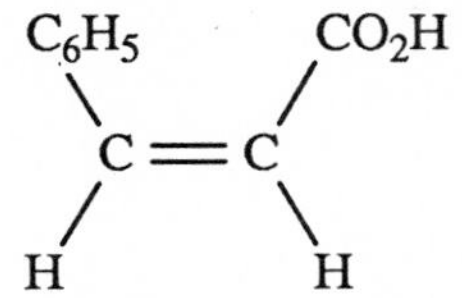

C_6H_5 / H on C=C with CO_2H / H (same side)

cis-cinnamic acid or
(*Z*)-3-phenylpropenoic acid
(H's on same side)

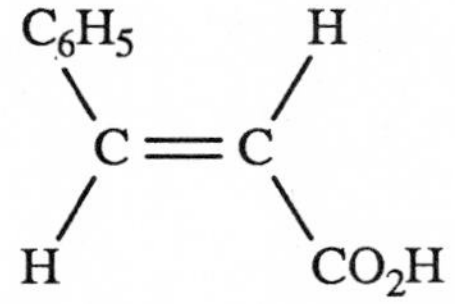

C_6H_5 / H on C=C with H / CO_2H (opposite sides)

trans-cinnamic acid or
(*E*)-3-phenylpropenoic acid
(H's on opposite sides)

4.21

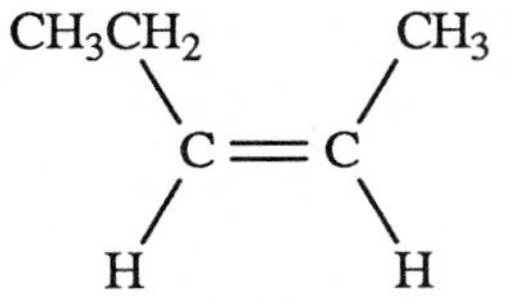

CH_3CH_2 / H on C=C with CH_3 / H

cis-2-pentene

CH_3CH_2 / H on C=C with H / CH_3

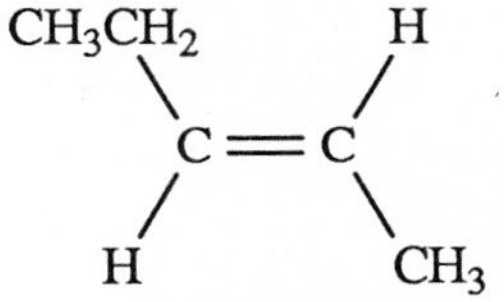

trans-2-pentene

1-Pentene, 2-methyl-1-butene, and 2-methyl-2-butene do not have geometric isomers.

4.22 In each case, the higher-priority atoms or groups are circled.

(a) CH_3, (Cl) on C=C with (Br), CH_3

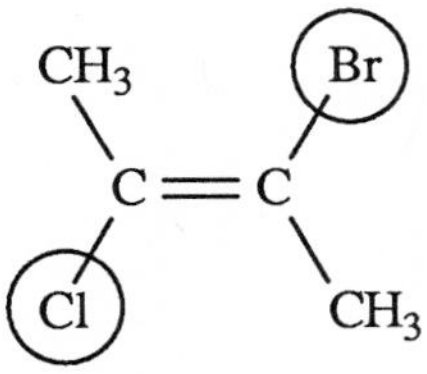

(*E*)

(b) (CH_3), H on C=C with (CO_2H), CH_3

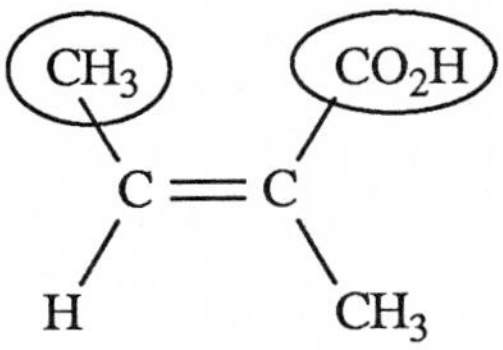

(*Z*)

(c) H, (CH_3) on C=C with CHO, (CO_2H)

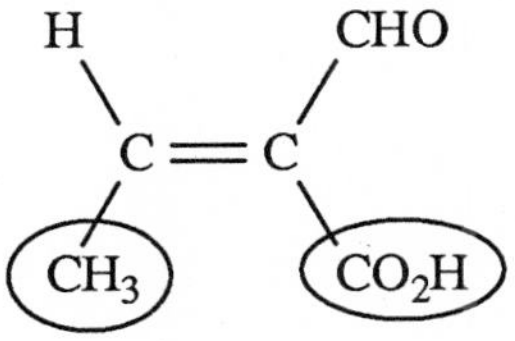

(*Z*)

(d) Cl, (Br) on C=C with H, (CH_3)

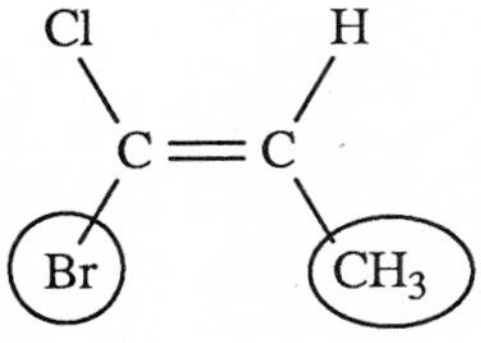

(*Z*)

(e) 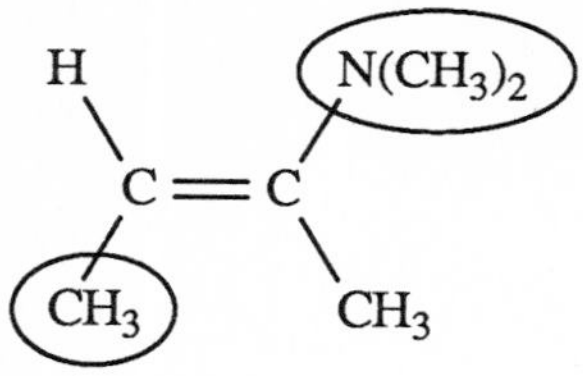 H, $N(CH_3)_2$ (circled), $C=C$, CH_3 (circled), CH_3

(*E*)

(f) 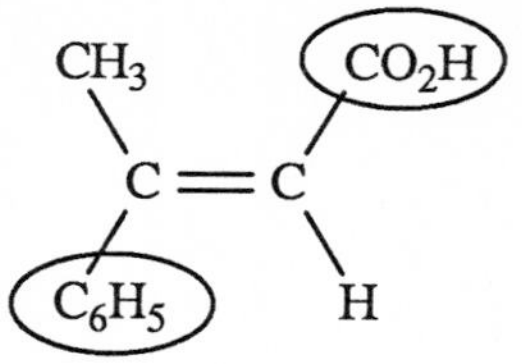 CH_3, CO_2H (circled), $C=C$, C_6H_5 (circled), H

(*E*)

4.23 (a) Only the double bond joining carbons 6 and 7 can lead to geometric isomerism. (You might find it helpful to write the formula of the compound first without regard to stereochemistry and then rewrite the formula to include stereochemistry.)

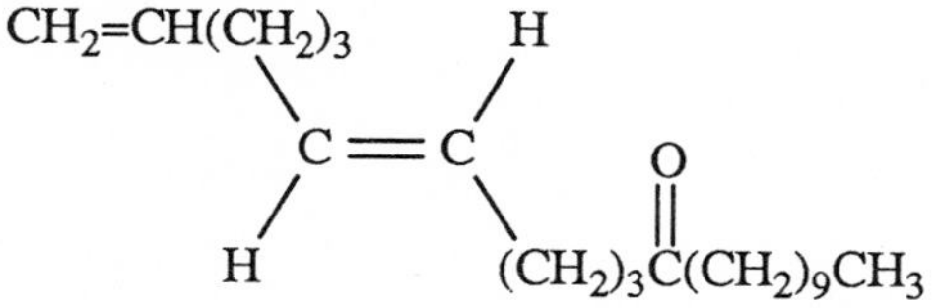

$CH_2{=}CH(CH_2)_3$, H, $C=C$, H, $(CH_2)_3\overset{O}{\overset{\|}{C}}(CH_2)_9CH_3$

(*E*)-1,6-heneicosadien-11-one

$CH_2{=}CH(CH_2)_3$, $(CH_2)_3\overset{O}{\overset{\|}{C}}(CH_2)_9CH_3$, $C=C$, H, H

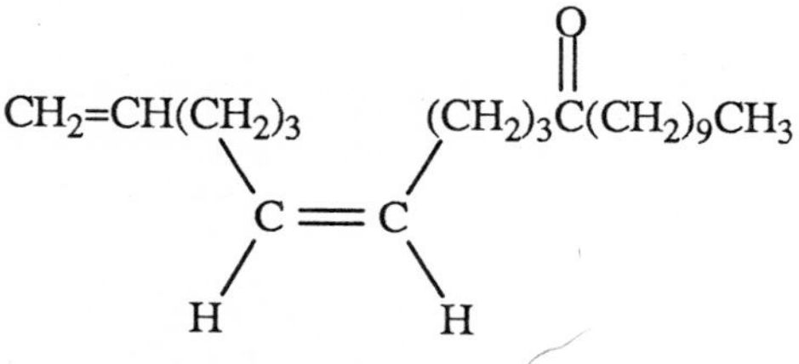

(*Z*)-1,6-heneicosadien-11-one

(b) 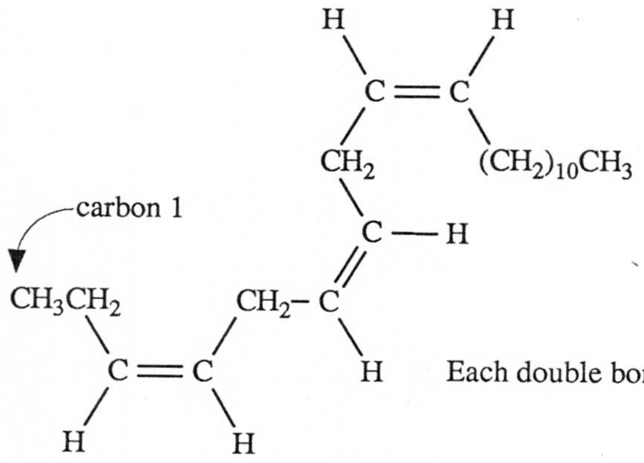 H, H, $C=C$, CH_2, $(CH_2)_{10}CH_3$, C—H, carbon 1, CH_3CH_2, CH_2– C, $C=C$, H, H, H

Each double bond is (*Z*).

(c) 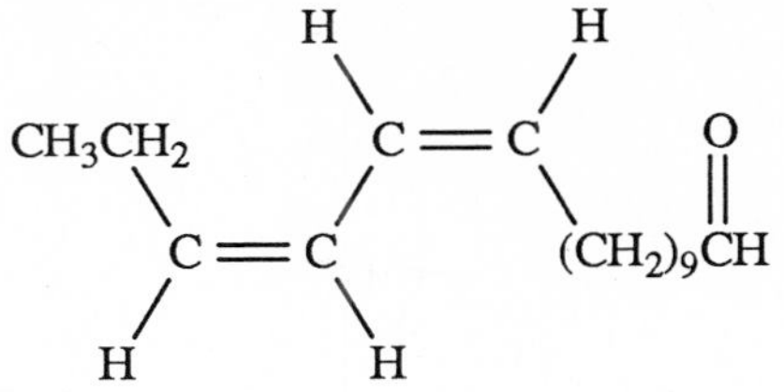 CH_3CH_2(H)C=C(H)–C(H)=C(H)–$(CH_2)_9$CH=O

4.24 (a) 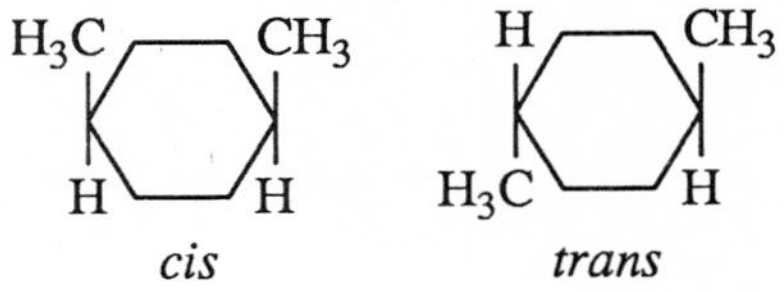 cis trans (b) none

(c) 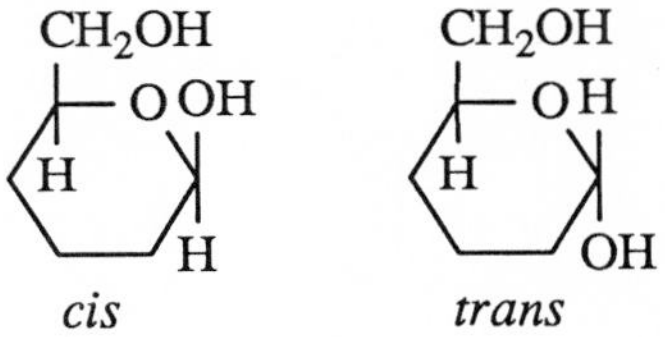 cis trans

In (b), there are no geometric isomers because the ring carbon atoms are sp^2 hybridized. This hybridization means that the ring carbons and the two methyl carbons all lie in the same plane--the methyl groups cannot be "on the same side" or "on opposite sides."

4.25

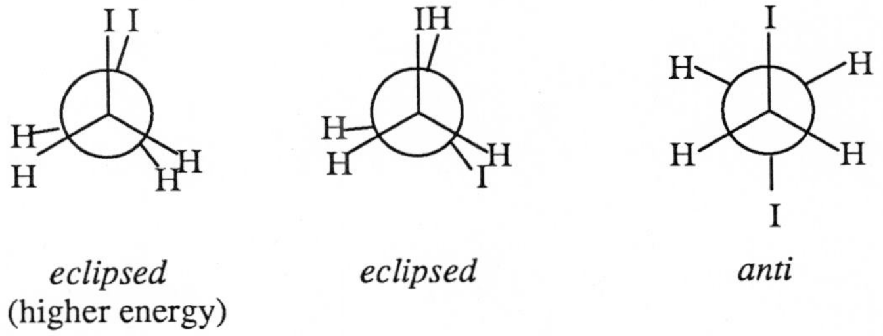

eclipsed (higher energy) eclipsed anti

In drawing different Newman projections for a compound, rotate only one of the two carbons of the projection. In our answer, the front carbon is fixed and the rear carbon is rotated. (Note that we have shown the eclipsed atoms not completely eclipsed so that all atoms can be seen.)

4.26 (a) Newman projection: front H, H; back CO_2H (top), H; bottom $CHCH_3$ with CO_2H — *anti*

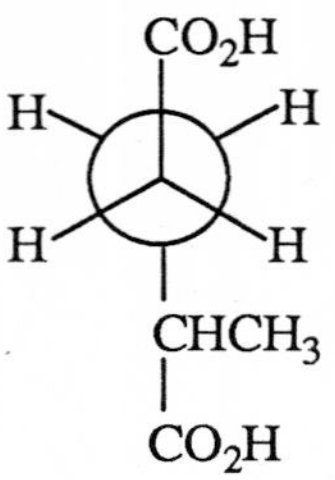

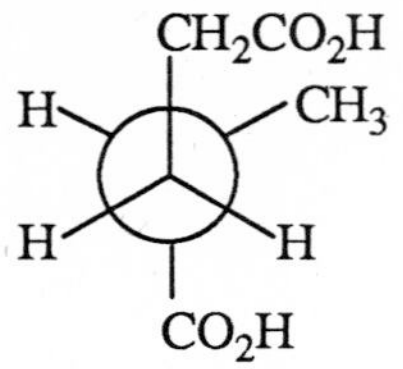

(b) Newman projection: CH_2CO_2H (top), CH_3, H, H, H; CO_2H (bottom) — *anti*

(c) Newman projection: H (top), CO_2H, H, H, H; $CH_2CH_2CO_2H$ (bottom) — not *anti*

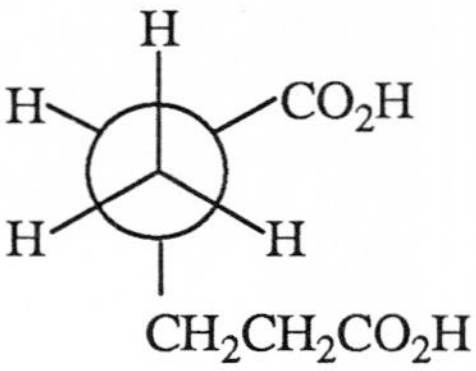

In (a) and (b), the two largest groups are *anti*. There is no *anti*-conformation for (c) around the two carbons shown because both large groups are on the same carbon.

4.27 (a) α-Pinene contains a four-membered ring and two six-membered rings (depending on how you view the structure). The four-membered ring is strained.

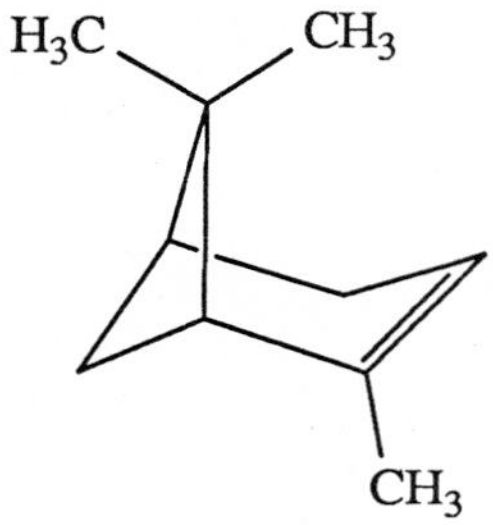

H_3C CH_3

CH_3

(b) Scopolamine contains a benzene ring (unstrained), a five-membered ring (unstrained), a six- or seven-membered ring (unstrained), and a three-membered oxygen-containing ring (strained).

4.28 (a), (b), (d), (e), and (g) are equatorial. (c) and (f) are axial.

4.29 (a) The second conformation of *cis*-1,4-dimethylcyclohexane is more stable because the methyl groups are farther apart. The first conformation shown would suffer from severe steric hindrance.

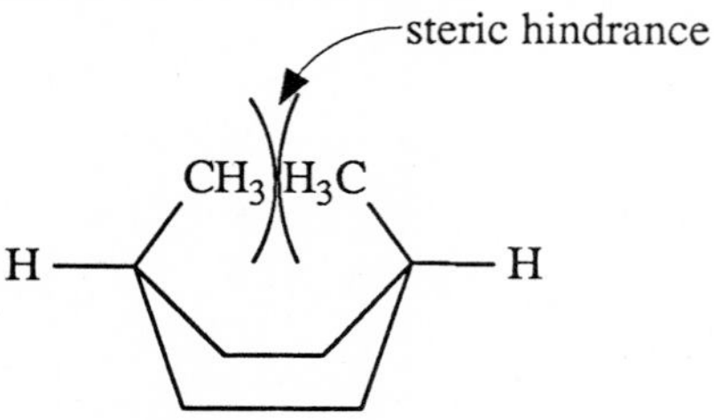

steric hindrance

CH_3 H_3C

H H

(b) The first isomer (*trans*-1,3-dimethylcyclohexane) is more stable than the conformation of the second isomer (*cis*) shown here. The first structure shows one methyl group equatorial and one axial while the second structure shows both methyl groups axial. (When the methyl groups in the *cis* isomer are both equatorial, that is the most stable chair-form conformer.)

(c) The *trans* isomer is more stable because its methyl groups can both be equatorial in one chair-form conformer. The *cis* isomer must have one methyl group axial in any chair-form conformer.

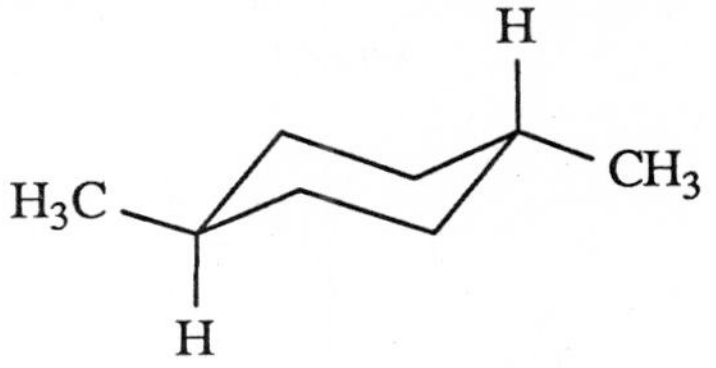

trans: *e,e* 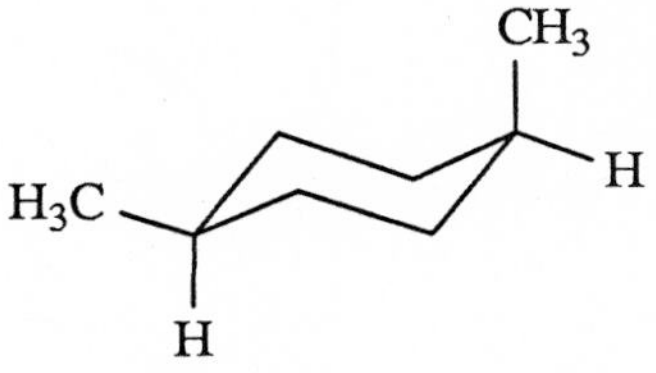 *cis*: *e,a*

4.30 (a), because the large *t*-butyl group is equatorial.

4.31 (a) In the *cis*-isomer, both substituents can be equatorial.

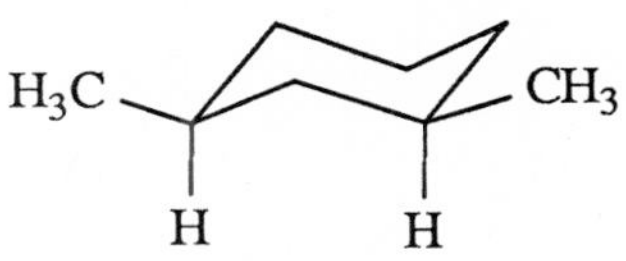

cis: *e,e* 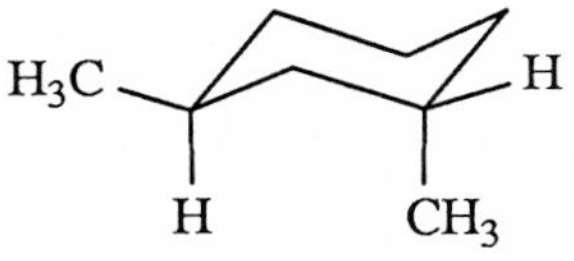 *trans*: *e,a*

(b) No, because any conformation of the *cis*-isomer has one group *e* and the other group *a*.

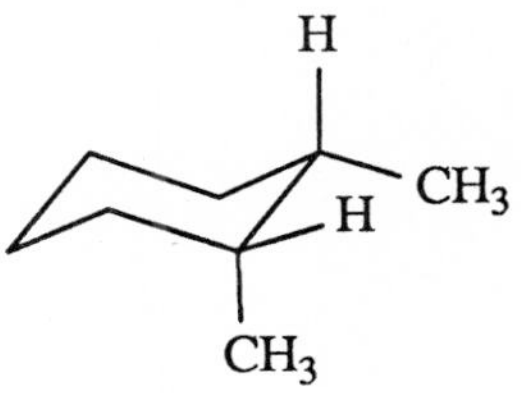

cis: *a,e* 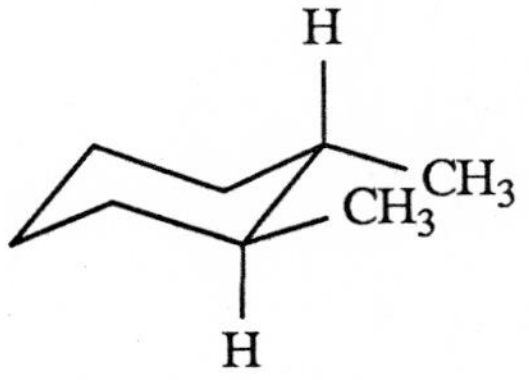 *trans*: *e,e*

4.32

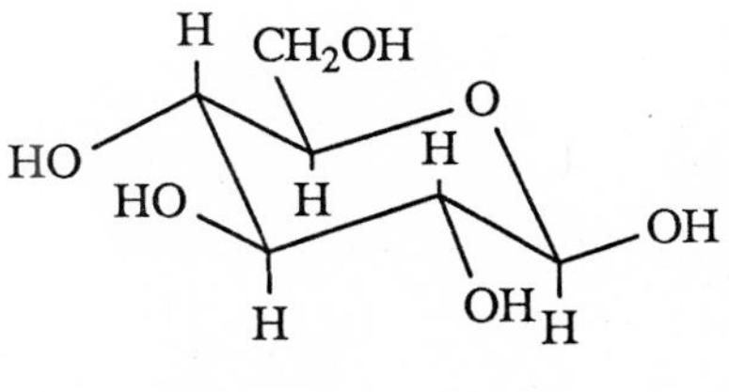

In this form of glucose, all the larger groups(-OH or -CH_2OH) are in equatorial positions. Therefore, this stereoisomer is the most stable. You may have drawn the mirror image (enantiomer) of the form of glucose that we have shown. This answer is also correct.

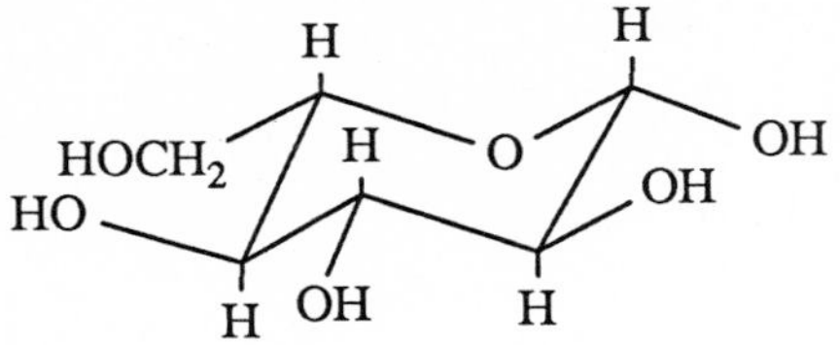

HOCH$_2$, H, O, OH, HO, OH, H, H, H OH, H

4.33 (a) 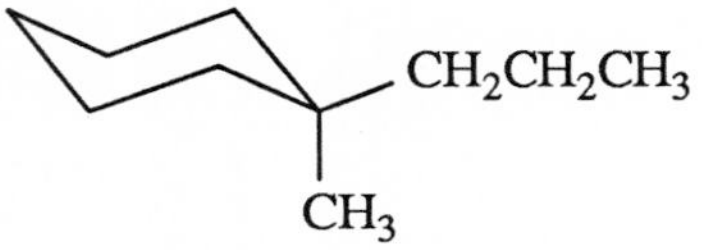 $CH_2CH_2CH_3$, CH_3 (b) 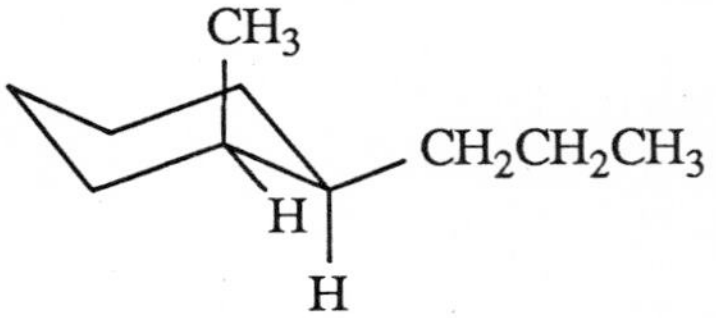 CH_3, $CH_2CH_2CH_3$, H, H

(c) 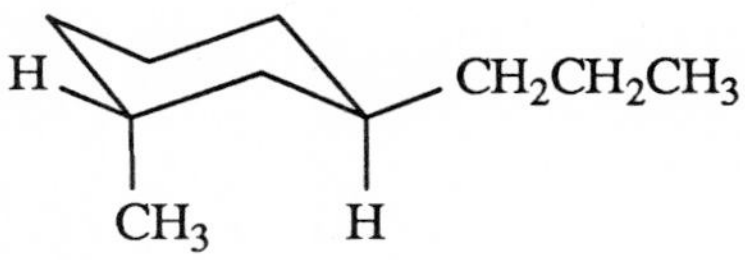 H, $CH_2CH_2CH_3$, CH_3, H

The most stable conformer is the one in which the largest group (propyl) is in an equatorial position.

4.34 (a) $(CH_3)_2CH\overset{*}{C}HBrCH_3$ (b) $CH_3CH_2CH_2\overset{*}{C}HOH$ with CH_3 (c) none

(d) $H_2N\overset{*}{C}HCO_2H$ with CH_3 (e) 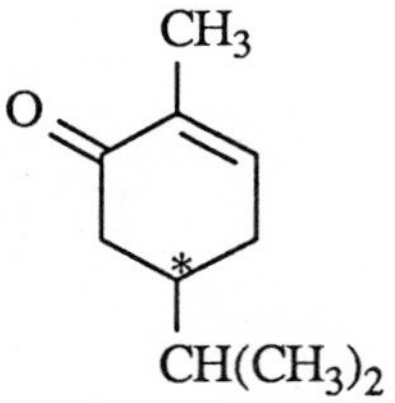 O, CH_3, $CH(CH_3)_2$ (f) 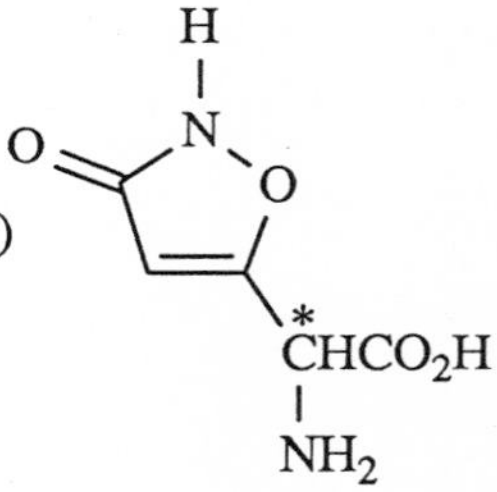 H, N, O, O, $\overset{*}{C}HCO_2H$, NH_2

4.35 (a)

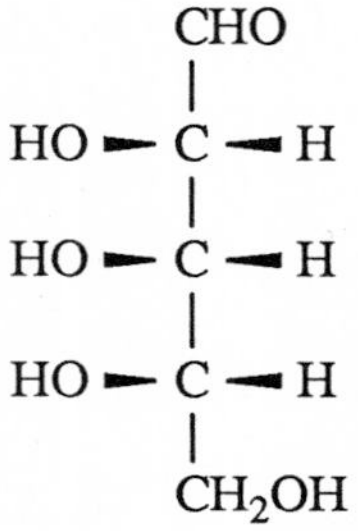

CHO
HO — C — H
HO — C — H
HO — C — H
CH_2OH

(b)

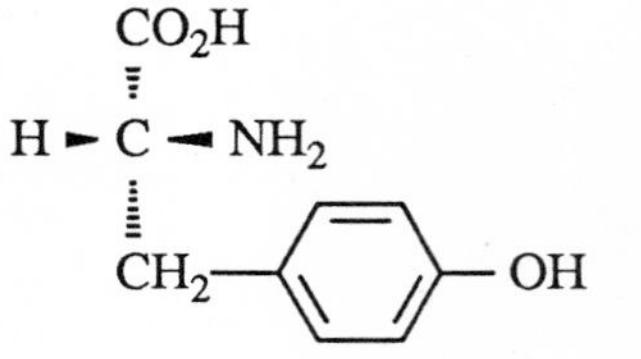

CO_2H
H — C — NH_2
CH_2— (benzene ring) —OH

The easiest way to solve this type of problem is to draw a line representing a mirror reflection at each chiral carbon. The result is that each group originally on the left is switched to the right and vice versa. (We need not reverse groups on achiral carbon atoms because the configuration is the same regardless of how we draw these groups.)

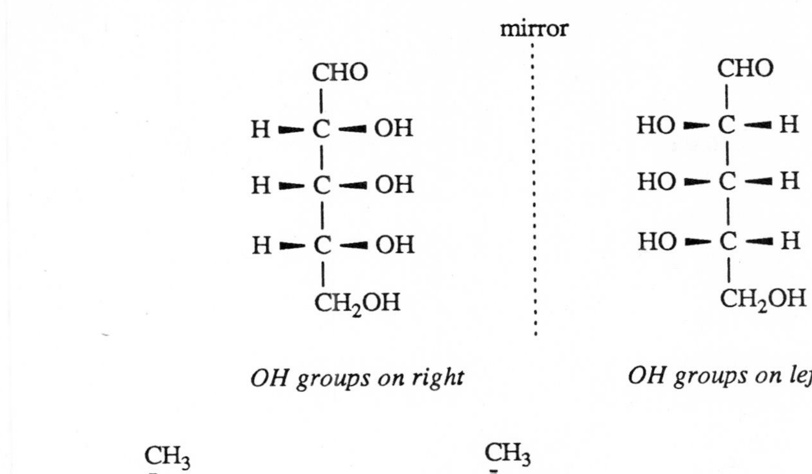

4.36

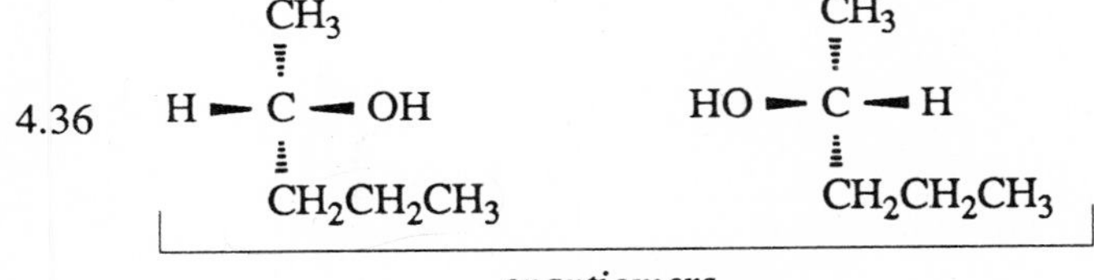

enantiomers

CH_3
H — C — OH
$CH(CH_3)_2$

CH_3
HO — C — H
$CH(CH_3)_2$

enantiomers

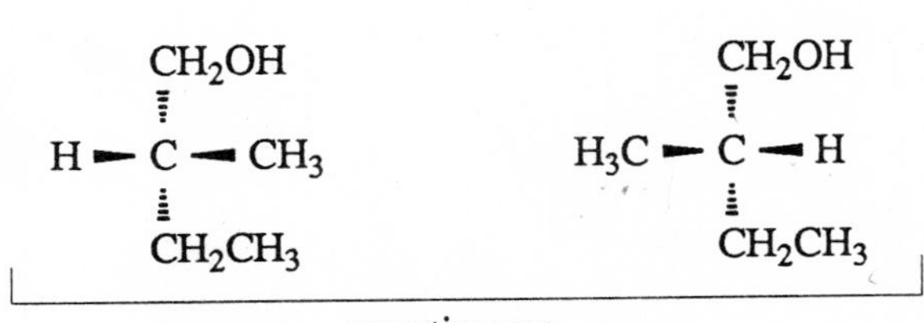

enantiomers

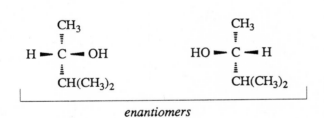

4.37 $CH_3CH(Br)CO_2H$, not $BrCH_2CH_2CO_2H$, because the optically active compound must have a chiral carbon atom.

4.38 Carbons 2 and 3 are chiral: $CH_3(CH_2)_6\overset{*}{C}H(OH)-\overset{*}{C}H(OH)CH_2OH$

Using R- to represent $CH_3(CH_2)_6$- and placing carbon 2 as the front carbon:

(2*R*,3*R*) (2*S*,3*S*) — enantiomers

(2*R*,3*S*) (2*S*,3*R*) — enantiomers

We have designated the configuration of each chiral carbon as (*R*) or (*S*); these terms are defined in Section 4.8.

4.39 (a) CO_2H top, H_2N left, H right, $CH_2CH(CH_3)_2$ bottom (b) CO_2H top, HO left, H right, CH_3 bottom (c) CHO top, Br left, CH_3 right, CH_2CH_3 bottom

(d) CHO top; H–C–OH; H_2N–C–H; CH_3 bottom (e) CO_2H top; H–C–Br; H–C–Br; CO_2H bottom ≡ CO_2H top; Br–C–H; Br–C–H; CO_2H bottom

In (e), construct models to verify that the two Fischer projections represent the same compound and not enantiomers.

4.40 (a) CO_2H top; H_2N–C–H; CH_2OH bottom (b) CHO top; H–C–OH; HO–C–H; CH_2OH bottom (c) CH_2OH top; C=O; H–C–OH; CH_2OH bottom

4.41 To convert Fischer projections to enantiomers, simply transpose the groups or atoms at each chiral carbon (at each intersection of lines). See Answer 4.35.

(a) CO_2H; H — — NH_2; CH_2OH

(b) CHO; HO — — H; H — — OH; CH_2OH

(c) CH_2OH; C=O; HO — — H; CH_2OH

4.42 None of the structures as a whole is chiral because each contains an internal plane of symmetry. Compounds (a) and (b) contain no chiral carbons because no carbon in these structures is bonded to four different substituents. Compounds (c) and (d) are meso.

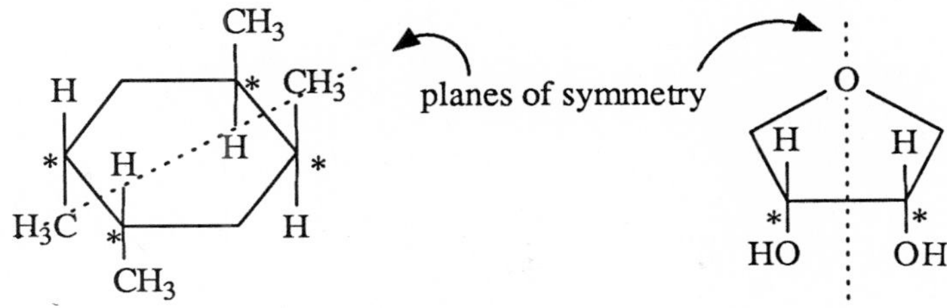

4.43 (b) and (d). The structures in (a) differ in the projection of only one chiral carbon. The structures in (c) are identical.

4.44 (a) is not a meso compound because of the configurations at carbons 3 and 4.

(b) is not meso because the functional groups at carbons 1 and 6 are different.

(c) is meso because it has an internal plane of symmetry in the conformation shown.

CO_2H
H — C — OH
HO — C — H
- - - - - - - - - - internal plane of symmetry
HO — C — H
H — C — OH
CO_2H

4.45 (a)

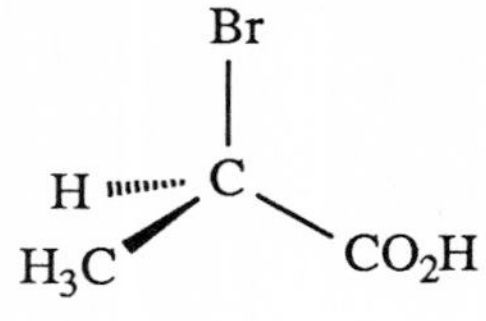

Br, H, H_3C, C, CO_2H or CO_2H, H, Br, CH_3

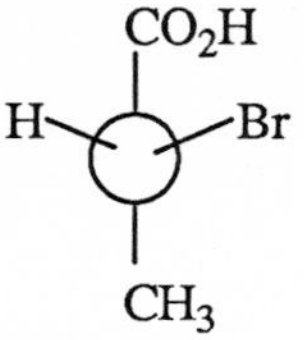

dimensional ball-and-stick

(b)

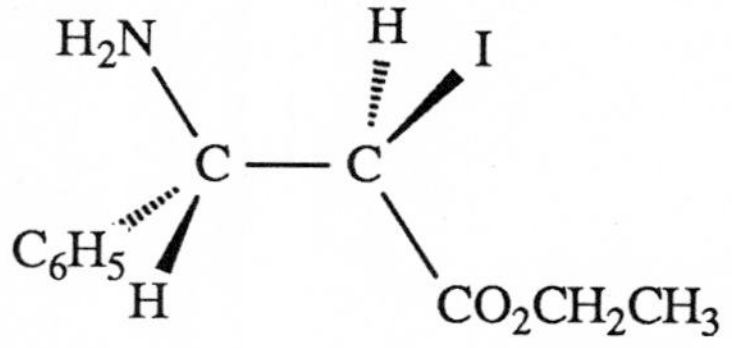

H_2N, C—C, C_6H_5, H, H, I, $CO_2CH_2CH_3$ or $CO_2CH_2CH_3$, H, I, H_2N, H, C_6H_5

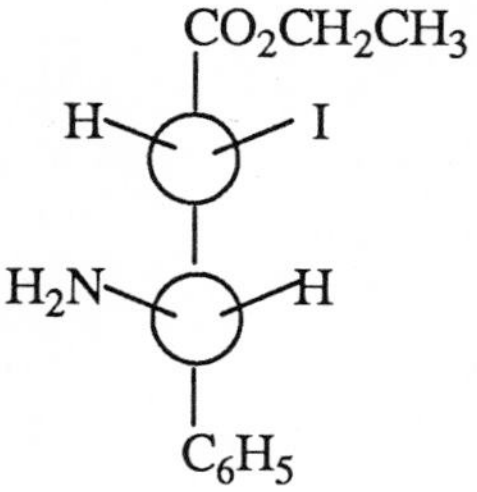

4.46 (a) 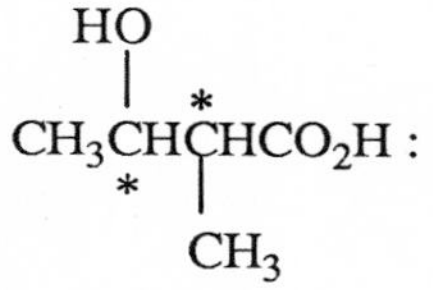 $CH_3\overset{*}{C}H(OH)\overset{*}{C}H(CH_3)CO_2H$: two chiral carbons, four stereoisomers

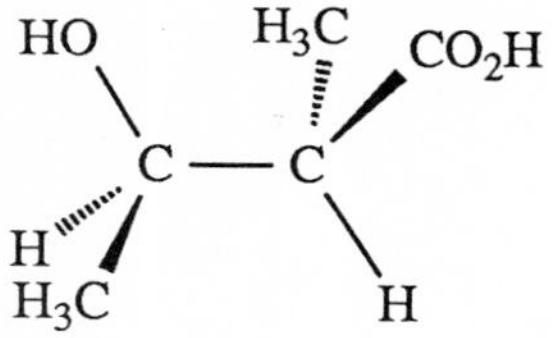

HO, H, H_3C, C—C, H_3C, CO_2H, H or

	CO_2H	
H		CH_3
H		OH
	CH_3	

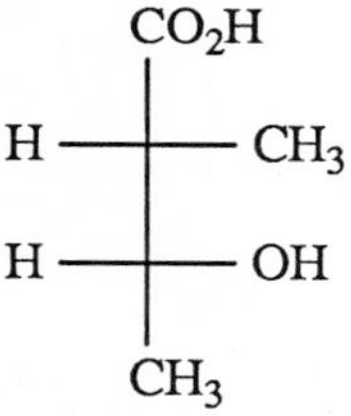

(2*R*,3*R*)-3-hydroxy-2-methylbutanoic acid

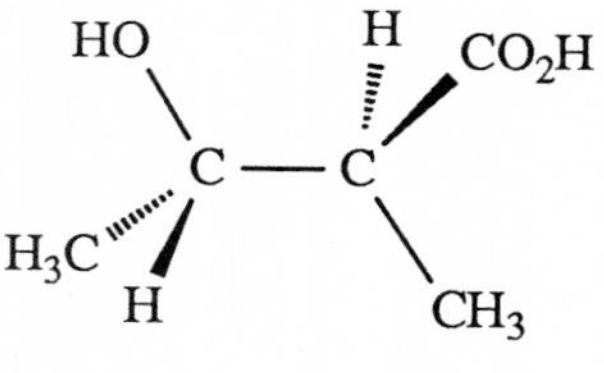

HO, H_3C, H, C—C, H, CO_2H, CH_3 or

	CO_2H	
H_3C		H
HO		H
	CH_3	

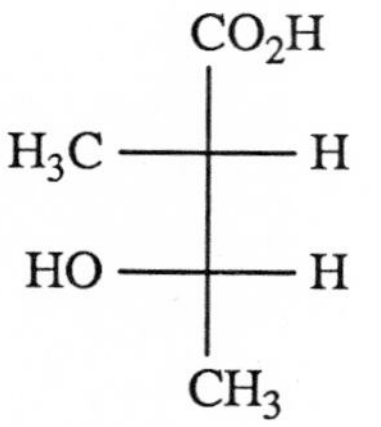

(2*S*,3*S*)

HO H$_3$C CO$_2$H
C — C
H$_3$C H H

or

CO$_2$H
H — CH$_3$
HO — H
CH$_3$

(2*R*,3*S*)

HO H CO$_2$H
C — C
H H$_3$C CH$_3$

or

CO$_2$H
H$_3$C — H
H — OH
CH$_3$

(2*S*,3*R*)

The priority assignments for the (2*R*,3*R*) stereoisomer follow (H in rear):

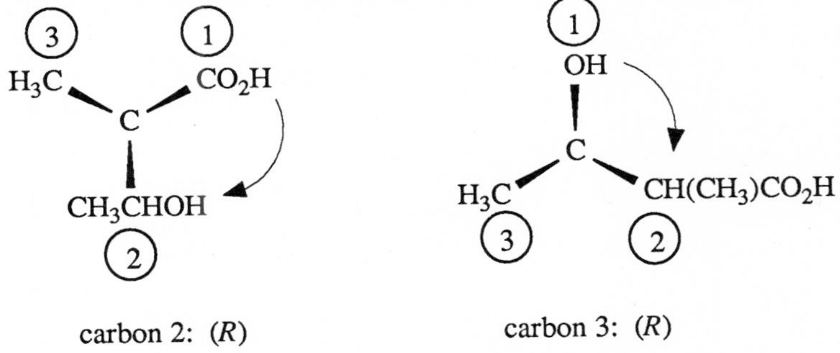

carbon 2: (*R*) carbon 3: (*R*)

When the configurations of one stereoisomer have been determined, the configurations of the other stereoisomers can be determined by comparison - for example, the enantiomer of the (2*R*,3*R*) isomer must be (2*S*,3*S*).

(b) $CH_3\underset{|}{C}HCH_2\overset{*}{C}HCH_2OH$ (CH$_3$ on C-2 and C-4): one chiral carbon, two stereoisomers

H$_3$C CH$_2$OH
(CH$_3$)$_2$CHCH$_2$— C
H

or

CH$_2$OH
H — CH$_3$
CH$_2$CH(CH$_3$)$_2$

(*R*)-2,4-dimethyl-1-pentanol

H, CH_2OH

$(CH_3)_2CHCH_2$— C   or H_3C — — H

CH_3 CH_2OH (top), $CH_2CH(CH_3)_2$ (bottom)

(*S*)-2,4-dimethyl-1-pentanol

4.47 (a) I and III are enantiomers (nonsuperposable mirror images).

(b) I and II are diastereomers, as are II and III (stereoisomers that are *not* enantiomers).

(c) II is a *meso* compound; a horizontal plane of symmetry may be drawn through carbon 3.

4.48

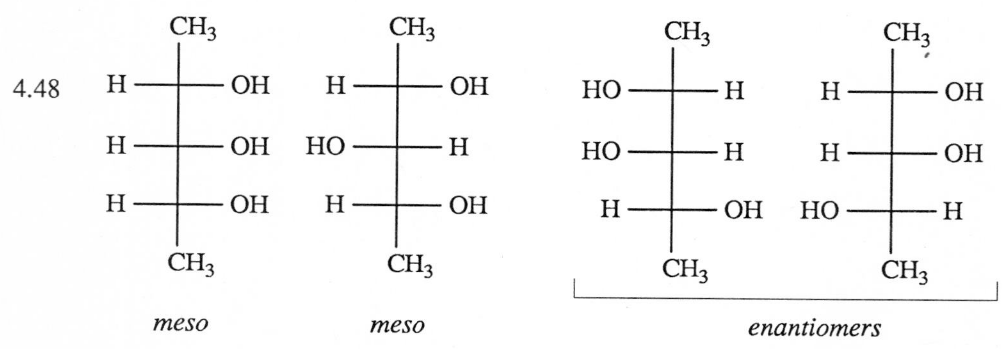

4.49 (a)

I	II	(b) III
CO_2H	CO_2H	CO_2H
Br — C — H	H — C — Br	H — C — Br
CH_2	CH_2	CH_2
H — C — Br	Br — C — H	H — C — Br
CO_2H	CO_2H	CO_2H

(c) Two pairs of diastereomers: I and III, II and III.

4.50 (a) (*R*) (b) (*R*) (c) (*S*) (d) (*S*) (e) (*S*)

(f) (2*S*,3*R*)

4.51 (b) is a meso form. The dashed line represents the internal plane of symmetry. No such plane of symmetry can be drawn for (c). Although (a) does have an internal plane of symmetry, it has no chiral carbons and therefore cannot be a meso compound.

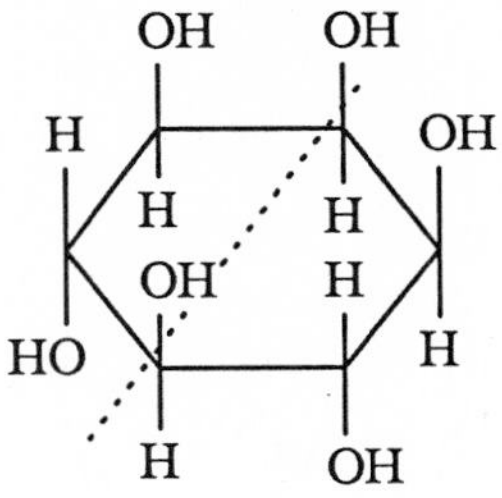

4.52 (a)-(c)

no chiral carbons:

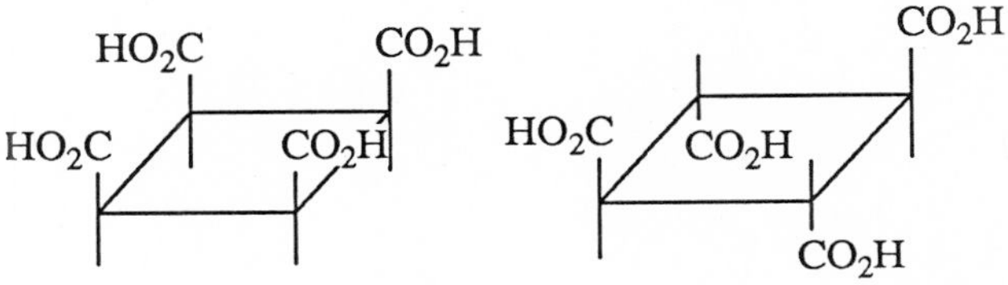

meso:

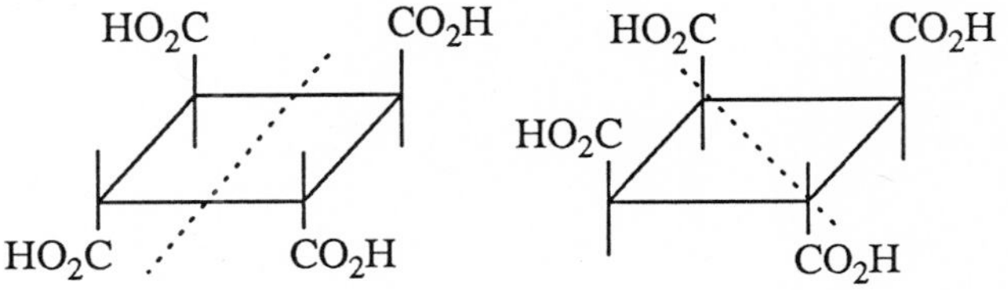

(d) None of these structures exists as an enantiomeric pair.

4.53 The most stable chair form of this compound would be the one in which a greater number of the large groups are in equatorial positions. Using R- to represent $-CH_2CH{=}CHCH_2CH{=}CHC{\equiv}CH$, let us draw the structures as follows:

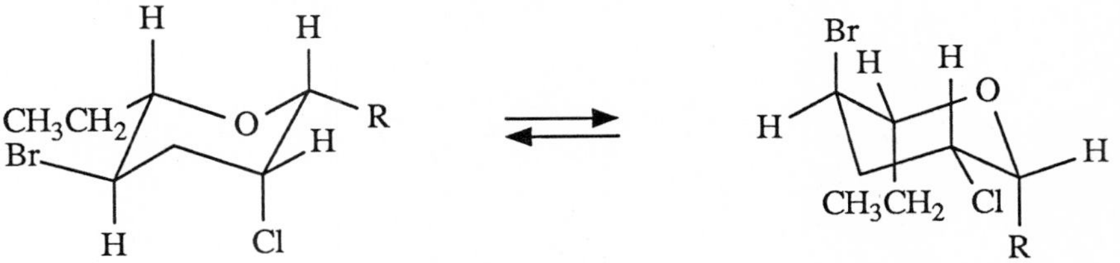

more stable

less stable

The first formula shows three equatorial substituents and one axial substituent. This chair form would be more stable than the second chair form in which only the Cl is equatorial.

4.54 (a) 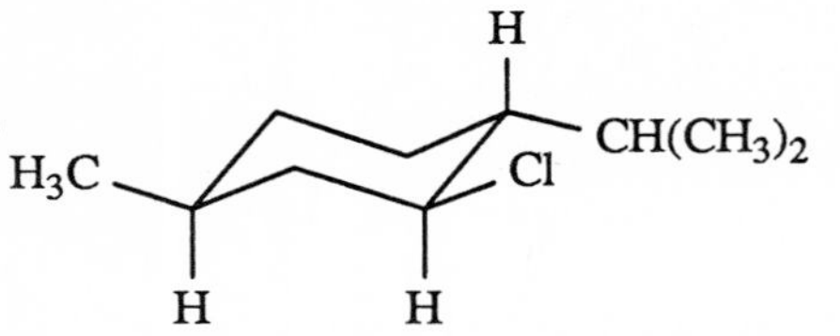 [chair structure: H_3C, H, Cl, H, $CH(CH_3)_2$, H] (b) 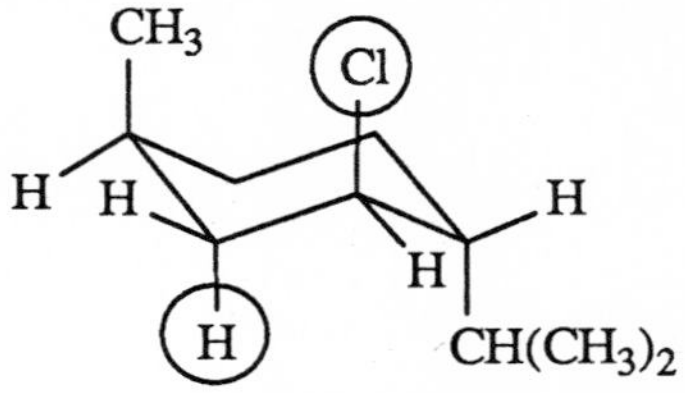 [chair structure: CH_3, Cl, H, H, H, H, H, $CH(CH_3)_2$]

Only a small percentage of molecules would be in the conformation shown in (b) at any given time because all the substituents are axial.

4.55 (a) 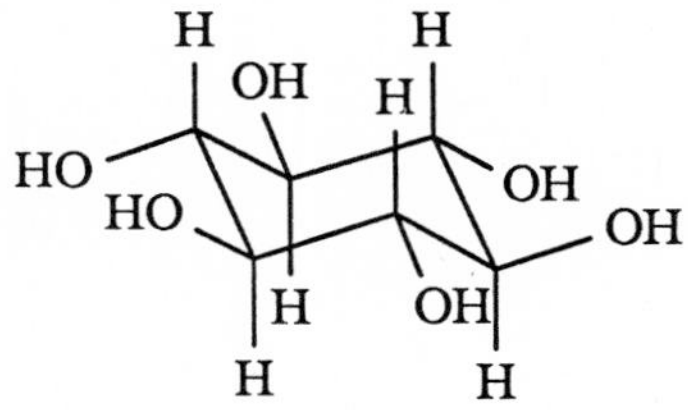 [chair structure: HO, HO, OH, OH, OH, OH, H, H, H, H, H, H]

(b) There is no simple way to solve a problem of this type. You may have to draw the formulas for all the stereoisomers to determine which may have internal planes of symmetry. The following formulas represent the enantiomeric pair. All other stereoisomers have an internal plane of symmetry.

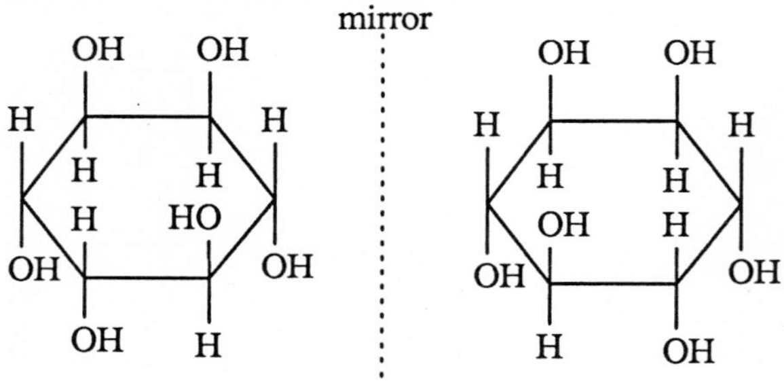

mirror

4.56 (a) Using the (1*S*) carbon as the front carbon:

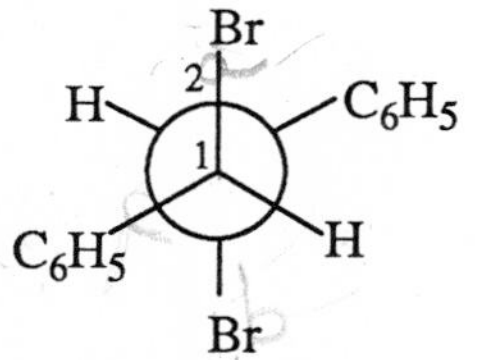

most stable
(Br's and phenyls *anti*)

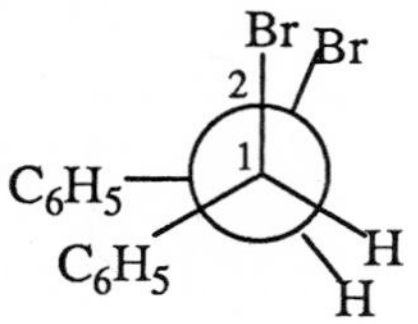

least stable

(b)

most stable *least stable*

4.57

4.58

(1*R*,2*R*) (1*S*,2*S*)

enantiomers

(1*R*,2*S*) (1*S*,2*R*)

enantiomers

The (1*S*,2*R*) and (1*R*,2*S*) isomers are diastereomers of either the (1*R*,2*R*) or the (1*S*,2*S*) isomer. Note the order of priority:

$$-Br > -\underset{C_6H_5}{CHBr} > -C_6H_5 > -\underset{CH_3}{CHC_6H_5} > -CH_3 > -H$$

4.59

```
     H       H
      \     /
       C = C          CH3
      /     \        /
   H3C       C  =  C
            /       \
           H         H
```

(2*Z*,4*Z*)-hexadiene

```
                 H         H
                  \       /
        H          C  =  C
         \        /       \
          C  =  C          CH3
         /       \
      H3C         H
```

(2*E*,4*Z*)-hexadiene
or (2*Z*,4*E*)-hexadiene

```
               H         CH3
                \       /
      H          C  =  C
       \        /       \
        C  =  C          H
       /       \
    H3C         H
```

(2*E*,4*E*)-hexadiene

4.60 (a) zero, because it is a racemic mixture.

(b) The solution is, in effect, half racemic and half (*S*)-enantiomer. The observed rotation is therefore half that of the pure (*S*)-enantiomer, or +8.0°.

4.61 (a) $$[\alpha]_D^{20} = \frac{\alpha}{lc} = \frac{+0.45^\circ}{0.10 \text{ dm} \times 0.200 \text{ g/mL}} = +22.5^\circ$$

(b) $$[\alpha]_D^{20} = \frac{-3.2^\circ}{1.0 \text{ dm} \times 0.10 \text{ g/mL}} = -32^\circ$$

4.62 (b), (c), and (e) would cause rotation. Compound (a) is meso and mixture (d) is racemic: neither of these would cause rotation.

4.63 In (a), the mixture is 50% optically active compound plus 50% optically inactive compound; therefore, the specific rotation is half that of the optically active compound, or -6°. In (b), the mixture is racemic; the specific rotation is therefore zero.

4.64 trans, racemic + (2R,3R)-tartaric acid →

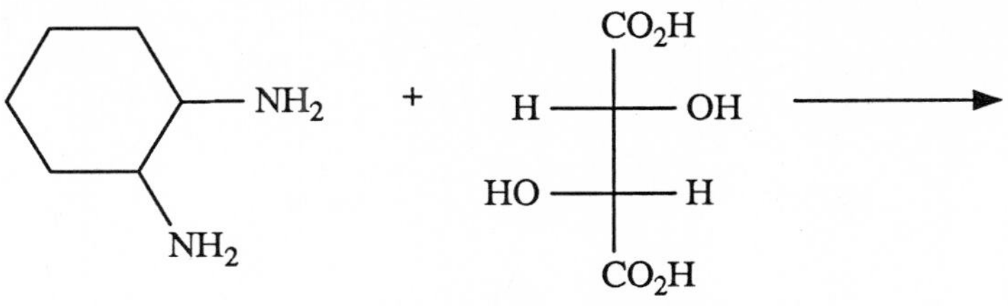

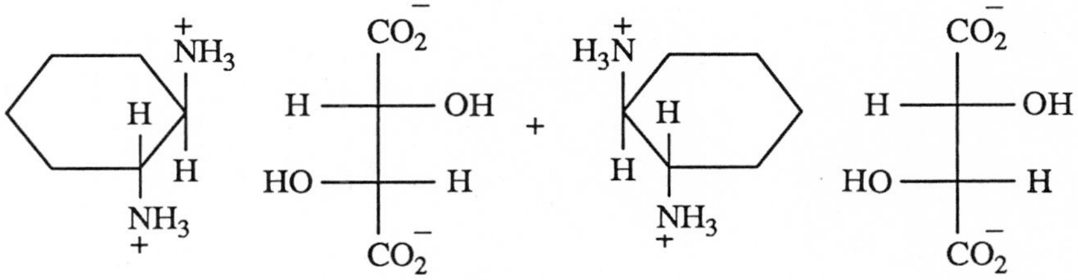

CO_2H
H — OH
HO — H
CO_2H

NH_2 NH_2

$\overset{+}{N}H_3$ H H NH_3^+ CO_2^- H OH HO H CO_2^- + $H_3\overset{+}{N}$ H H NH_3^+ CO_2^- H OH HO H CO_2^-

(R,R) (R,R) (S,S) (R,R)

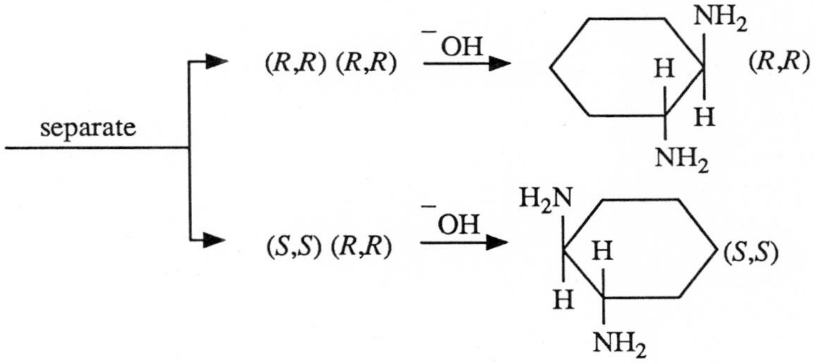

separate

→ (R,R) (R,R) $\xrightarrow{^-OH}$ NH_2 H H NH_2 (R,R)

→ (S,S) (R,R) $\xrightarrow{^-OH}$ H_2N H H NH_2 (S,S)

4.65 C C C

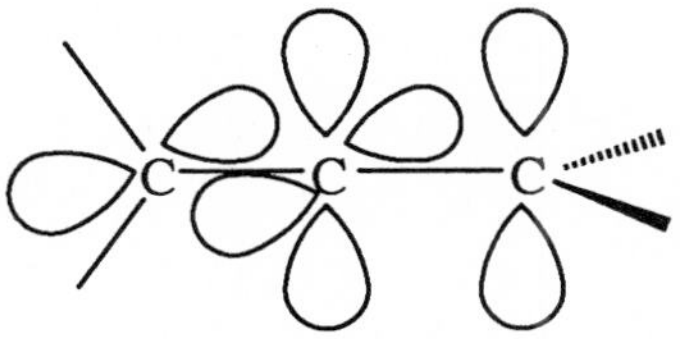

p orbitals

mirror

R_1 C=C=C R_2 H R_2 H C=C=C R_1 H

H

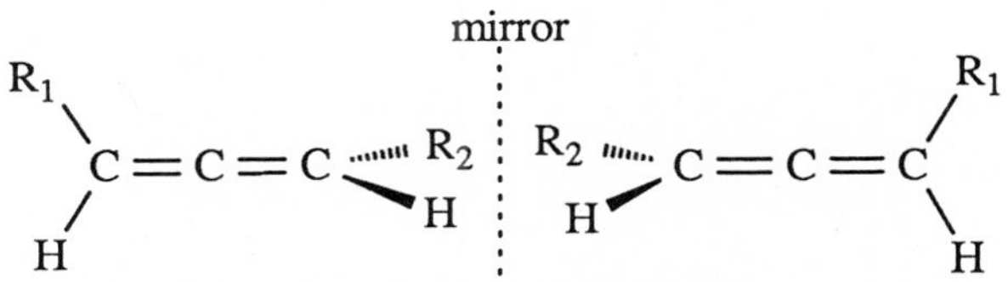

nonsuperposable mirror images,

or enantiomers

These molecules are chiral because they are nonsuperposable on their mirror images. Note that it is not necessary for a molecule to have a chiral carbon for the molecule to be chiral.

4.66 (a) α-Terpineol contains a chiral carbon and thus has an enantiomer.

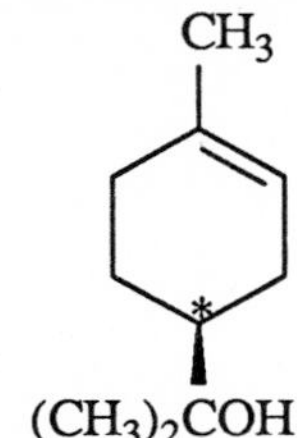

(b) A spiroheptadiene is chiral because its mirror images cannot be superposed. (See Answer 4.65.)

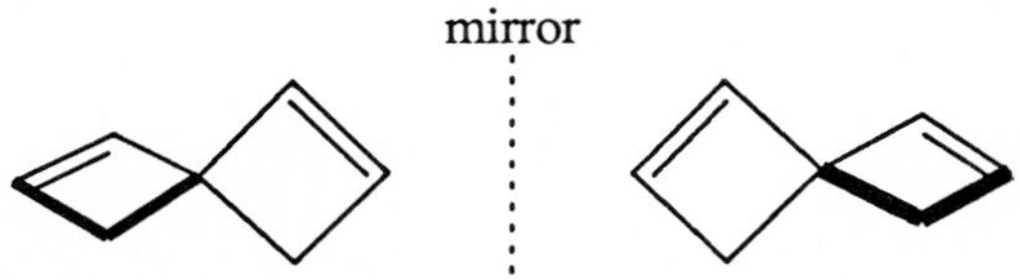

The nonsuperposability is easy to see, if the second structure is rotated 90°.

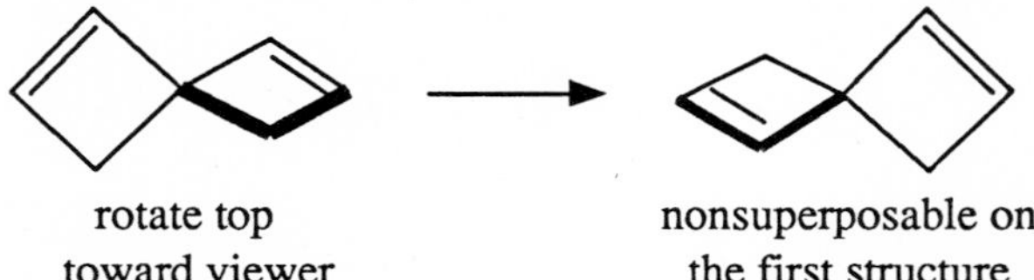

(c) contains a chiral carbon and has an enantiomer.

(d) has an enantiomer as well as a pair of diastereomers.

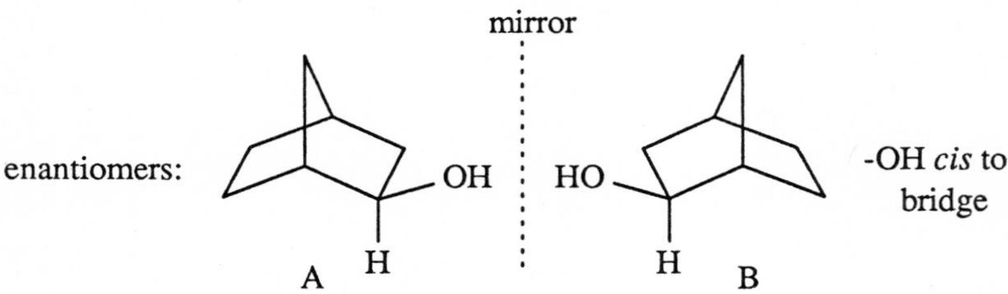

a different enantiomeric pair that are diastereomers of A and B:

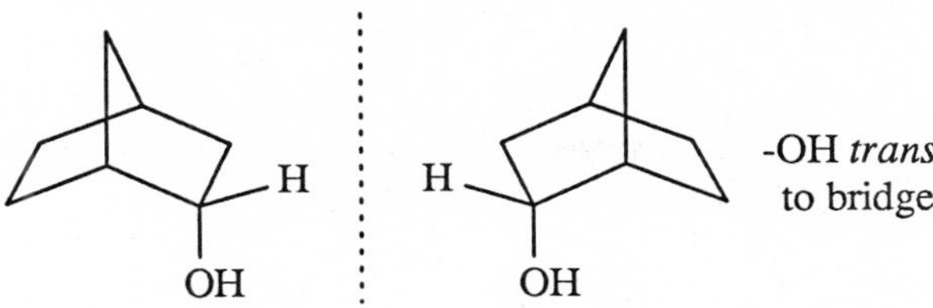

4.67 (a)

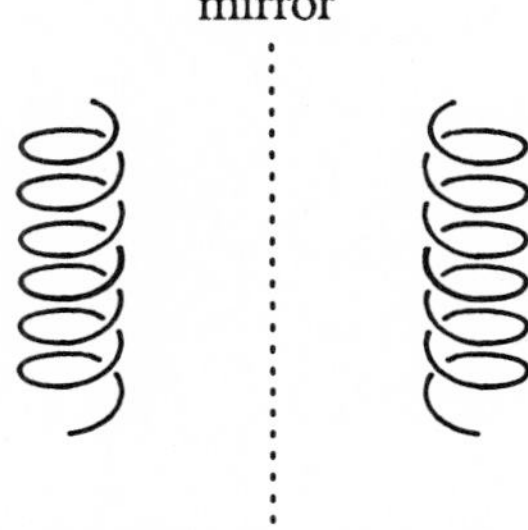

(b) Yes. The screw spiral can be threaded either to the right or to the left. One of these screws would require a clockwise twisting motion to screw it into the wood and the other, a counterclockwise motion.

4.68 (a) *cis:* (1*R*,2*S*) (1*S*,2*R*)

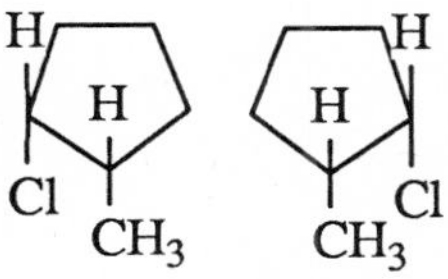

trans: (1*S*,2*S*) (1*R*,2*R*)

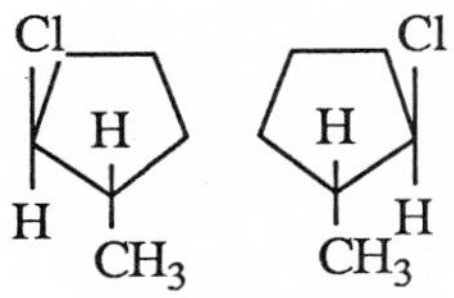

(b) *cis:* (1*R*,3*S*) (1*S*,3*R*)

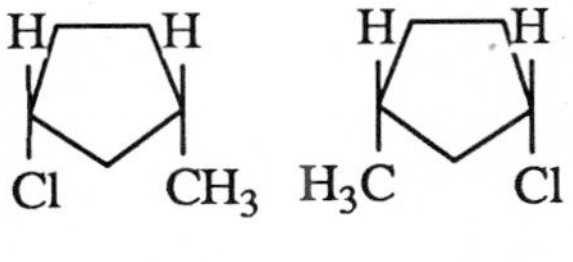

trans: (1*S*,3*S*) (1*R*,3*R*)

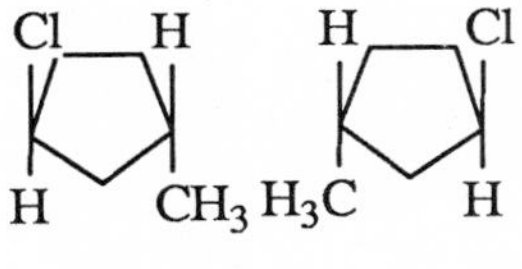

4.69 (a) The twisting of the ring can be either right- or left-handed. The two mirror images cannot be superposed, nor can they be interconverted without breaking bonds.

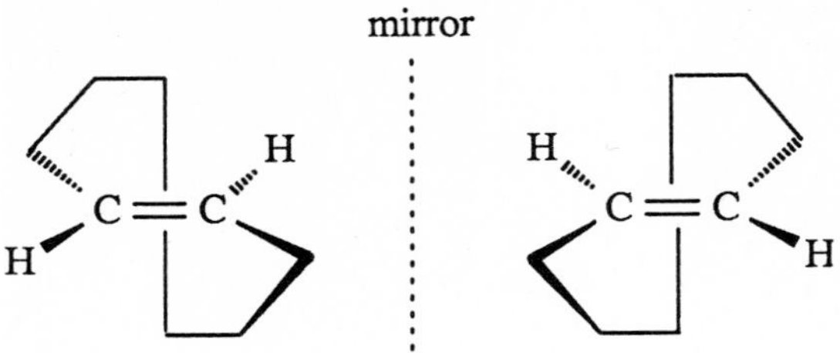

(b) The name (from *helix*) should give you a clue. Because of steric hindrance, the molecules are shaped like right- or left-handed helices (spirals), which cannot be superposed.

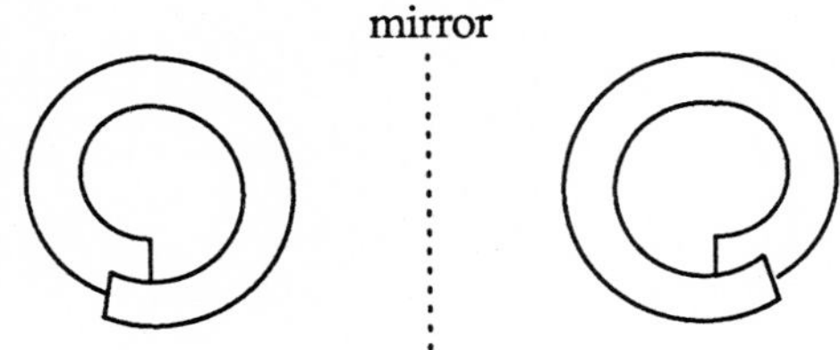

Chapter 5

Alkyl Halides; Substitution and Elimination Reactions

5.24 (a) 3,3,3-trichloropropene (b) 1,3-dibromobutane

(c) 1-bromo-1-methylcyclohexane

(d) *trans*-2-chlorocyclopentanol

(e) (*R*)-2-iodopropanoic acid

5.25 (a) and (b), $(CH_3)_2CHCH_2I$

(c)

H

H

Cl

Cl

(d) $(CH_3)_2CHCHBrCH_2OH$

(e)

Br H CH_3

C — C

H CH_3 Cl

or

CH_3

Br H

H Cl

CH_3

5.26 (a) 1° (b) 3° (c) 3°

5.27 (a) $(CH_3CH_2)_2CHI$, the 2° halide (b) $(CH_3)_2CHI$, the iodide

(c) CH_2Cl, the 1° halide (d) Cl, the 2° halide

5.28 (a) Reaction 1 (b) Reaction 2 because its E_{act} is lower

(c) Because Reaction 2 is faster, its products will predominate.

5.29 rate = k [CH_3I] [OH^-]

(a) The rate is multiplied by 3 x 2, or 6.

(b) The rate is halved. (c) The rate increases.

(d) The concentration of each reactant is halved; therefore, the rate is divided by four: (1/2) x (1/2) = 1/4.

5.30 (a) is preferred because (b) would lead to an alkene instead of the desired ether.

5.31 (a) 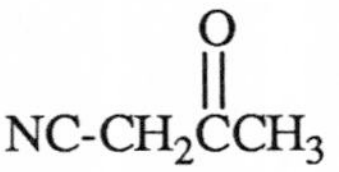 $NC\text{-}CH_2C(=O)CH_3$ (b) 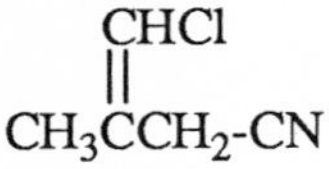 $CH_3C(=CHCl)CH_2\text{-}CN$

In (b), we expect displacement of the allylic Cl, as shown, but not displacement of the vinylic Cl, because these do not undergo ordinary S_N2 reactions (see Section 5.1).

(c) $HOCH_2CH_2CN$

In (c), only the Cl^- is a leaving group.

5.32 (a) 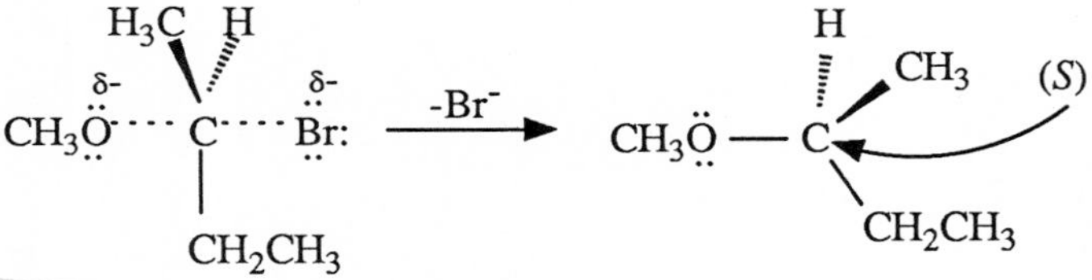 $CH_3\ddot{O}^{\delta-}\cdots C(H_3C)(H)(CH_2CH_3)\cdots Br:^{\delta-}$ $\xrightarrow{-Br^-}$ $CH_3\ddot{O}—C(H)(CH_3)(CH_2CH_3)$ (*S*)

(b) H_3C–cyclohexane–CH_2 with $:Cl:^{\delta-}$ and $\delta- :OCH_3$ $\xrightarrow{-Cl^-}$ H_3C–cyclohexane–$CH_2\ddot{O}CH_3$ still *trans*

(c) $CH_3\ddot{O}^{\delta-}\cdots C(H_3C)(H)(CH(CH_3)_2)\cdots Br:^{\delta-}$ $\xrightarrow{-Br^-}$ $CH_3\ddot{O}—C(H)(CH_3)(CH(CH_3)_2)$ (*S*)

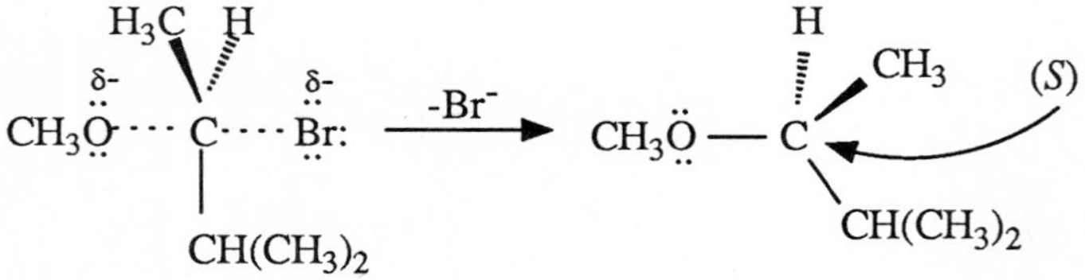

(d) 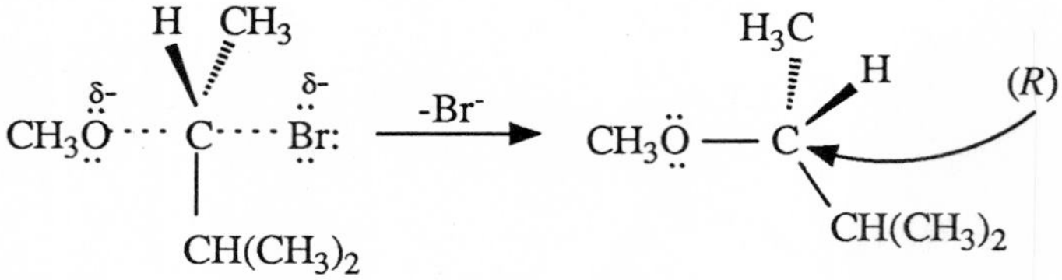 $CH_3\ddot{O}^{\delta-}\cdots C(H)(CH_3)(CH(CH_3)_2)\cdots Br:^{\delta-}$ $\xrightarrow{-Br^-}$ $CH_3\ddot{O}—C(H_3C)(H)(CH(CH_3)_2)$ (*R*)

5.33 (a) 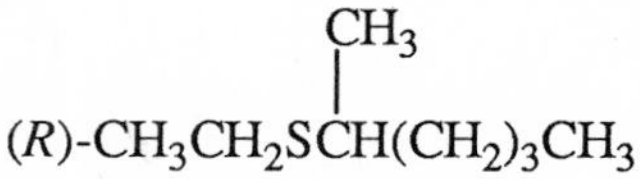 (*R*)-$CH_3CH_2SCH(CH_3)(CH_2)_3CH_3$ (b) 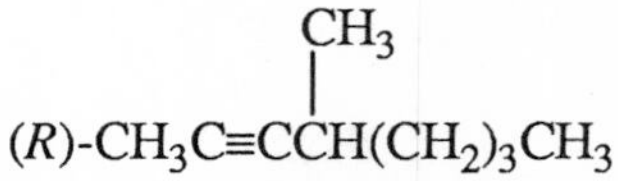 (*R*)-$CH_3C{\equiv}CCH(CH_3)(CH_2)_3CH_3$

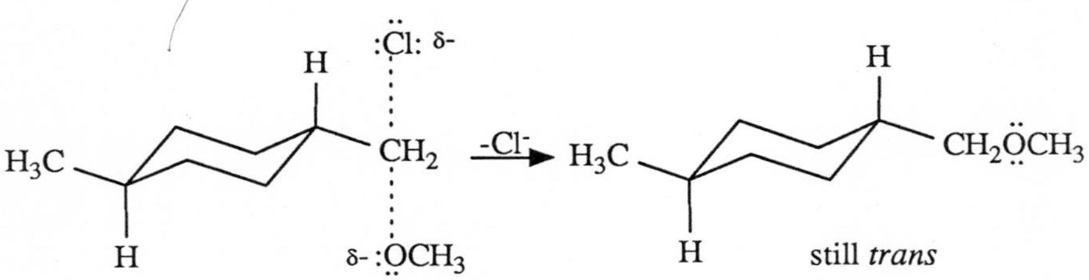

(c) 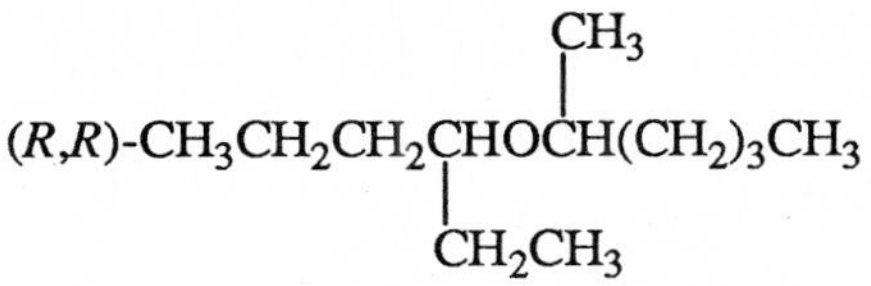 (R,R)-$CH_3CH_2CH_2CH(CH_2CH_3)OCH(CH_3)(CH_2)_3CH_3$

In each case, the configuration around the chiral carbon atom of (*S*)-2-iodohexane is inverted. In (c), the configuration of the attacking nucleophile is not changed in the reaction.

5.34 (a) 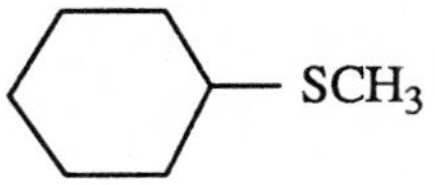 cyclohexyl—SCH_3 + Na^+ + Cl^- (b) 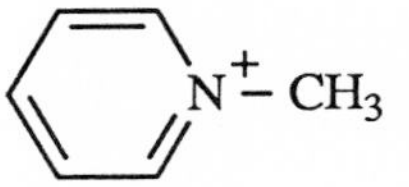 pyridinium $\overset{+}{N}-CH_3$ + I^-

(c) 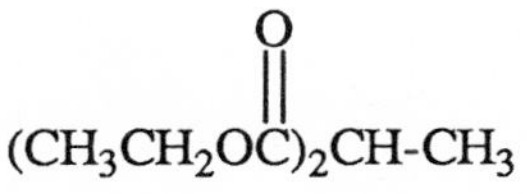 $(CH_3CH_2O\overset{O}{\overset{||}{C}})_2CH\text{-}CH_3$ + I^- (d) 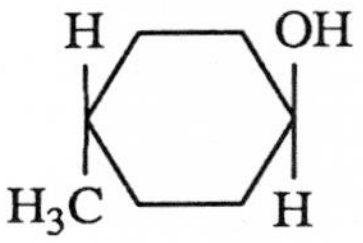 cyclohexane ring with H, OH, H_3C, H + Cl^-

In (d), the *trans* isomer is formed by way of the following transition state:

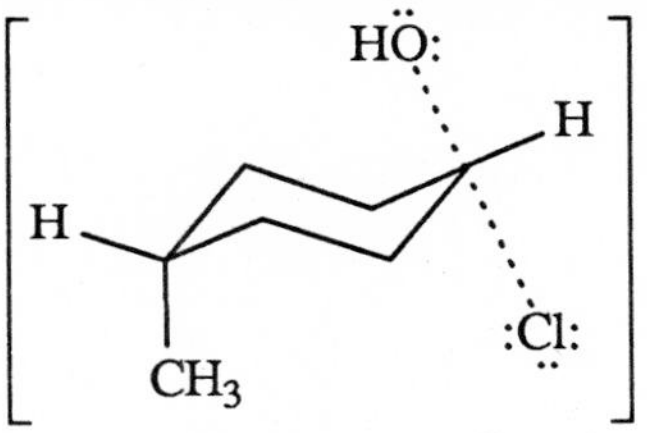

[H, HÖ:, H, :Cl:, CH_3]

Contrast this answer with Answer 5.32(b). In that reaction the nucleophile does not attack a ring carbon and the *cis, trans* relationship is not changed. Here, the nucleophile does attack a ring carbon, and a *cis* isomer is converted to a *trans* isomer.

5.35 (a) $(CH_3)_2CHBr$ + $C_6H_5O^-\ Na^+$

(b) $BrCH_2CH_2CH_2CH_2O^-\ Na^+$ ≡ :Br—CH_2 (H_2C—CH_2) CH_2 :Ö: $\xrightarrow{-NaBr}$ tetrahydrofuran (:O:)

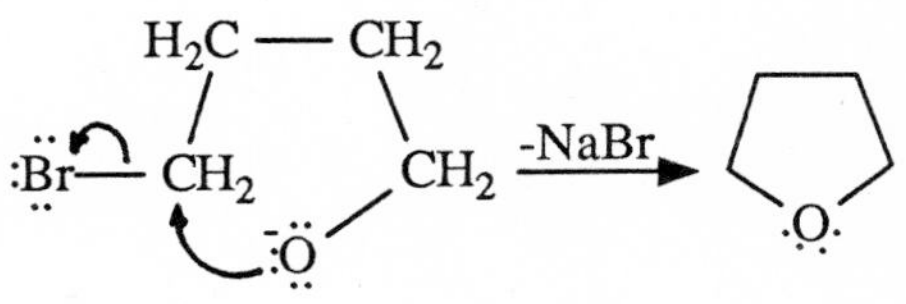

(c) 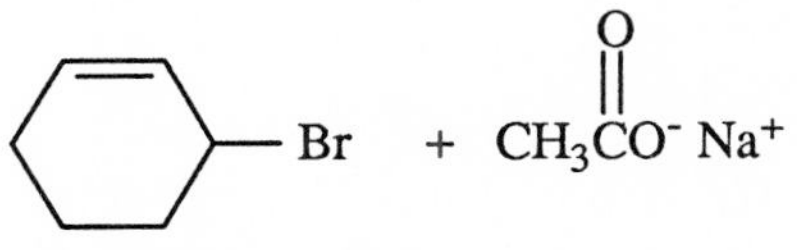 cyclohexenyl—Br + $CH_3\overset{O}{\overset{||}{C}}O^-\ Na^+$

an allylic halide

5.36 The most stable carbocation is (c) because it is 3°. The least stable is (a) because it is 1°.

5.37 (a) energy of activation

(b) energy of activation for the intermediate going to product

(c) the ΔH (change in enthalpy) of the reaction, or energy absorbed in this endothermic reaction

5.38 In aqueous ethanol, two nucleophiles are present - H_2O and CH_3CH_2OH. Therefore, two substitution products will be observed in each reaction mixture - an alcohol and an ether.

(a) $(CH_3)_2CHOH$ + $(CH_3)_2CHOCH_2CH_3$ + some $CH_3CH{=}CH_2$

(b) $(CH_3)_3COH$ + $(CH_3)_3COCH_2CH_3$ + some $(CH_3)_2C{=}CH_2$

(c) $(CH_3)_3COH$ + $(CH_3)_3COCH_2CH_3$ + some $(CH_3)_2C{=}CH_2$

Reaction (c) would be the fastest because RX is 3° and an iodide (a better leaving group).

5.39 (a) H_3C–cyclohexane–I (*cis*) $\xrightarrow[\text{-HI}]{H_2O}$ H_3C–cyclohexane–OH (*cis* and *trans*)

In (a), the *cis* and *trans* isomers would both be formed because solvolysis proceeds by an S_N1 path and thus the nucleophile can attack the intermediate carbocation from one side or the other.

(b) (*R*)-$CH_3CH(I)(CH_2)_5CH_3$ $\xrightarrow[\text{-HI}]{H_2O}$ racemic, or (*R*)(*S*)-$CH_3CH(OH)(CH_2)_5CH_3$

(c) $CH_3CH(I)CH_2CH(CH_3)CH_2CH_3$ [(*R*) at C–I, (*S*) at C–CH_3] $\xrightarrow[\text{-HI}]{H_2O}$ $CH_3CH(OH)CH_2CH(CH_3)CH_2CH_3$ [(*R*)(*S*) at C–OH; (*S*), unchanged at C–CH_3]

Racemization occurs at carbon 2. The product is a pair of diastereomers--(2*R*,4*S*) and (2*S*, 4*S*).

(d) 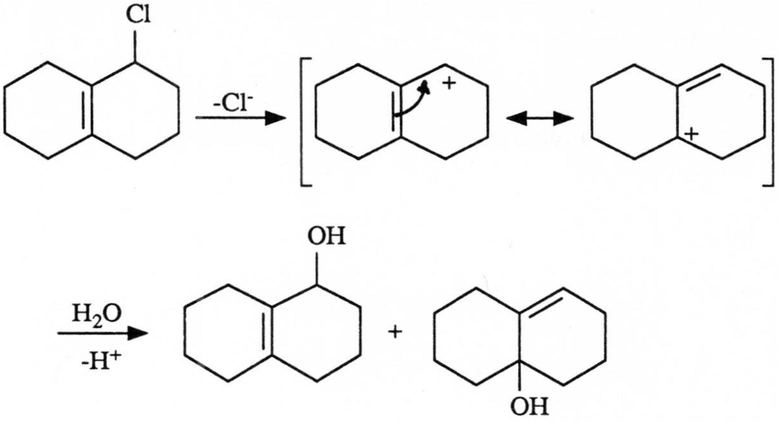 [structure: chloro-substituted bicyclic alkene] $\xrightarrow{-Cl^-}$ [carbocation $\longleftrightarrow$ carbocation]

$\xrightarrow[-H^+]{H_2O}$ [structure: OH] + [structure: OH]

5.40 $C_6H_5Se^- + CH_2{=}CHCH_2Cl \xrightarrow{S_N2} CH_2{=}CHCH_2SeC_6H_5 + Cl^-$

Se is directly below S in the periodic table; therefore, we would expect $C_6H_5Se^-$ to behave similarly to $C_6H_5S^-$, as a strong nucleophile.

5.41 (a) $(CH_3)_2\overset{+}{C}CHCH_2CH_3$ with CH_3 on the CH (methyl shift, 2° to 3° carbocation)

(b) $CH_2{=}CH\overset{+}{C}HCH_2CH_3$ (hydride shift, 2° to allylic carbocation)

(c) $(CH_3)_2\overset{+}{C}CH_2CH_2CH(CH_3)_2$ (hydride shift, 2° to 3°)

(d) 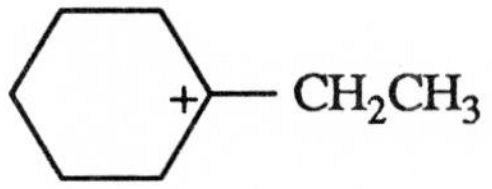 [cyclohexyl cation, + at ring carbon]— CH_2CH_3 (hydride shift, 2° to 3°)

5.42 (a) $(CH_3)_3CCHCH_3$ (with :I: on CH) $\xrightarrow{-\,:I:^-}$ $(CH_3)_3C\overset{+}{C}HCH_3$ $\longrightarrow$ $(CH_3)_2\overset{+}{C}CH(CH_3)_2$

$\xrightarrow{H_2\ddot{O}:}$ $(CH_3)_2CCH(CH_3)_2$ (with $:\overset{+}{O}H_2$ on C) $\underset{}{\overset{-H^+}{\rightleftharpoons}}$ $(CH_3)_2CCH(CH_3)_2$ (with $:\ddot{O}H$ on C)

(b) 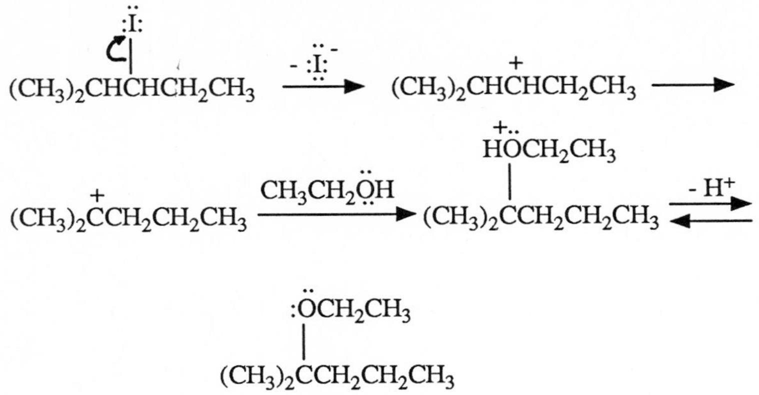 $(CH_3)_2CHCH(I)CH_2CH_3 \xrightarrow{-\ :I:^-} (CH_3)_2CH\overset{+}{C}HCH_2CH_3 \longrightarrow$

$(CH_3)_2\overset{+}{C}CH_2CH_2CH_3 \xrightarrow{CH_3CH_2\ddot{O}H} (CH_3)_2C(\overset{+}{H}\ddot{O}CH_2CH_3)CH_2CH_2CH_3 \underset{}{\overset{-\ H^+}{\rightleftharpoons}}$

$(CH_3)_2C(:\ddot{O}CH_2CH_3)CH_2CH_2CH_3$

5.43 (a) $CH_2{=}CH{-}CH{=}CH{-}\overset{+}{C}H_2 \longleftrightarrow CH_2{=}CH{-}\overset{+}{C}H{-}CH{=}CH_2$

$\longleftrightarrow \overset{+}{C}H_2{-}CH{=}CH{-}CH{=}CH_2$

(b) 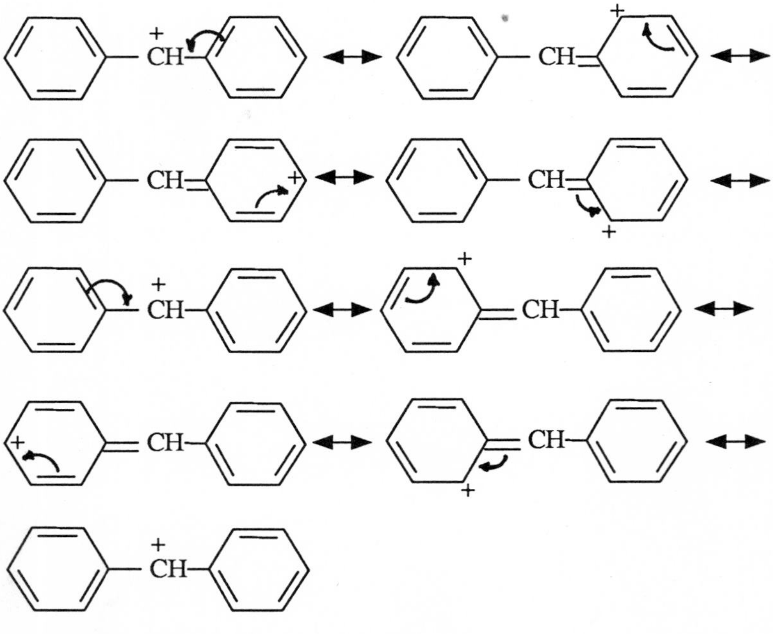 [Resonance structures of the diphenylmethyl cation, $C_6H_5\overset{+}{C}HC_6H_5$, with the positive charge delocalized onto the ortho and para positions of each ring]

plus other benzenoid resonance structures.

(c) $\overset{+}{}$ CH= CH_2 ⟷ =CH– $\overset{+}{C}H_2$

⟷ $\overset{+}{}$ CH= CH_2

(d) $\overset{+}{C}HCH_3$ ⟷ = $CHCH_3$ $^{+}$

In (d), note that the positive charge is delocalized by only one of the two double bonds. It is impossible to draw a resonance structure in which the positive charge is delocalized by the other double bond because that double bond is too far away. In each case, check your resonance structures to make sure that you have no five-bonded carbons.

5.44 (a) (*R*)(*S*) CN, = CH_2 + CH_2CN

(b) $CH_3CH{=}CHCH{=}CHCH_2OCH_3$ + $CH_3CH{=}CHCH(OCH_3)CH{=}CH_2$ (*R*)(*S*)

+ $CH_3CH(OCH_3)CH{=}CHCH{=}CH_2$ (*R*)(*S*)

(c) O, O, N – $CHCO_2CH_2$– , CH_2OH, NO_2

5.45 (a) $CH_3CHI(CH_2)_3CH_3 \xrightarrow[(1)]{-I^-} CH_3\overset{+}{C}H(CH_2)_3CH_3 \xrightarrow[(2)]{-H^+} CH_3CH{=}CH(CH_2)_2CH_3$ *trans*

(b) Step 1

(c) *cis*-$CH_3CH{=}CH(CH_2)_2CH_3$ and $CH_2{=}CH(CH_2)_3CH_3$

5.46 (a)

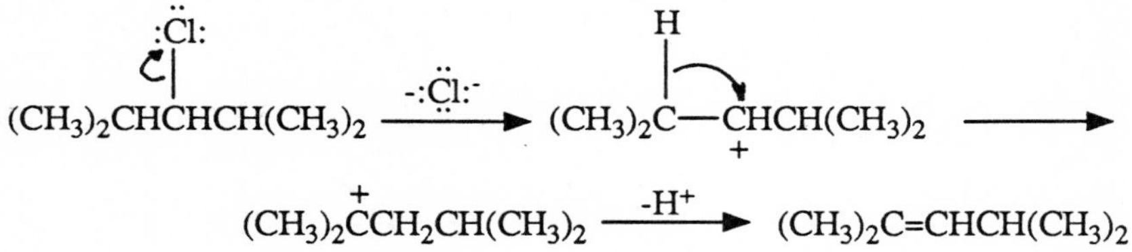

(b)

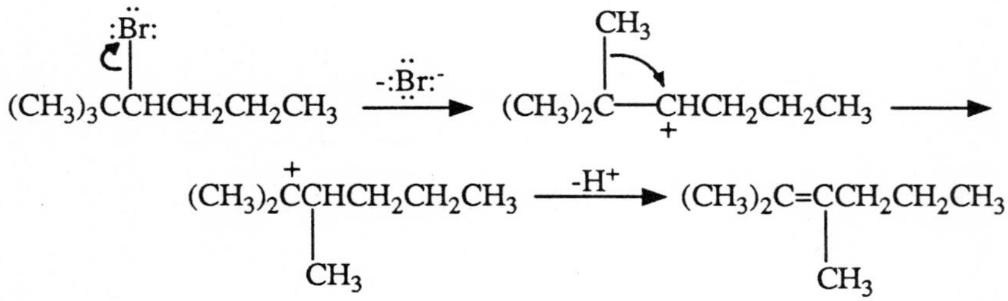

5.47 (a) $(CH_3)_2CBrCH_2CH_2CH_3$, the 3° halide

(b) $(CH_3)_2CHCHICH_3$, the 2° halide

(c) CH_3CHICH_3, the iodide

5.48 (a)

major product

(b)

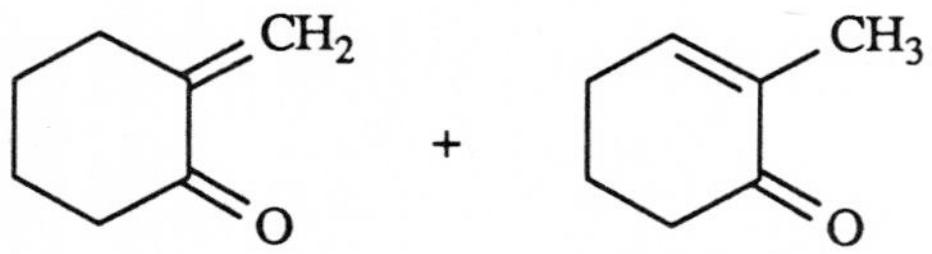

major product

5.49 (a) $(CD_3)_2CClCD_3$

(b)

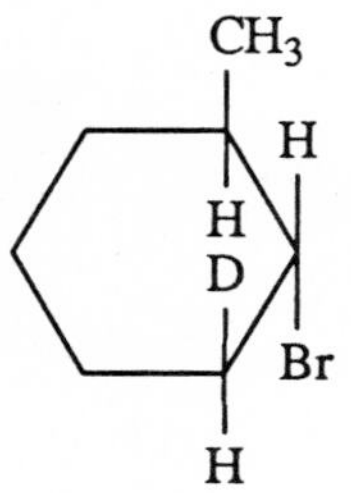

(c)

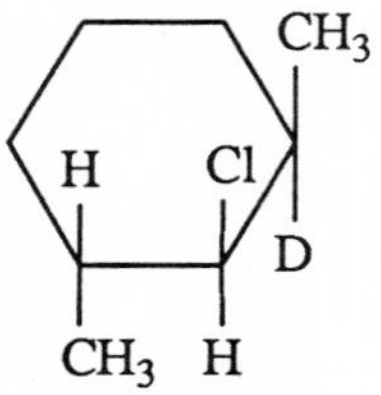

In (a), it is not necessary to replace all the indicated H atoms by D in order to observe a kinetic isotope effect that is less than maximum.

5.50 (a) Hofmann (b) and (c) Saytzeff

5.51 (a) (*S*)-$CH_3CH(Br)CH_2CH_2CH_3$ $\xrightarrow[-HBr]{^-OCH_3}$ $H_3C(H)C=C(H)CH_2CH_3$

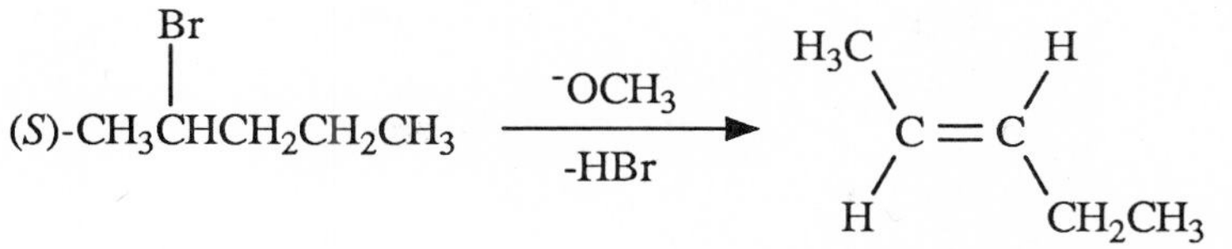

(*E*)- or *trans*-2-pentene

the most stable alkene

(b) $CH_3CH(Cl)CH_2CH_2CH_2CH(Cl)CH_3$ $\xrightarrow[-2\ HCl]{2\ ^-OCH_3}$ $H_3C(H)C=C(H)CH_2(H)C=C(H)CH_3$

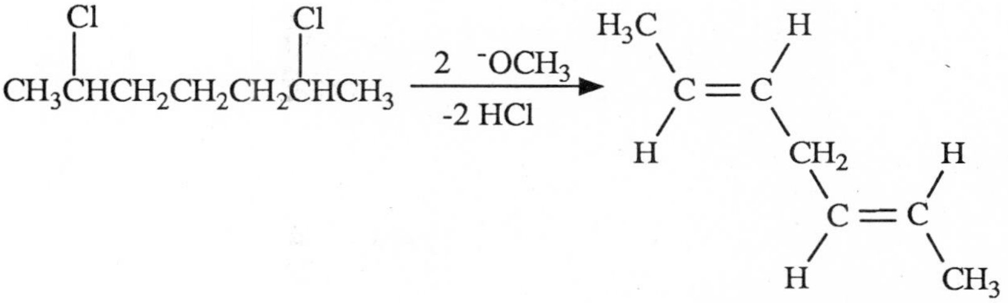

(2*E*,5*E*)-heptadiene

(c) $:\ddot{Br}:$ CH_3CH_2 C_6H_5 C—C H C_6H_5 H $^-:\ddot{O}CH_3$ ⟶ H(C_6H_5)C=C(CH_2CH_3)(C_6H_5) + $:\ddot{Br}:^-$ + CH_3OH

(*Z*)

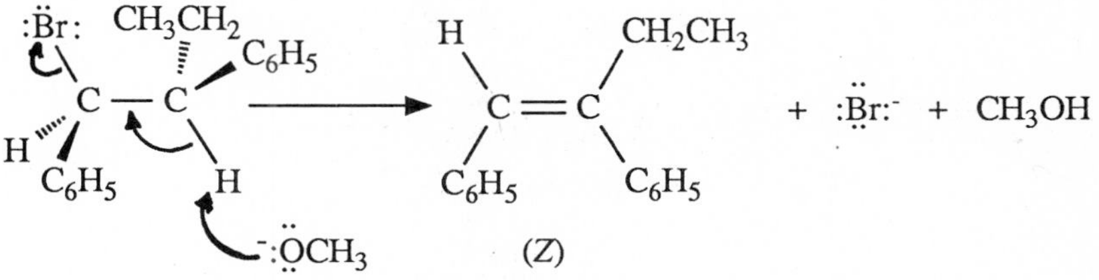

(d) (*S*) cyclohexyl–C(Cl)(H)(CH_3), H $\xrightarrow[-\ HCl]{^-OCH_3}$ cyclohexylidene=$CHCH_3$

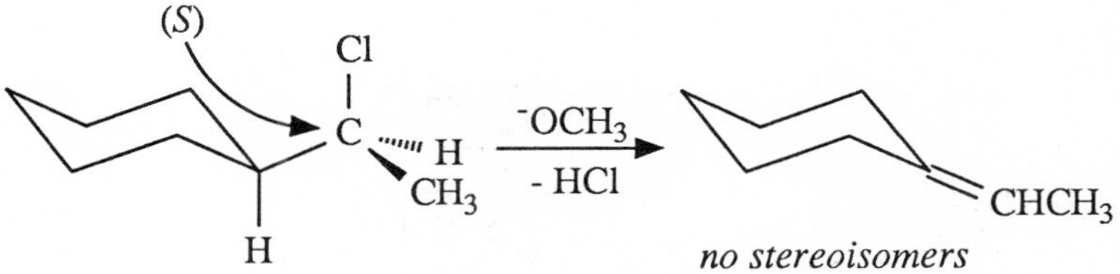

no stereoisomers

5.52 (a) $(CH_3)_2CHI$ (S_N2 because I^- is a weak base and a good nucleophile.)

(b) $CH_3CH{=}CH_2$ (E2 because OH^- is a strong base, the halide is secondary, and heat is applied.)

(c) $(CH_3)_2CHOCH_2CH_3$ (S_N1 because CH_3CH_2OH is a weak nucleophile.)

(d) $(CH_3)_2C{=}CHCH_3$ (E2 because $^-OCH_3$ is a strong base and the halide is tertiary. Because $^{-}OCH_3$ is small, the Saytzeff product is formed.)

(e) *trans*-$CH_3CH{=}CHCH_2CH_3$ (E2 because OH^- is a strong base, the halide is secondary, and heat is applied. The Saytzeff product is formed because OH^- is small.)

(f) $CH_2{=}C(CH_3)CH_2CH_3$ (E2 because the halide is tertiary and the alkoxide is a strong base. The Hofmann product would be formed preferentially because of steric hindrance.)

(g) $\xrightarrow[\text{acid-base reaction}]{Na^+\ ^-OCH_2CH_3,\ -\ CH_3CH_2OH} ClCH_2C(=O)O^-\ Na^+ \xrightarrow[S_N2]{Na^+\ ^-OCH_2CH_3,\ -\ Cl^-} CH_3CH_2OCH_2C(=O)O^-\ Na^+$

(Two reactions occur - a fast acid-base reaction, followed by an S_N2 reaction. The S_N2 reaction occurs because the halide is primary and because RO^- is a good nucleophile. No alkene can be formed because there are no beta hydrogens.)

5.53 (b), because the lower concentration of nucleophile causes a decrease in the rate of reaction by an S_N2 path. Therefore, reaction by an S_N1 path (with rearrangement) has a greater chance of occurring.

5.54 (a) $C_6H_5CH_2CHBrC_6H_5$ + KOH in $CH_3CH_2OH \xrightarrow{\text{heat}}$

(b) [cyclopentane ring: one carbon bearing CH_3 and H_3C; adjacent carbon bearing H (up) and H (down); carbon bearing Cl and H] + $Na^+\ ^-OCH_3 \longrightarrow$

(c) (2*R*,3*S*)-$C_6H_5CHBrCHBrCH_3$ + 2 $Na^+\ ^-CN \longrightarrow$ (Both chiral carbons undergo inversion.)

(d) 1 $ClCH_2CH_2CH_2I$ + 1 $K^+\ ^-CN \longrightarrow$ (RI reacts faster than Cl. We would use a 1:1 molar ratio of the reactants to minimize reaction of RCl.)

(e) 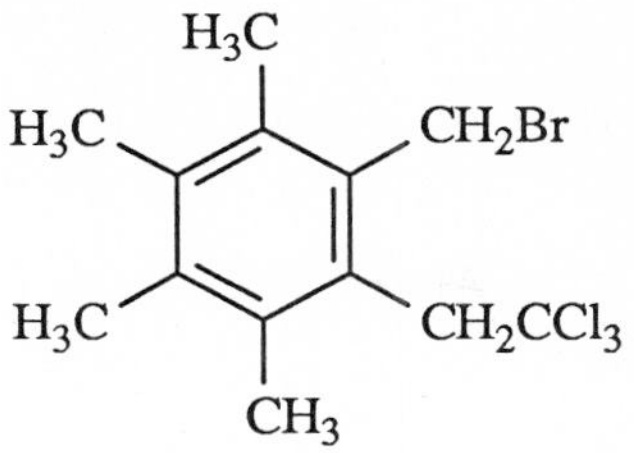 + CH_3OH ⟶ (Under S_N1 conditions, benzylic halides are very reactive.)

(f) 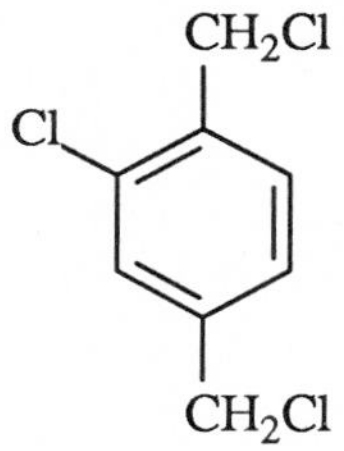 + 2 $Na^+ {}^-CN$ ⟶ (The Cl on the ring does not undergo displacement.)

5.55 (a) 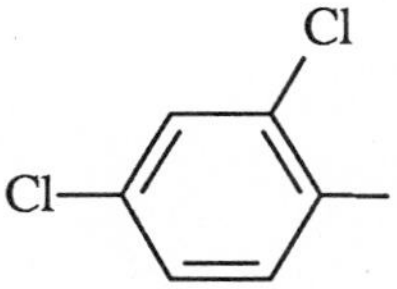 $O^- Na^+$ + $ClCH_2\overset{O}{\overset{\|}{C}}O^- Na^+$ $\xrightarrow[S_N2]{-NaCl}$

(b) 2 CH_3I + 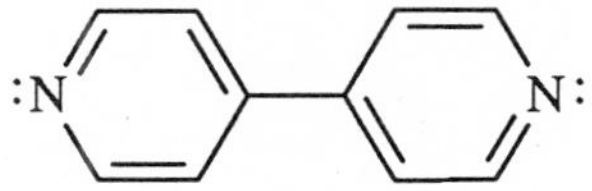 $\xrightarrow{S_N2}$

5.56 (a) $CH_2{=}CCH_2Br$ + $CH_3CH_2NH_2$ ⟶ $CH_2{=}CCH_2\overset{+}{N}H_2CH_2CH_3$ + Br^-

Br   Br

2,3-dibromopropene   initial product

(b) initial product + $CH_3CH_2NH_2$ ⟶

$CH_2{=}CCH_2NHCH_2CH_3$ + $CH_3CH_2\overset{+}{N}H_3$ + Br^-

Br

final product

5.57 A tertiary halide and a weak nucleophile (CH_3OH) in a polar solvent (CH_3OH) indicate a solvolysis (S_N1) reaction accompanied by elimination (E1).

$$C_6H_5CH_2C(Cl)(CH_3)CH_3 \xrightarrow{-Cl^-} C_6H_5CH(H)—\overset{+}{C}(CH_3)—CH_2(H)$$

a 3° carbocation

CH_3OH → $C_6H_5CH_2—\overset{+}{C}(CH_3)_2$ with $H—\ddot{O}CH_3$ bonded; $-H^+$ → $C_6H_5CH_2—C(CH_3)_2$ with $:\ddot{O}CH_3$

A

$-H^+$ → $C_6H_5CH=C(CH_3)_2$

B

(conjugated)

$-H^+$ → $C_6H_5CH_2C(CH_3)=CH_2$

C

The preceding flow equation shows the products that were actually reported. You might also have predicted a carbocation rearrangement to yield a second substitution product.

$$\underset{3°}{C_6H_5CH(H)—\overset{+}{C}(CH_3)_2} \longrightarrow \underset{\text{benzylic}}{C_6H_5\overset{+}{C}H—CH(CH_3)_2} \xrightarrow[(2)\ -H^+]{(1)\ CH_3OH} C_6H_5CH(OCH_3)CH(CH_3)_2$$

5.58 (a) The ring system prevents S_N2 backside attack on the 3° carbon bonded to the Br. The ring system does not allow a planar carbocation to form; thus the S_N1 path is also blocked.

(b) The transition state in the reaction with quinuclidine is less sterically hindered because the alkyl groups on the N are "tied back." Steric hindrance is less important in an acid-base reaction because of the small size of H^+.

(c) $$\underset{(S)}{CH_3CH=CHCH(Br)CH_3} \xrightarrow[-Br^-]{\text{heat}} \underset{achiral}{CH_3CH=CH\overset{+}{C}HCH_3} \xrightarrow{Br^-} \underset{(R)\ \text{or}\ (S)}{CH_3CH=CHCH(Br)CH_3}$$

an allylic halide, which undergoes ionization

an allylic cation

Note that the allylic cation can be attacked at two positions. Attack at either position yields the same racemic product.

$$CH_3\overset{+}{C}H—CH=CHCH_3 \longleftrightarrow CH_3CH=CH—\overset{+}{C}HCH_3$$

(d) Although $ClCH_2OCH_3$ is a primary alkyl halide, the carbocation is resonance stabilized.

$$\overset{+}{C}H_2-\ddot{O}CH_3 \longleftrightarrow CH_2=\overset{+}{O}CH_3$$

5.59 $CH_3CH_2CHClCH_3 \xrightarrow{\text{base}} CH_3CH{=}CHCH_3 \;+\; CH_3CH_2CH{=}CH_2$

the alkyl halide — *cis* and *trans*

All other butyl chlorides would yield only one alkene upon elimination.

5.60 Br and D *anti* —(-DBr)→ *trans* (no deuterium)

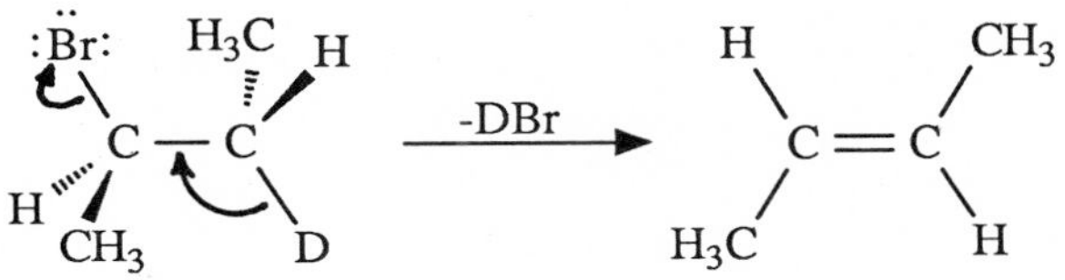

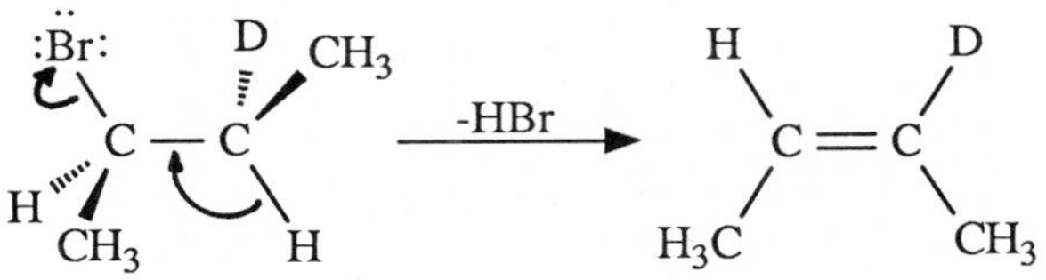

Br and H *anti* —(-HBr)→ *cis* (containing deuterium)

5.61 $\xrightarrow{-Br^-}$ (attack by $^-{:}\ddot{O}CH_3$) *meso*

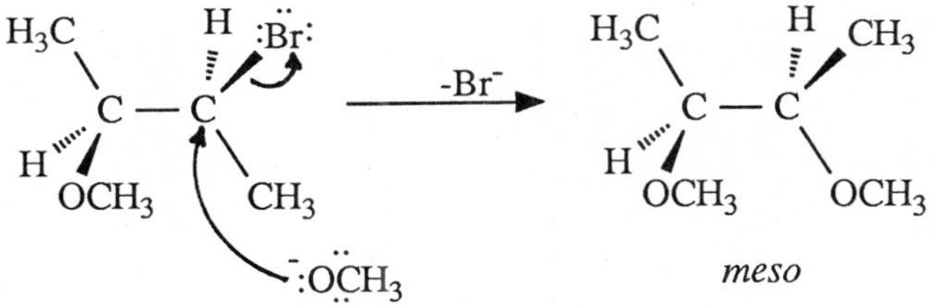

Inversion at carbon 3 leads to the meso (optically inactive) dimethoxy product.

5.62 (a) The slowest step is the rate-determining step. Although the rate of the overall reaction depends on the concentration of AB, this concentration is proportional to the concentrations of both A and B.

$$\text{overall rate} = k[AB] = k'[A]\,[B]$$

Therefore, the reaction would show *second-order kinetics*.

(b) Only one particle, AB, is involved in the transition state of the rate-determining step. Therefore, this reaction is *unimolecular* (even though it is second order in rate).

5.63 Under these strong conditions (molten KOH), mechanisms are not always predictable. A likely possibility (similar to an E2 mechanism) follows:

$$HO:^- + C_6H_5(H)C{=}C(H)Br \longrightarrow C_6H_5{-}C{\equiv}C{-}H + H_2O: + :Br:^-$$

5.64 (a) $(CH_3)_2CHCl \xrightarrow[-AgCl]{Ag^+} (CH_3)_2\overset{+}{C}H \xrightarrow{H_2O:} (CH_3)_2CH\overset{+}{O}H_2 \underset{}{\overset{-H^+}{\rightleftharpoons}} (CH_3)_2CHOH$

(b) $:Br:^- + CH_2{=}CH{-}CH(Br)CH_3 \xrightarrow{-Br^-} BrCH_2CH{=}CHCH_3$

or

$CH_2{=}CHCH(Br)CH_3 \xrightarrow{-Br^-} [CH_2{=}CH{-}\overset{+}{C}HCH_3 \leftrightarrow \overset{+}{C}H_2CH{=}CHCH_3] \xrightarrow{Br^-} BrCH_2CH{=}CHCH_3$

(c) $CH_3CH_2CH_2{-}Cl: + {}^-:O{-}N{=}O: \xrightarrow{-Cl^-} CH_3CH_2CH_2O{-}N{=}O:$

$CH_3CH_2CH_2{-}Cl: + {}^-:O{-}N{=}O: \xrightarrow{-Cl^-} CH_3CH_2CH_2{-}\overset{+}{N}(=O:)(O:^-)$

(d)

$$Ag_2O + H_2O \rightleftharpoons 2\ Ag^+ + 2\ ^-{:}\ddot{O}H$$

$$CH_2{=}CHC(Cl)(CH_3)_2 \xrightarrow[-AgCl]{Ag^+} \left[CH_2{=}CH{-}\overset{+}{C}(CH_3)_2 \leftrightarrow \overset{+}{C}H_2CH{=}C(CH_3)_2 \right]$$

allylic and 3° (major) — allylic and 1° (minor)

$\xrightarrow{^-:\ddot{O}H}$

$$CH_2{=}CHC(OH)(CH_3)_2 + HOCH_2CH{=}C(CH_3)_2$$

(e) Recall that nucleophilicity of anions, such as RCO_2^-, is increased in a nonprotic solvent such as DMSO.

$$BrCH_2(CH_2)_8CH_2CO_2H \xrightarrow[-HCO_3^-]{CO_3^{2-}} :\ddot{Br}{-}CH_2(CH_2)_8CH_2CO_2^-$$

$\xrightarrow{-Br^-}$ cyclic ester (lactone): $CH_2{-}(CH_2)_8{-}CH_2{-}C(=O){-}O$ ring ≡ 12-membered lactone

5.65 Redraw the ring undergoing reaction to show the conformation in which the Br atoms are in the necessary axial positions.

→ observed product

Neither of these H's can be lost.

The only *trans* axial H's are circled. These are the only atoms that can be eliminated with Br^-.

5.66 (a) Consider the conformations in which Cl and a beta H are *trans* and diaxial (the necessary conformations for elimination by an E2 path).

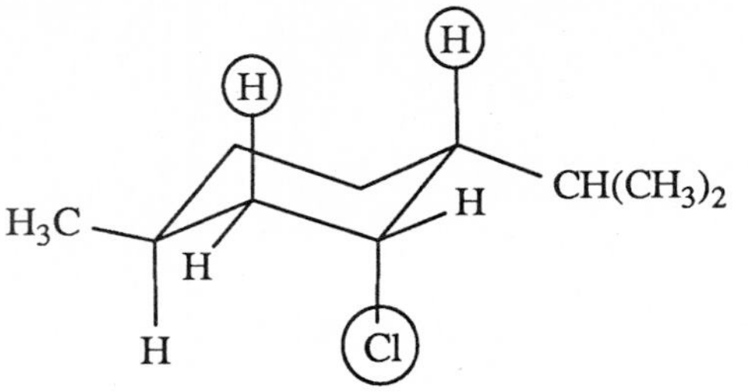

(1)

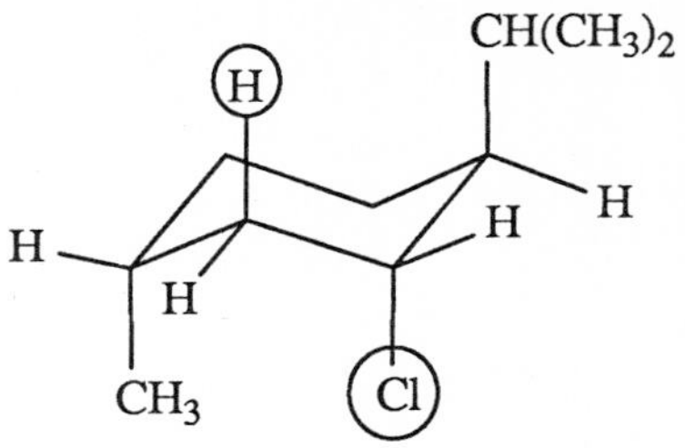

(2)

In the necessary conformation, structure (**1**) is also in a preferred conformation with CH_3 and $CH(CH_3)_2$ equatorial. Structure (**2**) is in a nonpreferred conformation with CH_3 and $CH(CH_3)_2$ axial. The concentration of the conformation that can undergo E2 reaction is greater for (**1**); therefore, (**1**) undergoes the faster reaction.

(b) When structure (**1**) undergoes reaction, either of two H's could be lost.

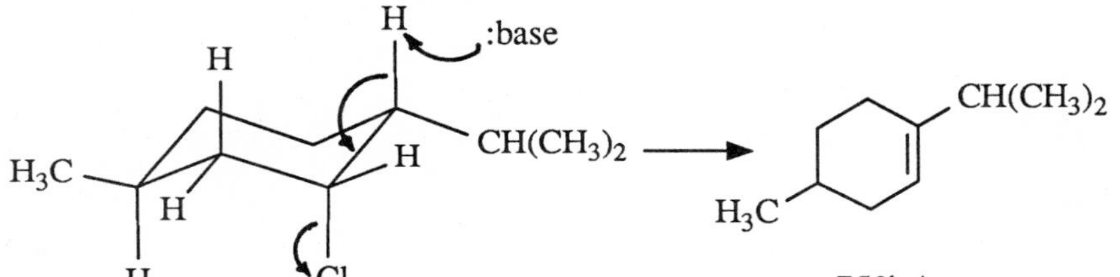

75% A

or

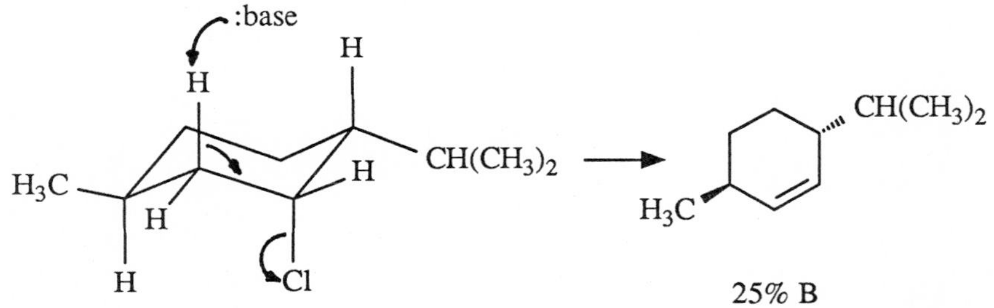

25% B

Alkene A predominates because it is the more highly substituted, more stable alkene (Saytzeff rule).

(c) When structure (**2**) undergoes reaction, only one alkene can be formed because it contains only one *trans*-axial beta H.

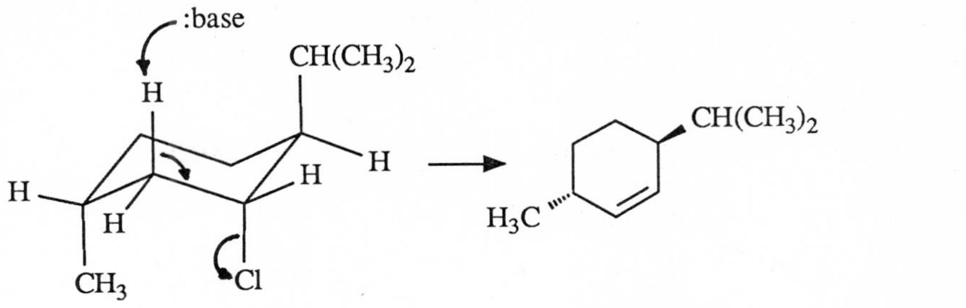

The reaction is slow both for the reason discussed in (a) and because the less substituted (higher-energy) alkene is the only possible product.

Chapter 6

Free-Radical Reactions

6.15 (a) $H:\underset{\underset{H}{\cdot\cdot}}{\overset{\overset{H}{\cdot\cdot}}{C}}:\underset{\underset{H}{\cdot\cdot}}{\overset{\overset{H}{\cdot\cdot}}{C}}:\underset{\cdot\cdot}{\overset{\cdot\cdot}{O}}\cdot$ (b) $\overset{H\cdot}{\underset{H^{\cdot}}{}}:C::\underset{\underset{H}{\cdot\cdot}}{C}:\underset{\underset{H}{\cdot\cdot}}{\overset{\cdot}{C}}:H$

6.16 (b) and (f) are initiation steps.

(a) and (e) are propagation steps.

(c) and (d) are termination steps.

These different steps can be distinguished as follows: If a radical is formed from a nonradical reactant, the reaction is an initiation step. If both a reactant and a product are radicals, the reaction is a propagation step. If radical reactants lead to nonradical products, the reaction is a termination step.

6.17 Initiation:

(1) $Cl_2 \xrightarrow{\text{light}} 2\ Cl\cdot$

Propagation:

(2) Cl· + cyclopentane (H, H) ⟶ HCl + cyclopentyl radical (H) [usually represented: cyclopentyl radical]

(3) cyclopentyl radical + Cl_2 ⟶ chlorocyclopentane (Cl) + Cl·

(4) chlorocyclopentane (Cl) + Cl· ⟶ 2-chlorocyclopentyl radical (Cl) + HCl

(5) 2-chlorocyclopentyl radical (Cl) + Cl_2 ⟶ 1,2-dichlorocyclopentane (Cl, Cl) + Cl·

Termination could occur by the combination of any two radicals; for example,

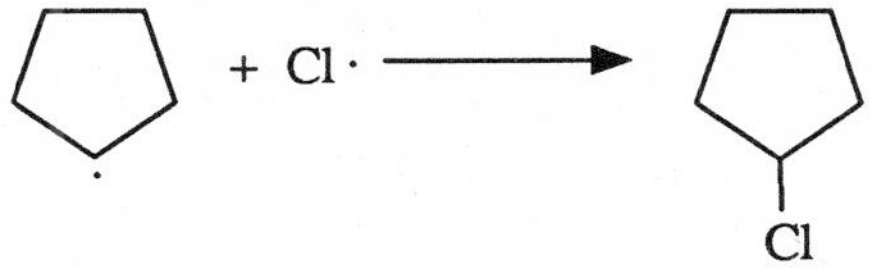

+ Cl· ⟶ Cl

6.18 The ratio would be the same as the ratio of the types of H being extracted: two propane CH_2 hydrogens, six propane CH_3 hydrogens, twelve cyclohexane CH_2 hydrogens, or 1:3:6. The product ratio would therefore be as follows:

1 part $CH_3CHClCH_3$ 3 parts $CH_3CH_2CH_2Cl$ 6 parts — Cl

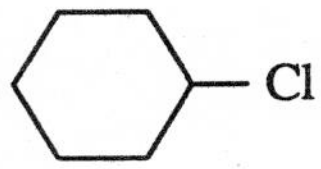

or 10% $CH_3CHClCH_3$ 30% $CH_3CH_2CH_2Cl$ 60% $C_6H_{11}Cl$

6.19 (a) $ClCH_2CHClCH_2Cl$, (R)-$Cl_2CH\overset{*}{C}HClCH_3$, $ClCH_2CCl_2CH_3$

(b) (S)-$ClCH_2\overset{*}{C}HClCH_2CH_3$, $CH_3CCl_2CH_2CH_3$, (R)-$CH_3\overset{*}{C}HClCH_2CH_2Cl$,

an equimolar mixture of (2R, 3S)-$CH_3\overset{*}{C}HCl\overset{*}{C}HClCH_3$ (*meso*) and

(2R, 3R)-$CH_3\overset{*}{C}HCl\overset{*}{C}HClCH_3$

The chiral carbon atoms have been starred in the products. In each case, chlorination at a chiral carbon yields a racemic or achiral product, while chlorination at an achiral carbon does not affect the stereochemistry of other chiral carbons. In (a), chirality can be lost by the formation of identical groups on a chiral carbon. In (b), the order of priority of the groups around the chiral carbon in the first compound has changed.

6.20 $(CH_3)_4C$ is the only isomer of C_5H_{12} in which all twelve hydrogens are equivalent to one another and, therefore, is the only isomer that can yield only one monochlorination product.

6.21 (d) < (b) < (a) < (c) < (e)

6.22 (b) < (a) < (c), ranked in the order of stability of the most stable intermediate radical that could be formed from the hydrocarbon.

6.23 (a) 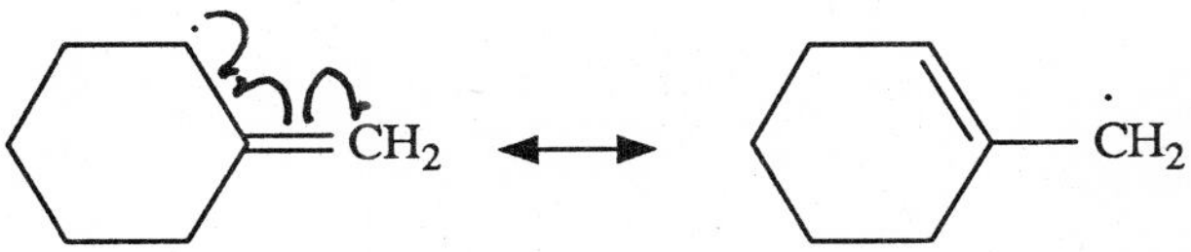 CH_2 ⟷ $\dot{C}H_2$

(b) [resonance structures: benzylic radical of ethylbenzene, CHCH_3 ↔ ring-delocalized forms with $=\text{CHCH}_3$]

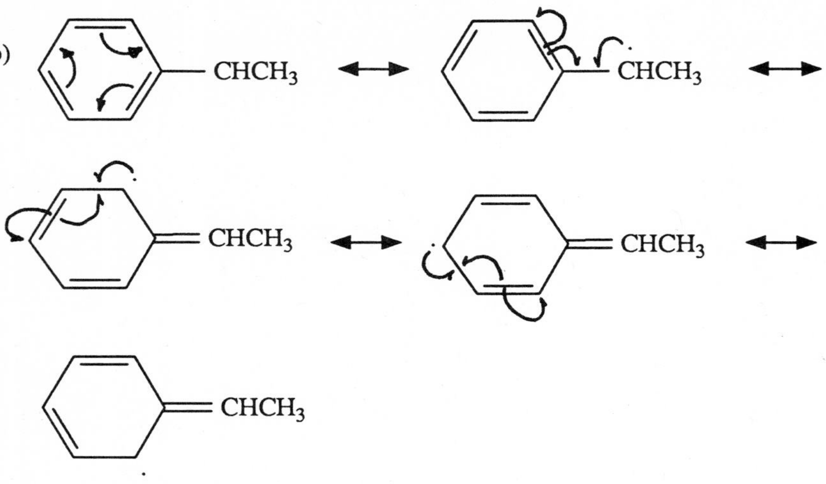

(c) [resonance structures: cyclohexanecarboxylate radical, $\text{C(=O)-O}\cdot$ ↔ $\text{C(-O}\cdot\text{)=O}$]

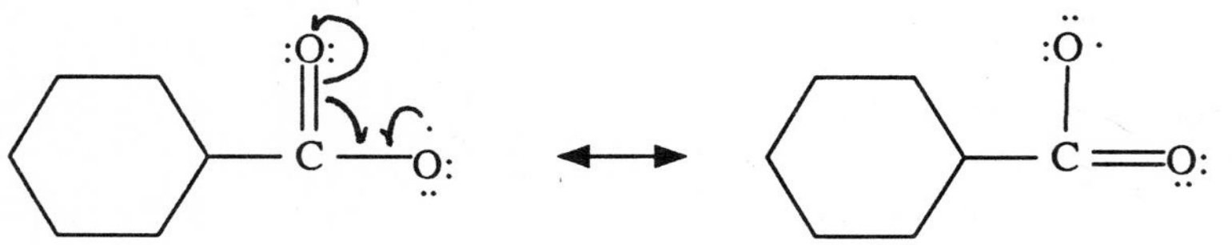

6.24 (a) allylic

$$H_3C-\underset{\large CH_3}{\overset{}{C}}=CH-\overset{O}{\overset{\|}{C}}-OCH_3 \longrightarrow H_3C-\underset{}{C}(CH_2Br)=CH-\overset{O}{\overset{\|}{C}}-OCH_3$$

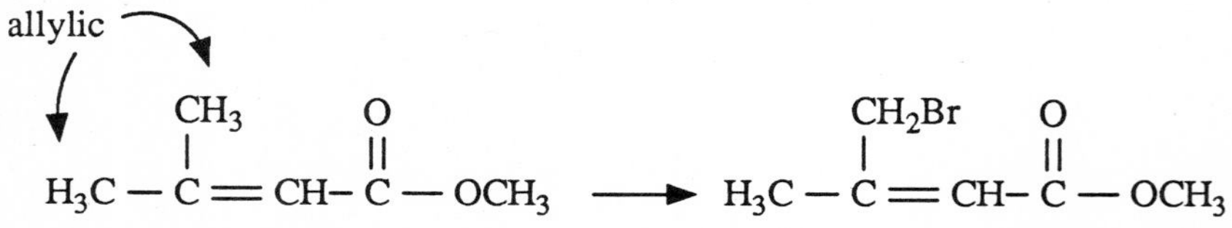

Replacement of any allylic H would yield the same product. You might also have predicted an allylic rearrangement to yield the following bromide.

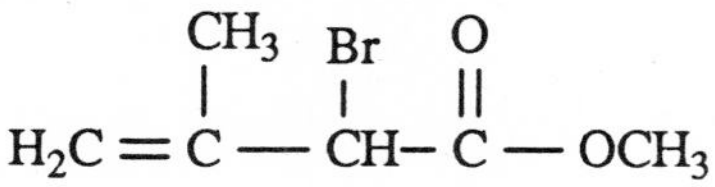

$$H_2C=C(CH_3)-CH(Br)-\overset{O}{\overset{\|}{C}}-OCH_3$$

(b) benzylic

[indane → 1-bromoindane ≡ 1-bromoindane (drawn with Br at other benzylic position)]

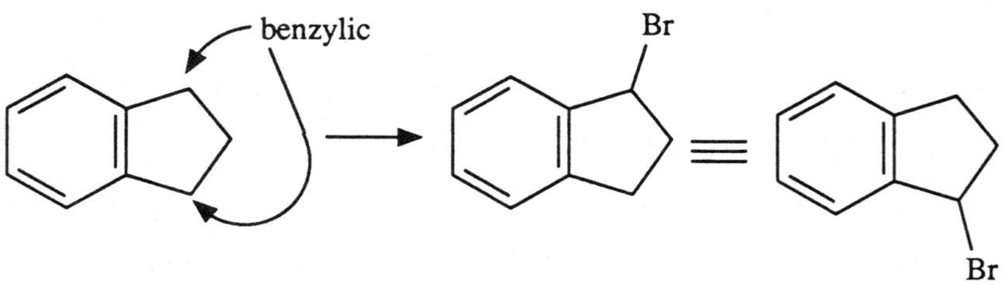

(c) 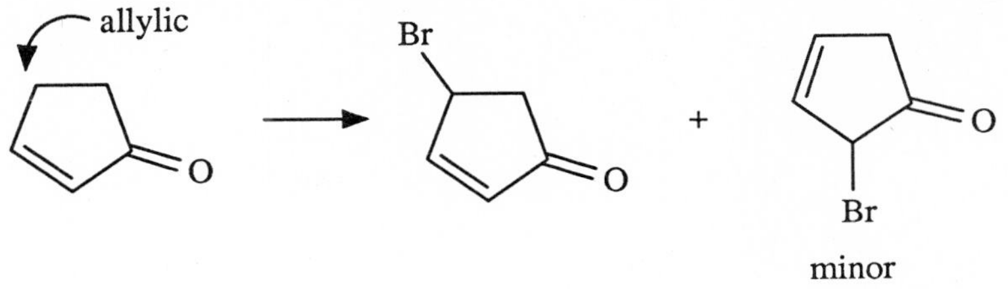 allylic

$\rightarrow$ + minor

Br O O Br

6.25

Br· / - HBr

Br_2 / - Br·

Br· / - HBr

Br_2 / - Br·

Br Br Br

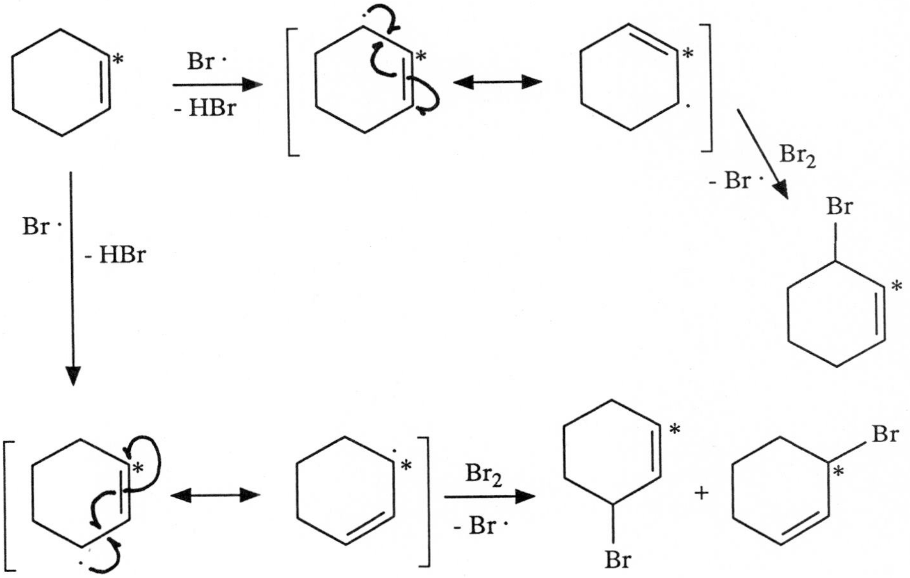

6.26 (a) 2 $(CH_3)_2CCH_2CH_2CH_3$ $\xrightarrow{\text{coupling}}$

$(CH_3)_2CCH_2CH_2CH_3$
|
$(CH_3)_2CCH_2CH_2CH_3$

disproportionation

$(CH_3)_2C{=}CHCH_2CH_3$ + $(CH_3)_2CHCH_2CH_2CH_3$ + $H_2C{=}C(CH_3)CH_2CH_2CH_3$ (minor)

(b) 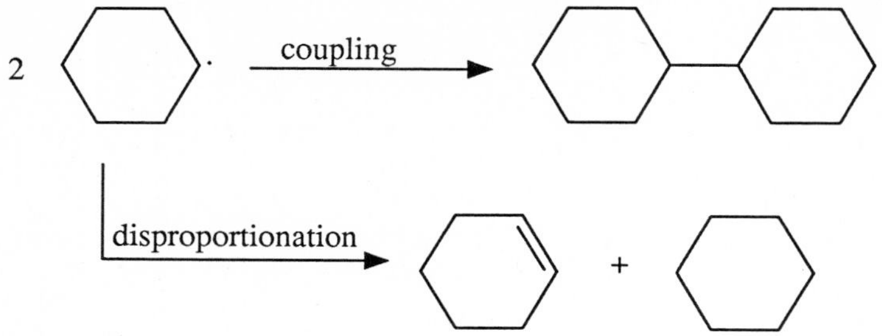 2 [cyclohexyl radical] —coupling→ [bicyclohexyl]

disproportionation → [cyclohexene] + [cyclohexane]

6.27 (a) Hexaphenylethane would suffer from extreme steric hindrance.

(b) The triphenylmethyl radical is resonance stabilized by three benzene rings. We show stabilization by one of the rings here; the other two rings are similarly involved in resonance stabilization. (For practice, draw the remaining resonance structures.)

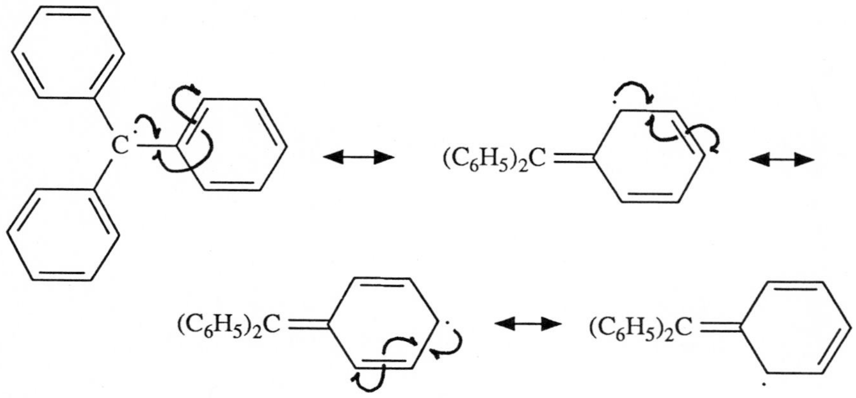

$(C_6H_5)_2C$ ↔ $(C_6H_5)_2C$ ↔ $(C_6H_5)_2C$ ↔ $(C_6H_5)_2C$

6.28 All can form hydroperoxides. (a) contains benzylic hydrogens -CH_3 and -$CH(CH_3)_2$. The isopropyl H is also tertiary. The probable structure of the hydroperoxide would be as follows:

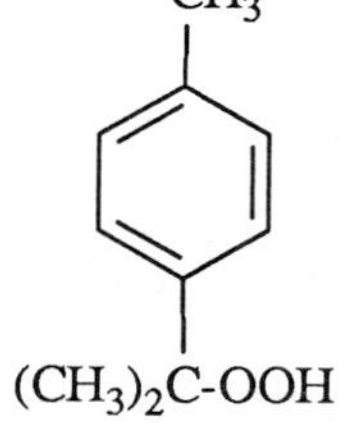

CH_3

$(CH_3)_2C$-OOH

(b) Vanillin is not oxidized as rapidly as is benzaldehyde because of the inhibitory effect of the phenol group, but it still can be oxidized in air. The peroxide structure follows:

$$\text{4-hydroxy-3-methoxyperoxybenzoic acid: } HO\text{-}C_6H_3(OCH_3)\text{-}C(=O)\text{-}OOH$$

(c) This compound contains hydrogen atoms bonded to carbons that are bonded to ether oxygen atoms. The formula for the peroxide follows:

$$\text{1,4-dioxane ring with OOH on C2}$$

6.29 (a) $\text{N-vinyl-2-pyrrolidone: } H_2C{=}CH\text{-}N(\text{pyrrolidin-2-one})$ (b) $CH_2{=}CHO\overset{O}{\overset{\|}{C}}CH_3$ (c) $CH_2{=}CHOCH_2CH_3$

6.30

$$C_6H_5CH{=}CH_2 + \text{maleic anhydride} \longrightarrow \left(\!\!-CH(C_6H_5)CH_2-CH\,CH-\!\!\right)_x \text{ (with anhydride ring O=C–O–C=O bridging the two CH)}$$

6.31 initiation:

$$C_6H_5\overset{O}{\overset{\|}{C}}O-O\overset{O}{\overset{\|}{C}}C_6H_5 \xrightarrow{\text{heat}} 2\ C_6H_5\overset{O}{\overset{\|}{C}}O\cdot$$

propagation:

$$C_6H_5\overset{O}{\overset{\|}{C}}O\cdot\ +\ H_2C{=}C(CH_3)_2 \longrightarrow C_6H_5\overset{O}{\overset{\|}{C}}O-CH_2\overset{CH_3}{\underset{CH_3}{C}}\cdot$$

$$C_6H_5\overset{O}{\overset{\|}{C}}OCH_2\dot{C}(CH_3)_2 + H_2C{=}C(CH_3)_2 \longrightarrow C_6H_5\overset{O}{\overset{\|}{C}}OCH_2C(CH_3)_2{-}CH_2\dot{C}(CH_3)_2$$

termination:

(a) coupling

$$2\ RCH_2\dot{C}(CH_3)_2 \longrightarrow RCH_2C(CH_3)_2{-}C(CH_3)_2CH_2R$$

(b) disproportionation

$$2\ RCH_2\dot{C}(CH_3)_2 \longrightarrow RCH_2CH(CH_3)_2 + RCH{=}C(CH_3)_2$$

6.32 If HBr is present in the reaction mixture, an exchange reaction leading to $C_6H_5CH_3$ occurs. The presence of $C_6H_5CH_3$ leads to a decreased isotope effect and a faster reaction.

$$C_6H_5\dot{C}H_2 + HBr \longrightarrow C_6H_5CH_3 + Br\cdot$$

no isotope effect

If HBr is removed, the reaction of $C_6H_5CH_2D$ alone is slower because of the following step:

$$C_6H_5CH_2D + Br\cdot \longrightarrow C_6H_5\dot{C}H_2 + DBr$$

shows isotope effect

6.33 *anode*:

$$CH_3\overset{O}{\overset{\|}{C}}\ddot{O}{:}^- \xrightarrow{-e^-} CH_3{-}\overset{O}{\overset{\|}{C}}{-}\dot{O}{:} \longrightarrow \dot{C}H_3 + {:}\dot{O}{=}C{=}\dot{O}{:}$$

$$2\ \dot{C}H_3 \longrightarrow CH_3CH_3$$

cathode: $2\ H_2O \longrightarrow H_2 + 2\ OH^-$

6.34 This type of problem is called a *road map problem* and is typical of what a chemist encounters in structure determination. To approach a road map problem, first write out a flow diagram.

$$CH_3CH_2CH_2CH_2CH_3 \xrightarrow[h\nu]{Br_2} A + B \xrightarrow[E2]{CH_3O^-} C$$

n-pentane — *two bromopentanes* — *a pentene*

We have added what is immediately apparent about the products of the two reactions: A and B are bromopentanes, while C must be an elimination product (and therefore a pentene). All we need to do is fill in the positions of substitution. *n*-Pentane is most likely to yield 2- and 3-bromopentane (A and B). Will these two compounds yield the same alkene (C)? The answer is yes. Now, we fill in the flow diagram with structures.

$$CH_3CH_2CH_2CH_2CH_3 \xrightarrow[h\nu]{Br_2} \underset{A}{CH_3CHBrCH_2CH_2CH_3} + \underset{B}{CH_3CH_2CHBrCH_2CH_3}$$

$$\xrightarrow[E2]{} \underset{C}{CH_3CH{=}CHCH_2CH_3}$$

6.35 Polypropylene $-\!\!(CH_2CH(CH_3))_x\!-$ has a 3° hydrogen that can be abstracted by atmospheric oxygen to yield a radical intermediate. This intermediate can cleave into lower-molecular-weight material. Repeated reaction and cleavage result in degradation of the polymer and loss of its strength.

$$\sim CH_2C(CH_3)(H){-}CH_2C(CH_3)(H)\sim \xrightarrow{\cdot O\text{-}O\cdot} \sim CH_2C(CH_3)(H){-}CH_2{-}\dot{C}(CH_3)\sim$$

$$\longrightarrow \sim CH_2\dot{C}(CH_3)(H) + H_2C{=}C(CH_3)\sim \xrightarrow[\text{steps}]{\text{more}} \text{smaller fragments}$$

Polyethylene, $-\!\!(CH_2CH_2)_x\!-$, does not have a 3° hydrogen; therefore, its rate of reaction with oxygen is very slow.

6.36 The initiator forms radicals. If present in excess, these radicals can undergo coupling with the growing polymer chain and prematurely terminate the radical polymerization reaction. The coupling reactions would thus limit the average molecular weight of the polymer.

6.37

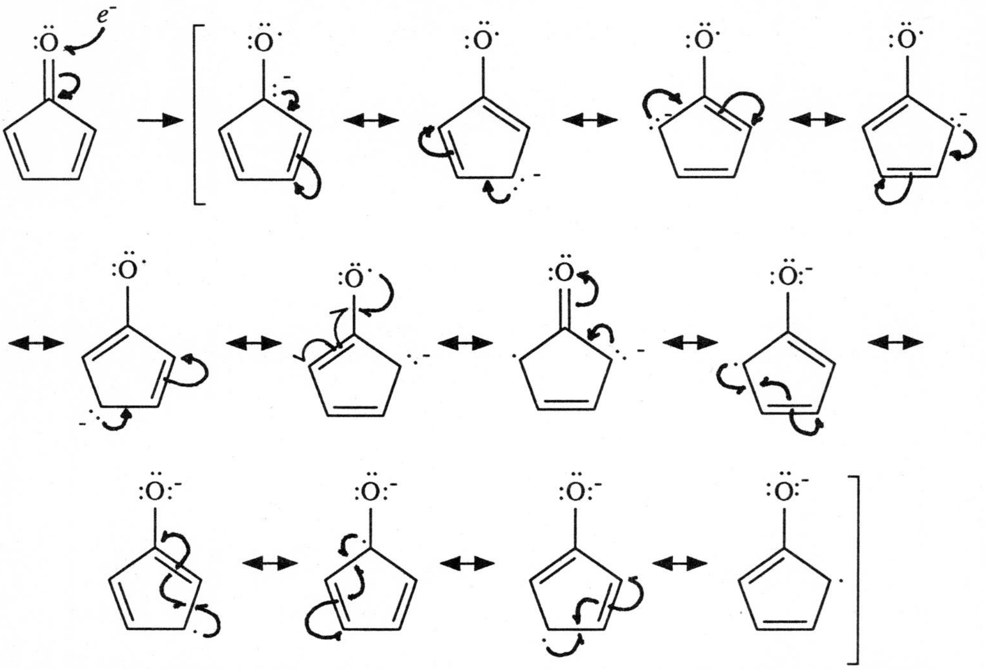

(Note that the sixth resonance structure is the same as the fourth. This was included for the sake of clarity.)

6.38 Initiator (as a radical) + $CH_3C(CH_3)_2OH$ ⟶ initiator-H + $\cdot CH_2C(CH_3)_2OH$

$$2\ \cdot CH_2C(CH_3)_2OH \longrightarrow HOC(CH_3)_2CH_2CH_2C(CH_3)_2OH$$

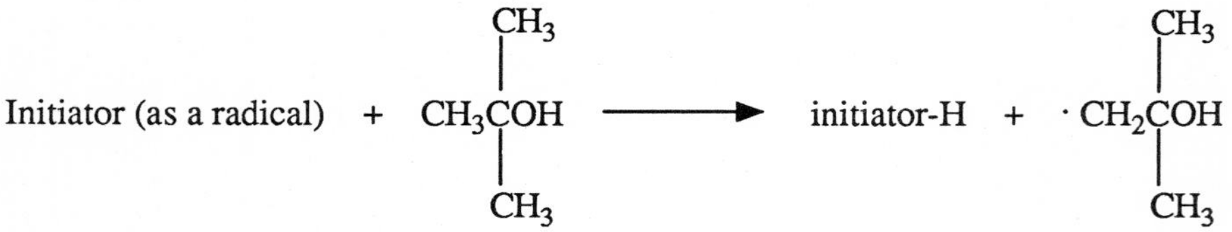

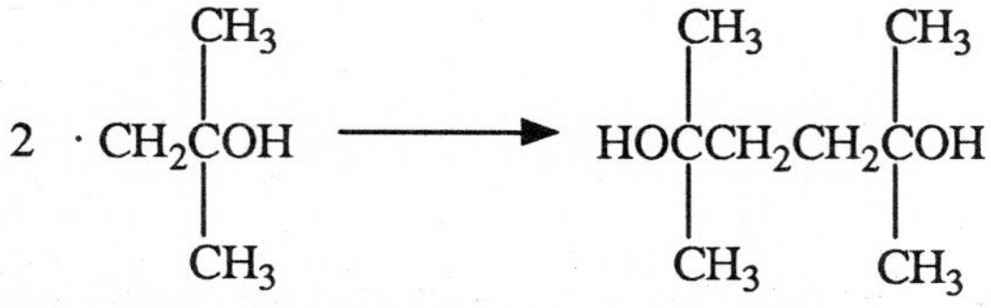

Note that the initiator is really a reactant and must be used in 1 : 1 molar ratio with the *t*-butyl alcohol.

6.39 (a) $CH_3CH_2CH_2CH(Br)CH{=}CHCH_3$ + $CH_3CH_2CH_2CH_2CH{=}CHCH_2Br$

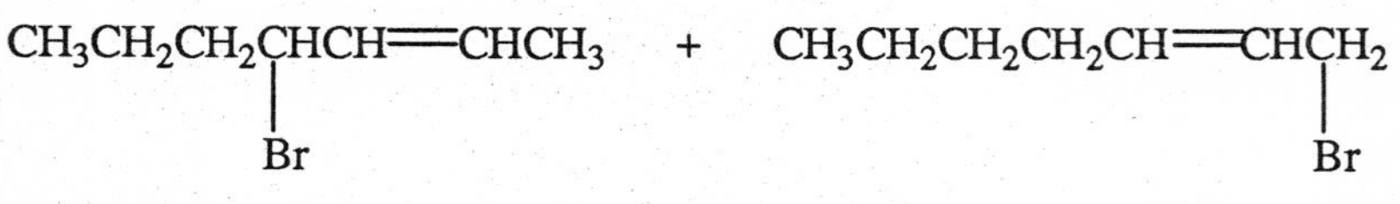

(b) The first product is formed in greater yield because an allylic secondary radical $CH_3CH_2CH_2\dot{C}HCH{=}CHCH_3$ would be an intermediate. The other isomer would require an allylic primary radical as the intermediate.

6.40 (a) (1) $-\!\!\left(CH_2\underset{}{\overset{C_6H_5}{\overset{|}{C}H}}\right)_x\!\!- \xrightarrow{\text{heat}} x\ CH_2{=}\overset{C_6H_5}{\overset{|}{C}H}$

(2) $-\!\!\left(CF_2CF_2\right)_x\!\!- \xrightarrow{\text{heat}} x\ CF_2{=}CF_2$

(b) The depolymerization releases possibly toxic and, in the case of polystyrene, flammable gases.

Chapter 7

Alcohols

7.23 (a) $(CH_3)_2NCH_2CH_2OCH(C_6H_5)_2$

ether

(b)

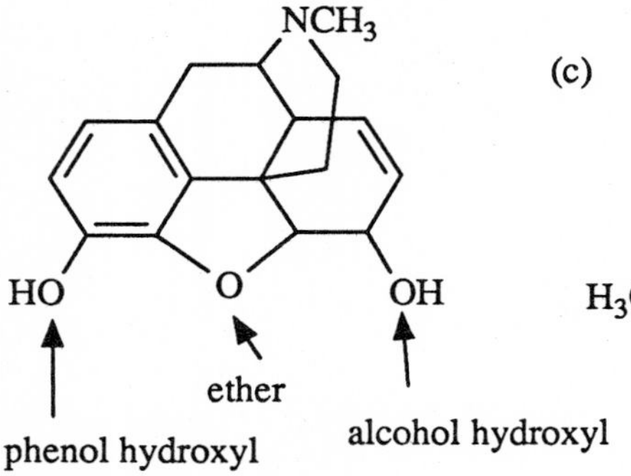
NCH3
HO
O
OH
ether
phenol hydroxyl
alcohol hydroxyl

(c)

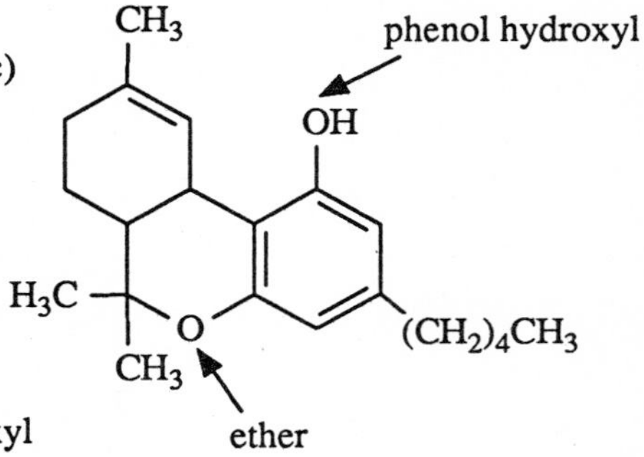
CH3
phenol hydroxyl
OH
H3C
O
CH3
(CH2)4CH3
ether

7.24

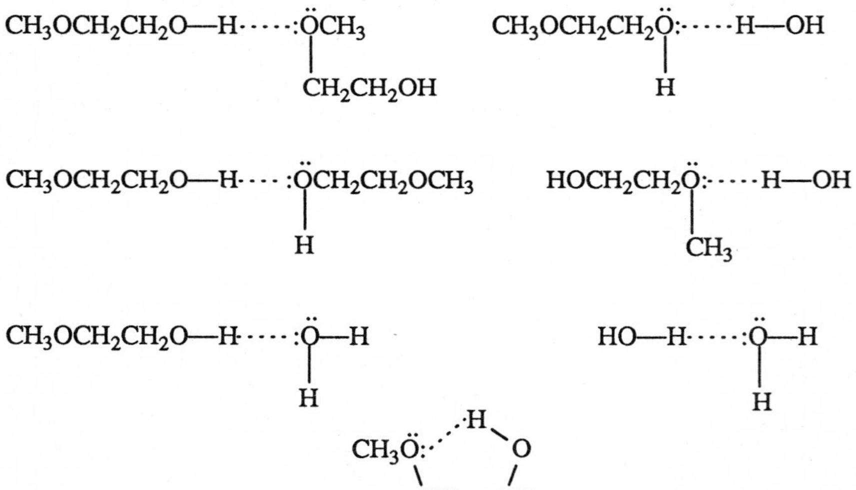
CH3OCH2CH2O—H·····:ÖCH3
CH2CH2OH
CH3OCH2CH2Ö:·····H—OH
H
CH3OCH2CH2O—H···· :ÖCH2CH2OCH3
H
HOCH2CH2Ö:·····H—OH
CH3
CH3OCH2CH2O—H·····:Ö—H
H
HO—H·····:Ö—H
H
H
CH3Ö:
O
CH2— CH2

7.25 (a)

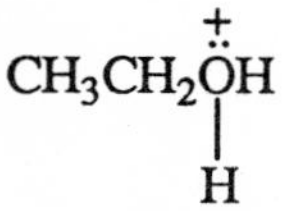
+
CH3CH2ÖH
H

(b)

O- K+ + H2

(c) $(CH_3)_3COH$ + ^{-}OH

(d) $CH_3CH_2CH_2CH_2O^{-}$ Na^{+} + H_2

(e) $CH_3\overset{+}{\underset{..}{O}}H_2 + HSO_4^-$ (f) $CH_3CH_2OH + H_3O^+$

7.26 (a) 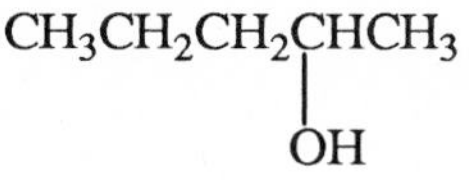

(b) 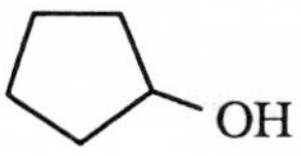

(c) 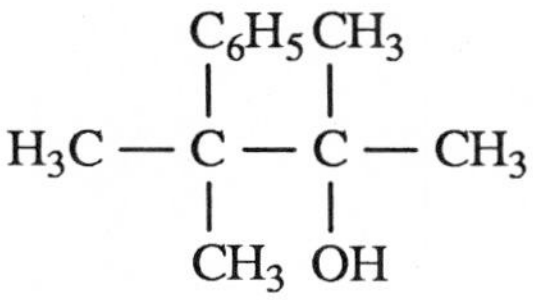

(d) 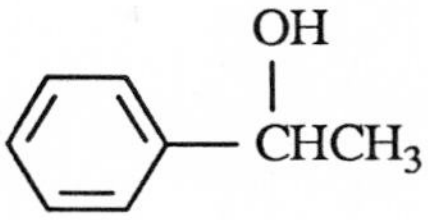

(e) 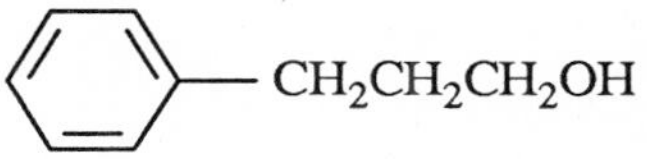

(f) 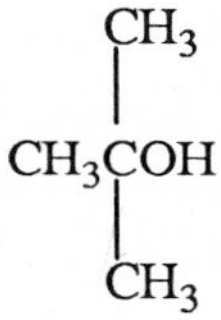

(g) $(CH_3)_2CHCH_2OH$

(h) 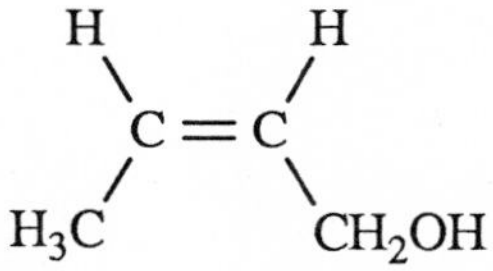

(i) 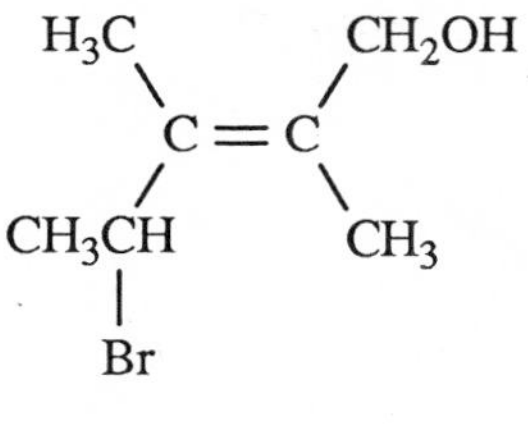

(j) 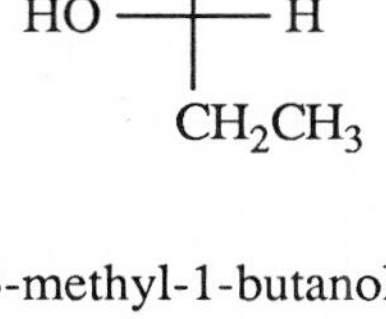 or 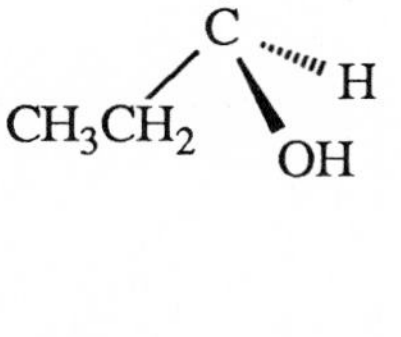

7.27 (a) 3-methyl-1-butanol, 1°

(b) *trans*-3-methylcyclohexanol or (1*S*,3*S*)-3-methylcyclohexanol, 2°

(c) 2,5-heptanediol, both 2°

(d) *cis*-3-penten-1-ol or *cis*-3-pentenol, 1°

In (d), you might have used (*Z*) instead of *cis*.

7.28 (a) propyl alcohol or *n*-propyl alcohol

(b) allyl alcohol (c) benzyl alcohol

(d) isopropyl alcohol (e) cyclobutyl alcohol

7.29 (a) C_6H_5—CH_2CH_2Br + ^-OH ⟶ product

(b) $CH_3CH_2CH_2CHCH_3$ (with Br on CH) + ^-OH ⟶ product

(c) (*S*)-$CH_3CH_2CHCH_3$ (with Br on CH) + ^-OH ⟶ product

(d) $(CH_3)_3C$-cyclohexane (chair; Br axial, H; H) + ^-OH ⟶ product

7.30 (a) (1) H_3C—C_6H_4—$\overset{O}{\overset{\|}{C}}H$ $\xrightarrow[(2)\ H_2O,\ H^+]{(1)\ NaBH_4}$

(2) H_3C—C_6H_4—MgBr $\xrightarrow[(2)\ H_2O,\ H^+]{(1)\ HCHO}$

(b) (1) $(CH_3)_2CHCH_2\overset{O}{\overset{\|}{C}}CH_3$ $\xrightarrow[(2)\ H_2O,\ H^+]{(1)\ NaBH_4}$

(2) $(CH_3)_2CHCH_2MgBr$ $\xrightarrow[(2)\ H_2O,\ H^+]{(1)\ CH_3CHO}$

7.31 The numbers over and under the reaction arrows in (b), (c), and (d) refer to a sequence of reactions. For example, in (b), cyclohexanone is first treated with methylmagnesium iodide, and then the product of that reaction is treated with aqueous acid.

(a) [cyclohexenyl]—MgI

(b) $\xrightarrow{(1)\ CH_3Li}$ [cyclohexane with CH_3 and $O^-\ Li^+$] $\xrightarrow{(2)\ H_2O,\ H^+}$ [cyclohexane with CH_3 and OH]

(c) $\xrightarrow{(1)\ Mg}$ [cyclohexyl]—MgBr $\xrightarrow{(2)\ HCHO}$ [cyclohexyl]—$CH_2O^-\ {}^+MgBr$

$\xrightarrow{(3)\ H_2O,\ H^+}$ [cyclohexyl]—CH_2OH

(d) $\xrightarrow{(1)\ HC{\equiv}CMgBr}$ $C_6H_5CH{=}CHCH(O^-\ {}^+MgBr)C{\equiv}CH$ $\xrightarrow{(2)\ H_2O,\ H^+}$

$C_6H_5CH{=}CHCH(OH)C{\equiv}CH$

7.32 (a) contains no acidic hydrogen

(b) $H\ddot{O}CH_2CH_2\ddot{O}H + 2\ CH_3MgI \longrightarrow {}^-{:}\ddot{O}CH_2CH_2\ddot{O}{:}^- + 2\ Mg^{2+}$

$+ 2\ I^- + 2\ CH_4$

(c) $(CH_3CH_2)_2\ddot{N}H + CH_3MgI \longrightarrow (CH_3CH_2)_2\ddot{N}{:}^- + Mg^{2+} + I^-$

$+ CH_4$

(d) $CH(CO_2H)_3 + 3\ CH_3MgI \longrightarrow CH(CO_2^-)_3 + 3\ Mg^{2+} + 3\ I^-$

$+ 3\ CH_4$

7.33 (a) $(CH_3)_2CHBr \xrightarrow[\text{ether}]{\text{Mg}} (CH_3)_2CHMgBr \xrightarrow[\text{(2) } H_2O,\ H^+]{\text{(1) } (CH_3)_2C{=}O} (CH_3)_2CHC(OH)(CH_3)_2$

(b) $(CH_3)_2CHMgBr \xrightarrow[\text{(2) } H_2O,\ H^+]{\text{(1) } CH_3CHO} (CH_3)_2CHCH(OH)CH_3$

7.34 (a) (1) $CH_3CH_2CHBrCH_3 + OH^- \longrightarrow$

(2) $CH_3CH_2MgBr \xrightarrow[\text{(2) } H_2O,\ H^+]{\text{(1) } CH_3CHO}$ *or* $CH_3MgI \xrightarrow[\text{(2) } H_2O,\ H^+]{\text{(1) } CH_3CH_2CHO}$

(3) $CH_3CH_2C(=O)CH_3 \xrightarrow[\text{(2) } H_2O,\ H^+]{\text{(1) } NaBH_4}$

(b) (1) $C_6H_5CH(Br)C_6H_5 + OH^- \longrightarrow$

(2) $C_6H_5MgBr \xrightarrow[\text{(2) } H_2O,\ H^+]{\text{(1) } C_6H_5CHO}$

(3) $C_6H_5C(=O)C_6H_5 \xrightarrow[\text{(2) } H_2O,\ H^+]{\text{(1) } NaBH_4}$

(c) (1) $CH_3CH_2CH_2CH_2Br + OH^- \longrightarrow$

(2) $CH_3CH_2CH_2MgBr \xrightarrow[\text{(2) } H_2O,\ H^+]{\text{(1) } HCHO}$

(3) $CH_3CH_2CH_2CHO \xrightarrow[\text{(2) } H_2O,\ H^+]{\text{(1) } NaBH_4}$

(d) (1) C_6H_{11}–CH_2Br + OH^- $\longrightarrow$

(2) C_6H_{11}–MgBr $\xrightarrow[\text{(2) } H_2O,\ H^+]{\text{(1) HCHO}}$

(3) C_6H_{11}–CHO $\xrightarrow[\text{(2) } H_2O,\ H^+]{\text{(1) } NaBH_4}$

7.35 (a) $(CH_3)_2CH\ddot{O}H \underset{}{\overset{H^+}{\rightleftharpoons}} (CH_3)_2CH\overset{+}{O}H_2 \underset{}{\overset{-H_2\ddot{O}:}{\rightleftharpoons}} (CH_3)_2CH^+$

2° alcohol, S_N1 $\xrightarrow{I^-}$ $(CH_3)_2CHI$

(b) $CH_3CH_2CH_2CH_2\ddot{O}H \overset{H^+}{\rightleftharpoons} CH_3CH_2CH_2CH_2\overset{+}{O}H_2 \xrightarrow{I^-}$

1° alcohol, S_N2

$$\left[\begin{array}{c} \overset{\delta+}{:OH_2} \\ \vdots \\ CH_3CH_2CH_2CH_2 \\ \vdots \\ I^{\delta-} \end{array}\right] \xrightarrow{-H_2\ddot{O}:} CH_3CH_2CH_2CH_2I$$

transition state

7.36 (a) C_6H_{11}–Cl + H_2O

(b) 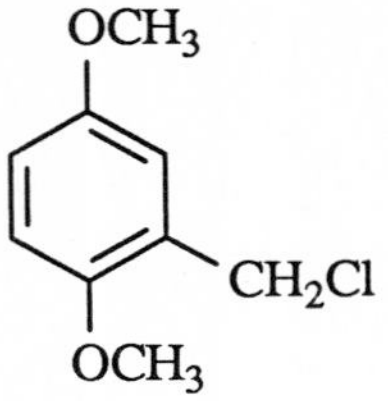 [2,5-dimethoxybenzyl chloride: benzene ring with OCH_3 at positions 1 and 4, CH_2Cl at position 2] + H_2O

Hydrochloric acid is not strong enough to cleave ether linkages. Benzylic alcohols are more reactive than other 1° alcohols and do react with HCl.

(c) 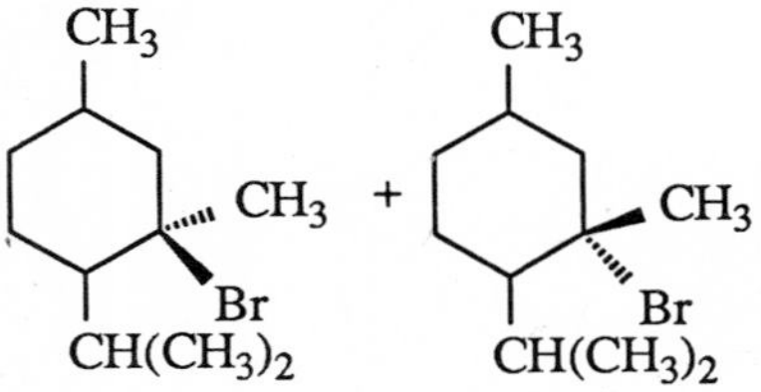 [cyclohexane ring with CH_3, CH_3/Br on the same carbon (two configurations), and $CH(CH_3)_2$] + [stereoisomer] + H_2O

In (c), two stereoisomers are obtained because the reaction proceeds by way of a carbocation.

7.37 (a) 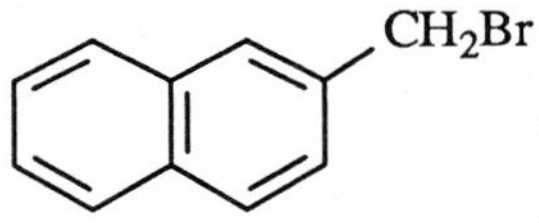 [2-naphthyl-CH_2Br] + H_2O

(fastest because it is a benzylic alcohol)

(b) $CH_3CH_2CH_2Br$ + H_2O (slowest because it is a 1° alcohol)

(c) 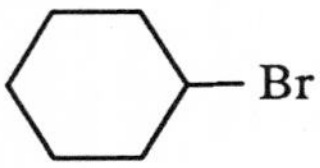 [cyclohexyl]—Br + H_2O

7.38 (a) 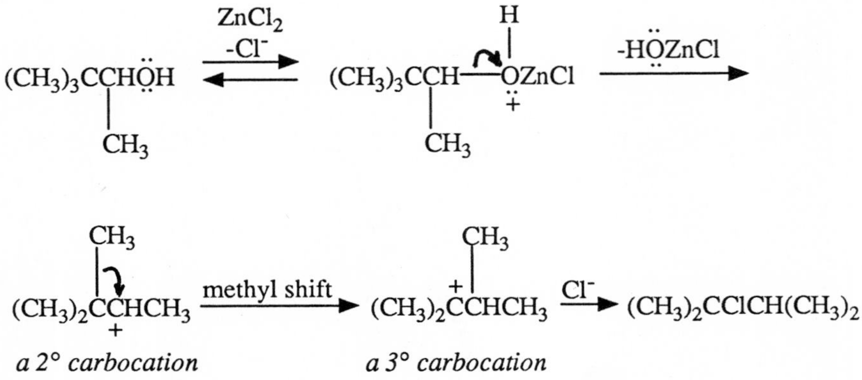 $(CH_3)_3CCH(CH_3)\ddot{O}H$ $\underset{}{\overset{ZnCl_2,\ -Cl^-}{\rightleftharpoons}}$ $(CH_3)_3CCH(CH_3)—\overset{+}{O}(H)ZnCl$ $\xrightarrow{-H\ddot{O}ZnCl}$

$(CH_3)_2\overset{CH_3}{C}\overset{+}{C}HCH_3$ $\xrightarrow{\text{methyl shift}}$ $(CH_3)_2\overset{+}{C}CH(CH_3)CH_3$ $\xrightarrow{Cl^-}$ $(CH_3)_2CClCH(CH_3)_2$

a 2° carbocation *a 3° carbocation*

(b) $(C_6H_5)_2CHCH_2\ddot{O}H \overset{H^+}{\rightleftharpoons} (C_6H_5)_2C(H)—CH_2—\overset{+}{O}H_2 \xrightarrow[- H_2O]{\text{hydride shift}}$

$(C_6H_5)_2\overset{+}{C}CH_3 \xrightarrow{I^-} (C_6H_5)_2CICH_3$

resonance stabilized by two rings

7.39 (a) a methyl alcohol, S_N2 (b) 2° ROH, S_N1

(c) 3° ROH, S_N1 (d) a benzylic ROH, S_N1

7.40 (a) $CH_3CH_2OCH_2CH_2Br$ (Ethers do not react with PBr_3.)

(b) H_3C CH_3 CH_2Br CH_3

Although the alcohol is allylic, we would expect no rearrangement with PBr_3.

(c) $(CH_3)_2NCH_2CH_2Cl + HCl \longrightarrow (CH_3)_2\overset{+}{N}HCH_2CH_2Cl\ Cl^-$

In each of the above reactions, inorganic esters would be intermediates. Under proper conditions, these inorganic esters could be isolated:

(a) $(CH_3CH_2OCH_2CH_2O)_3P$

(b) H_3C CH_3 CH_2O P CH_3 3

(c) $(CH_3)_2NCH_2CH_2OSCl$ (S=O) or $[(CH_3)_2NCH_2CH_2O]_2S{=}O$

(as their salts)

7.41 (a) (*E*), or *trans*-$C_6H_5CH{=}CHCH_3$ (b) $H_2C{=}CHCH_2CH_3$

(c) (*E*), or *trans*-$CH_3CH{=}CHCH_3$ (d) $HOCH_2CH_2CH{=}C(CH_3)_2$

In these cases, the more substituted, *trans* alkene is formed, if possible. In (d), the 4 -OH is eliminated preferentially because 3° alcohols undergo elimination more readily than 1° alcohols.

(e) (*E*) and (*Z*) $CHCO_2H{=}C(CO_2H)CH_2CO_2H$

$$\begin{array}{l} CHCO_2H \\ \| \\ CCO_2H \\ | \\ CH_2CO_2H \end{array}$$

Both isomers would probably be formed because the energy difference between them is small.

7.42 $H_2C{=}CHCH_2{-}OS(=O)Cl \xrightarrow{Cl^-} H_2C{=}CHCH_2Cl$

7.43 (a) $(S)\text{-}CH_3CH(OH)(CH_2)_2CH_3 + TsCl \xrightarrow{-HCl} (S)\text{-}CH_3CH(OTs)(CH_2)_2CH_3$

(b) $(R)\text{-}CH_3CH_2CH(OH)CH_3 \xrightarrow[\text{heat}]{H_2SO_4}$ (*E*)- or *trans*-$CH_3CH{=}CHCH_3 + H_2O$

(c) $(R)\text{-}CH_3CH_2CH(OH)CH_3 + ClSO_3H \longrightarrow (R)\text{-}CH_3CH_2CH(OSO_3H)CH_3 + HCl$

(d) $(R)\text{-}CH_3CH(OTs)CH_2CH_3 + CH_3CH_2OH \xrightarrow[\text{(solvolysis)}]{S_N1}$

$(R)(S)\text{-}CH_3CH(OCH_2CH_3)CH_2CH_3$

racemic

7.44 All three are phosphate esters (inorganic esters of phosphoric acid).

7.45 (a) $CH_3CH_2CO_2H$ contains more O atoms.

(b) CH_3CH_2CHO contains fewer H atoms.

(c) 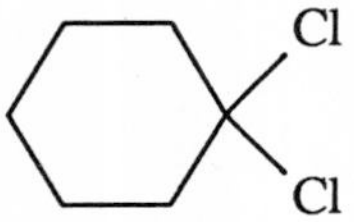 1,1-dichlorocyclohexane (cyclohexane ring with Cl, Cl on one carbon) contains fewer H atoms.

(c) 1,1-dichlorocyclohexane (cyclohexane ring with two Cl on one carbon) contains fewer H atoms.

7.46 (a) next higher: C_6H_5CHO (drawn as $C_6H_5\overset{O}{\overset{\|}{C}}H$) next lower: $C_6H_5CH_3$

(b) next higher: $CH_3C{\equiv}CH$ next lower: $CH_3CH_2CH_3$

(c) next higher: $CH_3CH_2\overset{O}{\overset{\|}{C}}OH$ next lower: $CH_3CH_2CH_2OH$

(d) next higher: $CH_3CH_2CBr_2$—$CHBrCH_3$ (Br above and below C; Br below CH)

next lower: $CH_3CH_2CH_2CHBrCH_3$ (Br below CH)

7.47 (a) oxidizing agent (b) oxidizing agent (c) neither

(d) reducing agent (e) neither (f) oxidizing agent

7.48 (a) $CH_3CH(OH)CH_2CH_3 + CrO_3 + H^+ \xrightarrow{\text{heat}} CH_3\overset{O}{\overset{\|}{C}}CH_2CH_3$

(b) $CH_3CH_2CH_2CH_2OH + CrO_3 \cdot 2$ pyridine $\longrightarrow CH_3CH_2CH_2\overset{O}{\overset{\|}{C}}H$

(c) $CH_3CH_2CH_2CH_2OH + CrO_3 + H^+ \xrightarrow{\text{heat}} CH_3CH_2CH_2\overset{O}{\overset{\|}{C}}OH$

(d) $ClCH_2CH(OH)CH_2Cl + CrO_3 + H^+ \xrightarrow{heat} ClCH_2C(=O)CH_2Cl$

(e) cyclooctanol $+ CrO_3 + H^+ \xrightarrow{heat}$ cyclooctanone

(f) $CH_3(CH_2)_8CH_2OH + CrO_3 \cdot 2$ pyridine $\longrightarrow CH_3(CH_2)_8CH(=O)$

7.49 (a) $H_2C\text{—}CHCH_2OH$ (cyclopropane ring, CH_2 bridge) $\xrightarrow{PBr_3}$ $H_2C\text{—}CHCH_2Br$ (cyclopropane ring, CH_2 bridge) $\xrightarrow[\text{ether}]{Mg}$ $H_2C\text{—}CHCH_2MgBr$ (cyclopropane ring, CH_2 bridge)

$\xrightarrow[\text{(2) } H_2O,\ H^+]{\text{(1) } CH_3CH_2CHO}$ $H_2C\text{—}CHCH_2CH(OH)CH_2CH_3$ (cyclopropane ring, CH_2 bridge) $\xrightarrow{SOCl_2}$

(b) $CH_3CH_2OH \xrightarrow{HBr} CH_3CH_2Br \xrightarrow[\text{ether}]{Mg} CH_3CH_2MgBr$

$\xrightarrow[\text{(2) } H_2O,\ H^+]{\text{(1) } CH_3C(=O)CH_3}$ $(CH_3)_2C(OH)CH_2CH_3 \xrightarrow[\text{heat}]{H_2SO_4}$

(c) $(CH_3)_2CHCH_2OH \xrightarrow{PBr_3} (CH_3)_2CHCH_2Br \xrightarrow[\text{ether}]{Mg} (CH_3)_2CHCH_2MgBr$

$\xrightarrow[\text{(2) } H_2O,\ H^+]{\text{(1) } H_2C\text{—}CH_2 \text{ (epoxide, O bridge)}}$

(d) $(CH_3)_2CHCH_2MgBr$ from (c) $\xrightarrow[\text{(2) } H_2O,\ H^+]{\text{(1) } CH_3CHO}$ $(CH_3)_2CHCH_2CH(OH)CH_3$ $\xrightarrow{HBr}$

$(CH_3)_2CHCH_2CHBrCH_3$ $\xrightarrow{CN^-}$

7.50 (a) $(CH_3)_2CHCH_2Cl$

(b) $CH_3CH_2CH_2CH_2OH$

(c) H_3C Cl

(*R*)

(d) (*S*)-2-chlorobutane, or (*S*)-$CH_3CHClCH_2CH_3$

(e) (cyclohexyl)–$CH_2CH_2O^-\ Na^+$

(f) $CH_3CH_2CH_2CH_2OH$

(g) $(C_6H_5)_2C{=}CHCH_3$

7.51 (a) $CH_3CH(OH)CH_2CH_3$ + Na ⟶ $CH_3CH(O^-\ Na^+)CH_2CH_3$

$(CH_3)_2CHCH_2OH$ + HBr ⟶ $(CH_3)_2CHCH_2Br$

} ⟶ product

or

$CH_3CH(OH)CH_2CH_3$ + HBr ⟶ $CH_3CHBrCH_2CH_3$

$(CH_3)_2CHCH_2OH$ + Na ⟶ $(CH_3)_2CHCH_2O^-\ Na^+$

} ⟶ product

(b) (cycloheptanone)=O $\xrightarrow[\text{(2) } H_2O,\ H^+]{\text{(1) } NaBH_4}$ (cycloheptyl)–OH

(c) (cycloheptyl)–OH $\xrightarrow[\text{heat}]{H_2CrO_4}$ (cycloheptanone)=O

(d) $(R)\text{-}CH_3\overset{\displaystyle OH}{\overset{|}{C}}HCH_2CH_3 + SOCl_2 \xrightarrow{\text{pyridine}} (S)\text{-}CH_3\overset{\displaystyle Cl}{\overset{|}{C}}HCH_2CH_3$

(e) $C_6H_5CH_2OH \xrightarrow{CrO_3 \cdot 2\text{ pyridine}} C_6H_5CHO$

(f) $CH_3CH_2CH_2Br \xrightarrow[\text{ether}]{Mg} CH_3CH_2CH_2MgBr \xrightarrow[(2)\ H_2O,\ H^+]{(1)\ HCHO} CH_3CH_2CH_2CH_2OH$

(g) $C_6H_5Br \xrightarrow[\text{ether}]{Mg} C_6H_5MgBr \xrightarrow[(2)\ H_2O,\ H^+]{(1)\ HCHO} C_6H_5CH_2OH$

(h) $C_6H_5CHO \xrightarrow[(2)\ H_2O,\ H^+]{(1)\ CH_3MgI} C_6H_5\overset{\displaystyle OH}{\overset{|}{C}}HCH_3$

(i) $CH_3\overset{\displaystyle OH}{\overset{|}{C}}HCH_2CH_2\overset{\displaystyle O}{\overset{\|}{C}}H \xrightarrow{CrO_3 \cdot 2\text{ pyridine}} CH_3\overset{\displaystyle O}{\overset{\|}{C}}CH_2CH_2\overset{\displaystyle O}{\overset{\|}{C}}H$

7.52 (a) $CH_4 + CH_3CH_2CH_2CH_2O^- \ {}^+MgI$

(b) [benzene] $+ CH_3CH_2CH_2CH_2O^- \ Li^+$

(c) $CH_3CH_2CH_2CH_2Br + H_2O$

(d) $CH_3CH_2CH_2CH_2O^- \ K^+ + H_2$

(e) $CH_3CH_2CH_2CH_2O^- \ Na^+ + H_2$

7.53 (a) achiral $CH_3(CH_2)_4\overset{\displaystyle O}{\overset{\|}{C}}CH_3$

(b) racemic $CH_3(CH_2)_4CHICH_3$ (by an S_N1 reaction)

(c) (*R*)-$CH_3(CH_2)_4\overset{O^-}{\overset{|}{C}}HCH_3$ Li^+ (chiral carbon unaffected)

(d) achiral (*E*)-, or *trans*-$CH_3(CH_2)_3CH{=}CHCH_3$

(e) (*R*)-$CH_3(CH_2)_4\overset{O^-}{\overset{|}{C}}HCH_3$ ^+MgI + CH_4 (chiral carbon unaffected)

(f) no reaction (g) no appreciable reaction

(h) (*S*)-$CH_3(CH_2)_4CHClCH_3$

7.54 Typical synthetic paths follow. There may be other correct answers.

(a) $(CH_3)_3CBr \xrightarrow[\text{ether}]{Mg} (CH_3)_3CMgBr \xrightarrow[(2)\ H_2O,\ H^+]{(1)\ CH_3CHO}$

or $CH_3I \xrightarrow[\text{ether}]{Mg} CH_3MgI \xrightarrow[(2)\ H_2O,\ H^+]{(1)\ (CH_3)_3CCHO}$

Because of steric hindrance, a substitution reaction is not practical. Rearrangement and elimination would probably be observed as side reactions.

(b) [bicyclic bromide, Br] $\xrightarrow[\text{ether}]{Mg}$ [bicyclic Grignard, MgBr] $\xrightarrow[(2)\ H_2O,\ H^+]{(1)\ HCHO}$

or [bicyclic, CH_2Br] $\xrightarrow{OH^-}$

(c) $CH_3I \xrightarrow[\text{ether}]{Mg} CH_3MgI \xrightarrow[(2)\ H_2O,\ H^+]{(1)\ \text{[bicyclic ketone, O]}}$

(d) $CH_3CH_2Br \xrightarrow[\text{ether}]{\text{Mg}} CH_3CH_2MgBr \xrightarrow[\text{(2) } H_2O,\ H^+]{\text{(1) cyclohexyl–}\overset{O}{\overset{\|}{C}}CH_3}$

or

$\text{cyclohexyl–Br} \xrightarrow[\text{ether}]{\text{Mg}} \text{cyclohexyl–MgBr} \xrightarrow[\text{(2) } H_2O,\ H^+]{\text{(1) } CH_3\overset{O}{\overset{\|}{C}}CH_2CH_3}$

(e) $(CH_3)_2CHBr \xrightarrow[\text{ether}]{\text{Mg}} (CH_3)_2CHMgBr \xrightarrow[\text{(2) } H_2O,\ H^+]{\text{(1) cyclopentyl–CHO}}$

or

$\text{cyclopentyl–Br} \xrightarrow[\text{ether}]{\text{Mg}} \text{cyclopentyl–MgBr} \xrightarrow[\text{(2) } H_2O,\ H^+]{\text{(1) } (CH_3)_2CHCHO}$

(f) $CH_2{=}CH\overset{Br}{\overset{|}{C}}{=}CH_2 \xrightarrow[\text{ether}]{\text{Mg}} CH_2{=}CH\overset{MgBr}{\overset{|}{C}}{=}CH_2 \xrightarrow[\text{(2) } H_2O,\ H^+]{\text{(1) } CH_3\overset{O}{\overset{\|}{C}}CH_3}$

or $CH_3I \xrightarrow[\text{ether}]{\text{Mg}} CH_3MgI \xrightarrow[\text{(2) } H_2O,\ H^+]{\text{(1) } CH_2{=}CH\overset{H_2C}{\overset{\|}{C}}{-}\overset{O}{\overset{\|}{C}}CH_3}$

In (b), (c), (d), and (e), you may have been tempted to use solvolysis of alkyl halides with water. Solvolysis reactions are not always good synthetic reactions; therefore, it is best to avoid these reactions where possible when you are working synthetic problems on paper.

7.55 The intermediate in the formation of the allylic Grignard reagent is resonance stabilized; therefore, *two* Grignard reagents (and two Grignard products) can be formed.

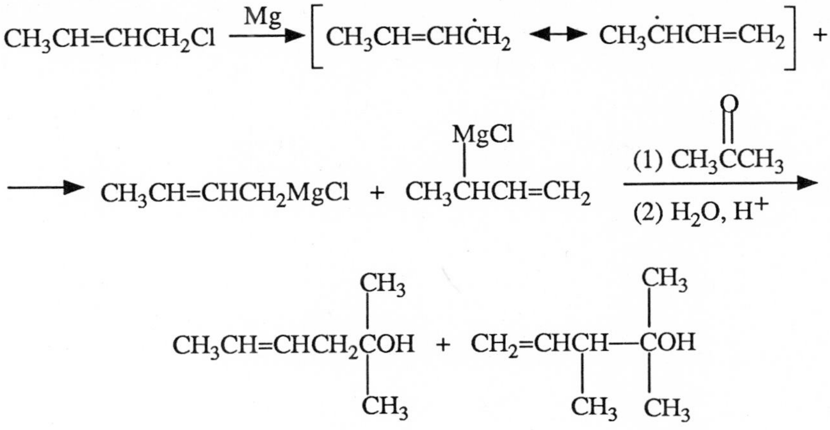

$$CH_3CH{=}CHCH_2Cl \xrightarrow{Mg} \left[CH_3CH{=}CH\dot{C}H_2 \longleftrightarrow CH_3\dot{C}HCH{=}CH_2\right] + \dot{M}gCl$$

$$\longrightarrow CH_3CH{=}CHCH_2MgCl + CH_3CH(MgCl)CH{=}CH_2 \xrightarrow[(2)\ H_2O,\ H^+]{(1)\ CH_3C(=O)CH_3}$$

$$CH_3CH{=}CHCH_2C(CH_3)_2OH + CH_2{=}CHCH(CH_3){-}C(CH_3)_2OH$$

7.56

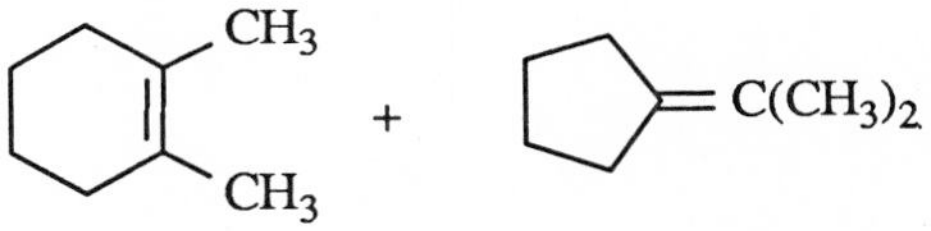

1,2-Dimethylcyclohexene + cyclopentylidene=$C(CH_3)_2$ The mechanisms follow:

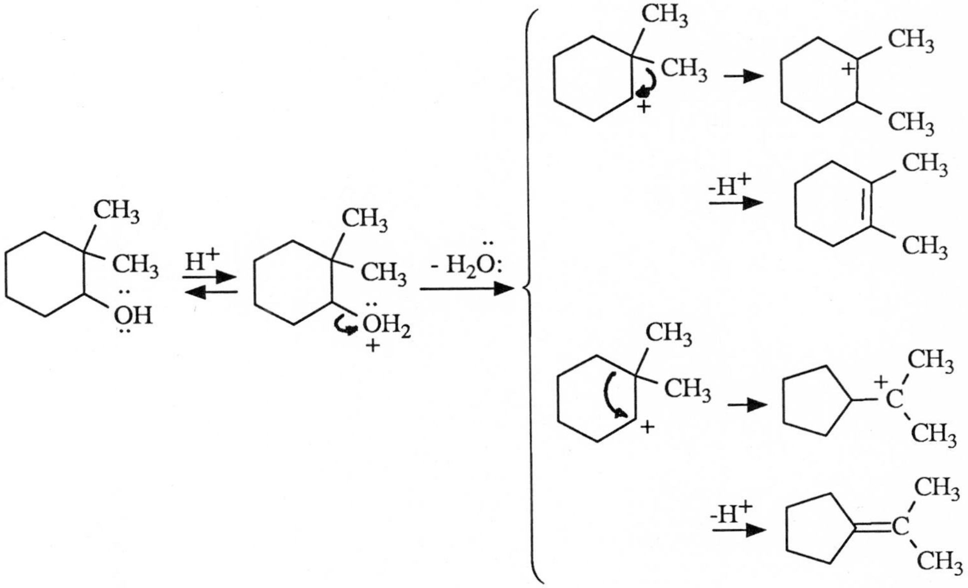

2,2-Dimethylcyclohexanol $\underset{}{\overset{H^+}{\rightleftharpoons}}$ protonated alcohol ($-\ddot{O}H_2^+$) $\xrightarrow{-H_2\ddot{O}:}$ 2,2-dimethylcyclohexyl cation, which reacts by two pathways:

- methyl shift → 1,2-dimethylcyclohexyl cation $\xrightarrow{-H^+}$ 1,2-dimethylcyclohexene
- ring contraction → cyclopentyl-$\overset{+}{C}(CH_3)_2$ $\xrightarrow{-H^+}$ cyclopentylidene=$C(CH_3)_2$

equatorial *t*-butyl group

7.57 $(CH_3)_3C$ [4-t-butylcyclohexyl tosylate] + $:\ddot{I}:^-$ $\xrightarrow[S_N2]{-\ ^-:\ddot{O}Ts}$ $(CH_3)_3C$ [4-t-butyliodocyclohexane] *cis* + $:\ddot{I}:^-$

$\xrightarrow[S_N2]{-\ :\ddot{I}:^-}$ $(CH_3)_3C$ [4-t-butyliodocyclohexane]

trans

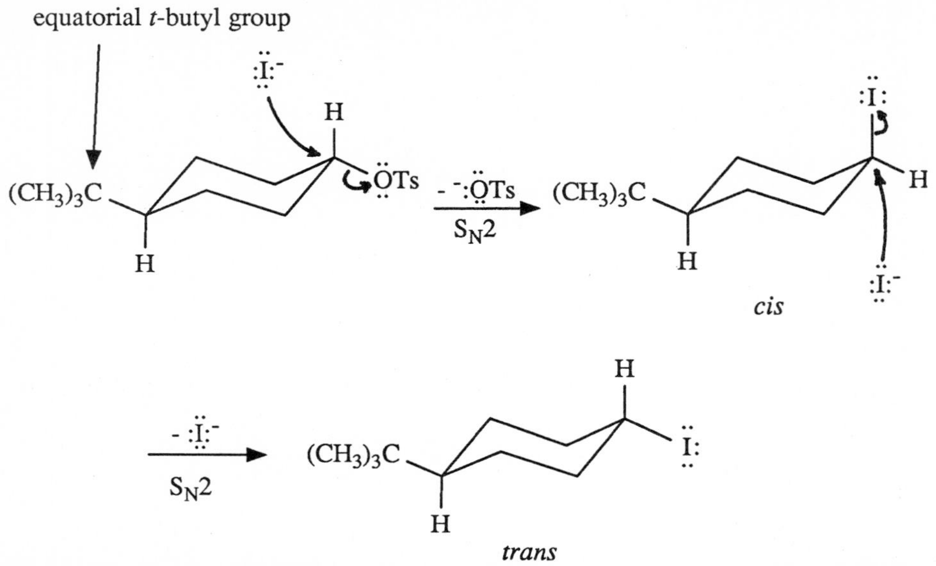

The iodide ion is a good leaving group and a good nucleophile as well. Iodide ion displaces the tosylate ion from the rear to yield the *cis*-4-*t*-butyliodocyclohexane. This *cis* product can be attacked, in turn, by iodide to yield the *trans* product. Therefore, the reactions yields a mixture of stereoisomers.

7.58 This reaction is an elimination (E2) reaction with the tosylate group as the leaving group. If possible, the most stable alkene will be formed--in this case, the most substituted conjugated alkene.

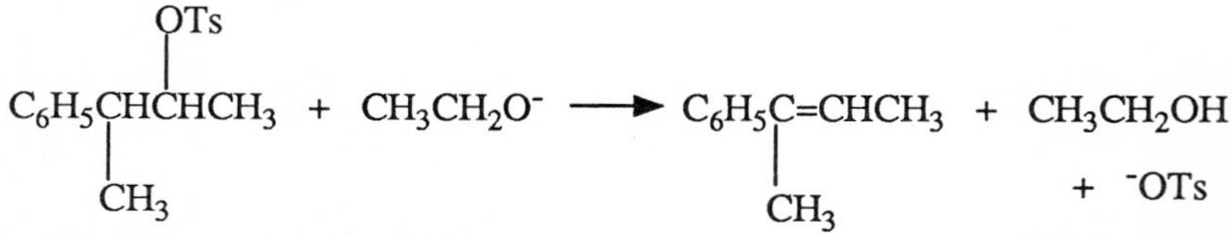

$$C_6H_5CH(CH_3)CH(OTs)CH_3 + CH_3CH_2O^- \longrightarrow C_6H_5C(CH_3){=}CHCH_3 + CH_3CH_2OH + {}^-OTs$$

This elimination is a stereospecific anti elimination. The (*E*) isomer is the product because the H being eliminated must be anti to the tosylate leaving group in the (2*S*,3*S*) stereoisomer.

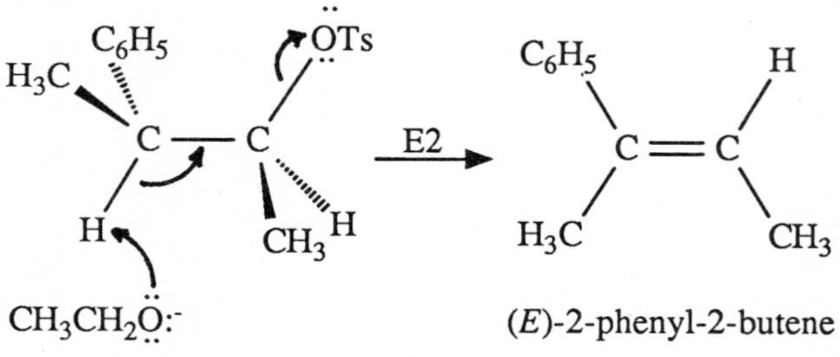

$CH_3CH_2\ddot{O}:^-$ + (2*S*,3*S*) tosylate $\xrightarrow{E2}$ (*E*)-2-phenyl-2-butene using Newman projections:

$$\text{Newman projection: front carbon } H_3C, H_3C, :\ddot{O}Ts;\ \text{back carbon } C_6H_5, H, H \ (H \text{ anti to OTs, removed by } CH_3CH_2\ddot{O}:^-) \xrightarrow{E2} H_3C(H_3C)C{=}C(C_6H_5)(H)$$

(*E*)-product

7.59 (a) $CH_3CH(OH)(CH_2)_5CH_3 \xrightarrow{[O]} CH_3C(=O)(CH_2)_5CH_3$ (2-octanone)

(b) $C_6H_5CH(OH)CH_3 \xrightarrow{[O]} C_6H_5C(=O)CH_3$ (phenylethanone, acetophenone)

(c) $C_6H_5CH{=}CHCH_2OH \xrightarrow{[O]} C_6H_5CH{=}CHCH{=}O$ (3-phenyl-2-propenal, cinnamaldehyde)

Note that the aldehyde is obtained in (c) because the oxidizing agent is a chromate-pyridine complex.

7.60 There are other possible synthetic schemes in each case.

(a) $CH_3CH_2Br \xrightarrow[\text{ether}]{Mg} CH_3CH_2MgBr \xrightarrow[(2)\ H_2O,\ H^+]{(1)\ CH_3CHO} CH_3CH(OH)CH_2CH_3$

$\xrightarrow[\text{heat}]{H_2SO_4} CH_3CH{=}CHCH_3 \xrightarrow{NBS} Br\text{-}CH_2CH{=}CHCH_3$

or $CH_3CH{=}CHI \xrightarrow[\text{ether}]{Mg} CH_3CH{=}CHMgI \xrightarrow[(2)\ H_2O,\ H^+]{(1)\ HCHO}$

$CH_3CH{=}CHCH_2OH \xrightarrow{PBr_3}$ product

(b) $CH_3CH_2CO_2H \xrightarrow[\text{heat}]{CH_3CH_2OH,\ H^+} CH_3CH_2CO_2CH_2CH_3$

$\xrightarrow[(2)\ H_2O,\ H^+]{(1)\ 2\ CH_3CH_2MgBr\ \text{from (a)}} CH_3CH_2C(OH)(CH_2CH_3)CH_2CH_3$

7.61 (a) $CH_3O-C_6H_4-CH_2OH \xrightarrow{HBr} CH_3O-C_6H_4-CH_2Br \xrightarrow{^-CN}$

(b) $C_6H_5CH_2Br \xrightarrow[\text{ether}]{Mg} C_6H_5CH_2MgBr \xrightarrow[(2)\ H_2O,\ H^+]{(1)\ C_6H_5CHO}$

$C_6H_5CH_2-CH(OH)C_6H_5 \xrightarrow[E1]{-H_2O}$ (*E*), or *trans*-$C_6H_5CH{=}CHC_6H_5$

(c) $C_6H_5CH_3 \xrightarrow{NBS} C_6H_5CH_2Br \xrightarrow[\text{ether}]{Mg} C_6H_5CH_2MgBr$

$\xrightarrow[(2)\ H_2O,\ H^+]{(1)\ CH_3CHO}$

(d) cyclopentyl$-CH_2CH_2OH \xrightarrow{PBr_3}$ cyclopentyl$-CH_2CH_2Br \xrightarrow[\text{ether}]{Mg}$

cyclopentyl$-CH_2CH_2MgBr \xrightarrow[(2)\ H_2O,\ H^+]{(1)\ HCHO}$ cyclopentyl$-CH_2CH_2CH_2OH$

$\xrightarrow{PBr_3}$ cyclopentyl$-CH_2CH_2CH_2Br \xrightarrow[\text{ether}]{Mg}$

cyclopentyl$-CH_2CH_2CH_2MgBr \xrightarrow[(2)\ H_2O,\ H^+]{(1)\ C_6H_5CHO}$

cyclopentyl$-CH_2CH_2CH_2CH(OH)C_6H_5$

7.62 (a) cyclohexyl–Br $\xrightarrow{^{-}OH}$ cyclohexyl–OH $\xrightarrow[H^+]{CrO_3}$ cyclohexanone (=O)

(b) cyclohexyl–OH $\xrightarrow[H^+]{CrO_3}$ cyclohexanone (=O)

$\xrightarrow[(2)\ H_2O,\ H^+]{(1)\ CH_2{=}CHCH_2MgBr}$ product

(c) cyclohexyl–Br $\xrightarrow{^{-}OH}$ cyclohexyl–OH $\xrightarrow[H^+]{CrO_3}$ cyclohexanone (=O)

$\xrightarrow[(2)\ H^+]{(1)\ C_6H_5MgBr}$ 1-phenylcyclohexanol (OH) $\xrightarrow[-H_2O]{H^+}$ 1-phenylcyclohexene

7.63 $(CH_3)_2C(\ddot{O}H){-}C(\ddot{O}H)(CH_3)_2 \xrightarrow{H^+} (CH_3)_2C(H_2\overset{+}{\ddot{O}}){-}C(\ddot{O}H)(CH_3)_2 \xrightarrow{-H_2\ddot{O}:}$

$(CH_3)_2\overset{+}{C}{-}C(\ddot{O}H)(CH_3)CH_3 \longrightarrow (CH_3)_2C(CH_3){-}C(=\overset{+}{\ddot{O}}H)CH_3 \xrightarrow{-H^+}$ product

a 3° carbocation

a protonated ketone (more stable than an ordinary 3° carbocation)

Chapter 8

Ethers, Epoxides, and Sulfides

8.8 (a) 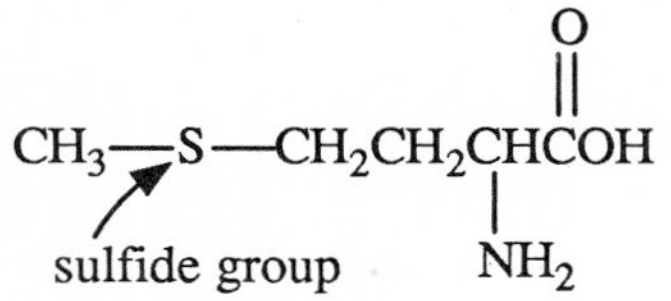 CH_3—S—$CH_2CH_2CH(NH_2)COOH$ (sulfide group indicated at S)

(b) 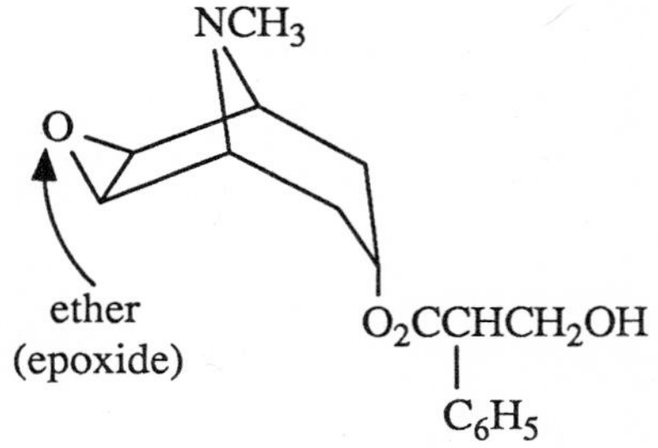 Epoxide-containing tropane structure with NCH3 bridge and ester group $O_2CCH(C_6H_5)CH_2OH$; the epoxide oxygen is labeled "ether (epoxide)"

(c) Isoquinoline bearing two CH_3O groups and a CH_2-linked 3,4-dimethoxyphenyl group (OCH_3, OCH_3); each CH_3O/OCH_3 is labeled "ether"

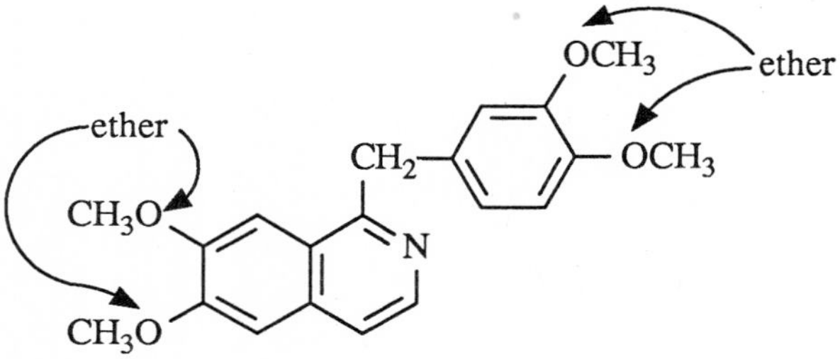

(d) 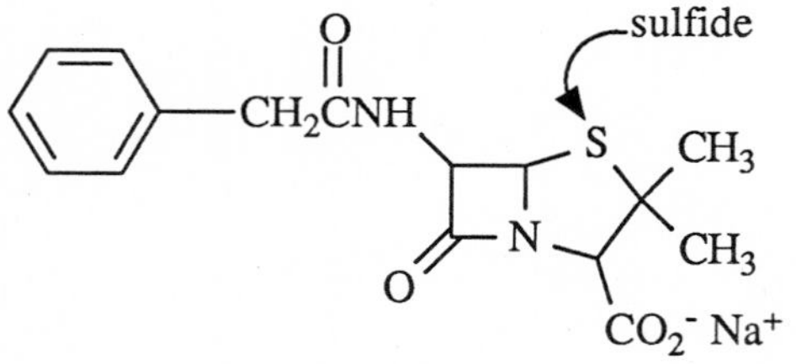 Penicillin structure: $C_6H_5CH_2C(=O)NH$– on the β-lactam ring, fused thiazolidine ring with S (labeled "sulfide"), two CH_3 groups, and CO_2^- Na^+

8.9 (R) and (S) enantiomers of 2-(chloromethyl)oxirane, drawn as mirror images:

(R): H_2C—C(O epoxide ring) with H and CH_2Cl on C ¦ mirror ¦ (S): H and $ClCH_2$ on C—CH_2 (O epoxide ring)

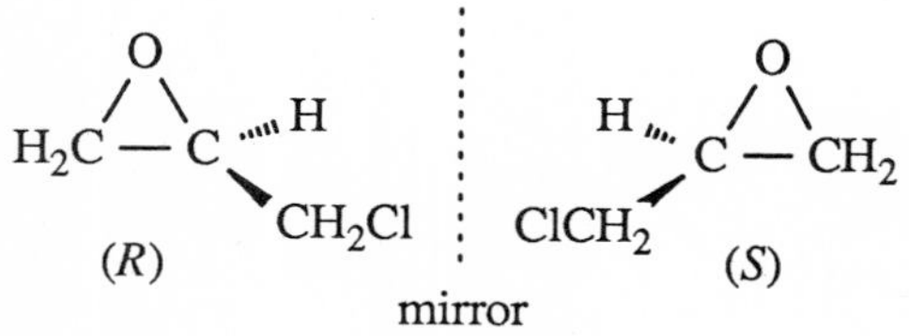

8.10 (a) ethyl isopropyl ether

(b) methyl phenyl ether, methoxybenzene, or anisole

(c) methoxyoxirane

(d) 1-methylcyclooctene oxide (Epoxides are often named after the alkene used for their synthesis; see Section 8.3C. The oxide prepared from cyclooctene, for example, is named cyclooctene oxide.)

8.11 (a) $C_6H_5-O-C(CH_3)_3$

(b) $(CH_3)_2CHCH_2OCH_2CH_2CH_2CH_2OCH(CH_3)CH_2CH_3$

(c) H, O, H; C–C; $CH(CH_3)_2$; cyclohexyl (epoxide: cyclohexyl–CH(O)CH–$CH(CH_3)_2$)

(d) $C_6H_5CH-CH_2$ (epoxide, O bridging)

8.12 2,6-dichlorophenol–OH $\xrightarrow{^-OH}$ 2,6-dichlorophenoxide–O^- $\xrightarrow{CH_3I}$ 2,6-dichloro–OCH_3

8.13 (a) $(C_6H_5)_2CHOCH_2CH_2Cl$ + $(C_6H_5)_2CHOCH(C_6H_5)_2$

+ $ClCH_2CH_2OCH_2CH_2Cl$ + $CH_2{=}CHCl$

(b) $CH_3CH_2CH(OCH_3)CH_2SCH_3$ + NaI

(c) o-$NO_2C_6H_4$–$OCH_2CH_2CH_2CH_3$ + NaBr

(d) 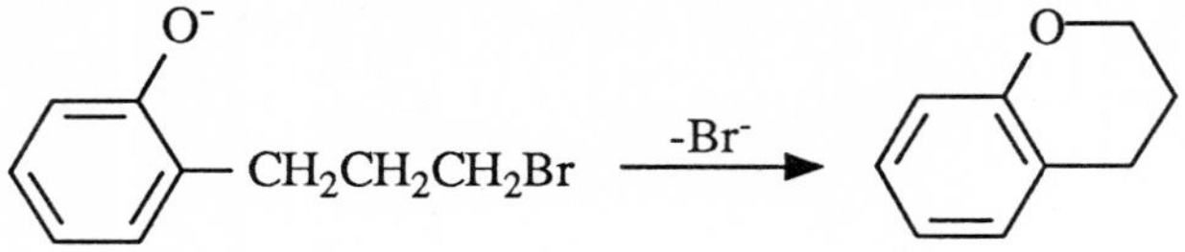 [phenoxide with O^- ortho to] $-CH_2CH_2CH_2Br$ $\xrightarrow{-Br^-}$ [chromane]

(e) $(C_6H_5)_2CHOCH_2CH_2N(CH_3)_2$

8.14 (a) 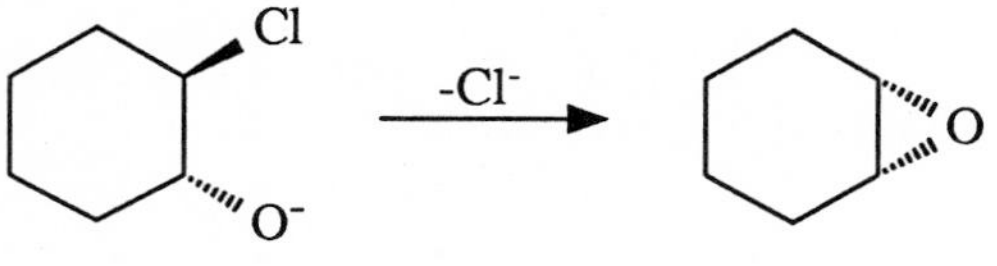 [cyclohexane with Cl and O^-] $\xrightarrow{-Cl^-}$ [cyclohexene oxide, O]

(b) 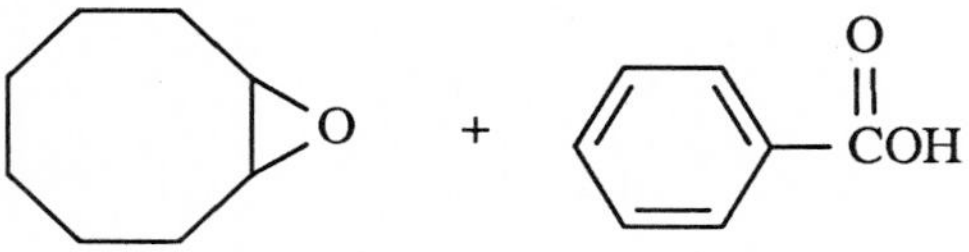 [cyclooctene oxide, O] + [benzene]$-\overset{O}{\overset{\|}{C}}OH$

(c) (2*R*,3*S*)-$CH_3CH_2\overset{CH_3}{\overset{|}{C}H}$—$\overset{O}{CH-CH_2}$

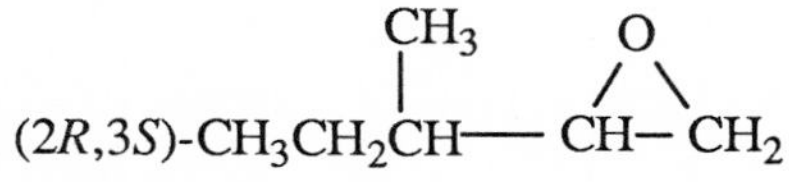

The intermediate is similar to that in Answer 8.14(a).

8.15 (a) 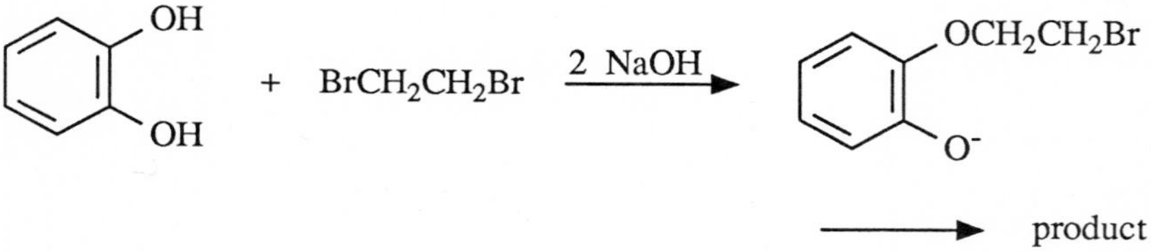 [benzene with OH, OH] + $BrCH_2CH_2Br$ $\xrightarrow{2\ NaOH}$ [benzene with OCH_2CH_2Br, O^-]

$\longrightarrow$ product

(b) 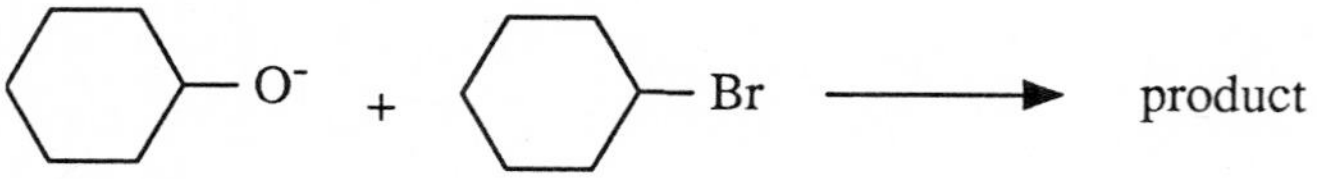 [cyclohexyl]$-O^-$ + [cyclohexyl]$-Br$ $\longrightarrow$ product

(c) (1) 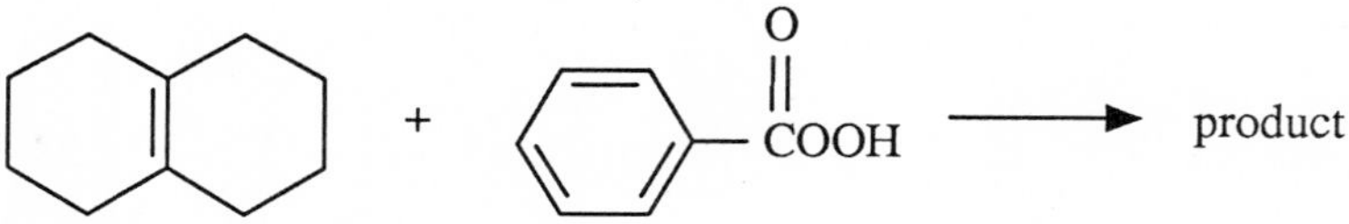 [octalin] + [benzene]$-\overset{O}{\overset{\|}{C}}OOH$ $\longrightarrow$ product

(2) 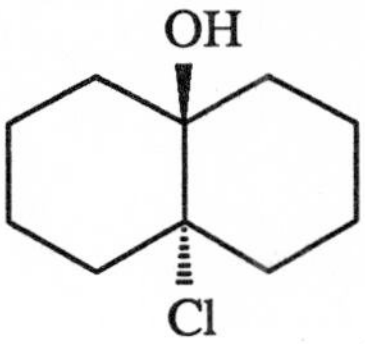 [decalin structure with OH (wedge) and Cl (hashed) at ring-fusion carbons] $\xrightarrow{NaOH}$ product

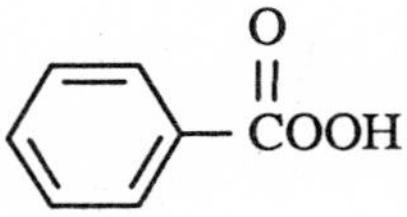

(d) (1) 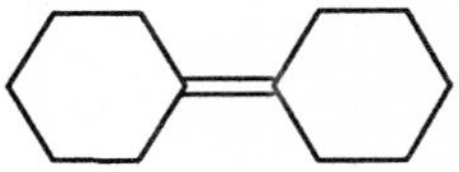 [cyclohexylidenecyclohexane] + [benzene ring]–$\overset{O}{\overset{\|}{C}}$OOH $\longrightarrow$ product

(2) 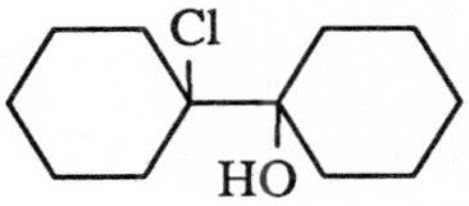 [1-chlorocyclohexyl attached to 1-hydroxycyclohexyl; Cl, HO] $\xrightarrow{NaOH}$ product

8.16 (a) $BrCH_2CH_2CH_2CH_2CH_2Br$ (b) 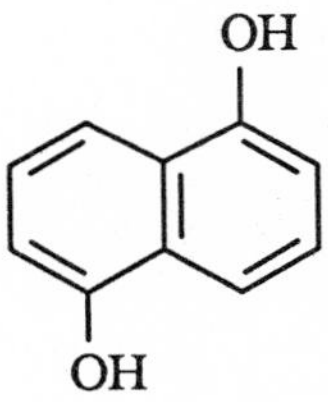 [1,5-dihydroxynaphthalene; OH, OH] + 2 CH_3I

(c) $CH_3CH_2CH_2Br$ + CH_3Br (d) $HOCH_2CH_2CH_2CH_2Cl$

(e) 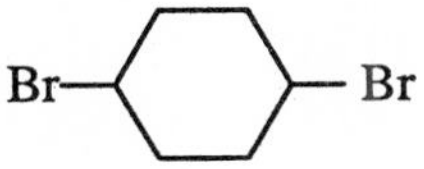 Br–[cyclohexane]–Br

8.17 (a) $(CH_3)_2C\overset{O}{—}CHCH_3$ (epoxide) + $CH_3(CH_2)_3CH_2OH$ $\xrightarrow{HCl}$

$(CH_3)_2C(OCH_2(CH_2)_3CH_3)—CH(OH)CH_3$

Depending on the quantity of HCl present, some chlorohydrin might also be observed:

$$(CH_3)_2C(Cl){-}CH(OH)CH_3$$

(b) $$(CH_3)_2\overset{O}{C{-}}CHCH_3 \xrightarrow[(2)\ H_2O,\ H^+]{(1)\ CH_2{=}CHCH_2MgBr} (CH_3)_2C(OH){-}CH(CH_2CH{=}CH_2)CH_3$$

(c) $$(CH_3)_2\overset{O}{C{-}}CHCH_3 \xrightarrow[(2)\ H_2O,\ H^+]{(1)\ C_6H_5Li} (CH_3)_2C(OH){-}CH(C_6H_5)CH_3$$

(d) $$(CH_3)_2\overset{O}{C{-}}CHCH_3 \xrightarrow{C_6H_5O^-\ Na^+} (CH_3)_2C(OH){-}CH(OC_6H_5)CH_3$$

Remember that oxiranes are attacked at the least hindered position in base. In acidic solution, the reaction proceeds according to the rules of carbocation stability.

8.18 (a) $HOCH_2CH_2Br$

(b) $$H_2\overset{:\ddot{O}:}{C{-}}CHCH_2N(CH_2CH_3)_2 \xrightarrow{S_N2} CH_2(:\ddot{S}H){-}CH(:\ddot{O}:^-)CH_2N(CH_2CH_3)_2$$

$^-:\ddot{S}H$
a strong nucleophile

$$\xrightarrow{H^+\ \text{from solvent}} CH_2(:\ddot{S}H){-}CH(:\ddot{O}H)CH_2N(CH_2CH_3)_2$$

(c) [oxetane] $\xrightarrow{\text{(1) } CH_3CH_2\text{—}MgI}$ $CH_2CH_2CH_2CH_3$ / $CH_2\ddot{O}:^{-}\ {}^{+}MgI$ $\xrightarrow{\text{(2) } H_2O,\ H^+}$

$CH_3(CH_2)_3CH_2OH$

In (c), the strained ring reacts similarly to an oxirane ring.

(d) $CH_3SCH_2CH(OH)CH_2Cl \xrightarrow[-Cl^-]{CH_3S^-} CH_3SCH_2CH(OH)CH_2SCH_3$

(e) [cyclopentane with OH (wedge) and $CH_2CH_2CH_3$ (hashed)]

(f) [naphthalene with $OCH_2CH(O^-\ Na^+)CH_2OH$]

(g) [naphthalene with $OCH_2\text{—}C(OH)(H)CH_2NHCH(CH_3)_2$]

8.19 (a) $CH_3CH_2\overset{O}{\overset{\|}{S}}CH_2CH_3$ (b) [cyclohexane]—SCH_2CH_3

(c) $CH_3CH_2CH_2SSCH_2CH_2CH_3$ + 2 HI

(d) $CH_3CH_2S(=O)_2\text{—}C(CH_3)(CH_2CH_3)\text{—}S(=O)_2CH_2CH_3$

8.20 (a) The epoxide formation requires backside (S_N2) attack by the intermediate alkoxide on the carbon bearing the chlorine. The oxygen and the chlorine of the *trans* isomer are positioned so that a backside attack can occur, while those of the *cis* isomer are not.

(b) *trans*:

$$\text{trans-2-chlorocyclohexanol} + OH^- \rightleftharpoons \text{alkoxide } (:\ddot{O}:^-) \xrightarrow{-Cl^-} \text{cyclohexene oxide}$$

cis:

$$\text{cis-2-chlorocyclohexanol} \xrightarrow[E2]{OH^-} \xrightarrow[-Cl^-]{-H_2O} \text{1-cyclohexenol (OH)} \xrightarrow{\text{proton transfer}} \text{cyclohexanone}$$

backside displacement of Cl by OH impossible

8.21 (a) $CH_2{=}CHCH_2OH \xrightarrow{PCl_3} CH_2{=}CHCH_2Cl \xrightarrow{\text{excess } Na^+ \ {}^-SH}$

$$CH_2{=}CHCH_2SH \xrightarrow{I_2 \text{ or } K_3Fe(CN)_6} \text{product}$$

8.22 (a) $CH_2{=}C(CH_3)CH_2CH_2OH \xrightarrow{H_3C-C_6H_4-SO_2Cl} CH_2{=}C(CH_3)CH_2CH_2CH_2OTs$ (OTs: good leaving group)

$$\xrightarrow[-\ {}^-OTs]{CH_3S^-\ Li^+} CH_2{=}C(CH_3)CH_2CH_2SCH_3$$

The preceding route is reported in the literature. You may have varied the reactants somewhat, using a halogenating agent instead of tosyl chloride, as in the following proposed synthesis.

(b) $C_6H_5CH_2CH_2OH \xrightarrow{PBr_3} C_6H_5CH_2CH_2Br \xrightarrow{CH_3S^-} C_6H_5CH_2CH_2SCH_3$

8.23 (a) cyclopentene $\xrightarrow{HBr}$ bromocyclopentane (Br) $\xrightarrow{(CH_3)_3CO^-}$ product

(b) cyclohexene oxide (O) $\xrightarrow{HCl}$ 2-chlorocyclohexanol (Cl, OH) $\xrightarrow[\text{pyridine}]{SOCl_2}$ product

(c) $H_2C\overset{O}{-}CH_2 \xrightarrow{C_6H_5O^-} C_6H_5OCH_2CH_2O^- \xrightarrow{\text{excess HBr}}$

$C_6H_5OCH_2CH_2Br \xrightarrow{C_6H_5O^-}$ product

(d) bromocyclopentane (Br) $\xrightarrow[\text{ether}]{Mg}$ cyclopentyl-MgBr $\xrightarrow[(2)\ H_2O,\ H^+]{(1) H_2C\overset{O}{-}CH_2}$ product

(e) $ClCH_2CH_2SCH_2CH_2Cl \xrightarrow{NH_3} ClCH_2CH_2SCH_2CH_2\overset{+}{N}H_3 \xrightarrow{-H^+}$

$ClCH_2CH_2SCH_2CH_2\ddot{N}H_2 \longrightarrow$ product

(f) $C_6H_5CH_2Br \xrightarrow[-Br^-]{C_6H_5S^-} C_6H_5CH_2S-C_6H_5$

$\xrightarrow[H^+,\ 100°]{H_2O_2}$ product

8.24 The attack by the alkoxide on the epoxide is an S_N2-type reaction that results in inversion of the chiral carbon.

(2R) $\rightleftarrows$ (2S)

8.25 Write out the information given:

A $\xrightarrow{CH_3OH}$ B $\xrightarrow[CH_3OH]{CH_3O^- Na^+}$ C

A $\xrightarrow[CH_3OH]{p\text{-toluenesulfonic acid}}$ C

Conversion of **A** to **B**: Methanol would attack at the rear of the most reactive position, which is a benzylic position. Attack at either benzylic position would yield the same product.

$CH_3\ddot{O}H$ A → $H-\overset{+}{\ddot{O}}CH_3$ $\ddot{O}:^-$ H^+ $\xrightarrow[-H^+]{H^+}$ $:\ddot{O}CH_3$ $\ddot{O}H$ B

Conversion of **B** to **C**: Attack of the methoxide ion would occur at the other benzylic carbon, again from the rear of the carbon-oxygen bond.

OCH_3 OH $CH_3\ddot{O}:^-$ B → OCH_3 OH $\ddot{O}:^-$ $CH_3\ddot{O}:$ $\xrightarrow[-CH_3O^-]{CH_3OH}$ OCH_3 $\ddot{O}H$ $\ddot{O}H$ $CH_3\ddot{O}:$ C

Conversion of **A** to **C** in acidic solution: Attack of methanol on the protonated oxirane rings occurs at the more positive benzylic position. (Remember carbocation stabilities.) If you make a model, you will also discover that the benzylic position is less hindered than the other carbon (carbon 2 relative to the benzene ring) - another reason for benzylic attack.

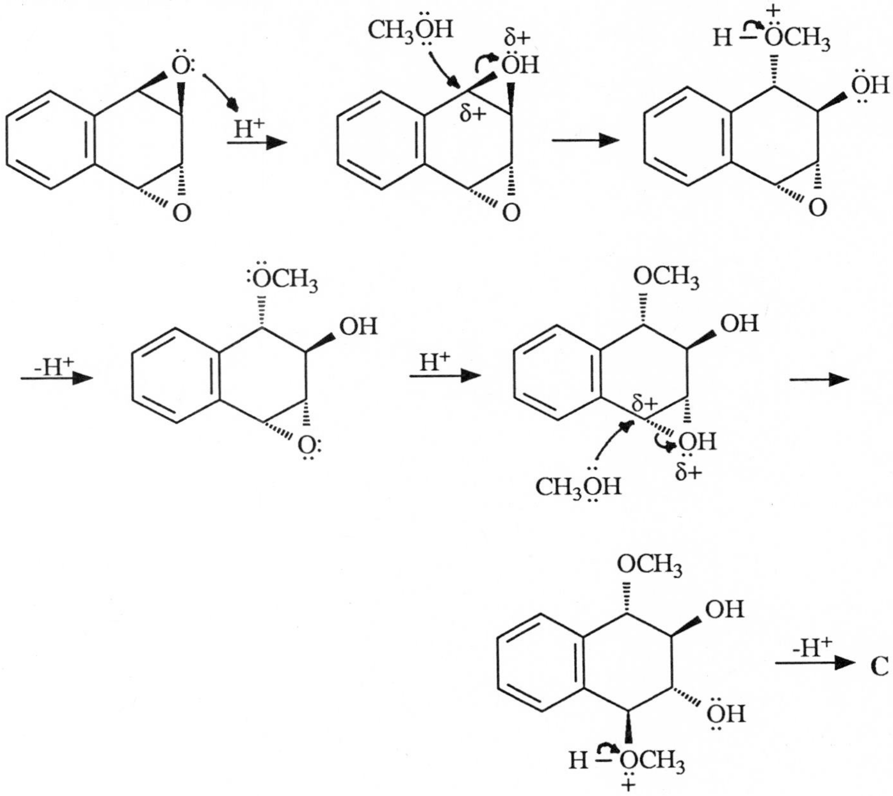

$CH_3\ddot{O}H$ δ+ $\ddot{O}H$ δ+ $H-\ddot{O}^{+}CH_3$ $\ddot{O}H$ H^+

$-H^+$ $:\ddot{O}CH_3$ OH H^+ OCH_3 OH δ+ $\ddot{O}H$ δ+ $CH_3\ddot{O}H$

OCH_3 OH $\ddot{O}H$ $H-\ddot{O}^{+}CH_3$ $\xrightarrow{-H^+}$ C

8.26 $H_2C-CHCH_2Cl$ (epoxide) $\longrightarrow$ $NCCH_2-CH(\ddot{O}:^-)-CH_2-\ddot{C}l:$ ($^-:CN$)

$\xrightarrow{-Cl^-}$ $NCCH_2CH-CH_2$ (epoxide) $\xrightarrow{^-:CN}$ $NCCH_2CH(\ddot{O}:^-)CH_2CN$

$\xrightarrow{H_2O}$ $NCCH_2CH(:\ddot{O}H)CH_2CN$

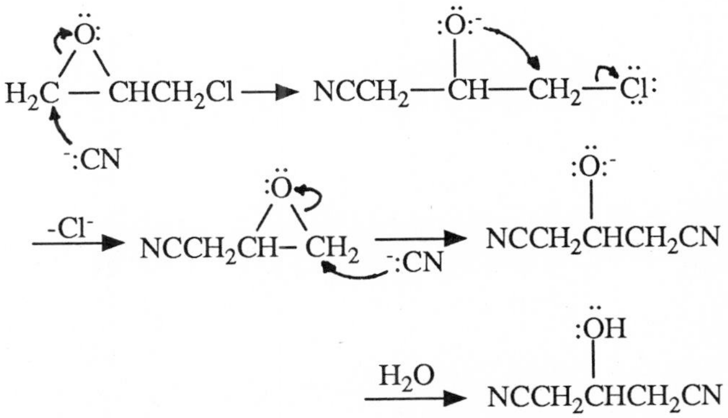

Chapter 9

Spectroscopy I: Infrared and Nuclear Magnetic Resonance

9.22 The equations for the interconversion of μm and cm^{-1} follow:

$$\text{wavenumber in cm}^{-1} = \frac{1}{\lambda \text{ in } \mu\text{m}} \times 10^4;$$

$$\lambda \text{ in } \mu\text{m} = \frac{1}{\text{wavenumber in cm}^{-1}} \times 10^4$$

The easiest way to make the conversion with a calculator is to enter the wavenumber value or the μm value and push the reciprocal (1/*x*) key. Then, either multiply by 10^4 (10,000) or adjust the decimal point yourself. Be sure to round your answer to the proper number of significant figures.

(a) $\lambda \text{ in } \mu\text{m} = \dfrac{1}{3000 \text{ cm}^{-1}} \times 10^4 = 3.33 \ \mu\text{m}$ (b) 5.68 μm

(c) 1786 cm^{-1} (d) 1220 cm^{-1}

9.23 (a) C=O (b) C=C-Cl (c) O-H

In each case, the vibration that results in the greater change in bond moment gives the stronger absorption. For example, in (a), the stretching of the very polar C=O bond results in a greater increase and decrease in bond moment than does the similar stretching of the nonpolar C=C bond.

9.24 (a) $CH_3CH_2CH_2NH_2$ shows NH absorption, while $CH_3CH_2CH_2N(CH_3)_2$ does not.

(b) $CH_3CH_2CH_2CO_2H$ shows characteristic broad, strong OH absorption, while $CH_3CH_2CO_2CH_3$ does not.

(c) $CH_3CH_2CO_2CH_3$ shows C-O absorption, while $CH_3CH_2C(=O)CH_3$ does not.

9.25 Cyclohexanol (–OH) $\xrightarrow{[O]}$ cyclohexanone (=O)

The organic content of the reaction mixture can be isolated and its infrared spectrum determined. As the reaction proceeds, the OH absorption of the reactant in the organic extract gradually disappears. The carbonyl absorption of cyclohexanone is not useful in this case because it appears early in the course of the reaction.

9.26 Any cyclic ether containing a total of five carbon atoms is consistent with the data. For example:

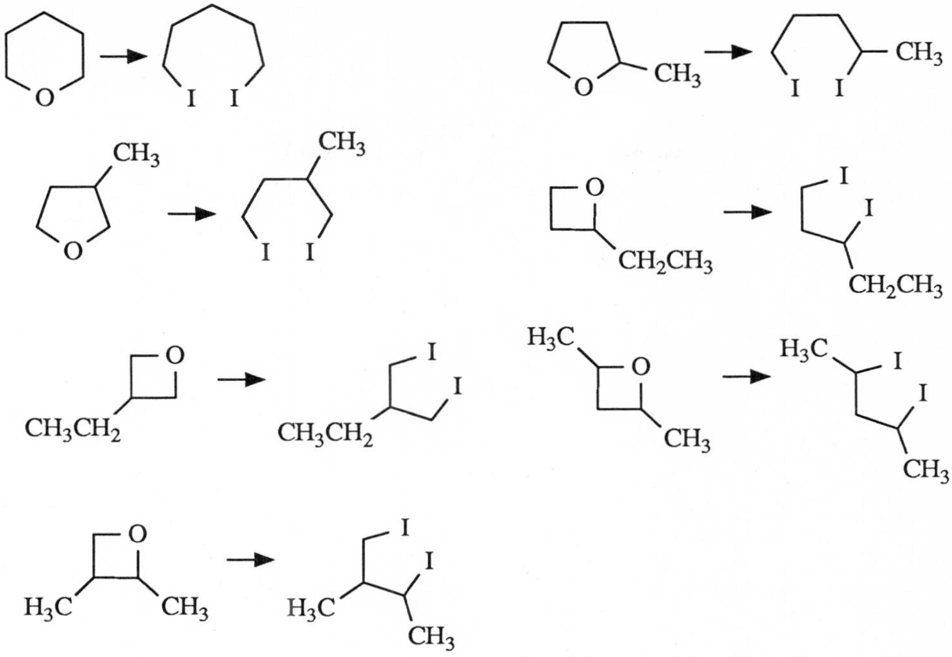

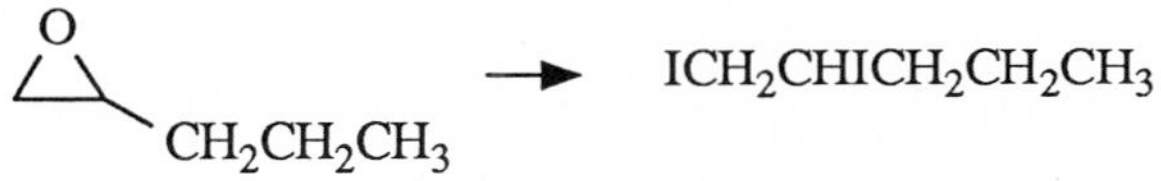

$\rightarrow$ $ICH_2CHICH_2CH_2CH_3$

$\rightarrow$ $CH_3CHICHICH_2CH_3$

$\rightarrow$ $ICH_2CHICH(CH_3)_2$

9.27 At 60 MHz, 1.0 ppm = 60 Hz and 7.5 ppm = 7.5 (60 Hz) = 450 Hz

9.28 (a) $CH_2ClCH_2C\underline{H}_3$ because it is farthest from the electronegative Cl atom.

(b) $CH_3C\underline{H}_2Cl$ because only one electronegative Cl atom is withdrawing electron density by the inductive effect.

(c) 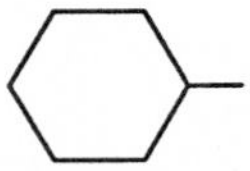 —$\underline{H}$ because it is an alkyl proton, not subjected to anisotropic effects of an aromatic ring.

(d) [cyclohexane]$=CH_2$ because it is an alkenyl proton, not subjected to electron withdrawal by the electronegative oxygen atom.

9.29 (a) two, $\overset{a}{CH_3}\overset{b}{CH_2}\overset{b}{CH_2}\overset{a}{CH_3}$ (b) two, $\overset{a}{CH_3}\overset{b}{CH_2}O\overset{b}{CH_2}\overset{a}{CH_3}$

(c) four, $\overset{a}{H_3C}(\overset{b}{H})C{=}C(\overset{d}{H})(\overset{c}{H})$ (d) one (e) one

(f) four, $\overset{a}{CH_3}\overset{b}{CH_2}\overset{O}{\overset{\|}{C}}O\overset{c}{CH_2}\overset{d}{CH_3}$ (g) five, $\overset{a}{CH_3}{-}\underset{b}{CH}(Cl){-}C(\overset{c}{H})(\underset{d}{H}){-}\overset{e}{CH_3}$

In (g), you may have counted only four sets of protons, but inspection of a molecular model or a Newman projection reveals that the protons labeled c and d cannot be made equivalent by bond rotation.

[Newman projection: front carbon Cl, H_c, H_d; back carbon H, CH_3, CH_3]

Another way of determining the lack of equivalence of H_c and H_d is considering the products of replacement of each of these by Cl.

two different compounds:

[Newman projection: Cl, Cl, H; H, CH_3, CH_3]
optically active

[Newman projection: Cl, H, Cl; H, CH_3, CH_3]
meso

(h) same as (g)

(i) three, $\overset{a}{(CH_3)_2}\overset{b}{CH}N\overset{b}{CH}\overset{a}{(CH_3)_2}$ with $\underset{c}{CH_3}$ on N

(j) four, $Br\overset{a}{CH_2}\overset{b}{CH_2}\overset{c}{CH}\overset{d}{(CH_3)_2}$

(k) four, I—C_6H_4—$\overset{c}{CH_2}\overset{d}{CH_3}$

(l) two, $\overset{a}{CH_3O}$—C_6H_4—$\overset{a}{OCH_3}$

(m) four, $\overset{a}{CH_3O}$—C_6H_4—$\overset{d}{CH_3}$

(n) one (all protons are equivalent)

9.30 The number of principal signals is equal to the number of equivalent protons. The relative areas are equal to the ratios of protons giving rise to the principal signals.

(a) two principal signals (area ratio, 3 : 2)

(b) two (3 : 2)

(c) four (3 : 1 : 1 : 1)

(d) one

(e) one

(f) four (3 : 2 : 2 : 3)

(g) five (3 : 1 : 1 : 1 : 3)

(h) five (3 : 1 : 1 : 1 : 3)

(i) three (12 : 2 : 3)

(j) four (2 : 2 : 1 : 6)

(k) four (2 : 2 : 2 : 3)

(l) two (3 : 2)

(m) four (3 : 2 : 2 : 3)

(n) one

9.31 (d) and (e). None of the other possibilities have chemically nonequivalent protons in the ratio of 3 : 1. Instead, the following ratios would be expected:

(a) 3 : 1 : 1 : 1

(b) 3 : 1 : 1 : 1

(c) 3 : 2 : 2

(f) 1 : 1

9.32 Divide all areas by the smallest one:

$$\frac{81.5}{28} = 2.9 \qquad \frac{28}{28} = 1.0 \qquad \frac{55}{28} = 2.0 \qquad \frac{80}{28} = 2.9$$

Rounding, the ratios are 3 : 1 : 2 : 3

9.33 Divide all areas by the smallest one:

$$\frac{4.0}{3.5} = 1.1 \qquad \frac{3.5}{3.5} = 1.0 \qquad \frac{5.4}{3.5} = 1.5 \qquad \frac{5.5}{3.5} = 1.6$$

Multiply by the integer that will give the smallest whole numbers (in this case, 2):

2.2 2.0 3.0 3.2

Round: 2 : 2 : 3 : 3

9.34 (a) $\overset{3}{CH_3}\overset{4}{CH_2}CO_2\overset{1}{CH_3}$ (*areas*, 3 : 2 : 3)

(b) $CH_3O\overbrace{CH_2CH_2}^{1}OCH_3$ (*areas*, 3 : 2) — the two CH_3 groups: 1

(c) $\overset{1}{CH_3}\overset{O}{\overset{\|}{C}}\overset{4}{CH_2}\overset{3}{CH_3}$ (*areas*, 3 : 2 : 3)

(d) $\overset{1}{CH_3O}$–C_6H_4(a, b, a, b)–Cl (*areas*, 3 : 2 : 2)
a = 2; b = 2

(e) $Cl_2\overset{3}{CH}\overset{2}{CH_2}Br$ (*areas*, 1 : 2)

In each case, the *n*+1 rule is used to determine the multiplicity, and the number of protons giving rise to the signal is used to determine the area. For example,

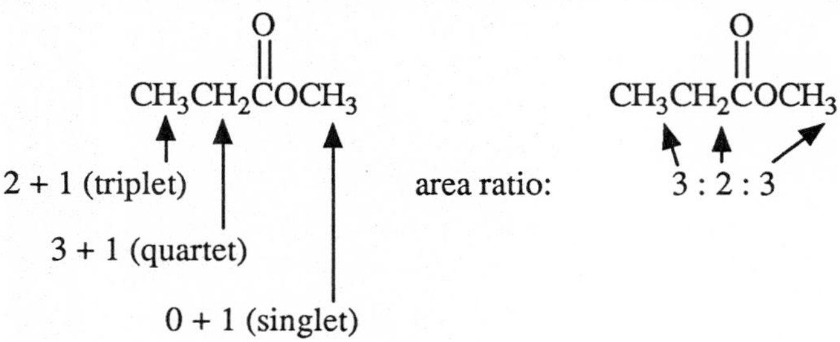

9.35 First, predict the peaks in the spectra of the compounds. Then, state the key distinguishing feature or features.

(a) $CH_3CH_2CH_2CH_2CH{=}CH_2$ would show a large number of peaks, while $(CH_3)_2C{=}C(CH_3)_2$ would show a singlet.

(b) CH_3CH_2CHO would show three principal signals (including offset absorption for an aldehyde proton), while CH_3COCH_3 would show a singlet.

(c) CH_3COCH_3 would show one singlet, while $CH_3CO_2CH_3$ would show two singlets.

(d) CH_3CH_2-C_6H_5 would show a quartet and a triplet between the TMS peak and aryl absorption, while *p*-CH_3-C_6H_4-CH_3 would show a singlet in this region.

9.36 (a) $CH_3CH_2CH_2OH$ and $CH_3CH{-}CH_2$ (with O bridging the two carbons) could be distinguished from each other on the basis of their nmr splitting patterns and chemical shifts. However, simply a glance at the infrared spectra would show which is which. 1-Propanol would show -OH absorption and propylene oxide would not.

(b) A comparison of the infrared spectra would not be useful because both compounds contain the same functional group. Because the compounds differ in their carbon-hydrogen bonding, the ^1H nmr spectra could be used to differentiate them. $(CH_3)_2CHOCH(CH_3)_2$ would show a doublet and septet, while $CH_3CH_2CH_2OCH_2CH_2CH_3$ would show two triplets and a probable sextet.

(c) Both compounds contain the same functional group; therefore, the infrared spectra would not be useful. However, in the ^1H nmr spectra, CH_3CH_2OH would show one singlet, one triplet, and one quartet; $HOCH_2CH_2OH$ would show two singlets.

(d) The NH bond shows characteristic absorption in the infrared spectrum. The first compound shown, *N*-methylpiperidine, would not show this absorption. Piperidine itself, with an NH bond, would show absorption in the NH stretching region.

(e) Although ^1H nmr spectroscopy could be used to distinguish between these compounds, infrared spectroscopy would be the technique of choice. CH_3CH_2OH would show strong OH absorption; CH_3CH_2Cl would not.

(f) Because of the difference in functional groups, infrared spectroscopy would, again, be the technique of choice. CH_3CO_2H would show the broad intense OH absorption characteristic of carboxylic acids; $(CH_3)_2C{=}O$ would not. (Both compounds would show carbonyl absorption.)

9.37

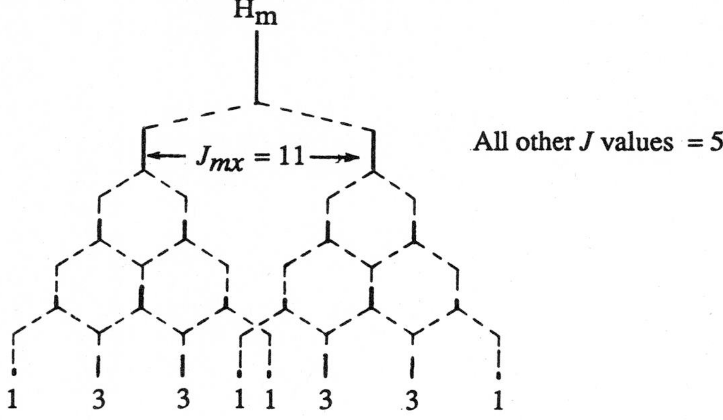

9.38 (a) seven

(b) aryl, doublet (area 2) plus triplet (area 1), but all aryl protons may appear as a singlet (area 3); CH_3, singlet (area 6); NH, singlet (area 1); CH_2, singlet (area 2); CH_3CH_2, triplet (area 6) plus quartet (area 4)

(c) NH absorption (~3500 cm^{-1}), C-H absorption (~3000 cm^{-1}), and C=O absorption (~1700 cm^{-1}) are the most characteristic peaks we have discussed in this chapter.

9.39

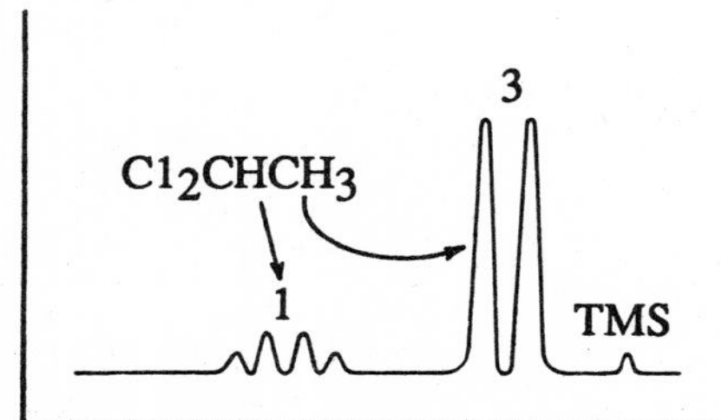

9.40 (a) $CH_3CHClCHClCH_3$

3

1

TMS

(b)

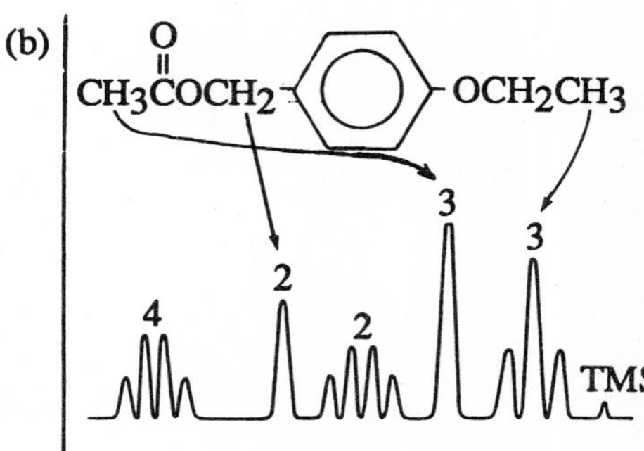

9.41 The compound contains four carbons (from the number of signals in the proton-decoupled ^{13}C spectrum). Therefore, from the molecular weight, we deduce the molecular formula to be C_4H_8O. The general formula $C_nH_{2n}O$ indicates one double bond or one ring. The peak at δ208 (a singlet) indicates C=O. The splitting patterns and chemical shifts all indicate the following formula:

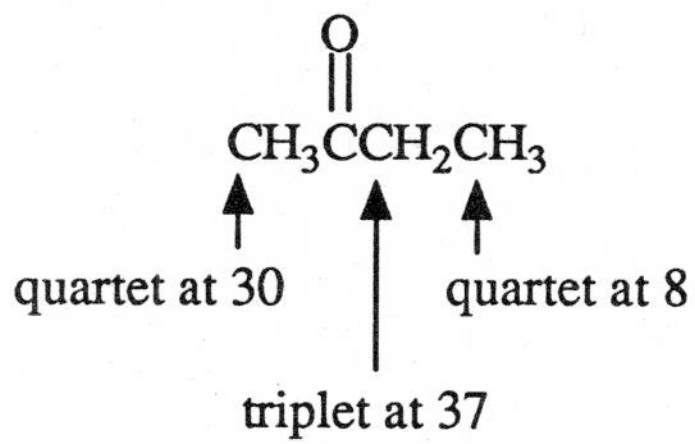

Another way to approach the problem is to interpret the spectra in more detail. Rough sketches of the spectra will help you visualize the data.

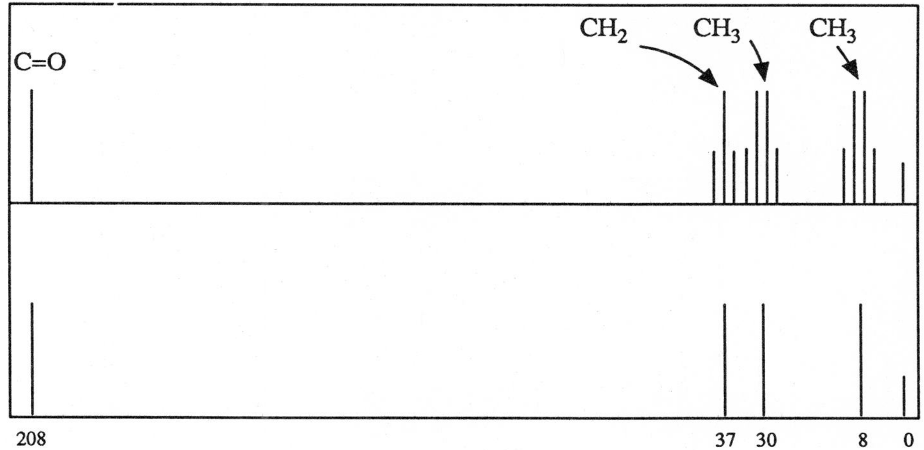

From the sketches, we can see that the peak at 208 is consistent with C=O. The peak at 8 arises from an alkane-like carbon (because it is so far upfield); this carbon must be bonded to another carbon. It is also bonded to three hydrogen atoms (because the coupled spectrum shows a quartet). The peaks at 30 and 37 must arise from $-CH_3$ and $-CH_2-$ bonded to an electronegative atom or, in this case, the carbonyl group. From these assignments and the molecular weight, we would also determine the structure to be that of $CH_3\underset{\underset{O}{\|}}{C}CH_2CH_3$.

9.42 The outside protons, like those of benzene, are deshielded and absorb downfield in the aromatic region. The inner protons are highly *shielded* and absorb upfield. (This absorption is observed upfield even of TMS.)

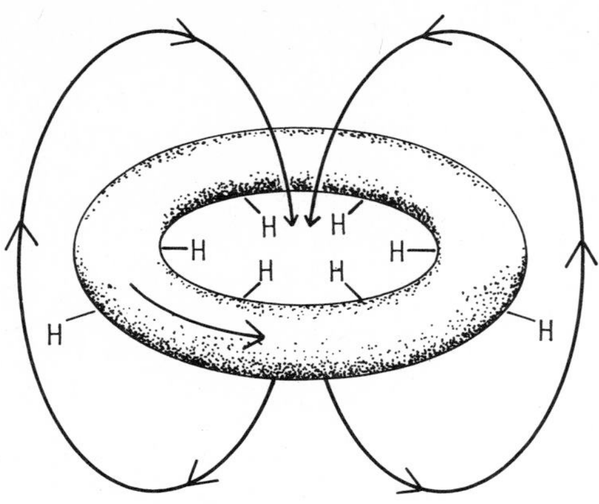

Not all outer H's are shown

The actual absorption for the outer protons is at δ10.75, while that for the inner protons is at δ -4.22.

9.43 As in [18]annulene, the CH_3 protons are shielded by the field induced by the ring current.

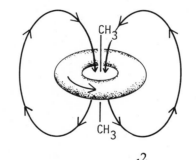

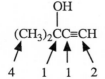

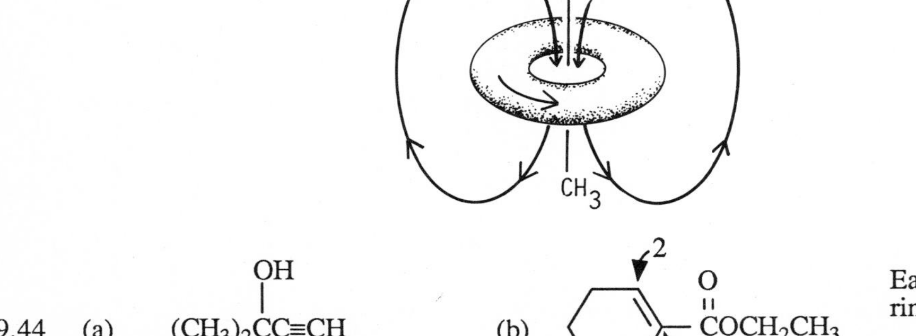

9.44 (a) $(CH_3)_2C(OH)C{\equiv}CH$ — arrows: $(CH_3)_2$ 4; C 1; C 1; CH 2

(b) 1-cyclohexenyl–$\overset{O}{\overset{\|}{C}}OCH_2CH_3$ — arrows: vinyl H 2; 1 1 3 4. Each other ring carbon: 3

9.45 (a) First let us consider the alkyl bromide. The molecular formula $C_5H_{11}Br$ indicates an open-chain saturated skeleton. A singlet (area 6) arises from $(CH_3)_2C\langle$. A quartet (area 2) and a triplet (area 3) arise from CH_3CH_2-. Therefore, the bromide is $CH_3CH_2CBr(CH_3)_2$.

The spectrum of the alcohol is not consistent with the carbon skeleton of this bromide; therefore, a carbocation rearrangement must have occurred. The singlet (area 1) must come from OH. The doublet (area 3) is from $\underline{CH_3}CH\langle$. The doublet (area 6) is from $(\underline{CH_3})_2CH$-. The multiplets (areas 1) are from -CH- and -CH-. The only feasible structure for the alcohol, plus the mechanism for its rearrangement, follows:

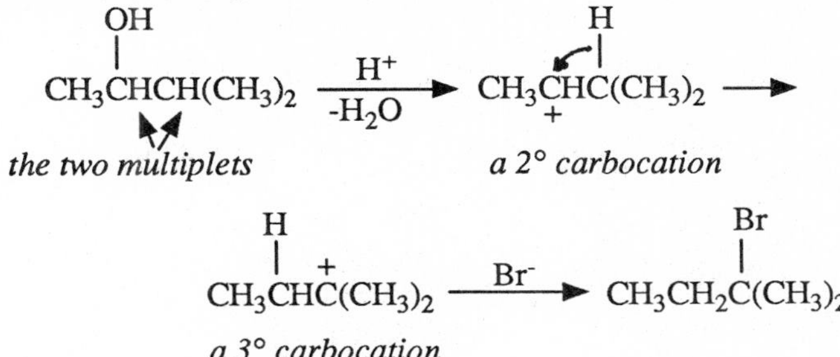

(b) The infrared absorption at 3000 cm^{-1} arises from C-H stretching. The absorption at 2240 cm^{-1} suggests the presence of C≡C or C≡N. The spectrum shows no OH or C=O absorption; therefore, the O atom is in an ether group. (Outlining or tabulating assignments in spectral problems is often helpful.)

infrared absorption	assignment
3000 cm^{-1} (3.3 μm)	C-H
2240 cm^{-1} (4.46 μm)	C≡N or C≡C
no C=O or OH	C-O-C

Let us consider the nmr spectrum:

$-CH_2-CH_2-$(a deshielding group)

triplet at 2.6; area 2 (first CH_2); the other triplet, at 3.5; area must be 2 (second CH_2)

The singlet must arise from $-CH_3$ because of its area (5 - 2 = 3).

nmr absorption	assignment
δ2.6 (triplet), area 2	$-\underline{CH_2}CH_2-$
δ3.3 (singlet), area 3	$-\underline{CH_3}$
δ3.5 (triplet), area 2	$-CH_2\underline{CH_2}-$(a deshielding group)

From the nmr data, we can rule out C≡C. Thus, we have the following pieces:

CH_3- $\qquad$ $-OCH_2CH_2-$ $\qquad$ -C≡N

The chemical shift of the $-CH_3$ proton and its lack of splitting suggest that this group is bonded to the O atom; therefore, the structural formula is $CH_3OCH_2CH_2C{\equiv}N$.

(c) Four peaks in the aryl region of the nmr spectrum with a relative area of 4 suggest a disubstituted benzene. Because only four peaks appear, let us assume a *p*-disubstituted ring.

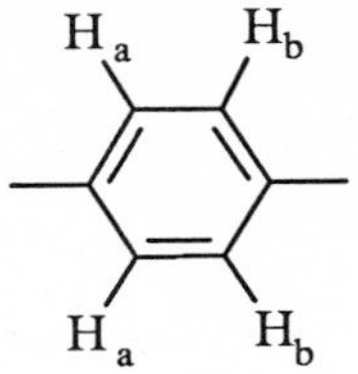

H_a H_b (benzene ring drawing)

H_a: doublet, area 2

H_b: doublet, area 2

sum: four peaks, area 4

The singlet (area 3) at δ2.15 must arise from $-CH_3$ with no neighboring protons. The downfield singlet (area 3) arises from another $-CH_3$, but this one must be bonded to an electronegative atom (oxygen). The fragments so far:

$-C_6H_4-$ $\quad -CH_3 \quad -OCH_3$

The sum of the formula weights of these fragments is 122, the molecular weight of the unknown. Therefore, the structure must be:

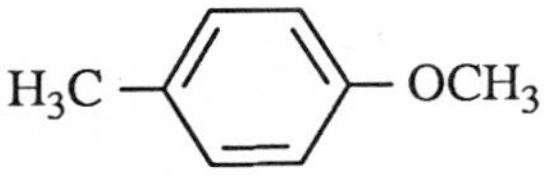

$H_3C-C_6H_4-OCH_3$

(d) The infrared spectrum shows carbonyl absorption. The nmr spectrum shows characteristic isopropyl absorption.

infrared absorption	assignment
1750 cm^{-1} (5.71 μm)	C=O

nmr absorption	assignment
δ1.2 (doublet), area 6	$(C\underline{H}_3)_2CH-$
δ2.0 (singlet), area 3	$-C\underline{H}_3$, possibly $-\overset{O}{\overset{\|\|}{C}}CH_3$
δ4.95 (septet), area 1	$(CH_3)_2C\underline{H}$-(a deshielding group)

The chemical shift of the septet proton suggests that its carbon is bonded to oxygen. The chemical shift of the singlet suggests CH_3 bonded to a carbonyl group. Thus, we have the following pieces for $C_5H_{10}O_2$.

$$(CH_3)_2CHO- \quad \text{and} \quad -\overset{O}{\overset{\|\|}{C}}CH_3$$

Putting them together, we arrive at the structure.

$$(CH_3)_2CHO\overset{O}{\overset{\|\|}{C}}CH_3$$

(e) Assign the peaks in the spectra.

infrared absorption	assignment
2900–3050 cm^{-1} (3.28–3.45 μm)	C-H
1760 cm^{-1} (5.68 μm)	C=O
1460 cm^{-1} (6.85 μm)	presumably from C-H bending
1380 cm^{-1} (7.25 μm)	C-H bend

nmr absorption	assignment
δ1.3 (singlet)	12 equivalent protons

For a compound with twelve protons to show only a singlet in the ^1H nmr spectrum, its structure must be highly symmetrical. The infrared spectrum shows carbonyl absorption, presumably from two carbonyl groups.

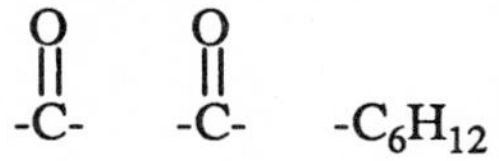

$$\overset{\displaystyle O}{\overset{\|}{-C-}} \quad \overset{\displaystyle O}{\overset{\|}{-C-}} \quad -C_6H_{12}$$

The alkyl portion of the molecule (C_nH_{2n}) contains one ring or double bond. From the chemical shift, none of the protons is alpha to a carbonyl group. The following structure fits these data.

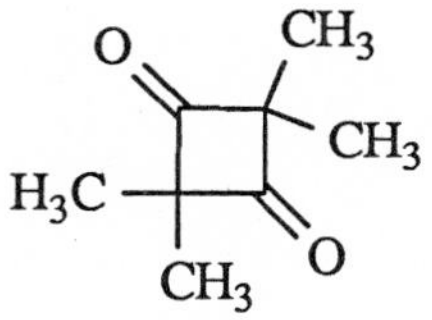

O CH$_3$ CH$_3$ H$_3$C CH$_3$ O

9.46 The spectrum was carried out on a 220-MHz instrument. Therefore, a delta value of 1.0 ppm corresponds to 220 Hz, not 60 Hz as with a 60-MHz instrument. When sketching the spectrum, we must use this 220-Hz value in determining the J values. In the sketched spectrum, we have not shown Nx splitting because J_{Nx} is so small.

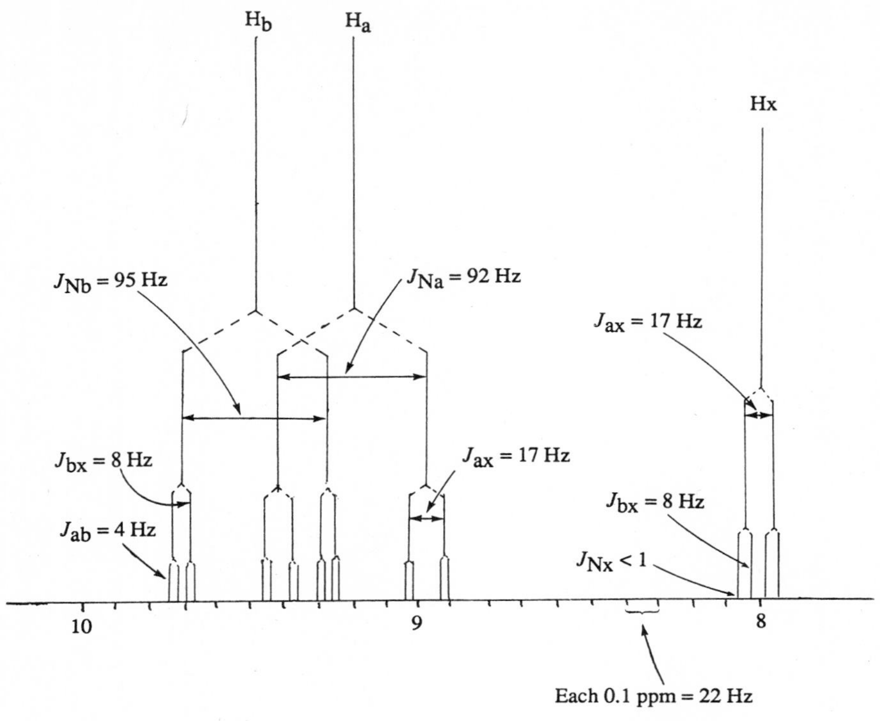

9.47 (a) The formula $C_8H_{10}O_2$ indicates extensive unsaturation or ring(s). In such a case, check the aryl region of the ^{1}H nmr spectrum. This compound does contain an aromatic ring. The infrared spectrum shows OH absorption at about 3370 cm^{-1} (3.0 μm), but no C=O absorption. At least one of the oxygens in the formula is accounted for. The second oxygen could be in a second OH group or in an OR group (an ether).

The nmr spectrum indicates a 1,4-disubstituted benzene ring (the two doublets in the δ7.0 region). The singlet at δ3.7 must arise from a CH_3 group (area 3) bonded to O (because the signal is shifted so far downfield). The pieces determined so far are:

-OH (1,4-disubstituted benzene ring) CH_3O-

Because of the CH_3O- group, there is only one OH in the molecule. In the nmr spectrum, this is the peak at δ3.1. The remaining CH_2 in the formula gives rise to the singlet at δ4.4. It must be bonded to O (because of its chemical shift).

The only way the pieces can be fit together is as follows:

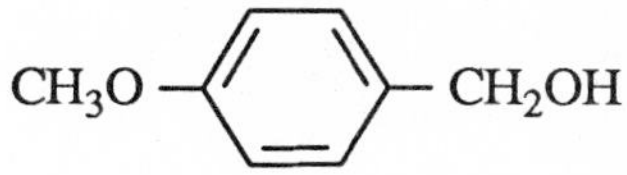

CH_3O–⟨benzene ring⟩–CH_2OH

(b) This is another example of a compound containing an aromatic ring. The infrared spectrum shows C=O absorption (1680 cm^{-1}, 5.95 μm). Also note the CH peak at about 2700 cm^{-1} (3.7 μm). This absorption suggests an aldehyde. Indeed, the nmr spectrum shows offset absorption characteristic of RCHO or RCO_2H. (This compound, however, could not be RCO_2H. Why not?) The nmr spectrum also indicates a 1,4-disubstituted benzene ring (the pair of doublets in the aryl region). The fragments determined so far are:

–⟨benzene ring⟩– and $-\overset{O}{\overset{\|}{C}}H$

This leaves only C_3H_7O to account for. The triplet at δ1.0 must arise from a CH_3 group (area 3) bonded to a CH_2 group (because its multiplicity is 3). One CH_2 group in the molecule exhibits a sextet in the nmr spectrum: $CH_3\underline{CH_2}CH_2$-. The CH_2 group that shows a triplet is shifted downfield; therefore, it must be attached to the other O: $CH_3CH_2\underline{CH_2}O$-. Now a complete formula may be written:

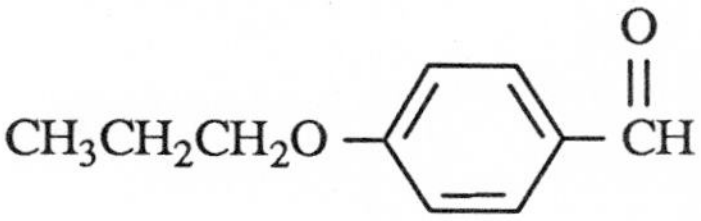

$CH_3CH_2CH_2O$–⟨benzene ring⟩–$\overset{O}{\overset{\|}{C}}H$

(c) The infrared spectrum shows strong carbonyl absorption at about 1730 cm^{-1} (5.8 μm), which accounts for one of the two oxygens. The second oxygen could be involved in a second C=O or in OR (as an ether or an ester), but not in an OH group. (Why not?)

Let us consider the nmr spectrum. The downfield septet arises from an isopropyl group, $(CH_3)_2\underline{CH}$-, which must be bonded to O (not to C=O because it is so far downfield). The upfield doublet is therefore from $(\underline{CH_3})_2CHO$-. The two triplets must come from $-CH_2CH_2$-. Because one of them is shifted downfield to δ3.75, the group must be $-CH_2CH_2Cl$. (The downfield CH_2 group is not bonded to O, because the two O's have been accounted for.) The fragments of the molecule are:

$-\overset{O}{\overset{\|}{C}}-$ $\qquad -OCH(CH_3)_2 \qquad -CH_2CH_2Cl$

The formula for the compound is $ClCH_2CH_2\overset{O}{\overset{\|}{C}}OCH(CH_3)_2$.

(d) The infrared spectrum shows C=O, but no OH. The nmr spectrum shows CH_3 and CH_2, accounting for *half* of the ten protons in the molecule. Thus, the fragments are:

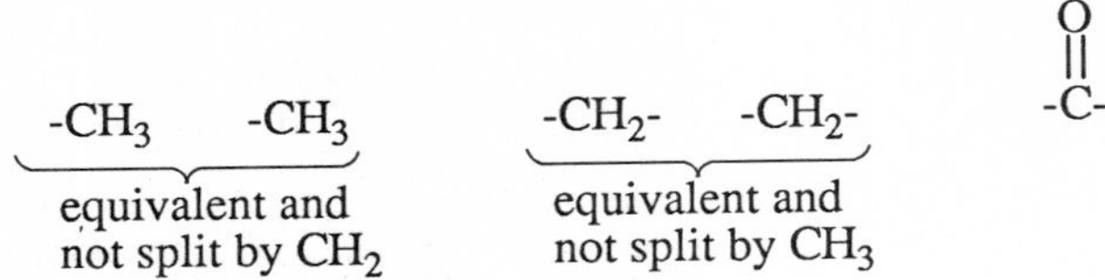

The remaining C and O must be in a second C=O group, and the compound must be:

$$CH_3\overset{\overset{\displaystyle O}{||}}{C}CH_2CH_2\overset{\overset{\displaystyle O}{||}}{C}CH_3$$

Any other arrangement of the fragments would result in splitting between the CH_2 and CH_3 groups.

(e) The infrared spectrum shows strong OH absorption and strong C=O absorption. The compound is not a carboxylic acid (or an aldehyde) because the ^1H nmr spectrum shows no offset absorption. The nmr spectrum shows a singlet fairly far downfield at δ3.75. This singlet (area 3) must be due to CH_3O- because of its large chemical shift. Now the three oxygens in the molecule have been identified: -OH, >C=O, and -OCH_3 . The singlet at δ3.3 represents the OH proton. The upfield doublet (area 3) arises from ($\underline{CH_3}CH<$). The downfield quartet is thus from $CH_3\underline{CH}<$. This CH must also be bonded to O. (Why?) The fragments are:

$$\overset{\overset{\displaystyle O}{||}}{-C-} \qquad -OCH_3 \qquad CH_3\overset{\overset{\displaystyle OH}{|}}{C}H-$$

The compound is thus $CH_3\overset{\overset{\displaystyle HO}{|}}{C}H-\overset{\overset{\displaystyle O}{||}}{C}OCH_3$.

(f) The infrared spectrum shows C=O absorption as the only truly significant peak. The CH region of the spectrum indicates that the compound is not an aldehyde. The nmr spectrum confirms this. Thus, the compound must be a ketone.

The nmr spectrum indicates a phenyl group, C_6H_5- (the singlet at δ7.4, area 5). The nmr spectrum also shows a CH_3 group not split by neighboring protons. The chemical shift of the CH_3 signal implies that the CH_3 group is bonded to the C=O group. The parts of the molecule determined so far are:

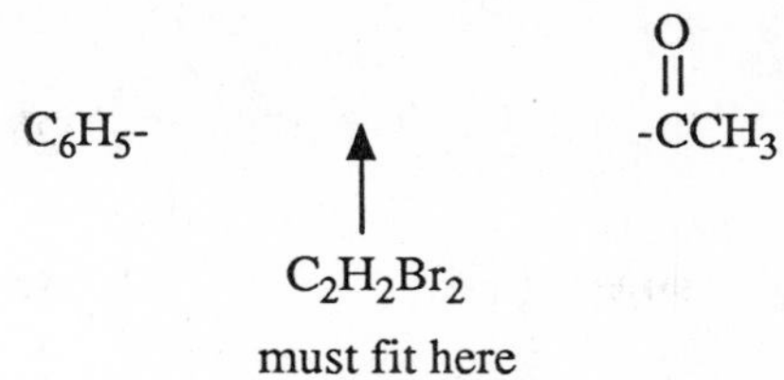

Because only two protons are unaccounted for, the absorption in the δ5.0 region must be two doublets and not a quartet. The middle of the molecule is thus -CHBrCHBr-, and the entire structure is:

```
    Br  Br  O
    |   |   ||
C6H5CH-CH-CCH3
```

(g) The infrared spectrum shows NH_2 absorption, leaving us with a partial structural formula of $C_4H_9NH_2$. Because of the number of hydrogens ($2n + 1$), we know that the alkyl portion of the molecule contains no double bonds or rings. The nmr spectrum is sufficiently complex that it cannot be completely analyzed. The downfield sextet arises from $CH_3\underline{C}\underline{H}CH_2$- (with a bond up from the CH). Because of its shift, we can assume that the CH carbon is bonded to N. Therefore, the remaining CH_3 group is bonded to the CH_2:

```
   NH2
   |
CH3CHCH2CH3
```

9.48 (a) C_8H_{12} (C_nH_{2n-4}) is a hydrocarbon with three double bonds, one double bond and one triple bond, two double bonds and a ring, one double bond and two rings, or three rings. The ^{13}C nmr spectra show four types of carbon atoms, two types of which are alkenyl. (Because the formula shows eight carbon atoms in a molecule, we must multiply the number of fragments by 2.)

2 =CH- + 2 =CH-, or 2 -CH=CH-

↑ ↑

doublet means one H

The spectra show two types of alkyl carbons, each as a poorly resolved triplet.

2 $-CH_2-$ + 2 $-CH_2-$

↑ ↑

triplet means two H's

Because no terminal $=CH_2$ or $-CH_3$ are indicated, the structure must be a cyclooctadiene. The correct answer follows, with the four types of carbon atoms (showing the four signals) indicated.

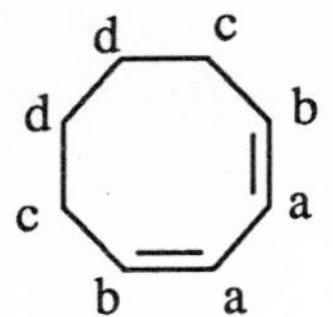

1,3-cyclooctadiene

The other cyclooctadienes would not show four signals.

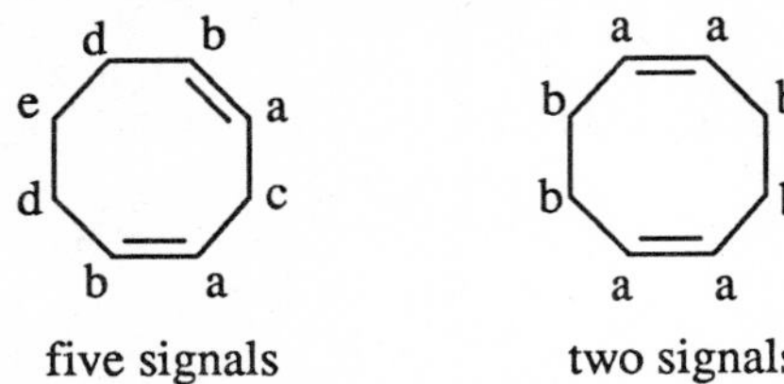

five signals two signals

(b) The compound is an ester that contains one double bond (besides the one in the carbonyl group) or one ring.

$\overset{O}{\overset{\|}{\text{-CO-}}}$ $\quad$ $\text{-}C_4H_8$ $\leftarrow C_nH_{2n}$

^{13}C nmr absorption	assignment
δ19 (triplet)	$-CH_2-$
δ22 (triplet)	$-CH_2-$
δ30 (triplet)	$-CH_2-$
δ70 (triplet)	$-OCH_2-$
δ172 (singlet)	$>C=O$

No alkenyl carbons are indicated by the spectra; therefore, the molecule is cyclic. The only way to combine the fragments is as a six-membered-ring cyclic ester.

(c) The formula $C_7H_{16}O_4$ shows a sufficient number of hydrogen atoms that the structure must be open-chain and saturated. The ^{13}C nmr spectrum shows no carbonyl groups and indicates *three* types of carbons.

^{13}C nmr absorption	assignment
δ36 (triplet)	$-CH_2-$
δ52 (quartet)	$-CH_3$ or $-OCH_3$
δ101 (doublet)	$>CHO-$ (farther downfield than usual)

^{1}H nmr absorption	assignment
δ1.95 (triplet), area 1	(see discussion)
δ3.3 (singlet), area 6	-OC<u>H</u>$_3$
δ4.5 (triplet), area 1	-CH$_2$C<u>H</u>O- (with a vertical bond on C)

By default, the -CH$_2$- group must give rise to the triplet at δ1.95; therefore, this group must be bonded to two -CHO- groups:

-OCHCH$_2$CHO-

The ^{13}C nmr spectra, showing three types of carbon atoms in the structure, suggests a symmetrical molecule with this -CH$_2$- in the center:

$C_3H_7O_2$ — CH_2 — $C_3H_7O_2$

The sum of the relative areas of the proton signals is 8 although the formula shows sixteen protons. Multiplying the relative areas by 2 allows us to list all the pieces. (Note that two oxygen atoms are repeated in the following list of fragments: the formula shows only four O atoms.)

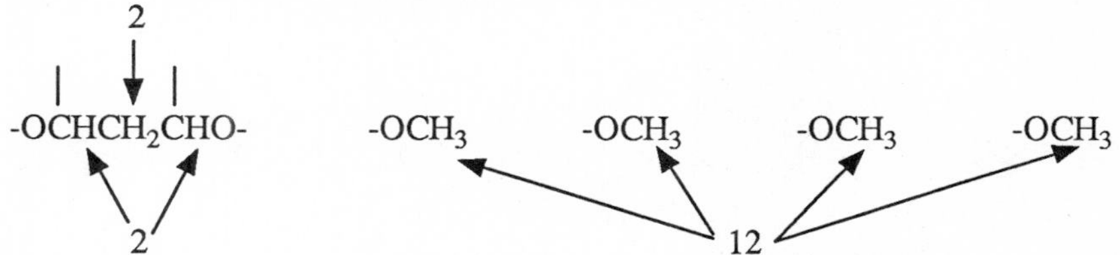

Now we can put the pieces together to form the structure of the compound:

CH$_3$O OCH$_3$
CH$_3$O-CHCH$_2$CH-OCH$_3$

This signal is far downfield in the ^{13}C nmr spectrum because each C is bonded to *two* O atoms.

9.49 CH$_3$CH$_2$CH$_2$C(=O)CH$_2$CH$_2$CH$_3$, 4-heptanone

The formula contains one oxygen. The infrared spectrum shows a carbonyl group. The ^{1}H nmr spectrum shows no evidence that the compound is an aldehyde; therefore, the compound is a ketone.

The ^{13}C nmr spectrum shows four peaks, one of which is the carbonyl carbon (δ210.2). Because the formula contains seven carbons, the structure must be symmetrical around the carbonyl carbon.

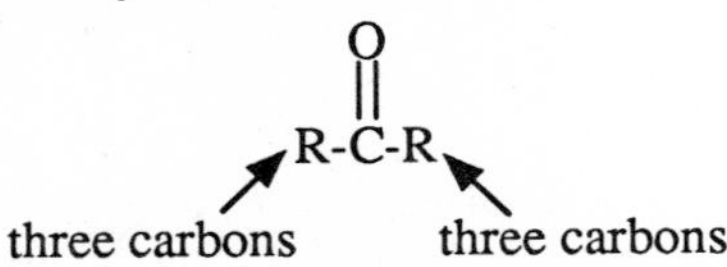

The ^{1}H nmr spectrum also shows symmetry – the integration accounts for only seven of the fourteen protons. The proton area ratio 2:2:3 suggests that R is $CH_2CH_2CH_3$ (*n*-propyl). Had the three-carbon unit been an isopropyl group, the area ratio would have been 1:6 [$-CH(CH_3)_2$].

Chapter 10

Alkenes and Alkynes

10.40 (a) 3-bromo-1-propene or 3-bromopropene

(b) 3-methyl-1-butyne

(c) 5,5-dichloro-1,3-cyclohexadiene (d) propenoic acid

(e) (4*R*)(2*E*)-6-methyl-2-hepten-4-ol or (4*R*)-*trans*-6-methyl-2-hepten-4-ol

10.41 (a)

$$\begin{array}{c} CH_3 \\ | \\ CH_2{=}CCH_2CH_3 \end{array}$$

(b) $CH_3CH_2C{\equiv}CCH_3$

(c) $CH_2{=}CHCH{=}CHCH_2CH_3$

(d)

```
H3C        CH2CH2CH3
   \      /
    C = C
   /      \
  H        H
```

(e)

```
C6H5   H
   \  /
   C = C
  /     \
 H       C6H5
```

(f)

```
Br       Br
  \     /
   C = C
  /     \
 H       H
```

(g)

```
     H       H
      \     /
       C = C        CH3
      /     \      /
   H3C       C = C
            /     \
           H       H
        (Z)
```

(h)

```
H3C        CO2H
   \      /
    C = C
   /      \
C6H5       H
```

10.42

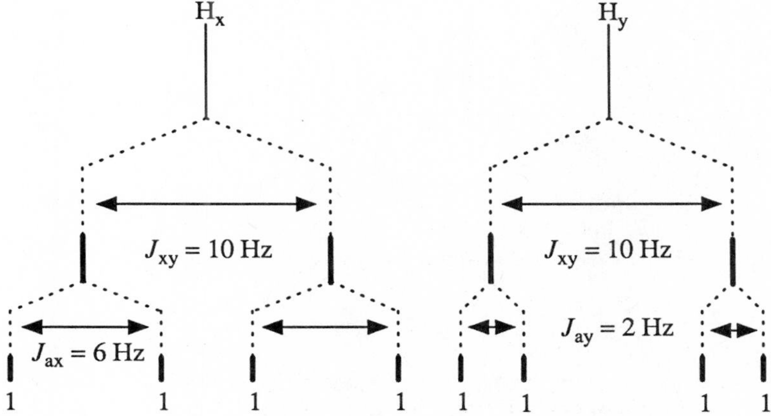

10.43 The infrared absorption of the terminal triple bond is easier to identify than that of a carbon-carbon double bond. Therefore, the spectrum should be inspected for a C≡C-H carbon-hydrogen absorption near 3300 cm^{-1} (3.0 μm) and for C≡C absorption near 2100 cm^{-1} (4.8 μm).

If these bands are absent, the unknown is probably 1-hexene. Confirmation that the unknown contains a terminal double bond should be obtained by inspecting the spectrum for bands near 3100 cm^{-1} (3.2 μm), C=C-H; 1640 cm^{-1} (6.1 μm), C=C; and 900 cm^{-1} (11.4 μm), $C=CH_2$.

In the following sketches, we have shown only the absorption you would check for.

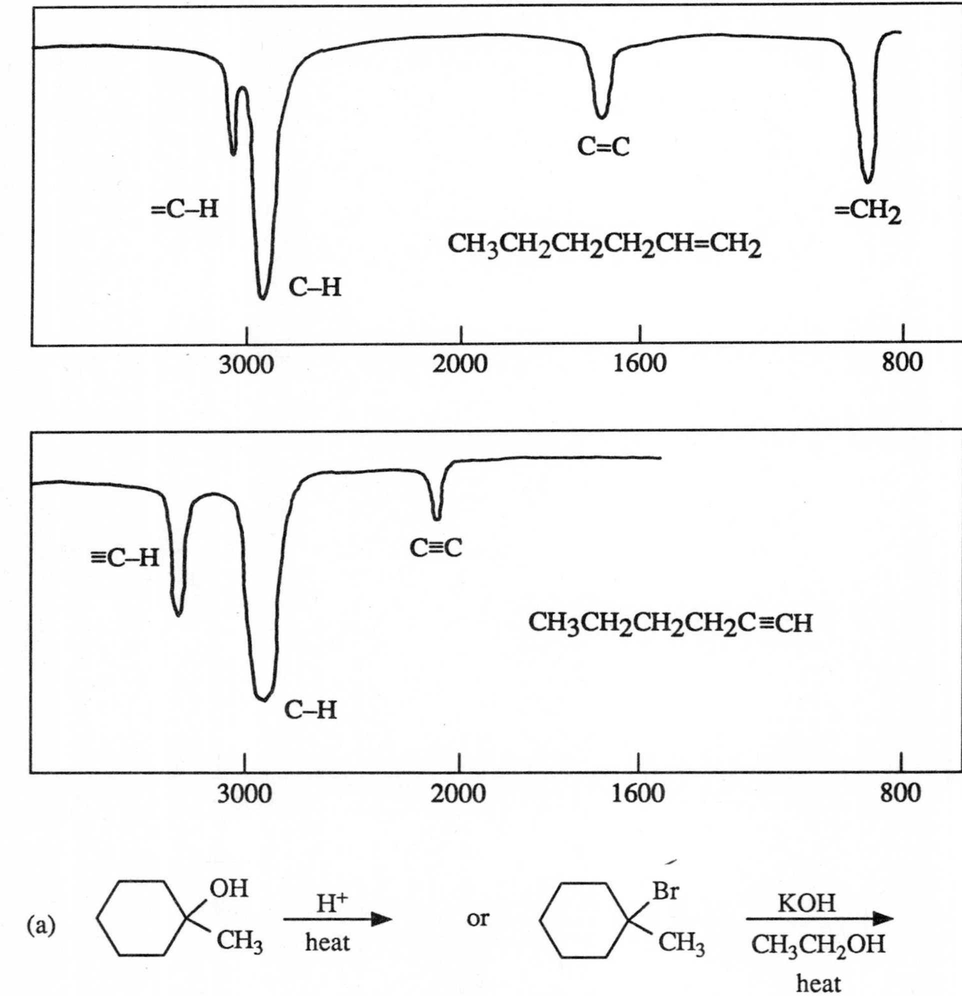

10.44 (a)

Because the most stable alkene is the desired product, either an E1 reaction of an alcohol or an E2 reaction of an alkyl halide would be satisfactory. In either case, the functional group should be at the tertiary position to maximize yield of the desired product.

(b) Cyclohexyl–CH_2Br $\xrightarrow[\text{warm}]{K^+ \, ^-OC(CH_3)_3 \,/\, (CH_3)_3COH}$ product

In (b), the functional group should be at the primary position, and an E2 reaction should be used to avoid rearrangement.

(c) carbon 1

[dimensional formula: C_6H_5, H_3C, H on C — C bearing Br, H, C_6H_5] or [Newman projection: Br, CH_3, H, H, C_6H_5, C_6H_5] $\xrightarrow[\text{(E2)}]{^-OH \,/\, CH_3CH_2OH}$ product

(1*R*,2*R*)-1-bromo-1,2-diphenylpropane
or its (1*S*,2*S*) enantiomer

(d) [dimensional formula: C_6H_5, H_3C, H on C — C bearing Br, H, C_6H_5] or [Newman projection: Br, C_6H_5, H, H, H_3C, C_6H_5] $\xrightarrow[\text{(E2)}]{^-OH \,/\, CH_3CH_2OH}$ product

(1*R*,2*S*)-1-bromo-1,2-diphenylpropane
or its (1*S*,2*R*) enantiomer

In (c) and (d), only an E2 reaction would yield the desired stereochemistry. In solving this type of problem, use dimensional formulas or Newman projections in order to check the stereochemistry. Also, note that the halogen is bonded to the carbon that gives it only one *anti* hydrogen. (Otherwise, mixtures would result.)

10.45 (a) $CH_3C{\equiv}CH \xrightarrow{NaNH_2} CH_3C{\equiv}C{:}^- \, Na^+ \xrightarrow{CH_3I} CH_3C{\equiv}CCH_3$

(b) $CH_3C{\equiv}CH \xrightarrow{CH_3MgI} CH_3C{\equiv}CMgI \xrightarrow[\text{(2) } H_2O,\, H^+]{\text{(1) cyclohexanone}}$ product

(c) $CH_3C{\equiv}CH \xrightarrow{NaNH_2} CH_3C{\equiv}C{:}^- \, Na^+ \xrightarrow{D_2O}$ product

or

$$CH_3C{\equiv}CH \xrightarrow{CH_3MgI} CH_3C{\equiv}CMgI \xrightarrow{D_2O} \text{product}$$

10.46 (a) $\xrightarrow{H^+} (CH_3)_2\overset{+}{C}CH_3 \xrightarrow{Cl^-} (CH_3)_3C\text{-}Cl$ by the more stable 3° carbocation

(b) $\xrightarrow{H^+}$ [indanyl cation, + at benzylic carbon] $\xrightarrow{Cl^-}$ [1-chloroindane] by the more stable benzylic carbocation

racemic

(c) $\xrightarrow{H^+} CH_3(CH_2)_3\overset{+}{C}{=}CH_2 \xrightarrow{Br^-} CH_3(CH_2)_3CBr{=}CH_2$

(d) $\xrightarrow{H^+}$ [cyclopentyl cation]$^+$–CH_2CH_3 $\xrightarrow{Br^-}$ [cyclopentane with Br and CH_2CH_3 on same carbon] by the more stable 3° carbocation

(e) $\xrightarrow{Br\cdot}$ [cyclopentyl radical]$\cdot$–$\underset{}{CHCH_3}$ (with Br on CH) $\xrightarrow[-Br\cdot]{HBr}$ [cyclopentane with H and $CHBrCH_3$ on same carbon] by the more stable 3° radical

racemic

10.47 (1) (a) < (c) < (b). The order is based upon the stabilities of the carbocations formed by protonation. Although (a) and (c) both form 2° carbocations, the one from (c) is slightly more stable.

(2) (a) $(CH_3)_2\overset{+}{C}H \longrightarrow (CH_3)_2CHOSO_3H$

(b) $(CH_3)_3\overset{+}{C} \longrightarrow (CH_3)_3COSO_3H$

(c) $CH_3\overset{+}{C}HCH_2CH_3 \longrightarrow CH_3\underset{}{\overset{OSO_3H}{C}}HCH_2CH_3$

10.48 (a) $(CH_3CH_2)_2\overset{OH}{C}CH_3$ (b) $(CH_3)_3COCH_2CH_3$

(c) 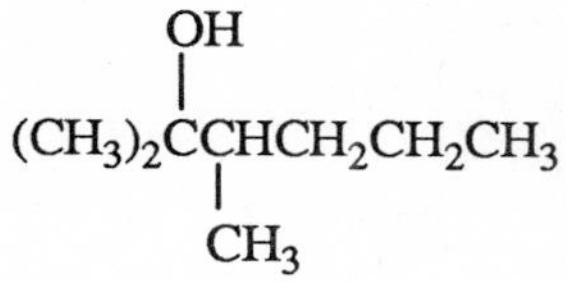 $(CH_3)_2C(OH)(CH_3)CHCH_2CH_2CH_3$ — structure:
$$\overset{\displaystyle OH}{\underset{\displaystyle CH_3}{(CH_3)_2C\!\!-\!CHCH_2CH_2CH_3}}$$
racemic

(d) 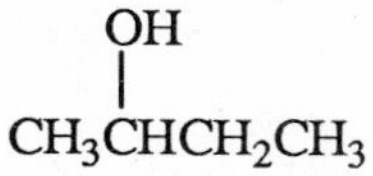 $CH_3\overset{OH}{C}HCH_2CH_3$
racemic

(e) 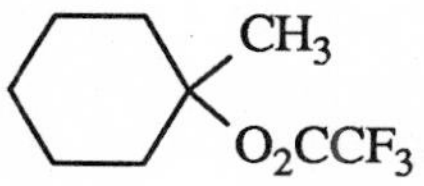 1-methylcyclohexyl with CH_3 and O_2CCF_3

In (c), rearrangement occurs. In (d), $CH_3CHICH_2CH_3$ will be formed in small amounts. In (e), the alkene is protonated and then attacked by the nucleophilic $CF_3CO_2^-$.

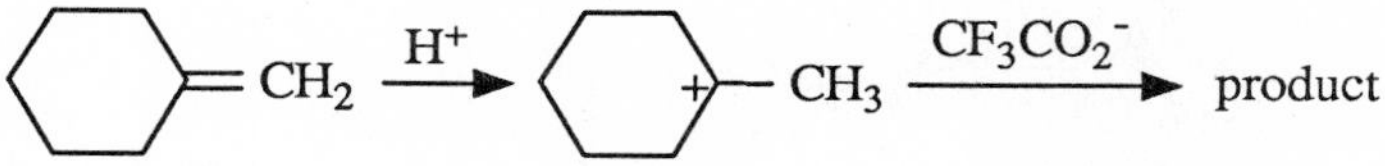

$$\text{cyclohexylidene}{=}CH_2 \xrightarrow{H^+} \text{cyclohexyl}^{+}{-}CH_3 \xrightarrow{CF_3CO_2^-} \text{product}$$

10.49 (a) racemic $(CH_3)_3C\overset{OH}{C}HCH_2CH_3$ + $(CH_3)_3CCH_2\overset{OH}{C}HCH_3$
major product

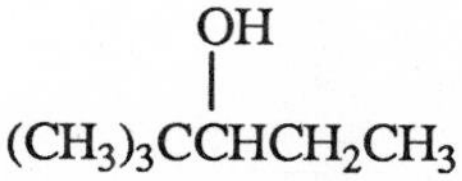

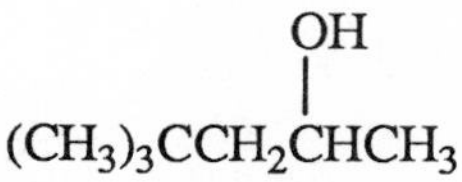

(b) racemic cyclohexyl$-\overset{OCH_2CH_3}{C}HCH_3$

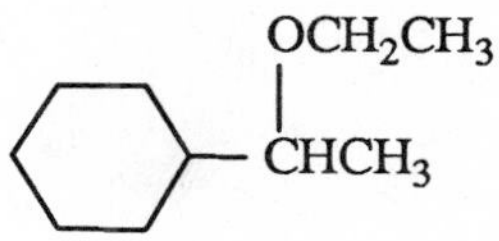

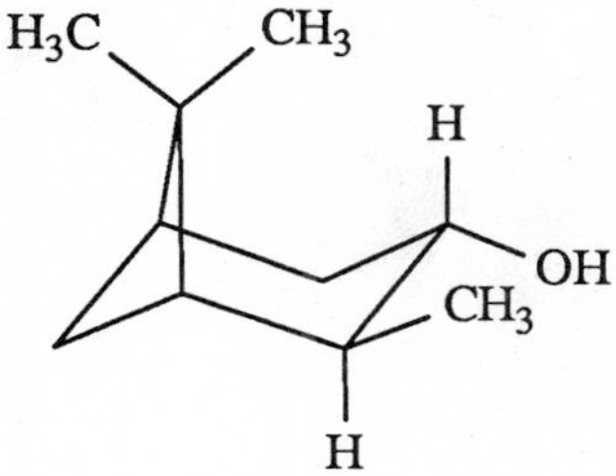

(c)

H₃C CH₃ ... H, OH, CH₃, H by way of H₃C CH₃ ... H, BR₂, CH₃, H as an intermediate

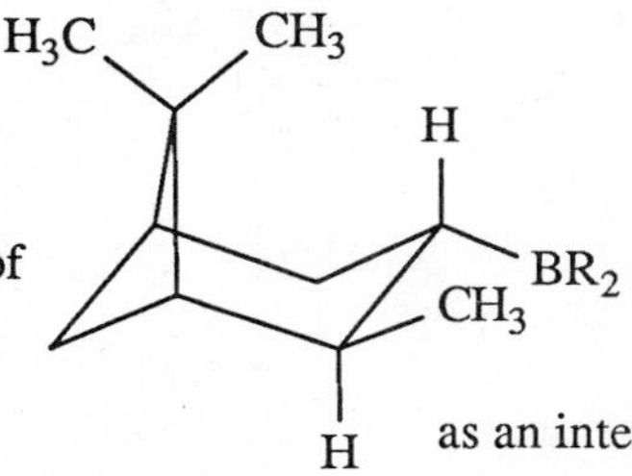

(d) racemic cyclopentane with Br and CH_3

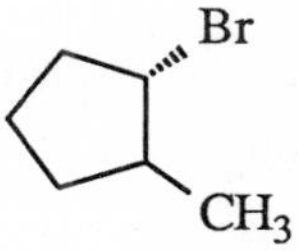

10.50 $Hg(OCOCH_3)_2 \rightleftharpoons {}^{+}HgOCOCH_3 + {}^{-}OCOCH_3$

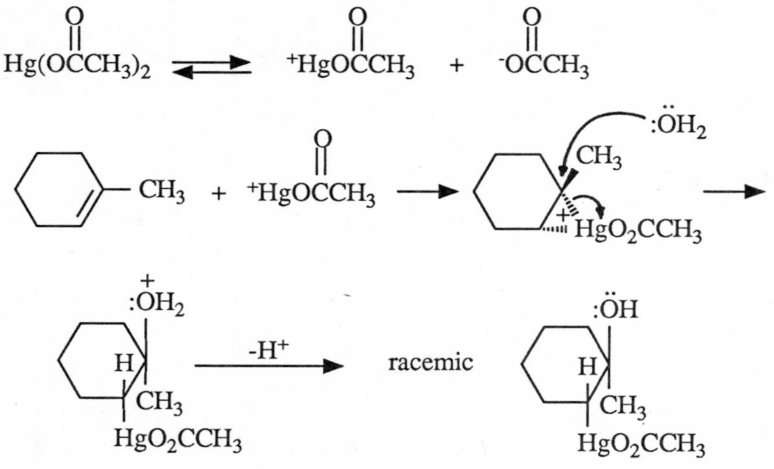

CH_3 + $^{+}HgOCOCH_3$ → CH_3 $:\ddot{O}H_2$ HgO_2CCH_3 →

$:\overset{+}{O}H_2$ H CH_3 HgO_2CCH_3 $\xrightarrow{-H^+}$ racemic $:\ddot{O}H$ H CH_3 HgO_2CCH_3

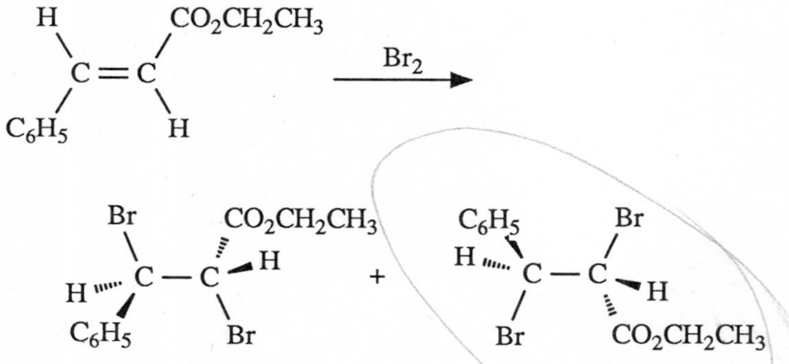

10.51 (a) $C_6H_5CH{=}CHCO_2CH_2CH_3$ (H, $CO_2CH_2CH_3$ / C_6H_5, H) $\xrightarrow{Br_2}$

Br, $CO_2CH_2CH_3$, H, C—C, H, C_6H_5, Br + C_6H_5, Br, H, C—C, H, Br, $CO_2CH_2CH_3$

(2*S*,3*R*) (2*R*,3*S*)

(b) $CH_3(CH_2)_7CH{=}CH(CH_2)_7CO_2CH_3$ (H, H / $CH_3(CH_2)_7$, $(CH_2)_7CO_2CH_3$) $\xrightarrow{Br_2}$

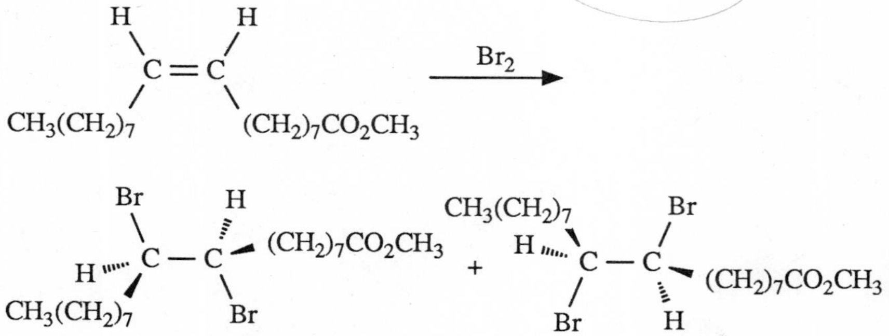

Br, H, C—C, $(CH_2)_7CO_2CH_3$, H, $CH_3(CH_2)_7$, Br + $CH_3(CH_2)_7$, Br, H, C—C, $(CH_2)_7CO_2CH_3$, Br, H

(9*R*,10*R*) (9*S*,10*S*)

(c) $BrCH_2CHBrCH_2Br$ (achiral)

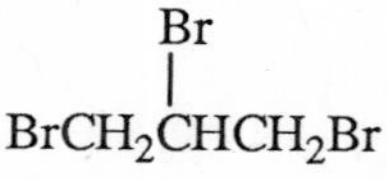

(d) trans-$C_6H_5CH{=}CHC_6H_5$ $\xrightarrow{Br_2}$

meso

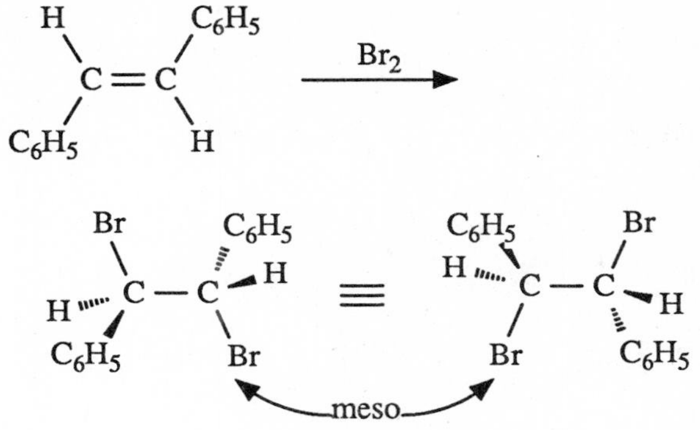

10.52 (a) In the laboratory, the addition of Cl_2 is not as stereospecific as the addition of Br_2, probably because Cl_2 is more likely to yield an open carbocation. Thus, a mixture of stereoisomers might result.

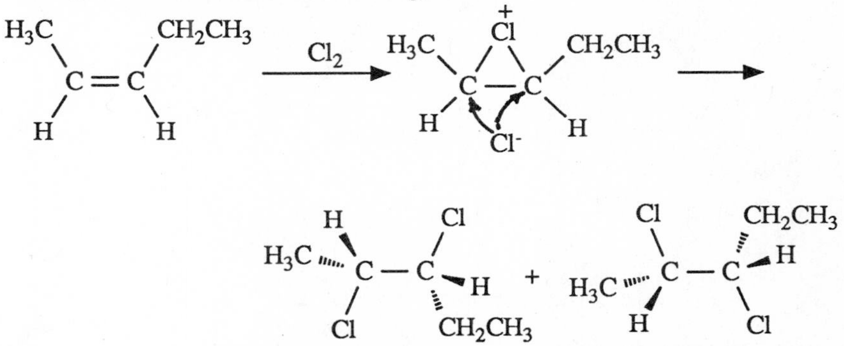

(2*R*,3*R*)-2,3-dichloropentane (2*S*,3*S*)-2,3-dichloropentane

(b) $\xrightarrow{Br_2}$

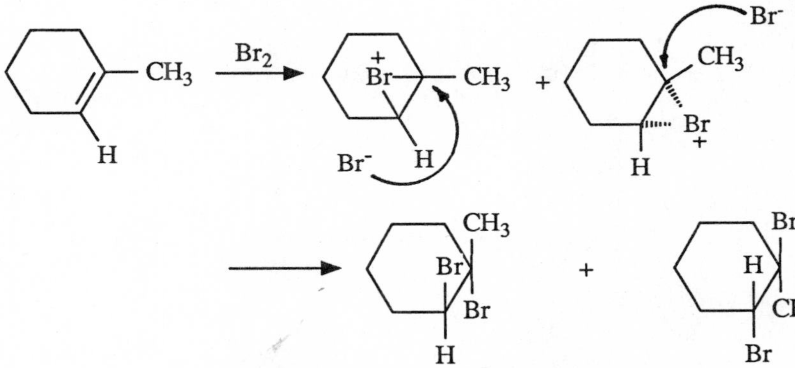

racemic 1,2-dibromo-1-methylcyclohexane

(c) 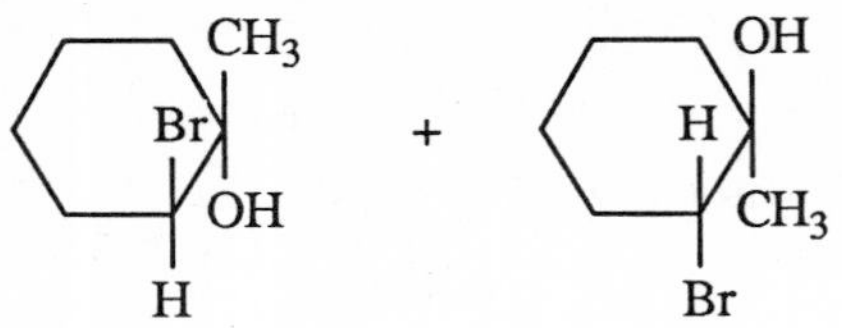 [cyclohexane structure with CH₃, Br, OH, H] + [cyclohexane structure with OH, H, CH₃, Br] by a route similar to that in (b)

10.53 $HO_2C(H)C{=}C(H)CO_2H \xrightarrow{Br_2}$ [bromonium ion: HO_2C, H, Br^+, H, CO_2H; attacked by Br^-]

$\longrightarrow$ HO_2C–C(Br)(H)–C(H)(Br)–CO_2H $\equiv$ (HO_2C)(H)C(Br)–C(Br)(H)(CO_2H)

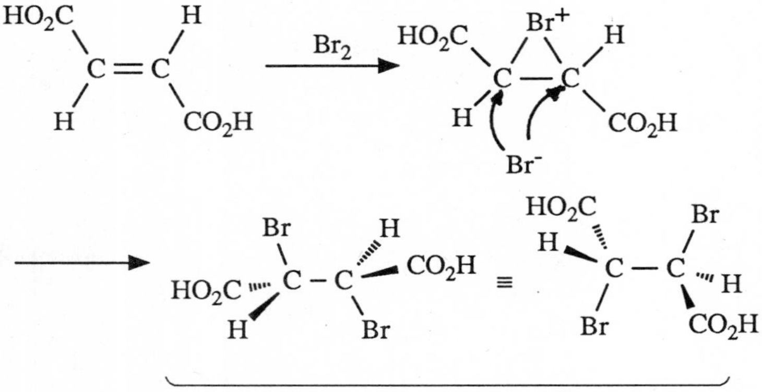

The meso product is obtained. (Make models to verify this conclusion.)

Another way to verify that the product is meso is to convert the formulas to Fischer projections:

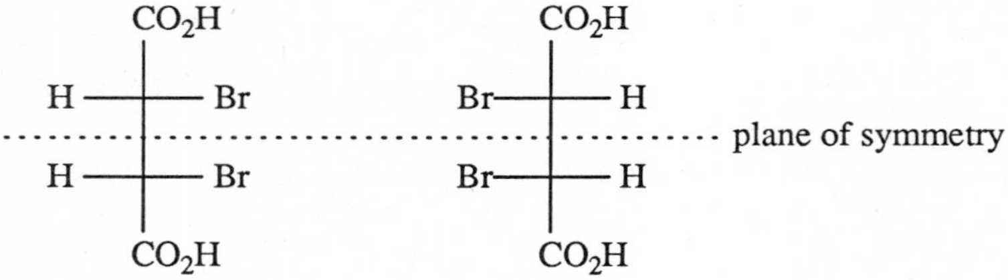

	CO_2H				CO_2H	
H	—	Br		Br	—	H
	plane of symmetry					
H	—	Br		Br	—	H
	CO_2H				CO_2H	

10.54 Either bromide ion or acetic acid can act as a nucleophile and attack the bromonium ion to yield either a dibromo compound or a bromo ester.

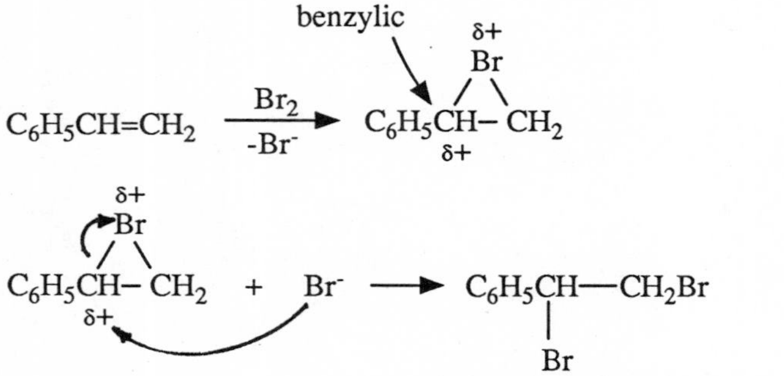

$C_6H_5CH{=}CH_2 \xrightarrow[-Br^-]{Br_2} C_6H_5\overset{\delta+}{C}H{-}CH_2$ (bridged $\overset{\delta+}{Br}$; benzylic)

$C_6H_5\overset{\delta+}{C}H{-}CH_2$ (bridged $\overset{\delta+}{Br}$) $+ \; Br^- \longrightarrow C_6H_5CH(Br){-}CH_2Br$

1,2-dibromo-1-phenylethane

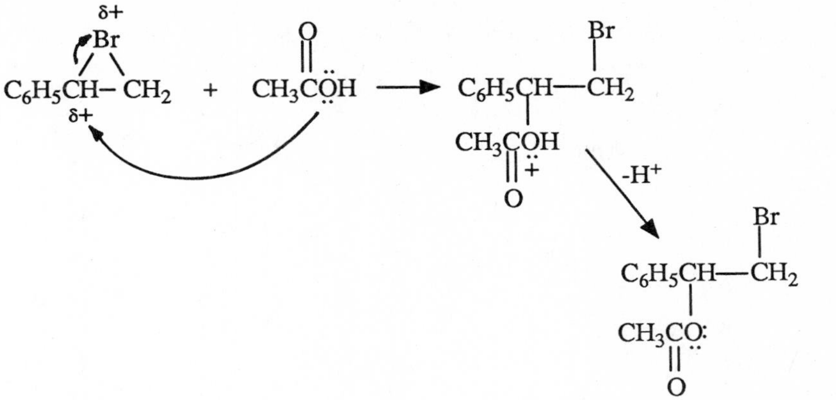

$$C_6H_5\overset{\delta+}{CH}-CH_2 \text{ (bromonium, } \overset{\delta+}{Br}) + CH_3\overset{O}{\overset{\|}{C}}\ddot{O}H \longrightarrow C_6H_5CH(Br\text{-}CH_2)\text{—}O^+(H)C(=O)CH_3 \xrightarrow{-H^+} C_6H_5CH(OC(=O)CH_3)\text{—}CH_2Br$$

2-bromo-1-phenylethyl acetate

10.55 (a) *cis*-$CH_3CH_2CH{=}CHCH_3$ + $CHCl_3$ $\xrightarrow{K^+\ ^-OC(CH_3)_3}$

(b) $CH_3(CH_2)_4CH{=}CH_2$ + Zn(Cu) + CH_2I_2 $\xrightarrow{\text{ether}}$

(c) 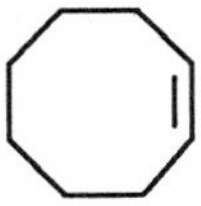 cyclooctene + $CHBr_3$ $\xrightarrow{K^+\ ^-OC(CH_3)_3}$

(d) $(CH_3)_2C{=}C(CH_3)_2$ + CH_2Cl_2 $\xrightarrow{K^+\ ^-OC(CH_3)_3}$

10.56 (a) 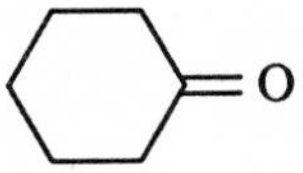 cyclohexanone (b) 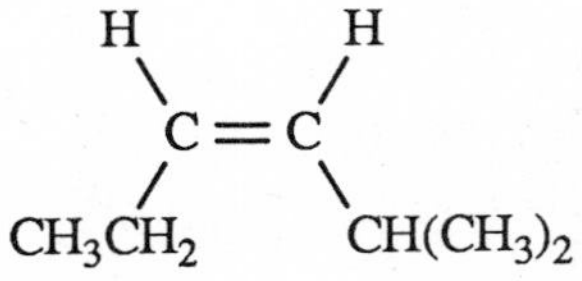 cis-$CH_3CH_2(H)C{=}C(H)CH(CH_3)_2$

(c) H_3C CH_3 H H CH_3 H

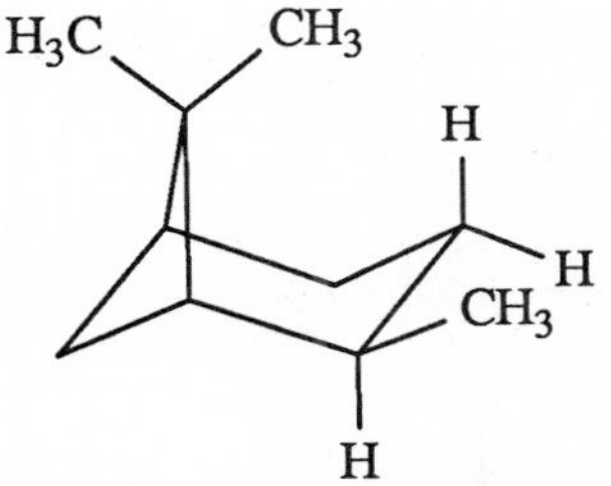

In (c), the α-pinene molecule would approach the catalyst from the less hindered side. This approach would cause H_2 to add *trans* to the bridge and would force the methyl group to become *cis* to the bridge.

(d) 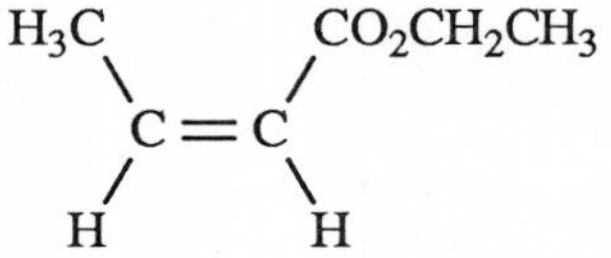 $H_3C(H)C=C(H)CO_2CH_2CH_3$

(e) 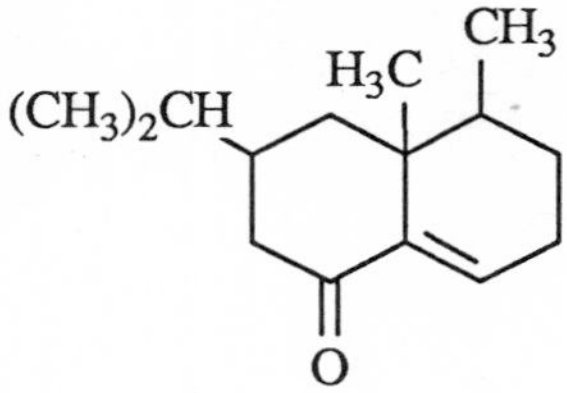 $(CH_3)_2CH$ ... H_3C ... CH_3 ... O

10.57 (a) 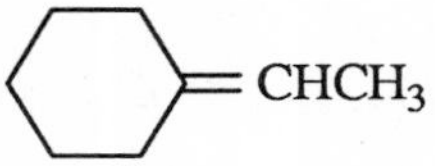 cyclohexylidene=$CHCH_3$ (more substituted) (b) *trans*

(c) (*E*) (d) CH_2=$CHCO_2CH_2CH_3$ (conjugated)

10.58 (a) racemic H_3C ... O ... CH_3C-CH_2 ... O

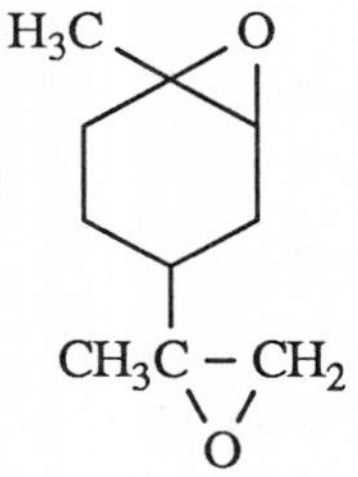

(b) 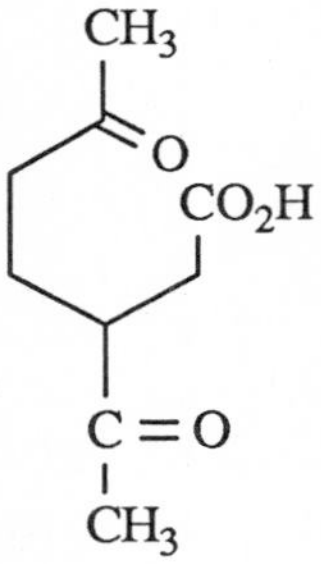 CH_3 ... O ... CO_2H + CO_2 (from HCO_2H) ... C=O ... CH_3

(c) racemic H_3C OH OH *cis* ... CH_3C-CH_2 OH OH

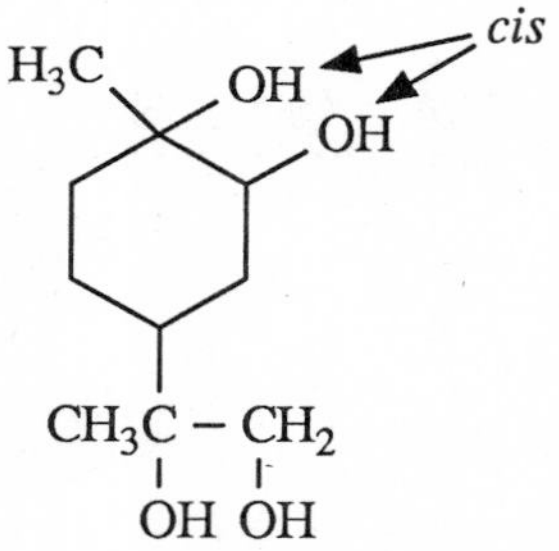

10.59 The product is a *cis* diol, which arises from the intermediate osmate ester. The osmate ester group would form on the less hindered side of the molecule to yield the following product:

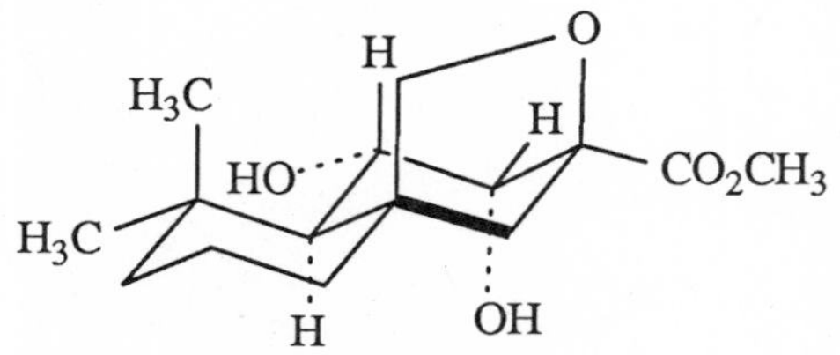

H O H_3C H HO CO_2CH_3 H_3C H OH

10.60 (a) 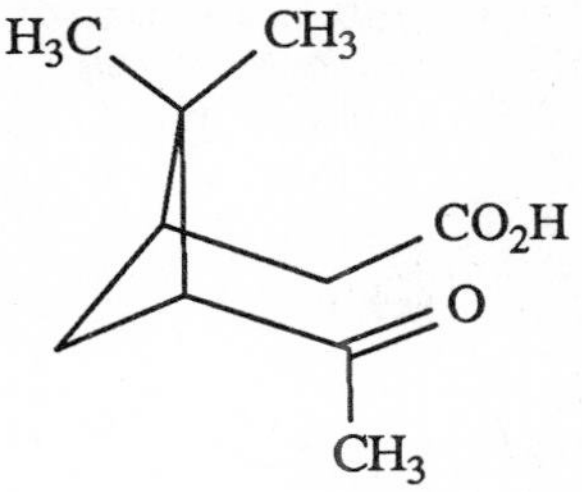 [bicyclic structure: H_3C, CH_3, CO_2H, O, CH_3] (b) 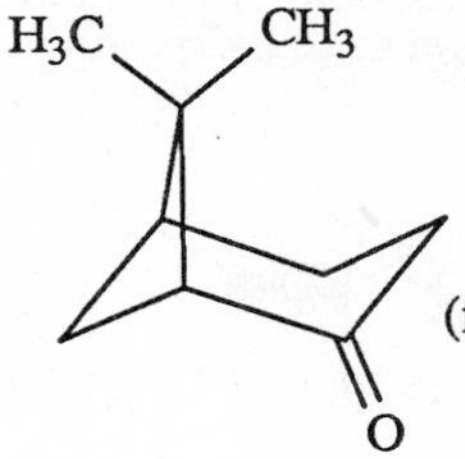 [bicyclic ketone: H_3C, CH_3, O] + CO_2 (from HCO_2H)

10.61 (a)  [cyclopentene] or [cyclic diene: $HC=CH$ linked by $(CH_2)_3$ bridges] (b) $CH_3CH_2CH{=}C(CH_3)_2$

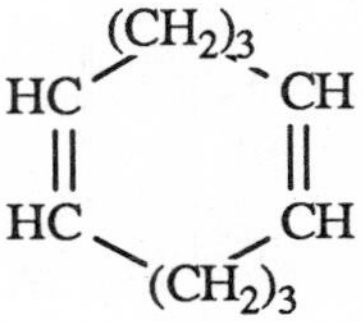

(c) (*E*) or (*Z*)-$CH_3CH{=}CHCH_3$ (d) 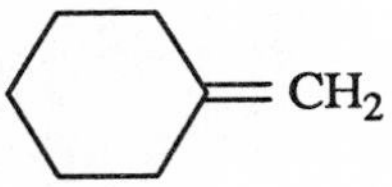 [cyclohexane]$=CH_2$

(e) 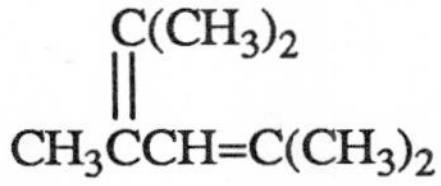 $CH_3C(=C(CH_3)_2)CH{=}C(CH_3)_2$ (f) (*E*) or (*Z*)-$CH_3CH{=}CHCH_2CH{=}CH_2$

To solve this type of problem, redraw the products so that the carbonyl groups are close together (because two carbonyl groups result from one double bond). For example,

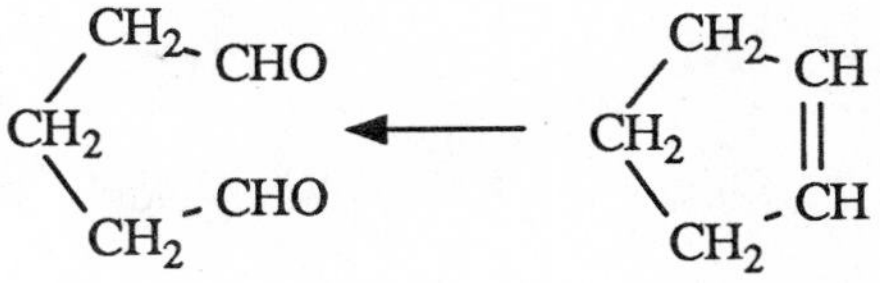

$CH_2(CHO)$–CH_2–$CH_2(CHO)$ ← CH_2–$CH{=}CH$–CH_2 (ring with CH_2)

product *reactant*

10.62 (a) 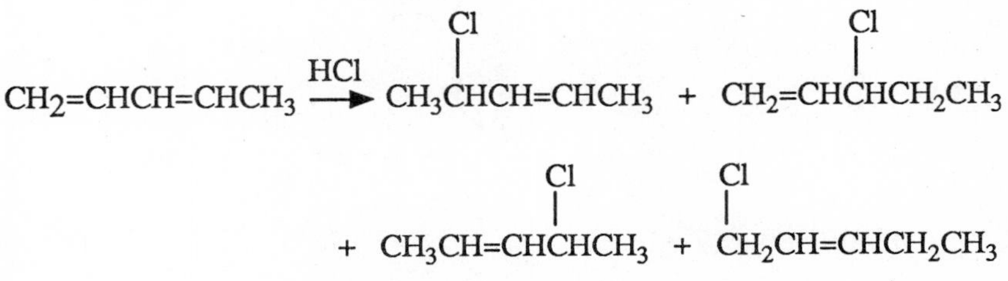 $CH_2{=}CHCH{=}CHCH_3 \xrightarrow{HCl} CH_3CHClCH{=}CHCH_3 + CH_2{=}CHCHClCH_2CH_3$

$+ CH_3CH{=}CHCHClCH_3 + CH_2ClCH{=}CHCH_2CH_3$

same as first structure

The first two structures shown arise from 1,2-addition, while the last two structures arise from 1,4-addition. The products shown here are those that arise from the most stable carbocations (allylic, in each case). Under acidic conditions, equilibration of (*E*) and (*Z*) double bonds would probably occur, regardless of the stereochemistry of the starting material.

(b)

$$(CH_3)_2C(CH_3)\text{—}CH{=}CH(CH_2)_2CH_3 \xrightarrow{H^+}$$

$$\left[(CH_3)_2C(CH_3)\text{—}CH_2\text{—}\overset{+}{C}H(CH_2)_2CH_3\right] + \left[(CH_3)_2C(CH_3)\text{—}\overset{+}{C}H\text{—}CH_2(CH_2)_2CH_3\right]$$

First cation $\xrightarrow{Cl^-}$ $(CH_3)_3CCH_2CH(Cl)(CH_2)_2CH_3$

Second cation $\xrightarrow{Cl^-}$ $(CH_3)_3CCH(Cl)CH_2(CH_2)_2CH_3$

Second cation $\xrightarrow{\text{methyl shift}}$ $\left[(CH_3)_2\overset{+}{C}\text{—}CH(CH_3)\text{—}CH_2(CH_2)_2CH_3\right]$ $\xrightarrow{Cl^-}$ $(CH_3)_2C(Cl)\text{—}CH(CH_3)\text{—}CH_2(CH_2)_2CH_3$

(c) cyclopentadiene $\xrightarrow[-Cl^-]{1\ HCl}$ [allylic cyclopentenyl cation ↔ resonance form] $\xrightarrow{Cl^-}$ 3-chlorocyclopentene (Cl)

In (c), 1,2-addition and 1,4-addition yield the same product.

10.63 $CH_2{=}C(CH_3)\text{-}CH{=}CH_2 \xrightarrow{H^+} \left[CH_3\text{-}\overset{+}{C}(CH_3)\text{-}CH{=}CH_2 \leftrightarrow CH_3\text{-}C(CH_3){=}CH\text{-}\overset{+}{C}H_2\right]$

more stable (first resonance form); Cl^- attacks each cationic carbon

A (75 parts) **B** (25 parts)

10.64 The copolymer could be represented by any number of formulas. A correct answer should contain (a) three units of butadiene and one unit of styrene; (b) two of the butadiene units joined by 1,4-addition and one unit joined by 1,2-addition and (c) most or all the double bonds as *trans*. For example:

$$\text{~}CH_2\overset{H}{\underset{|}{C}}=\underset{H}{C}CH_2\text{—}CH_2\underset{C_6H_5}{CH}\text{—}CH_2\underset{CH=CH_2}{CH}\text{—}CH_2CH=CHCH_2\text{~}$$

trans (arrow to the C=C double bond)

1,4-addition (80%) 1,2-addition (20%)

ratio of butadiene to styrene, 3 : 1

10.65 (a) Water provides a proton for the initial electrophilic addition.

$$BF_3 + H_2\ddot{O}: \rightleftharpoons [H_2\ddot{O}:\cdots BF_3] \rightleftharpoons H^+ + HO\overline{B}F_3$$

(b)

$$CH_2{=}C(CH_3)_2 + H^+ \rightleftharpoons CH_3\overset{+}{C}(CH_3)_2 \xrightarrow{CH_2=C(CH_3)_2}$$

$$CH_3C(CH_3)_2\text{—}CH_2\text{—}\overset{+}{C}(CH_3)_2 \xrightarrow{\text{etc.}} \text{polyisobutylene}$$

(c) The addition of the intermediate carbocation to isobutylene forms the most stable carbocation (Markovnikov addition). For $\text{~}CH_2C(CH_3)_2\text{—}C(CH_3)_2CH_2\text{~}$ to be a product, the reaction would proceed by way of a 1° carbocation rather than a 3° carbocation.

$$RCH_2\overset{+}{C}(CH_3)_2 + (CH_3)_2C{=}CH_2 \longrightarrow RCH_2C(CH_3)_2\text{—}C(CH_3)_2\text{—}\overset{+}{C}H_2$$

a primary carbocation: not formed

(d) No. For ethylene $CH_2{=}CH_2$ to be polymerized by a cationic mechanism, a 1° carbocation would have to be an intermediate.

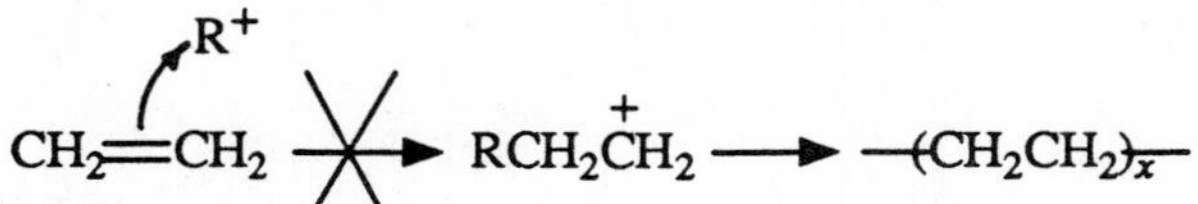

$CH_2{=}CH_2$ + R^+ ⟶ (crossed out) $R CH_2\overset{+}{C}H_2$ ⟶ $-(CH_2CH_2)_x-$

not formed

10.66 (a) 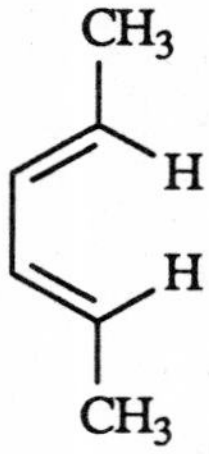 CH_3, H, H, CH_3 (b) no

(c) 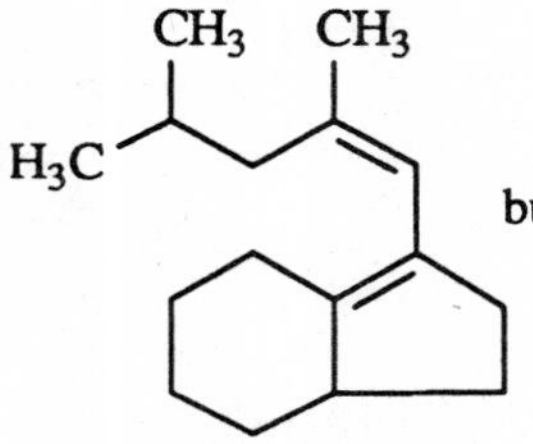 H_3C, CH_3, CH_3 but not H_3C, CH_3, CH_3 (*E*) instead of (*Z*)

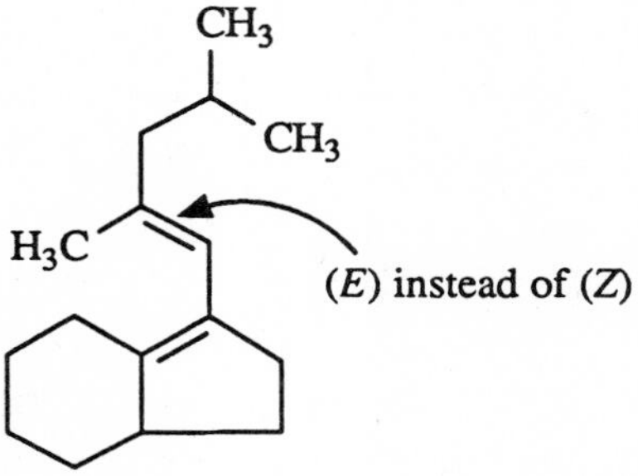

10.67 (a) 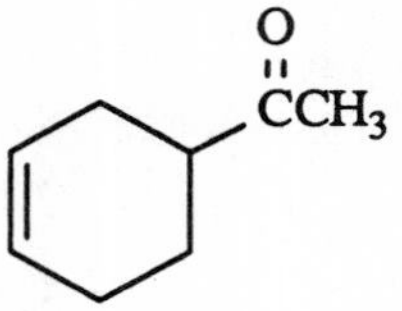 O, $\overset{\|}{C}CH_3$ (b) 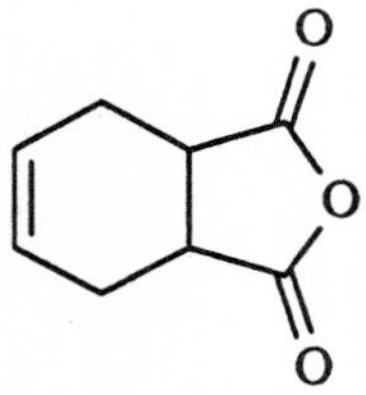 O, O, O

(c) CO_2H, OCH_3 + CO_2H, OCH_3

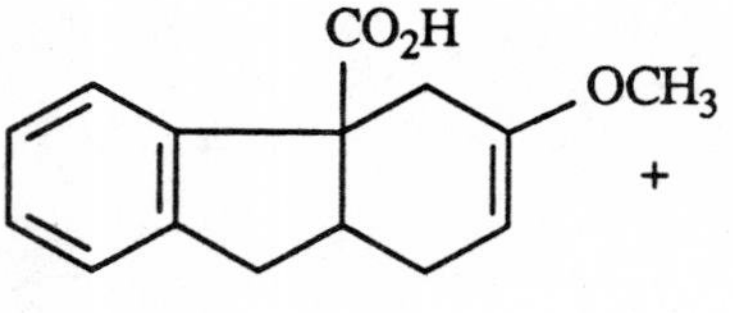

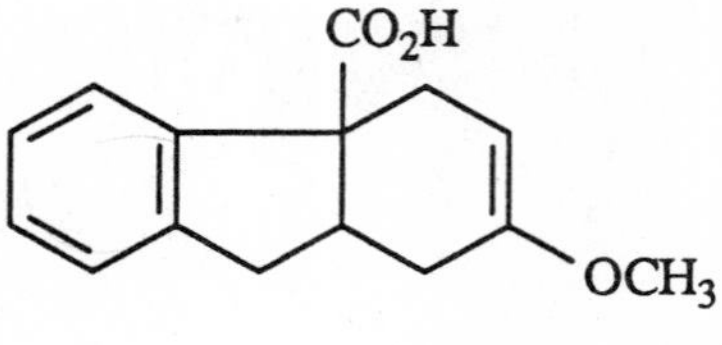

(d)  O + O

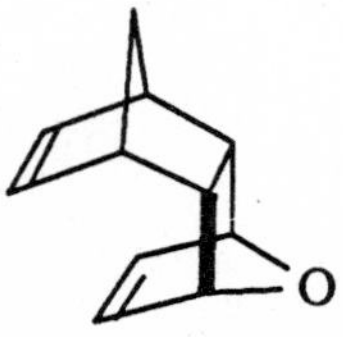

10.68 (a) [1,4-naphthoquinone] + [2,3-dimethyl-1,3-butadiene, CH_3, CH_3] $\xrightarrow{\text{heat}}$

(b) [cyclopentadiene] + $N{\equiv}C$–$CH{=}CH_2$ (c) [1,3-cyclohexadiene] + $CH_3CH_2O_2C$–$CH{=}CH_2$

(d) [2H-thiopyran, S] + [N-phenylmaleimide, NC_6H_5]

10.69 (a) *trans*-$CH_3CH{=}CHCH_3$ $\xrightarrow{\text{HI}}$ racemic $CH_3\overset{\text{I}}{C}HCH_2CH_3$

(b) $\xrightarrow[\text{(2) } Na_2SO_3]{\text{(1) } OsO_4}$ [H_3C, H, HO on C; H, CH_3, OH on C] + [HO, H_3C, H on C; OH, H, CH_3 on C]

(c) $\xrightarrow{\text{(1) } C_6H_5CO_3H}$ [H, H_3C on C; H, CH_3 on C; O bridge] + [H_3C, H on C; H, CH_3 on C; O bridge]

A B

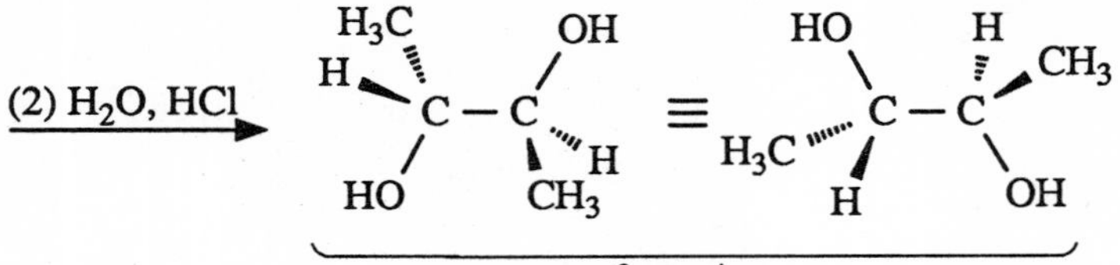

from A

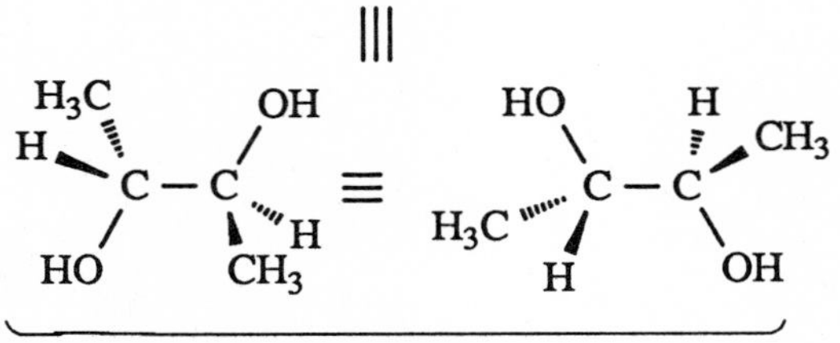

from B

Make models to verify that all the structures are identical (*meso*). If the concentration of HCl is sufficiently high, some 1,2-chlorohydrin would also be observed (a mixture of stereoisomers):

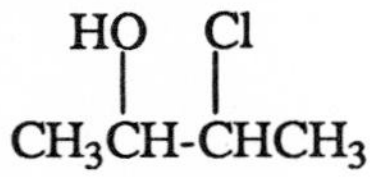

(d) racemic 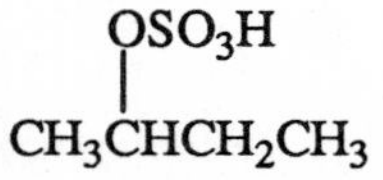 (e) 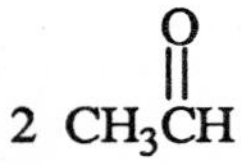

(f) racemic 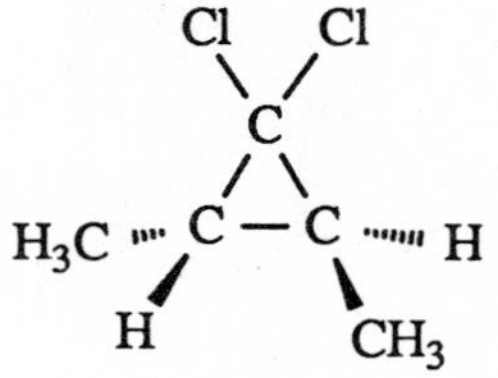

10.70 In each part, racemic or achiral products would be obtained:

(a) 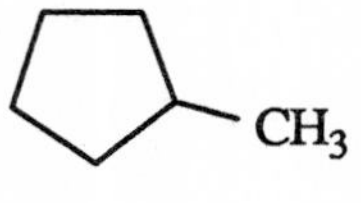 (b) 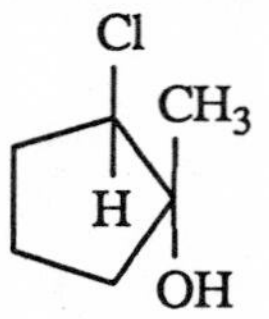 (c) 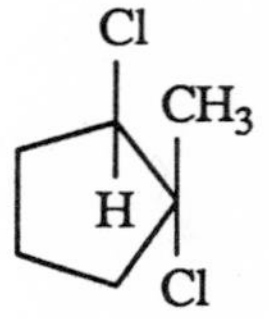

(d) 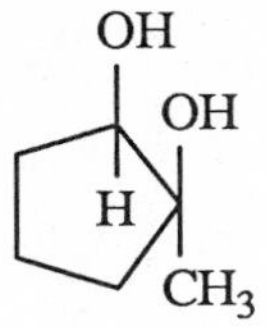 [cyclopentane structure: OH, OH, H, CH_3]

(e) 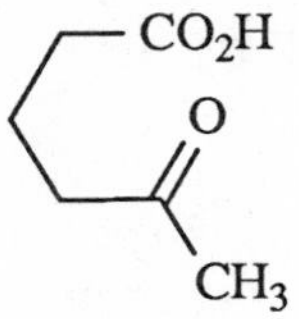 [structure: CO_2H, O, CH_3]

(f) 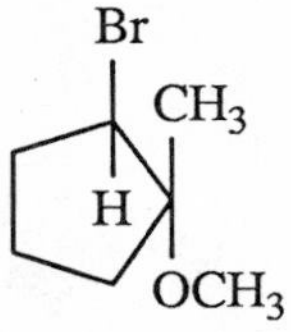 [cyclopentane structure: Br, CH_3, H, OCH_3]

(g) 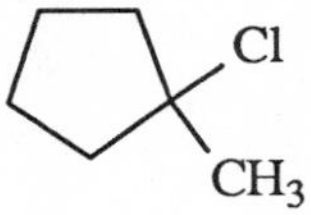 [cyclopentane structure: Cl, CH_3]

(h) *cis* and *trans* [cyclopentane structure: Br, CH_3]

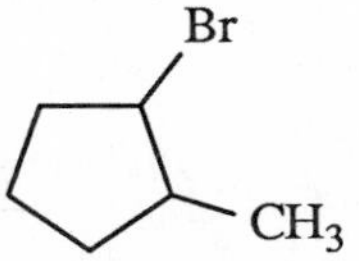

(i) 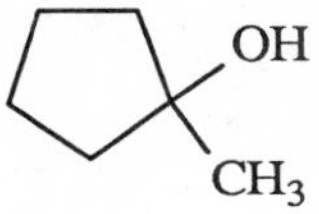 [cyclopentane structure: OH, CH_3]

(j) 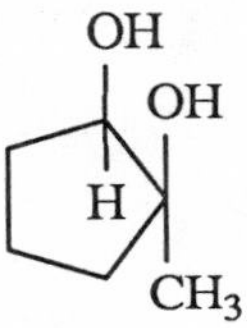 [cyclopentane structure: OH, OH, H, CH_3]

10.71 (a) 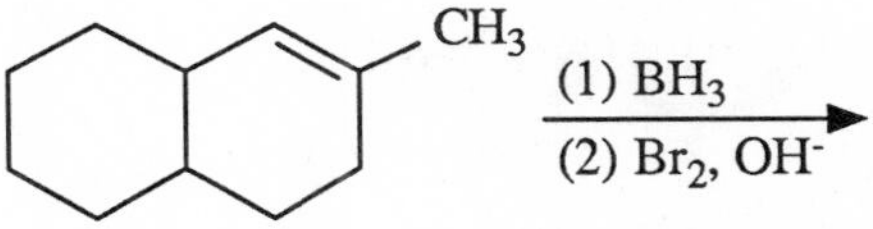 [structure with CH_3] $\xrightarrow[\text{(2) } Br_2,\ OH^-]{\text{(1) } BH_3}$

(b) 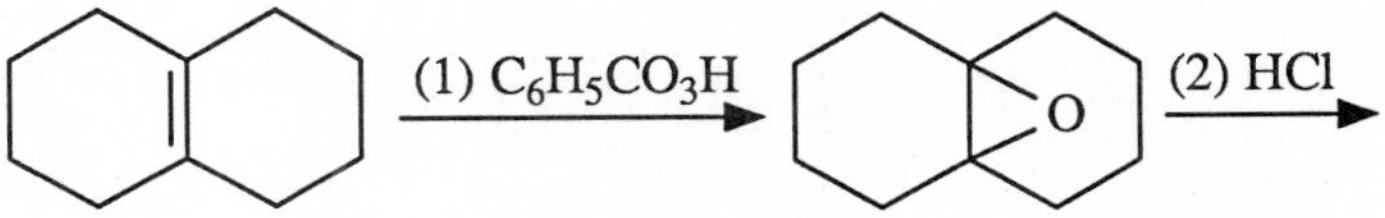 [structure] $\xrightarrow{\text{(1) } C_6H_5CO_3H}$ [epoxide structure, O] $\xrightarrow{\text{(2) HCl}}$

10.72 $CH_3CH_2CH_2C{\equiv}CCH_2CH_2CH_3 \xrightarrow[H_2O]{KMnO_4} 2\ CH_3CH_2CH_2\overset{O}{\overset{\|}{C}}OH$

$\xrightarrow[\text{phase transfer agent}]{KMnO_4,\ CH_2Cl_2} CH_3CH_2CH_2\overset{O}{\overset{\|}{C}}\text{-}\overset{O}{\overset{\|}{C}}CH_2CH_2CH_3$

10.73 (a) 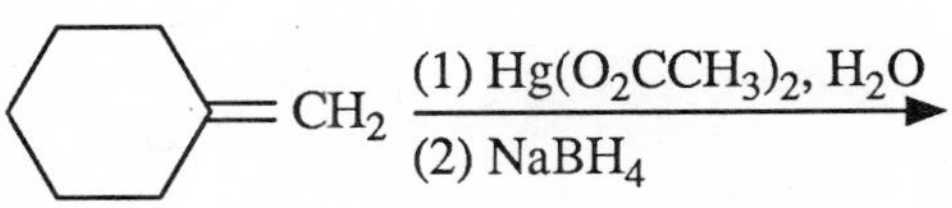 [cyclohexane]$=CH_2 \xrightarrow[\text{(2) } NaBH_4]{\text{(1) } Hg(O_2CCH_3)_2,\ H_2O}$

(b) [cyclohexane]$=CH_2$ $\xrightarrow[(2)\ H_2O_2,\ OH^-]{(1)\ BH_3}$

(c) [cyclohexane]$=CH_2$ $\xrightarrow{C_6H_5CO_3H}$ [cyclohexane spiro-epoxide, O, CH_2] $\xrightarrow[CH_3OH]{Na^+\ ^-OCH_3}$

10.74 (a) $CH_3C{\equiv}CH \xrightarrow{H_2O,\ Hg^{2+}} \left[CH_3\overset{OH}{\overset{|}{C}}{=}CH_2\right] \longrightarrow CH_3\overset{O}{\overset{\|}{C}}CH_3$

(b) $CH_3C{\equiv}CH \xrightarrow[(2)\ CH_3I]{(1)\ NaNH_2} CH_3C{\equiv}CCH_3 \xrightarrow{H_2O,\ Hg^{2+}}$

(c) $CH_3C{\equiv}CH \xrightarrow{CH_3MgI} CH_3C{\equiv}CMgI \xrightarrow[(2)\ H_2O,\ H^+]{(1) CH_3CHO}$

$CH_3C{\equiv}C\overset{OH}{\overset{|}{C}}HCH_3 \xrightarrow{K} CH_3C{\equiv}C\overset{O^-\ K^+}{\overset{|}{C}}HCH_3 \xrightarrow{CH_3I}$

(d) $CH_3C{\equiv}CH\ +\ 1\ H_2 \xrightarrow{Pt} CH_3CH{=}CH_2 \xrightarrow{NBS}$

$BrCH_2CH{=}CH_2 \xrightarrow{Br_2}$

10.75 (a) $CH_3CH_2CH_3 \xrightarrow[h\nu]{Br_2} CH_3\overset{Br}{\overset{|}{C}}HCH_3 \xrightarrow[heat]{KOH,\ CH_3CH_2OH}$

$CH_3CH{=}CH_2 \xrightarrow[Ni]{D_2} CH_3CHDCH_2D$

(b) $CH_3CH_2CH_2Br \xrightarrow{KOH} CH_3CH_2CH_2OH \xrightarrow[-H_2O]{H_2SO_4,\ heat}$

$CH_3CH{=}CH_2 \xrightarrow[heat]{KMnO_4}$

(c) 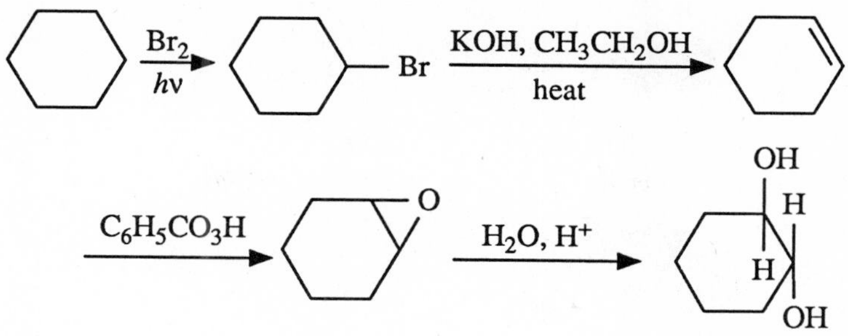 cyclohexane $\xrightarrow[h\nu]{Br_2}$ bromocyclohexane (–Br) $\xrightarrow[\text{heat}]{KOH,\ CH_3CH_2OH}$ cyclohexene

$\xrightarrow{C_6H_5CO_3H}$ cyclohexene oxide (O) $\xrightarrow{H_2O,\ H^+}$ trans-1,2-cyclohexanediol (OH, H / H, OH)

(d) 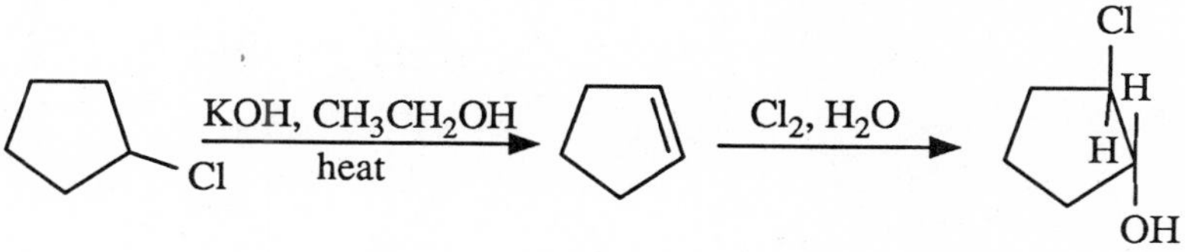 chlorocyclopentane (Cl) $\xrightarrow[\text{heat}]{KOH,\ CH_3CH_2OH}$ cyclopentene $\xrightarrow{Cl_2,\ H_2O}$ trans-2-chlorocyclopentanol (Cl, H / H, OH)

(e) $C_6H_5CH{=}CHC_6H_5 \xrightarrow{Br_2} C_6H_5\underset{}{CH}(Br)\text{-}CH(Br)C_6H_5 \xrightarrow{2\ {}^{-}\!:\ddot{N}H_2} C_6H_5C{\equiv}CC_6H_5$

10.76 Write a flow equation showing the information given:

a 1,2-hydroquinone (A) $\xrightarrow[(2)\ 2\ CH_3I]{(1)\ 2\ Na^{+}\ {}^{-}OH}$ B
($C_{21}H_{34}O_2$) dimethylated A ($C_{23}H_{38}O_2$)

$\xrightarrow[(2)\ Zn,\ H_2O]{(1)\ O_3}$ benzene ring with OCH_3, OCH_3, $(CH_2)_7\overset{O}{\overset{\|}{C}}H$ + $H\overset{O}{\overset{\|}{C}}(CH_2)_5CH_3$

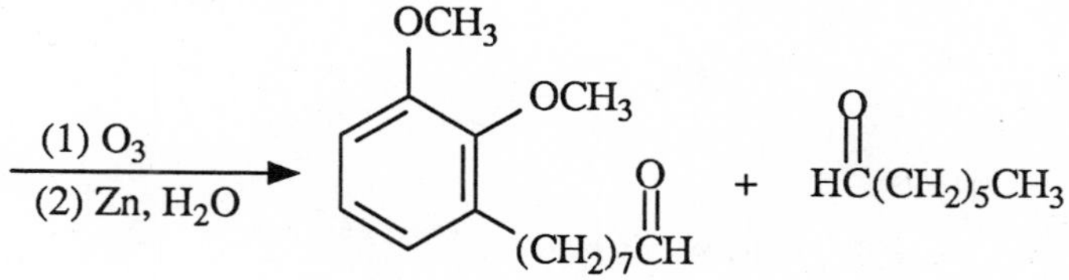

Working backwards:

B: benzene ring with OCH_3, OCH_3, $(CH_2)_7CH{=}CH(CH_2)_5CH_3$

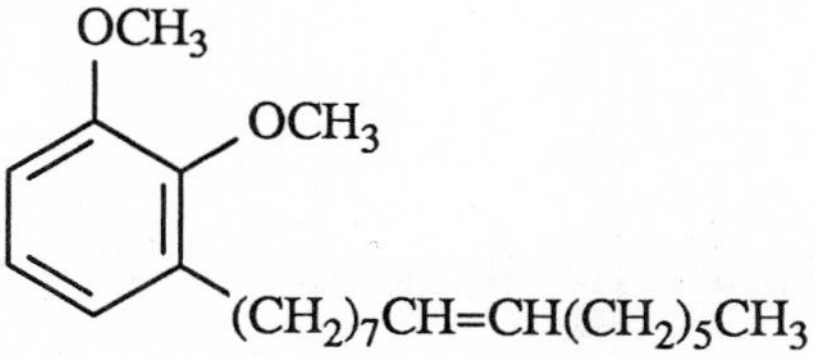

Therefore, the urushiol structure is as follows:

A: (*E*) or (*Z*) - [benzene ring with OH, OH, and $(CH_2)_7CH{=}CH(CH_2)_5CH_3$ substituents]

10.77 [1-ethynylcyclohexanol: cyclohexane with $C{\equiv}CH$ and OH] $\xrightarrow[HgO]{H_2O,\ H^+}$ [cyclohexane with $C(OH){=}CH_2$ and OH] $\longrightarrow$ [cyclohexane with $C(=O){-}CH_3$ and OH]

the product

10.78 (a) $HC{\equiv}CH \xrightarrow{1\ NaNH_2} HC{\equiv}C{:}^-\ Na^+ \xrightarrow{CH_3I} HC{\equiv}CCH_3$

$\xrightarrow[\text{deact. Pd}]{1\ H_2} CH_3CH{=}CH_2 \xrightarrow[\text{or (1) } BH_3 \text{ (2) } Br_2,\ OH^-]{HBr,\ \text{peroxide}} CH_3CH_2CH_2Br$

$\xrightarrow{HC{\equiv}C{:}^-\ Na^+} CH_3CH_2CH_2C{\equiv}CH \xrightarrow[(2)\ CH_3CH_2CH_2Br]{(1)\ NaNH_2}$

$CH_3CH_2CH_2C{\equiv}CCH_2CH_2CH_3 \xrightarrow[\text{deact. Pd}]{1\ H_2}$

cis-$CH_3CH_2CH_2CH{=}CHCH_2CH_2CH_3$

(b) $CH_2{=}CH_2 \xrightarrow{HBr} CH_3CH_2Br \xrightarrow[\text{from (a)}]{HC{\equiv}C{:}^-\ Na^+} CH_3CH_2C{\equiv}CH \xrightarrow[\text{deact. Pd}]{1\ H_2}$

$CH_3CH_2CH{=}CH_2$

(c) $CH_3CH_2CH{=}CH_2$ from (b) $\xrightarrow[H_2O]{Cl_2} CH_3CH_2CH(OH)CH_2Cl$ (racemic)

(d) 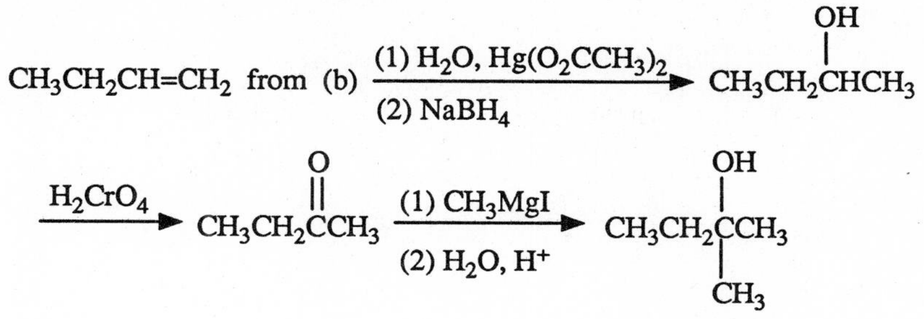 $CH_3CH_2CH{=}CH_2$ from (b) $\xrightarrow[\text{(2) NaBH}_4]{\text{(1) H}_2\text{O, Hg(O}_2\text{CCH}_3)_2}$ $CH_3CH_2CH(OH)CH_3$

$\xrightarrow{H_2CrO_4}$ $CH_3CH_2C(=O)CH_3$ $\xrightarrow[\text{(2) H}_2\text{O, H}^+]{\text{(1) CH}_3\text{MgI}}$ $CH_3CH_2C(OH)(CH_3)CH_3$

(e) 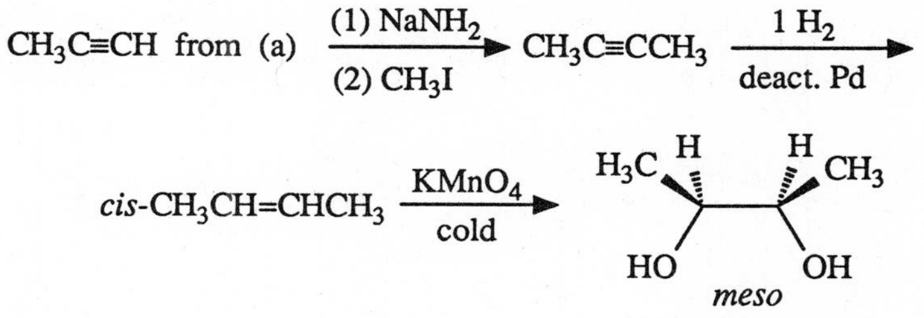 $CH_3C{\equiv}CH$ from (a) $\xrightarrow[\text{(2) CH}_3\text{I}]{\text{(1) NaNH}_2}$ $CH_3C{\equiv}CCH_3$ $\xrightarrow[\text{deact. Pd}]{1\ H_2}$

cis-$CH_3CH{=}CHCH_3$ $\xrightarrow[\text{cold}]{KMnO_4}$ H_3C, H, H, CH_3, HO, OH

meso

(f) 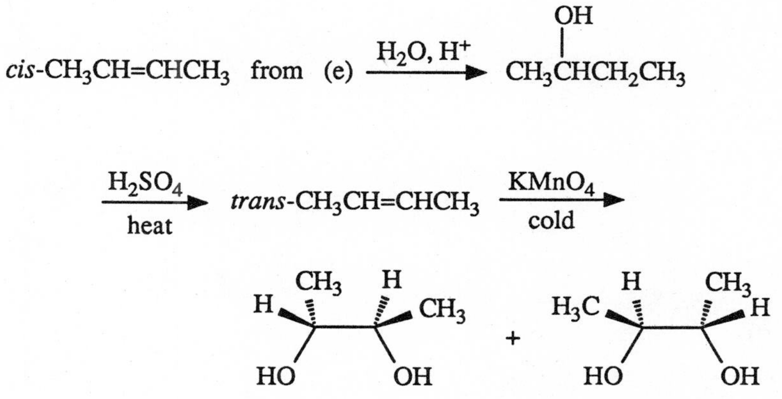 *cis*-$CH_3CH{=}CHCH_3$ from (e) $\xrightarrow{H_2O,\ H^+}$ $CH_3CH(OH)CH_2CH_3$

$\xrightarrow[\text{heat}]{H_2SO_4}$ *trans*-$CH_3CH{=}CHCH_3$ $\xrightarrow[\text{cold}]{KMnO_4}$

H, CH_3, H, CH_3, HO, OH + H_3C, H, CH_3, H, HO, OH

10.79 (a) 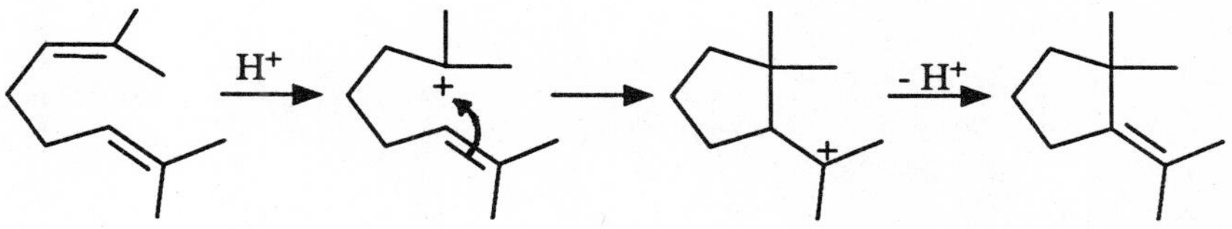 $\xrightarrow{H^+}$ + $\longrightarrow$ + $\xrightarrow{-\ H^+}$

(b) [cyclooctadiene] $\xrightarrow[-\ I^-]{I_2}$ [iodonium ion, H, I^+, H] attack from top → [bicyclic cation, H, I, H, +] ≡ [bicyclo[3.3.0]octyl cation, I, +] $\xrightarrow{I^-}$ 2 products

(c) [diene] $\xrightarrow{H^+}$ [cation +] → [cyclohexyl cation +] $\xrightarrow{-\ H^+}$ [cyclohexene]

10.80 (a) [cyclododecanone] $\xrightarrow[(2)\ H_2O,\ H^+]{(1)\ CH_3MgI}$ [1-methylcyclododecanol, HO, CH_3] $\xrightarrow[heat]{H^+}$

[1-methylcyclododecene, CH_3] $\xrightarrow[(2)\ Zn,\ HCl]{(1)\ O_3}$ [keto aldehyde, CH_3, O, CH=O]

(b) [1-methylcyclohexene, CH_3] $\xrightarrow[(2)\ Zn,\ HCl]{(1)\ O_3}$ $HC(O)(CH_2)_4C(O)CH_3$ $\xrightarrow[(2)\ H_2O,\ NH_4^+]{(1)\ 2\ CH_2{=}CHMgBr}$

$CH_2{=}CHCH(OH)(CH_2)_4C(OH)(CH_3)CH{=}CH_2$

In (b), NH_4^+ is used in place of a strong acid for hydrolyzing the magnesium alkoxide--the weakly acidic NH_4^+ ion is less likely to cause dehydration of the product tertiary alcohol.

10.81 (a) cyclohexene $\xrightarrow{HBr}$ bromocyclohexane $\xrightarrow[\text{ether}]{Mg}$ cyclohexyl-MgBr $\xrightarrow[(2)\ H_2O,\ H^+]{(1)\ CH_3CH_2CHO}$

cyclohexyl-$CH(OH)CH_2CH_3$ $\xrightarrow{H_2CrO_4}$ cyclohexyl-$C(=O)CH_2CH_3$

(b) cyclohexyl-MgBr from (a) $\xrightarrow[(2)\ H_2O,\ H^+]{(1)\ HCHO}$ cyclohexyl-CH_2OH

(c) cyclohexene $\xrightarrow{C_6H_5CO_3H}$ cyclohexene oxide $\xrightarrow[(2)\ H_2O,\ H^+]{(1)\ CH_3MgI}$ trans-2-methylcyclohexanol (OH, CH_3)

(d) cyclohexene $\xrightarrow[\text{cold}]{KMnO_4}$ cis-1,2-cyclohexanediol (H, H, OH, OH) $\xrightarrow{PBr_3}$ trans-1,2-dibromocyclohexane (Br, Br, H, H)

(e) cyclopentadiene + $CH_2{=}CHC(=O)CH_3$ $\xrightarrow{heat}$ bicyclic adduct (H, H, H, $C(=O)CH_3$) $\xrightarrow[\text{heat}]{KMnO_4}$ cyclopentane with H, CO_2H, H, CO_2H, H, $C(=O)CH_3$

endo adduct

(If the stereochemistry eludes you, make models.)

10.82 $C_{23}H_{46}$ $\xrightarrow{H_2,\ Ni}$ $C_{23}H_{48}$

$C_{23}H_{46}$ $\xrightarrow{\text{hot } KMnO_4}$ $CH_3(CH_2)_{12}CO_2H$ + $HO_2C(CH_2)_7CH_3$

$C_{23}H_{46}$ $\xrightarrow{Br_2}$ $C_{23}H_{46}Br_2$ (a pair of enantiomers)

The formula $C_{23}H_{46}$ (C_nH_{2n}) tells us that the structure contains one double bond or one ring. The H_2 reaction and the $KMnO_4$ reaction tell us that the structure contains one double bond. The $KMnO_4$ reaction shows the location of the double bond:

$$CH_3(CH_2)_{12}CH{=}CH(CH_2)_7CH_3$$

The fact that the Br_2 reaction yields a pair of enantiomers tells us that the housefly sex attractant is either *cis*-$CH_3(CH_2)_{12}CH{=}CH(CH_2)_7CH_3$ or the *trans* isomer, not a mixture of both. (A mixture would lead to *two* pairs of enantiomers.) It has been determined by other means that this alkene is the *cis* compound.

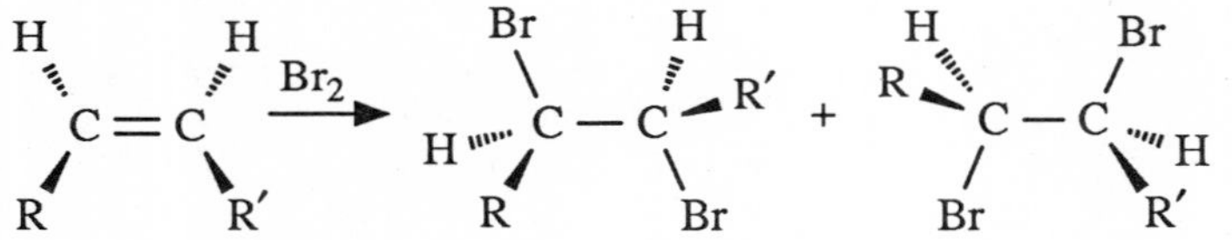

cis $\xrightarrow{Br_2}$ *enantiomers*

trans $\xrightarrow{Br_2}$ *enantiomers*

10.83 First, let us summarize the chemical information given.

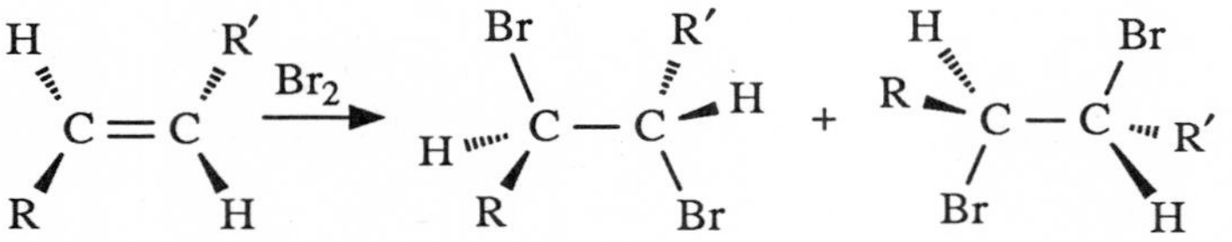

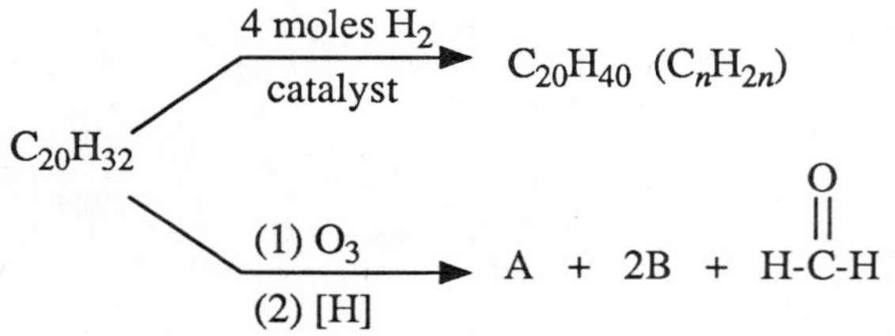

$$C_{20}H_{32} \xrightarrow[\text{catalyst}]{\text{4 moles } H_2} C_{20}H_{40}\ (C_nH_{2n})$$

$$C_{20}H_{32} \xrightarrow[(2)\ [H]]{(1)\ O_3} A + 2B + H\text{-}\overset{\overset{\displaystyle O}{\|}}{C}\text{-}H$$

Next, let us summarize the nmr data.

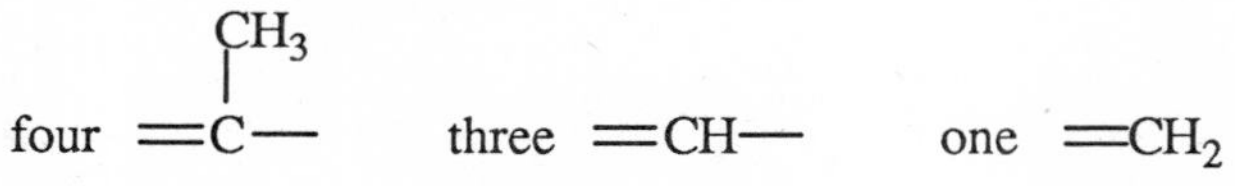

four $=\overset{\overset{\displaystyle CH_3}{|}}{C}-$ three $=CH-$ one $=CH_2$

no tetrasubstituted C=C

Then, we will determine the structure of compound A.

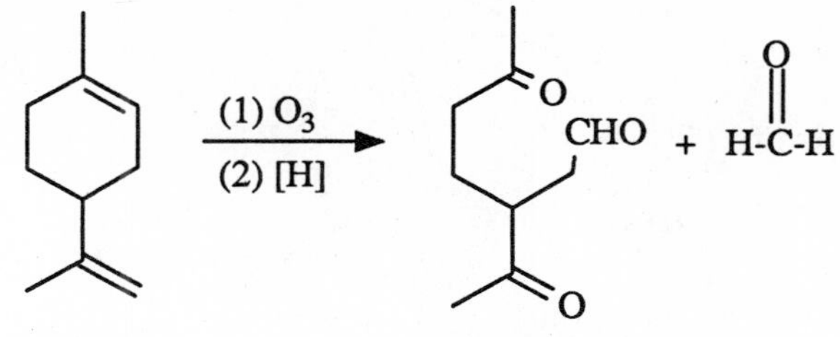

limonene → (1) O_3 (2) [H] → A (CHO) + H-C-H (=O)

limonene A

From the hydrogenation data and the molecular formula, we can deduce that cembrene A contains four carbon-carbon double bonds and one ring:

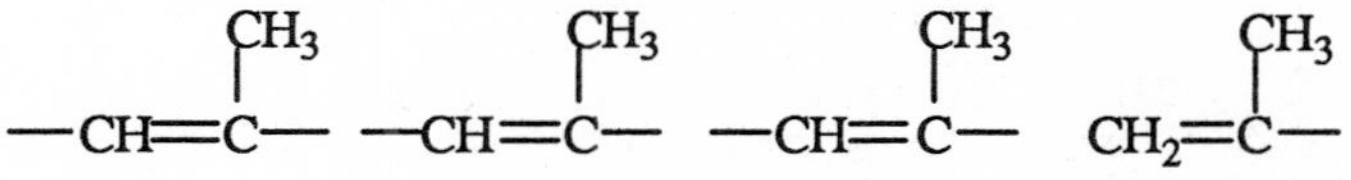

$-CH{=}C(CH_3)-$ $-CH{=}C(CH_3)-$ $-CH{=}C(CH_3)-$ $CH_2{=}C(CH_3)-$ one ring

From the ozonolysis data, we deduce the following fragments:

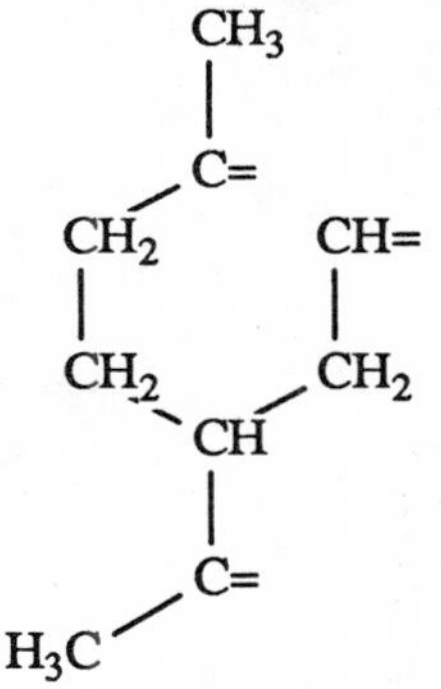

CH_3–C=, CH_2, CH=, CH_2, CH_2, CH_2, CH, C=, H_3C

This fragment leads to A.

$=CH_2$

This fragment leads to formaldehyde.

Ten carbon atoms are now accounted for. Compound B, therefore, must contain five carbon atoms (10 + 5 + 5 = 20) and the remaining double-bond fragments.

$CH_3\overset{\|}{C}CH_2CH_2CH=$ $CH_3\overset{\|}{C}CH_2CH_2CH=$

These fragments lead to 2 B.

Putting the fragments together gives us the structure of cembrene A:

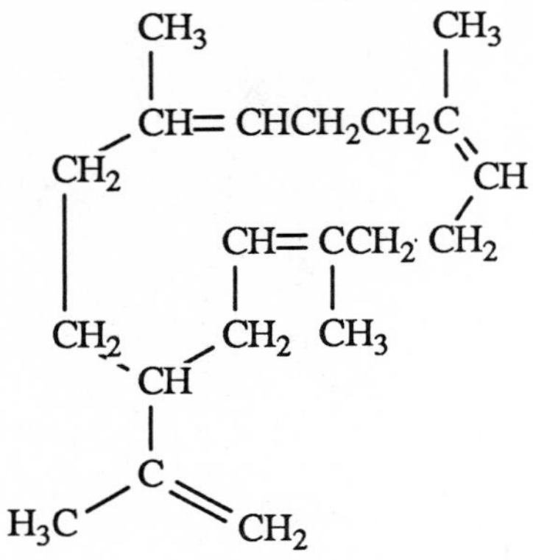

10.84 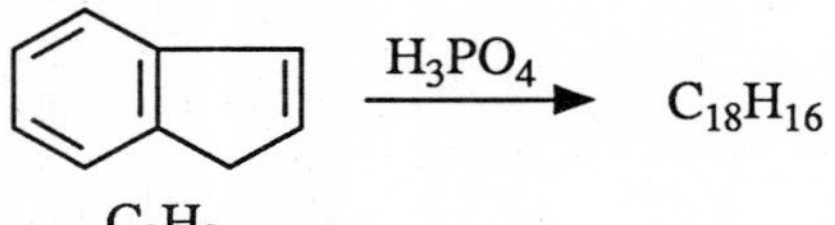 C_9H_8 $\xrightarrow{H_3PO_4}$ $C_{18}H_{16}$

^{1}H nmr absorption for $C_{18}H_{16}$:	assignment
δ 2.22 (multiplet), area 2	?-CH_2-?
δ 2.90 (multiplet), area 2	?-CH_2-?
δ 3.21 (singlet), area 2	-CH_2-
δ 4.75 (triplet), area 1	>C<u>H</u>-CH_2-
δ 6.53 (singlet), area 1	>CH—
δ 7.12 (multiplet), area 8	8 aryl protons

The product, with twice as many C's and H's as the starting material, is a dimer. Let us predict a likely dimerization product and then compare this product with the spectral data.

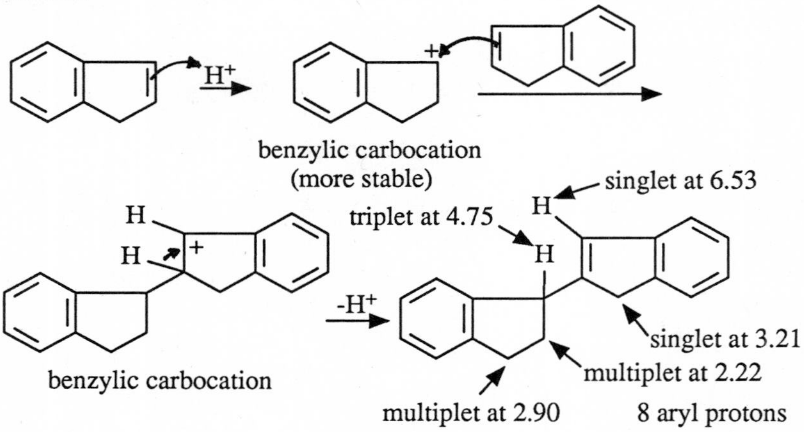

benzylic carbocation (more stable)

benzylic carbocation

singlet at 6.53

triplet at 4.75

singlet at 3.21

multiplet at 2.22

multiplet at 2.90

8 aryl protons

10.85 There can be four possible structures for the unknown:

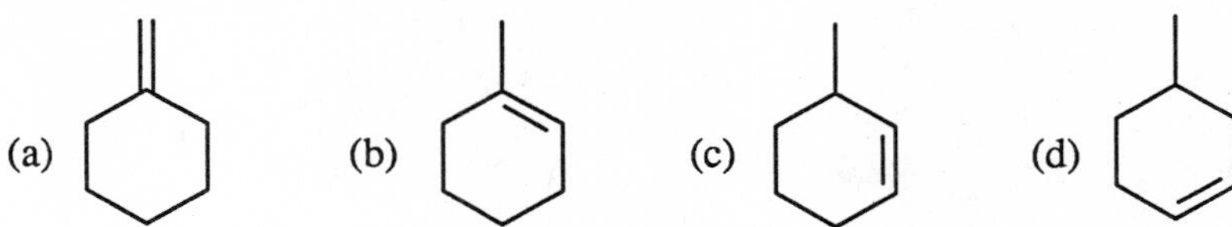

Of these, only (a) has five nonequivalent carbons, because of its symmetry. All the others have a greater number of nonequivalent carbons. Therefore, the unknown is (a), methylenecyclohexane.

10.86 $(CH_3)_3CC{\equiv}CH \xrightarrow[\text{catalyst}]{H_2} (CH_3)_3CCH{=}CH_2$

A → B

Note that the infrared cell was "overloaded." (The CH absorption runs off the bottom of the spectrum.) This fact should warn you that *all* the absorption in the spectrum is much stronger than would normally be observed. The peak at 1650 cm^{-1} (6.1 μm) is <u>not</u> a C=O peak, this peak is actually weak compared to the CH absorption.

10.87 (a) A: $(CH_3)_3CCH{=}C(CH_3)_2$

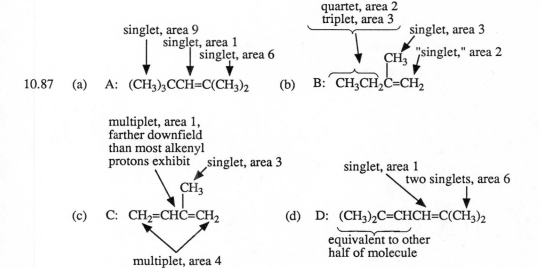

singlet, area 9 ($(CH_3)_3$); singlet, area 1 (CH); singlet, area 6 ($C(CH_3)_2$)

(b) B: $CH_3CH_2C(CH_3){=}CH_2$

quartet, area 2; triplet, area 3 (CH_3CH_2); singlet, area 3 (CH_3); "singlet," area 2 ($=CH_2$)

(c) C: $CH_2{=}CHC(CH_3){=}CH_2$

multiplet, area 1, farther downfield than most alkenyl protons exhibit (CH); singlet, area 3 (CH_3); multiplet, area 4 (CH_2 groups)

(d) D: $(CH_3)_2C{=}CHCH{=}C(CH_3)_2$

singlet, area 1; two singlets, area 6; $(CH_3)_2C{=}CH$ equivalent to other half of molecule

Chapter 11

Aromaticity and Benzene; Electrophilic Aromatic Substitution

11.14 (a) 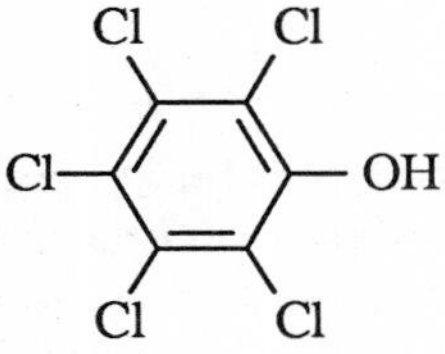 Cl, Cl, Cl, Cl, Cl, OH (pentachlorophenol structure) (b) 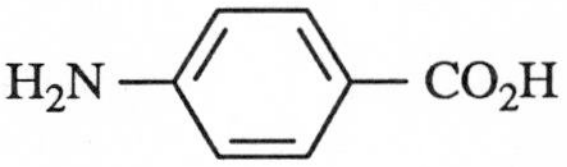 H_2N—C₆H₄—CO_2H

(c) 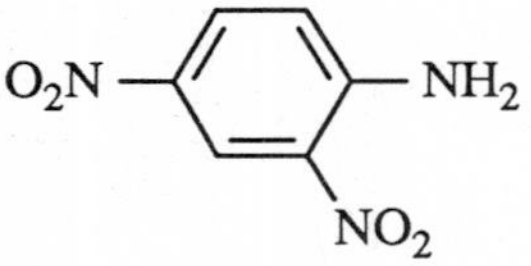 O_2N—, NH_2, NO_2 (d) 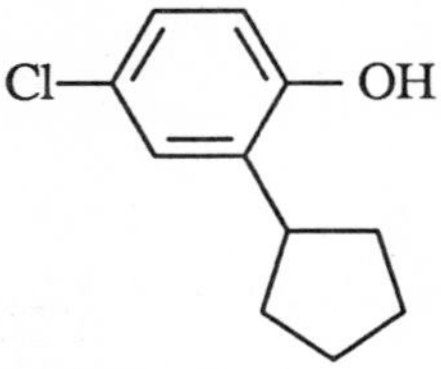 Cl—, OH, cyclopentyl

(e) 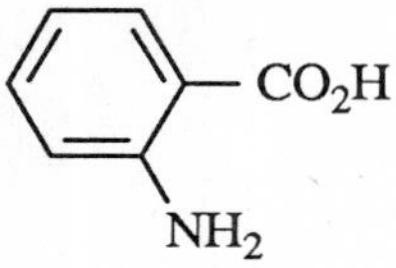 CO_2H, NH_2 (f) 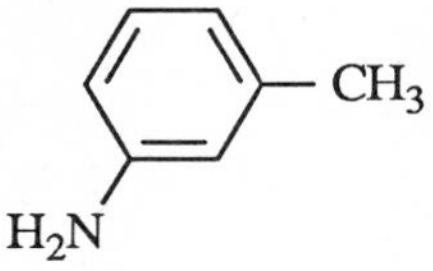 CH_3, H_2N

11.15 (a) phenylethene or styrene (b) *o*-bromostyrene

(c) *m*-methylbenzoic acid (*m*-toluic acid) (d) *o*-nitrotoluene

(e) 3,5-diiodotoluene (f) 1,4-diphenyl-1,3-butadiene

It is also correct to use numbers for disubstituted benzenes -- for example, 2-bromostyrene in (b).

11.16 (a) 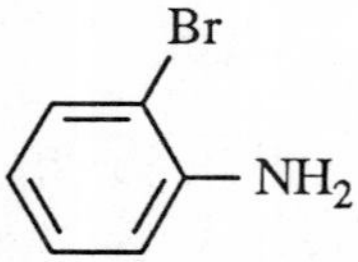 *o*-bromoaniline 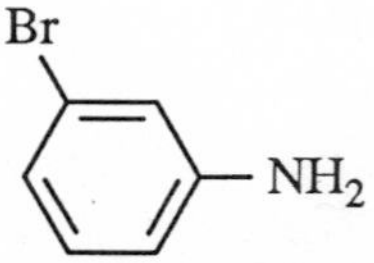 *m*-bromoaniline 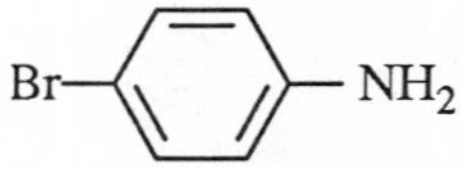 *p*-bromoaniline

(b) 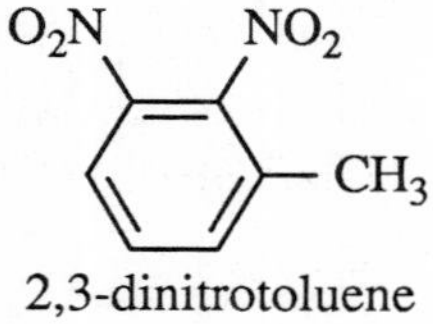 2,3-dinitrotoluene 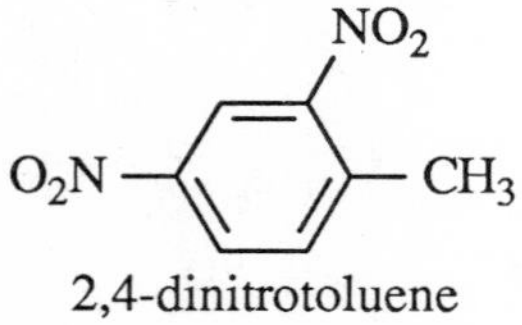 2,4-dinitrotoluene 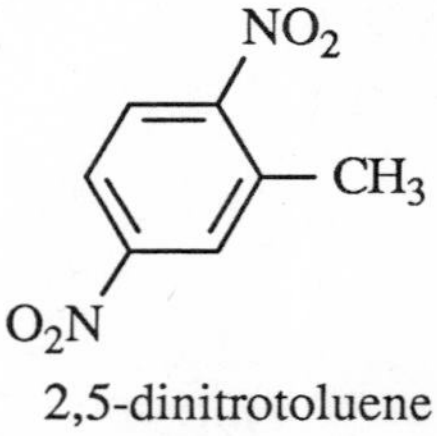 2,5-dinitrotoluene

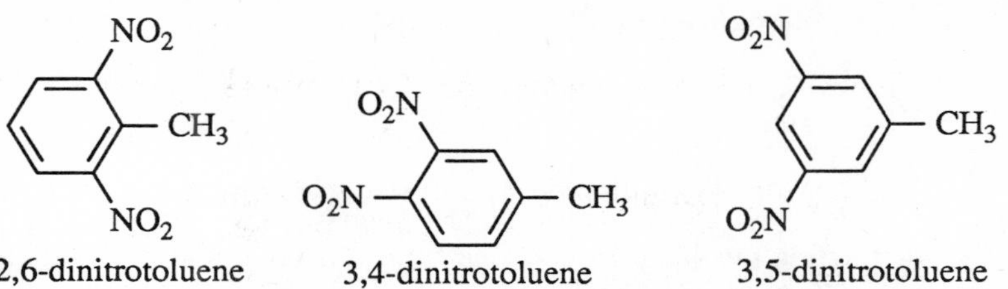

11.17 All the examples shown are aromatic. Compound (a) has 14 pi electrons, compound (b) has ten, and compound (c) has ten.

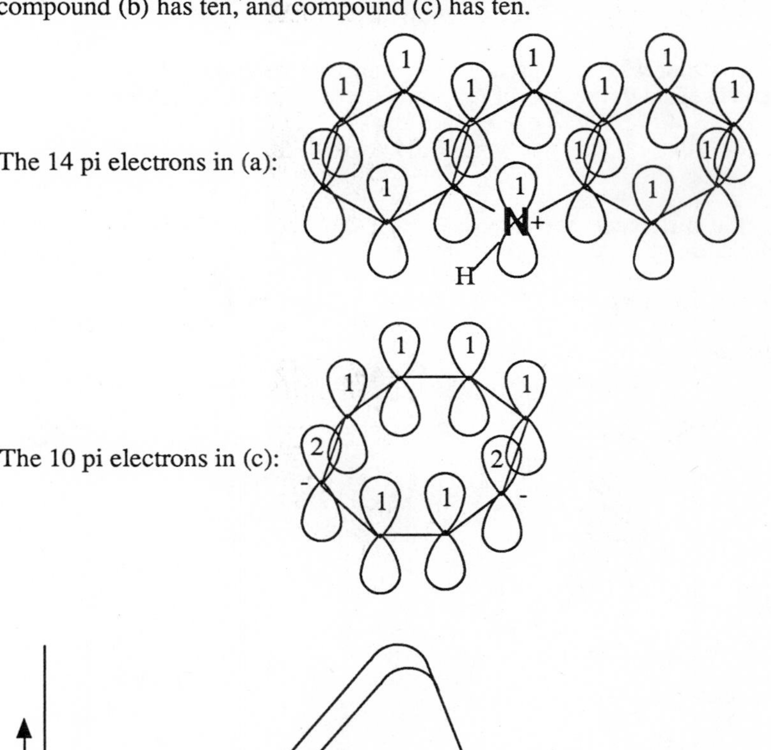

11.18

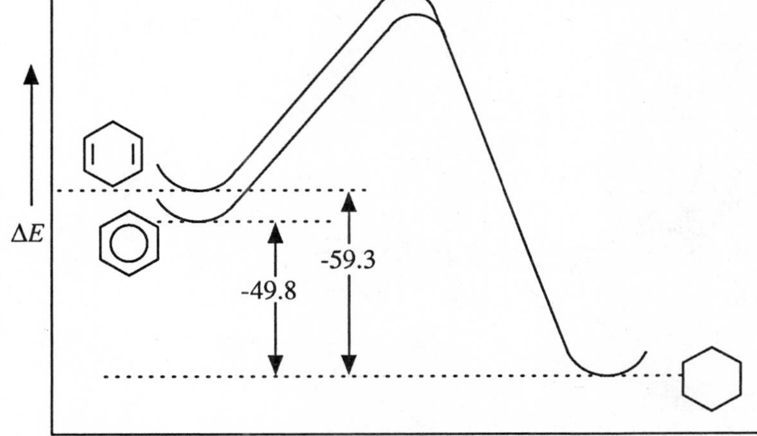

ΔE = -49.8 kcal/mol - (-59.3 kcal/mol) = +9.5 kcal/mol

Therefore, 9.5 kcal/mol must be supplied in the conversion of benzene to

1,4-cyclohexadiene.

11.19 Subtracting the ΔH of hydrogenation of benzene from that of styrene gives the ΔH of hydrogenation of C=C in styrene.

$$\Delta H \text{ for C=C in styrene} = -76.9 \text{ kcal/mol} - (-49.8 \text{ kcal/mol}) = -27.1 \text{ kcal/mol}$$

Subtracting the value for propene from this value shows the added stabilization of C=C in styrene by conjugation.

$$\Delta H \text{ of stabilization} = -27.1 \text{ kcal/mol} - (-28.6 \text{ kcal/mol}) = 1.5 \text{ kcal/mol}$$

Another way to approach this problem is to assume that a calculated ΔH for styrene is equal to the sum of ΔH for benzene and ΔH for propene:

$$\Delta H \text{ (theory) for styrene} = -49.8 \text{ kcal/mol} + (-28.6 \text{ kcal/mol}) = -78.4 \text{ kcal/mol}$$

Then the difference between the actual ΔH and this calculated ΔH is the added stabilization of the double bond in styrene.

$$-76.9 \text{ kcal/mol} - (-78.4 \text{ kcal/mol}) = 1.5 \text{ kcal/mol}$$

11.20 *the products:*

(a) Br (b) NO_2 (c) and (d) CH_2CH_3

(e) $\overset{O}{\|}{C}CH_3$

generation of the electrophiles:

(a) $:\ddot{Br}-\ddot{Br}: + FeBr_3 \rightleftharpoons :\ddot{Br}^+ + FeBr_4^-$

(b) $H\ddot{O}-NO_2 + H^+ \rightleftharpoons H_2\overset{+}{O}-NO_2 \rightleftharpoons H_2\ddot{O}: + {}^+NO_2$

(c) $CH_3CH_2-\ddot{Cl}: + AlCl_3 \rightleftharpoons \overset{\delta+}{CH_3CH_2} \rightarrow \ddot{Cl} \cdot\cdot \rightarrow \overset{\delta-}{AlCl_3}$ (polarized)

(d) $CH_2{=}CH_2 + HCl \longrightarrow CH_3CH_2Cl$

$CH_3CH_2-\ddot{Cl}: + AlCl_3 \rightleftharpoons \overset{\delta+}{CH_3CH_2} \rightarrow \ddot{Cl} \cdot\cdot \rightarrow \overset{\delta-}{AlCl_3}$

(e) $CH_3\overset{O}{\overset{\|}{C}}-\ddot{Cl}: + AlCl_3 \rightleftharpoons \left[CH_3\overset{O}{\overset{\|}{C}}_{\delta+} \cdots \ddot{Cl} \cdots \overset{\delta-}{AlCl_3} \right] \rightleftharpoons$

$[CH_3\overset{+}{C}=\ddot{O} \longleftrightarrow CH_3C\equiv O^+] + AlCl_4^-$

11.21 (a) $C_6H_6 + HNO_3 \xrightarrow{H_2SO_4} C_6H_5NO_2$

(b) $C_6H_6 + Cl_2 \xrightarrow{FeCl_3} C_6H_5Cl$

(c) $C_6H_6 + ClCCH_2CH_2CH_2CH_3$ (with C=O) $\xrightarrow{AlCl_3} C_6H_5CCH_2CH_2CH_2CH_3$ (with C=O)

(d) $C_6H_6 + SO_3 \xrightarrow{H_2SO_4} C_6H_5SO_3H$

(e) $C_6H_6 \xrightarrow[AlCl_3]{ClCCH_2CH_2CH_3 \text{ (C=O)}} C_6H_5CCH_2CH_2CH_3$ (C=O) $\xrightarrow[HCl]{Zn(Hg)}$

$C_6H_5CH_2CH_2CH_2CH_3$

(f) $C_6H_6 + Br_2 \xrightarrow{FeBr_3} C_6H_5Br$

(g) C_6H_6 + $(CH_3)_2CHCl$ $\xrightarrow{AlCl_3}$ $C_6H_5CH(CH_3)_2$

C_6H_6 + $CH_3CH{=}CH_2$ $\xrightarrow{H^+}$ $C_6H_5CH(CH_3)_2$

11.22 $H\ddot{O}{-}NO_2 + H^+ \rightleftharpoons H_2\overset{+}{O}{-}NO_2 \rightleftharpoons H_2\ddot{O}: + {}^+NO_2$

$C_6H_6 + {}^+NO_2 \longrightarrow [C_6H_6^+(H)(NO_2)] \xrightarrow{-H^+} C_6H_5{-}NO_2$

11.23 $CH_3CH_2CH_2CH_2Cl + AlCl_3 \rightleftharpoons \overset{\delta+}{CH_3CH_2CH_2CH_2}\cdots\overset{\delta-}{ClAlCl_3}$

$C_6H_6 + \overset{\delta+}{CH_3CH_2CH_2CH_2}\cdots\overset{\delta-}{ClAlCl_3} \xrightarrow{-AlCl_4^-}$

$[C_6H_6^+(H)(CH_2CH_2CH_2CH_3)] \xrightarrow{-H^+} C_6H_5{-}CH_2CH_2CH_2CH_3$

$CH_3CH_2CH(H)CH_2\cdots ClAlCl_3 \xrightarrow{-AlCl_4^-} CH_3CH_2\overset{+}{C}HCH_3$

$C_6H_6 + CH_3CH_2\overset{+}{C}HCH_3 \longrightarrow [C_6H_6^+(H)(CH(CH_3)CH_2CH_3)]$

$\xrightarrow{-H^+} C_6H_5{-}CH(CH_3)CH_2CH_3$

11.24 The starting acid chloride has an (*R*) configuration. Because the reaction does not affect the configuration of the chiral carbon, the product has the same stereochemistry as the reactant.

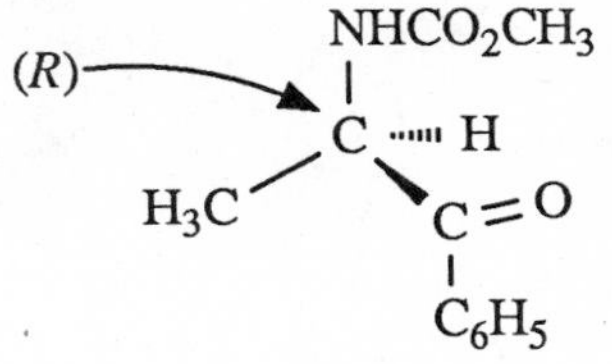

(*R*) → $NHCO_2CH_3$ / H_3C–C····H / C=O / C_6H_5

11.25 (a) $C_6H_5\overset{O}{\overset{\|}{C}}CH_2CH_2CO_2H$ (b) [benzene ring with ortho $\overset{O}{\overset{\|}{C}}C_6H_5$ and CO_2H]

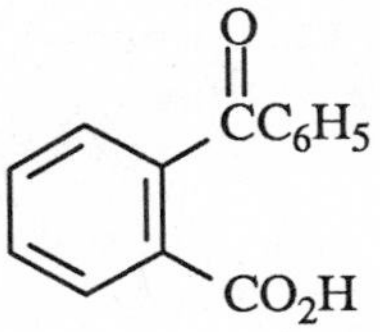

11.26 (a) acetanilide, because the N has unshared electrons that help stabilize the intermediate by resonance.

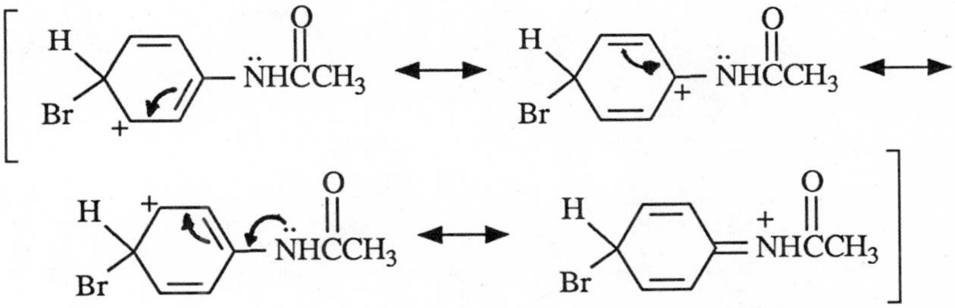

[resonance structures of the bromo-substituted arenium ion with $-\ddot{N}H\overset{O}{\overset{\|}{C}}CH_3$ ↔ ... ↔ $=\overset{+}{N}H\overset{O}{\overset{\|}{C}}CH_3$]

N helps delocalize the positive charge

(b) toluene, because the CH_3 group is electron releasing by the inductive effect.

(c) *p*-xylene, because it has *two* electron-releasing groups on the ring.

(d) *m*-nitrotoluene, because it has an electron-releasing CH_3 group.

(e) chlorobenzene, because it has only one electron-withdrawing group.

11.27 (a) *o,p* (b) *m* (c) *o,p* (d) *m* (e) *m*

(f) *o,p* (g) *o,p* (h) *m* (i) *o,p*

11.28 (a)

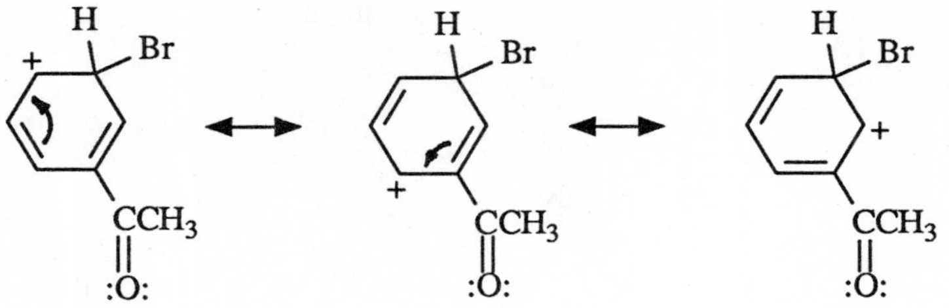

$\leftrightarrow$ $\leftrightarrow$

H, Br, + ; CCH$_3$, :O:

(b) HÖ— ... NO$_2$, H $\leftrightarrow$ HÖ— ... NO$_2$, H $\leftrightarrow$ HÖ— ... NO$_2$, H

$\leftrightarrow$ $H\overset{+}{O}$= ... NO$_2$, H

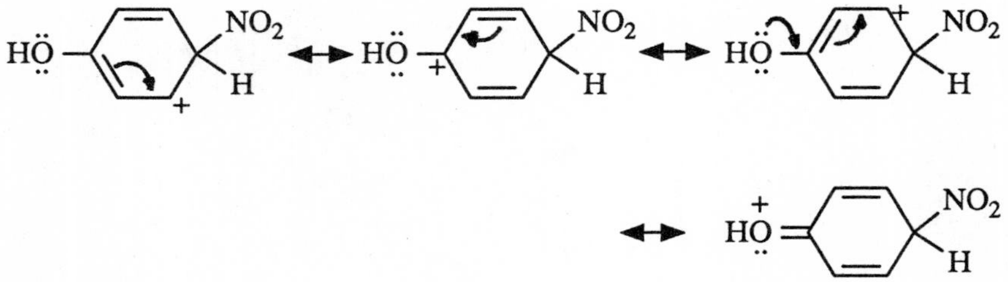

(c) Br, H, CH$_3$ $\leftrightarrow$ Br, H, CH$_3$ $\leftrightarrow$ Br, H, CH$_3$

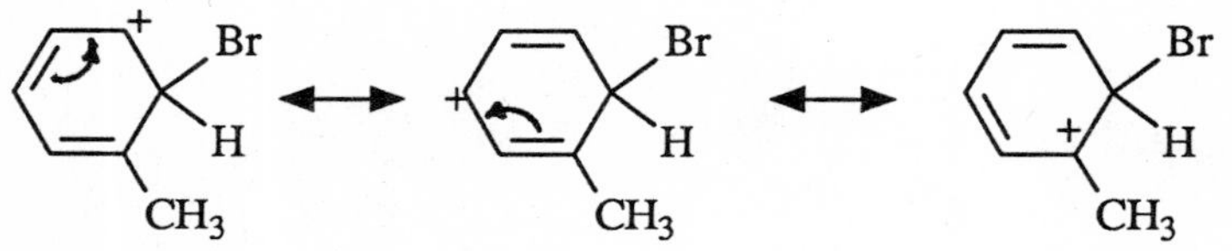

(d) NO$_2$, H, NO$_2$ $\leftrightarrow$ NO$_2$, H, NO$_2$ $\leftrightarrow$ NO$_2$, H, NO$_2$

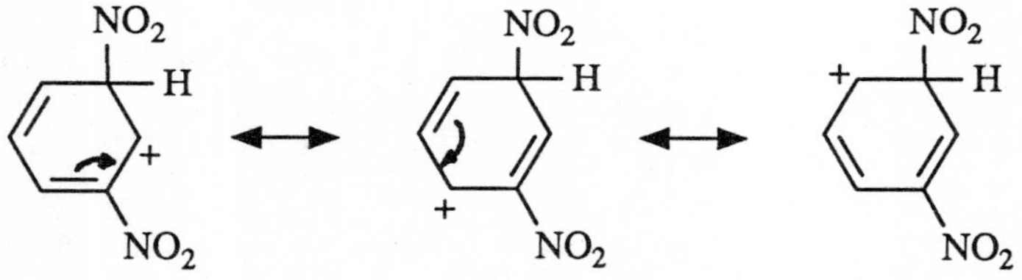

11.29 (a) HÖ–NO + H–ÖNO $\xrightarrow{-NO_2^-}$ $H_2\overset{+}{O}$–NO $\xrightarrow{-H_2O}$ $^+$NO

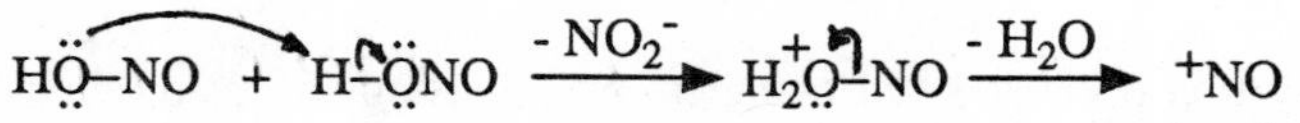

the electrophile

(b) $^+$NO + (benzene ring)—OH $\longrightarrow$ [H, ON, +, OH] $\xrightarrow{-H^+}$

ON—(benzene ring)—OH

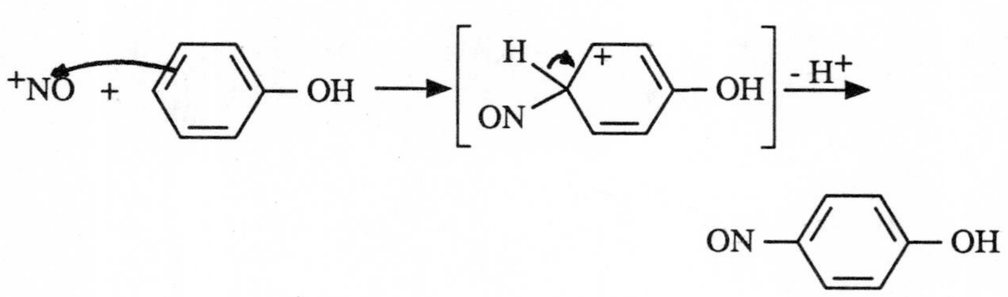

or

$C_6H_5OH + {}^+NO \longrightarrow [\text{arenium ion: H, NO, OH}] \xrightarrow{-H^+}$ o-nitrosophenol (NO, OH)

11.30 (a) 1-tetralone (structure)

(b) $(CH_3)_3C-C_6H_4-CH_3$ (para) + $C(CH_3)_3$ / CH_3 (ortho isomer)

(c) $C_6H_5CH_2CH{=}CHCH_3$ + $C_6H_5CH(CH_3)CH{=}CH_2$

mostly (*E*)

11.31 (a) The presence of *two* deactivating NO_2 groups deactivates the benzene ring sufficiently that a third substitution does not occur.

$C_6H_6 \xrightarrow{\text{moderate rate}} C_6H_5NO_2 \xrightarrow{\text{slow}}$

$1,3-(O_2N)_2C_6H_4 \xrightarrow{\text{negligible rate}} 1,3,5-(O_2N)_3C_6H_3$

(b) The activating CH_3 group counteracts the deactivation by the two NO_2 groups so that a third substitution can occur.

$C_6H_5CH_3 \xrightarrow{\text{fast}} O_2N-C_6H_4-CH_3 \xrightarrow{\text{moderate rate}}$

$O_2N-C_6H_3(NO_2)-CH_3 \xrightarrow{\text{slow}} O_2N-C_6H_2(NO_2)_2-CH_3$

11.32 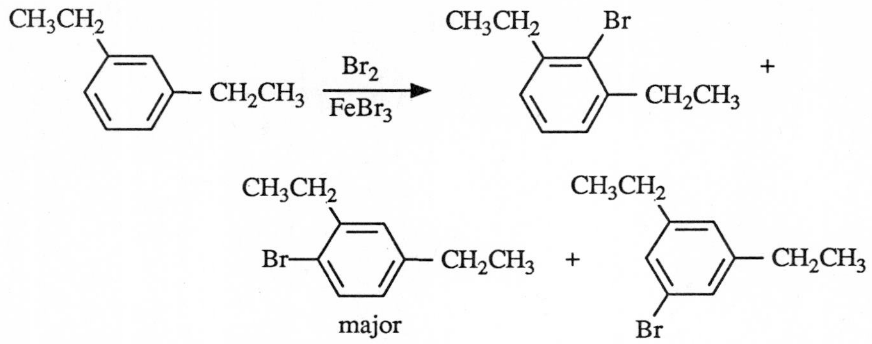

CH_3CH_2
CH_2CH_3
Br_2
$FeBr_3$
CH_3CH_2
Br
CH_2CH_3
+
CH_3CH_2
Br
CH_2CH_3
+
CH_3CH_2
CH_2CH_3
major
Br

By contrast, *o*-diethylbenzene yields two monobromo products, and *p*-diethylbenzene yields only one.

11.33 (a) 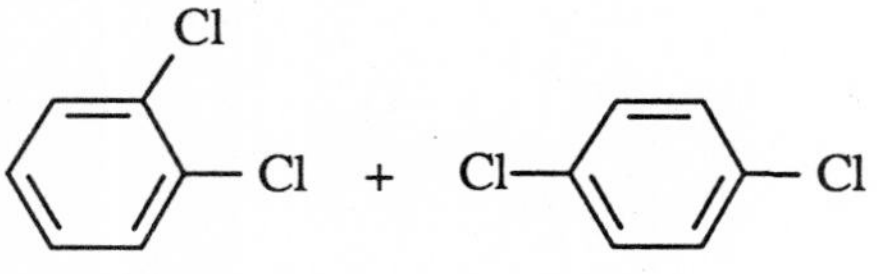

Cl
Cl
+
Cl
Cl

(b) 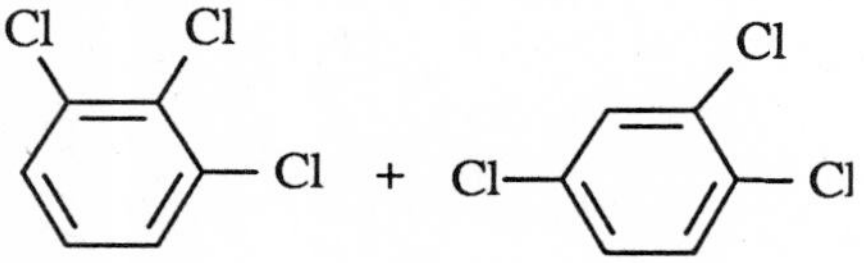

Cl
Cl
Cl
+
Cl
Cl
Cl

(c) 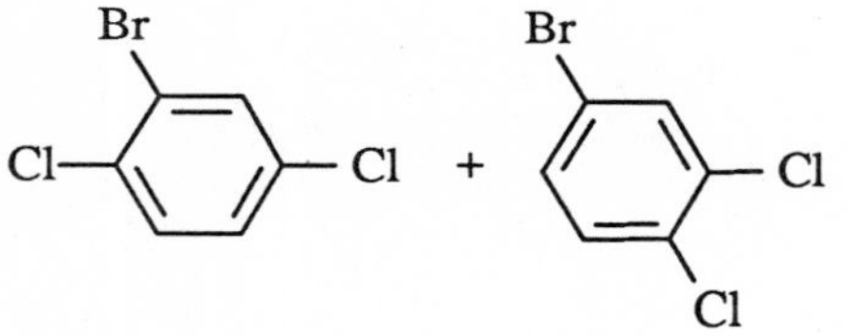

Br
Cl
Cl
+
Br
Cl
Cl

(d) 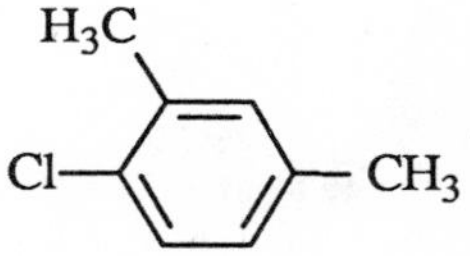

H_3C
Cl
CH_3

(e) 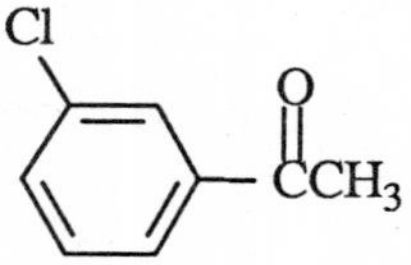

Cl
O
CCH_3

(f) 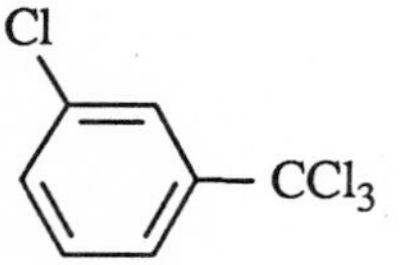

Cl
CCl_3

In (f), the $-CCl_3$ group is *m* directing because the three Cl atoms withdraw electron density by the inductive effect.

11.34 (a) 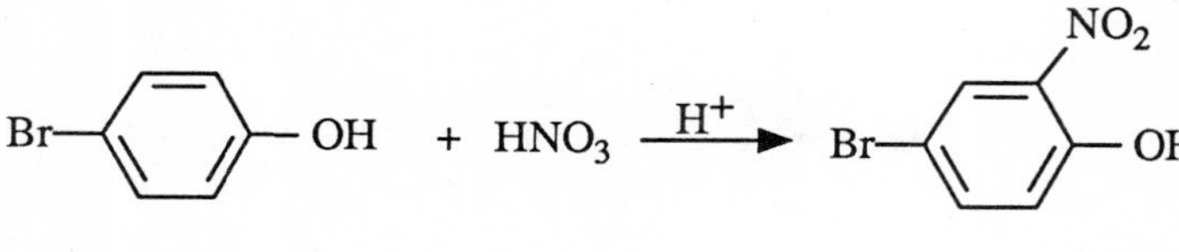

Br
OH
+ HNO_3
H^+
NO_2
Br
OH

(b) $O_2N-C_6H_4-CH_3 + HNO_3 \xrightarrow{H^+} O_2N-C_6H_3(NO_2)-CH_3$

(c) $CH_3O-C_6H_4-CO_2H + HNO_3 \xrightarrow{H^+}$

$CH_3O-C_6H_3(NO_2)-CO_2H + O_2N-C_6H_3(OCH_3)-CO_2H$

(d) $Br-C_6H_4-OCH_3 + HNO_3 \xrightarrow{H^+}$

$Br-C_6H_3(NO_2)-OCH_3 + O_2N-C_6H_3(Br)-OCH_3$

11.35 Fe is oxidized by halogens to FeX_3. The reagent ICl must be ionized by the catalyst into positive and negative ions. Because Cl is more electronegative than I, Cl is the negative end of the reagent and I is the positive end. Therefore, I^+ is the attacking electrophile, and the product is C_6H_5I.

$$C_6H_6 \xrightarrow[- ClFeX_3^-]{\overset{\delta+}{I}\cdots\overset{\delta-}{Cl}\cdots FeX_3} \left[C_6H_6I^+\right] \xrightarrow{-H^+} C_6H_5-I$$

11.36 Bromobenzene $\xrightarrow[FeCl_3]{Cl_2}$ A (1-bromo-2-chlorobenzene) + B (1-bromo-4-chlorobenzene)

A $\xrightarrow[FeBr_3]{Br_2}$ (1,3-dibromo-2-chlorobenzene) + (2,4-dibromo-1-chlorobenzene) + C (2,5-dibromo-1-chlorobenzene) + (1-chloro-2,3-dibromobenzene... Br, Br)

B $\xrightarrow[FeBr_3]{Br_2}$ C + D

11.37 (a) $FCH(=O) + BF_3 \rightleftharpoons HC^{\delta+}(=O)\cdots F \cdots BF_3^{\delta-}$

$C_6H_6 + HC^{\delta+}(=O)\cdots F \cdots BF_3^{\delta-} \xrightarrow{-BF_4^-}$ [arenium ion with H and CHO] $\xrightarrow{-H^+}$ C_6H_5CHO

(b)

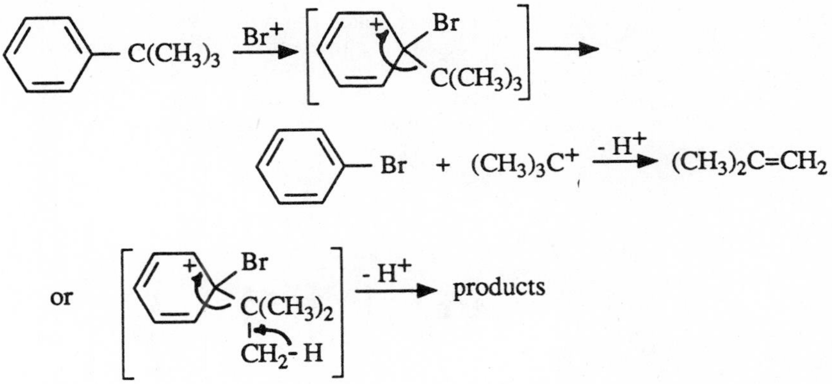
C(CH3)3
Br+
Br
C(CH3)3
Br
+ (CH3)3C+
- H+
(CH3)2C=CH2
or
Br
C(CH3)2
CH2- H
- H+
products

(c)

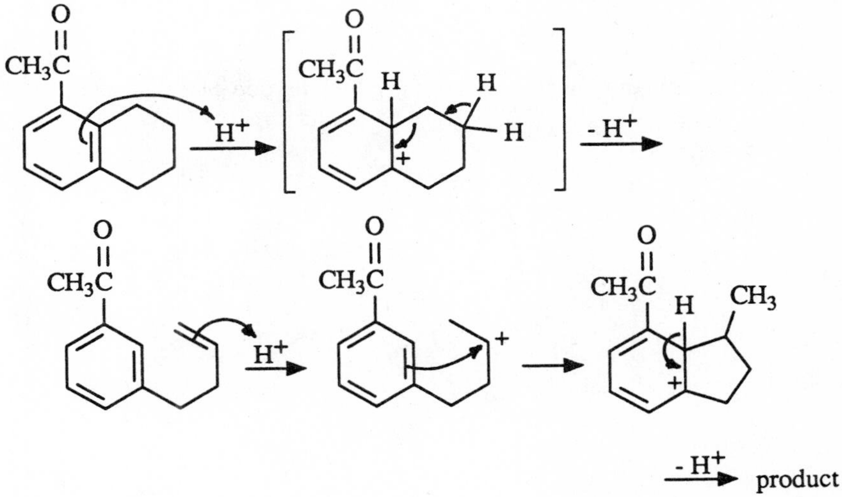
CH3C
H+
- H+
product

11.38 (a)

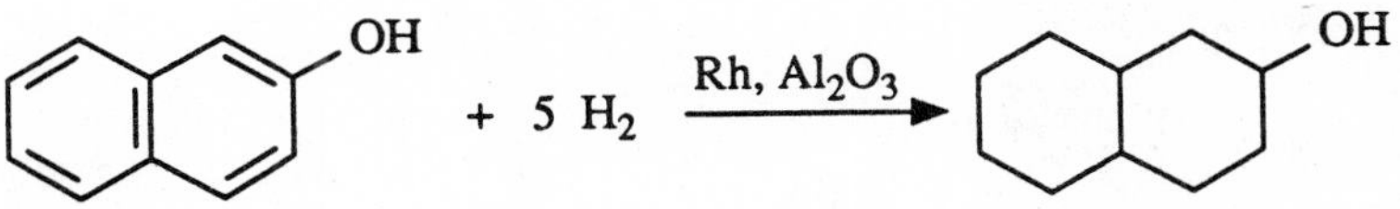
OH
+ 5 H2
Rh, Al2O3
OH

(b)

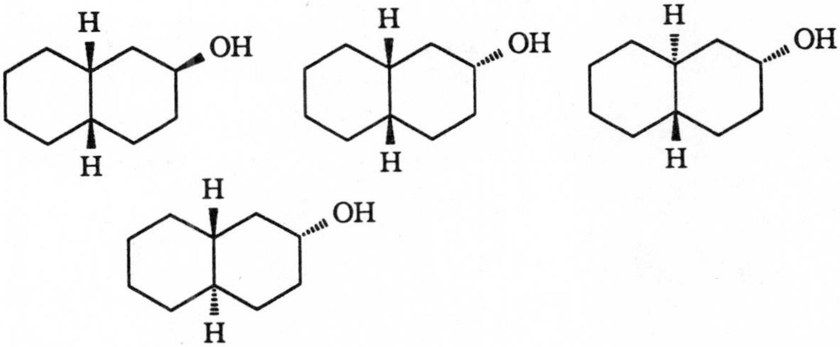
OH

11.39 The cation is aromatic. It is cyclic; each atom in the ring has a *p* orbital (each is sp^2 hybridized); and the ring has $4n + 2$ pi electrons ($n = 0$, 2 electrons).

11.40 The ring system is aromatic. The circulating pi electrons create a field that deshields the protons on the outside of the ring but shields the protons on the bridging $-CH_2-$ group.

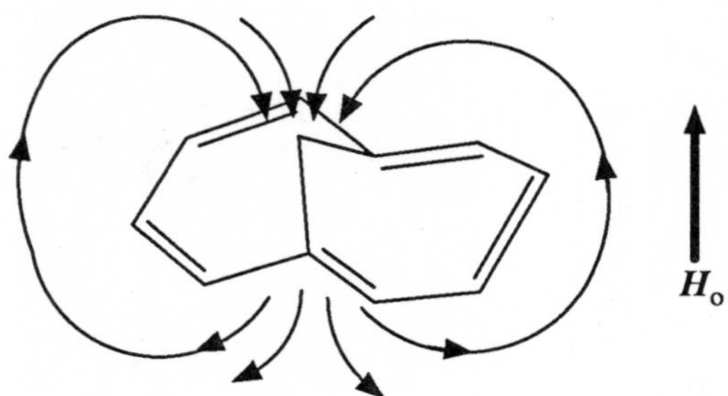

11.41 The circulating pi electrons of the ethynyl group cause the additional deshielding.

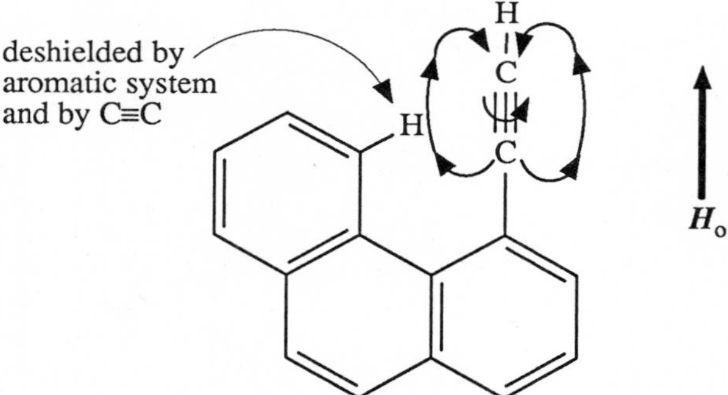

11.42 The areas of the peaks in the nmr spectrum account for only half the protons; therefore we double the areas.

10 aryl protons

4 protons alpha to C=O
or to an electronegative atom

Because the infrared spectrum shows C=O absorption, we can deduce the following fragments:

$-CH_2C(=O)CH_2-$ $\qquad$ $-C_{10}H_{10}-$ (aryl groups)

Because of the simplicity of the spectrum, denoting a symmetrical molecule, we determine the structure to be:

$C_6H_5-CH_2C(=O)CH_2-C_6H_5$

Chapter 12

Substituted Benzenes

12.10 (a) 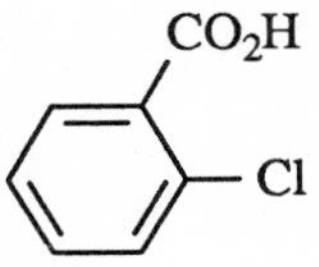 CO_2H, Cl (b) 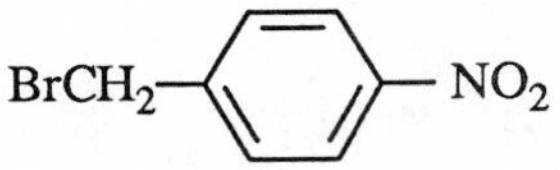 $BrCH_2$—, —NO_2

(c) 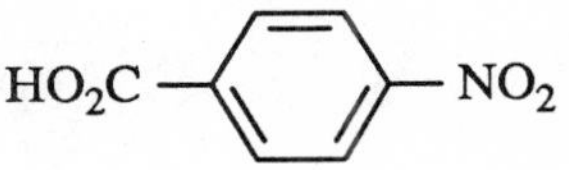 HO_2C—, —NO_2 (d) 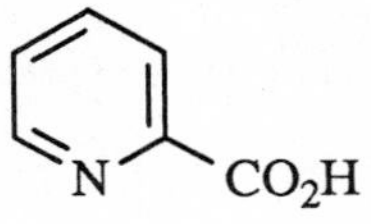 N, CO_2H

12.11 NO_2, O_2N—, —CH_3, NO_2 $\xrightarrow[H_2SO_4]{Na_2Cr_2O_7}$ NO_2, O_2N—, —CO_2H, NO_2

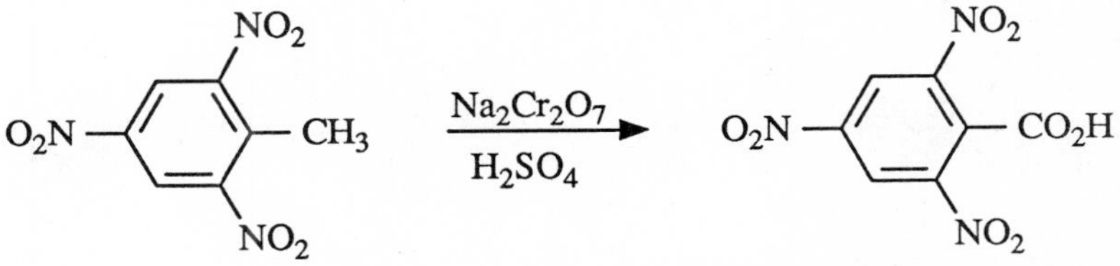

2,4,6-trinitrobenzoic acid

12.12 (a) 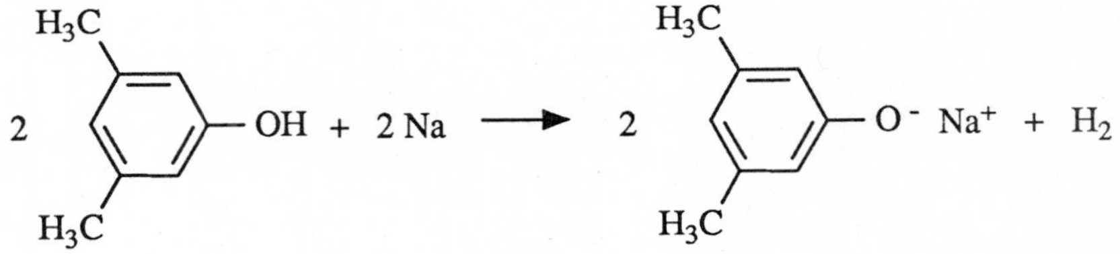 2 H_3C, H_3C, —OH + 2 Na ⟶ 2 H_3C, H_3C, —O^- Na^+ + H_2

(b) 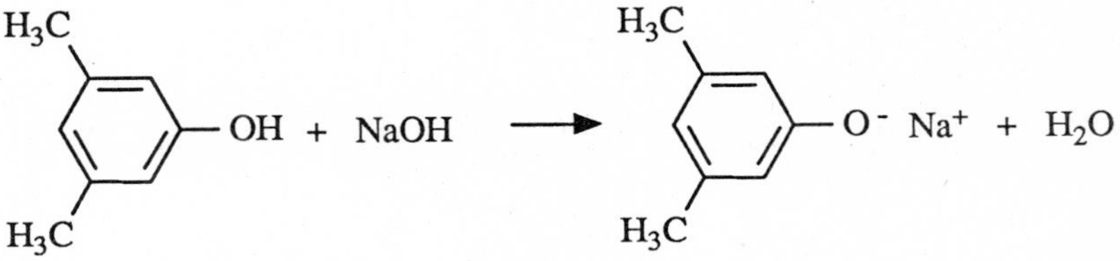 H_3C, H_3C, —OH + NaOH ⟶ H_3C, H_3C, —O^- Na^+ + H_2O

(c) 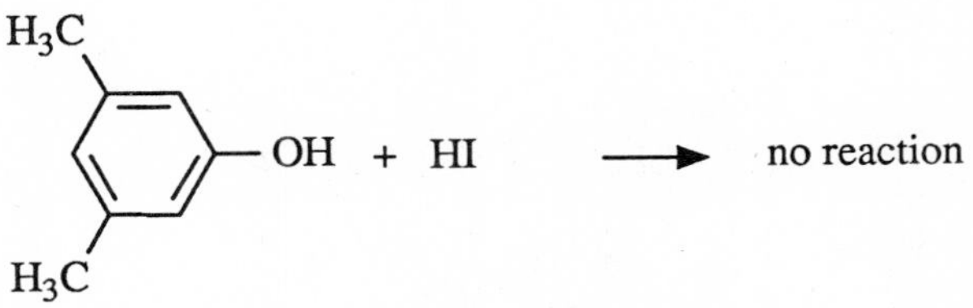 H_3C, H_3C, —OH + HI ⟶ no reaction

(d) 3,5-dimethylphenol: H_3C, H_3C, OH + H_2SO_4 $\xrightarrow{\text{heat}}$ HO_3S (ring with two H_3C) OH

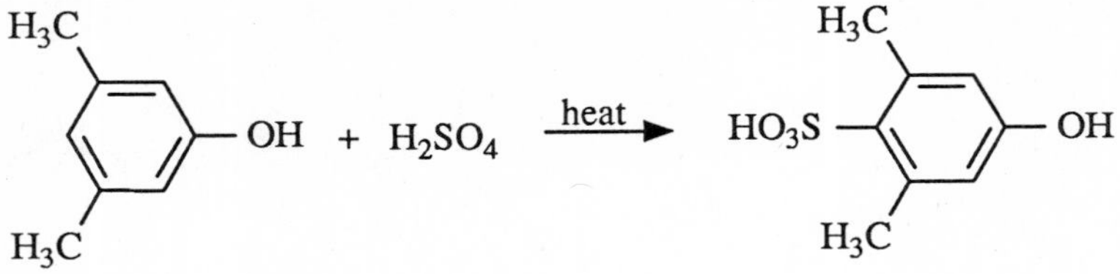

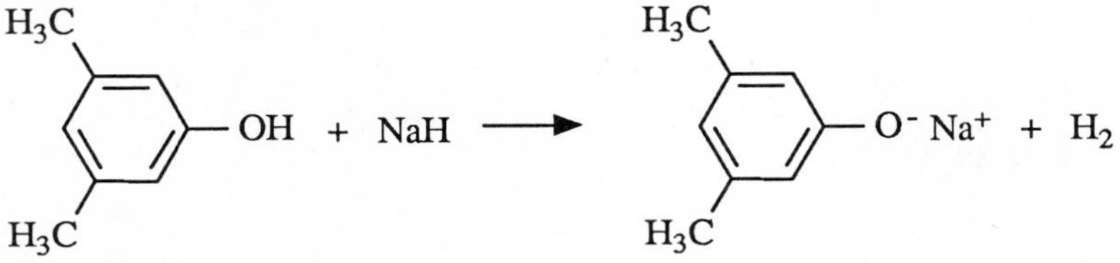

(e) 3,5-dimethylphenol: H_3C, H_3C, OH + NaH ⟶ (H_3C, H_3C ring) $O^-\ Na^+$ + H_2

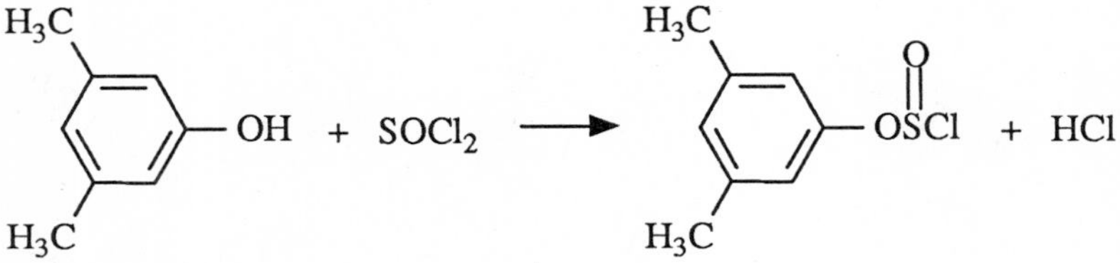

(f) 3,5-dimethylphenol: H_3C, H_3C, OH + $SOCl_2$ ⟶ (H_3C, H_3C ring) O–S(=O)–Cl + HCl

(g) 3,5-dimethylphenol: H_3C, H_3C, OH + NaCN ⟶ no reaction

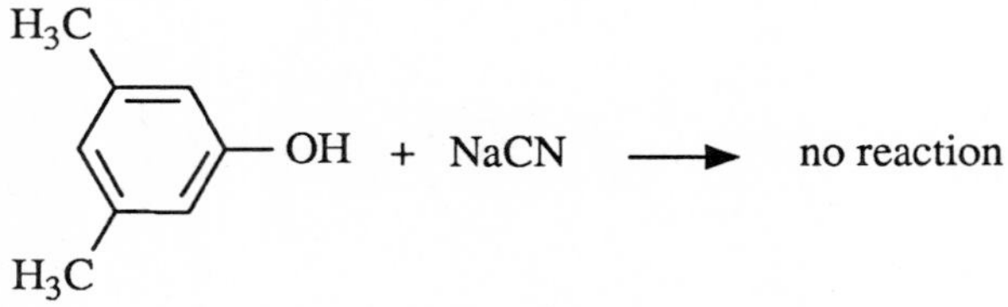

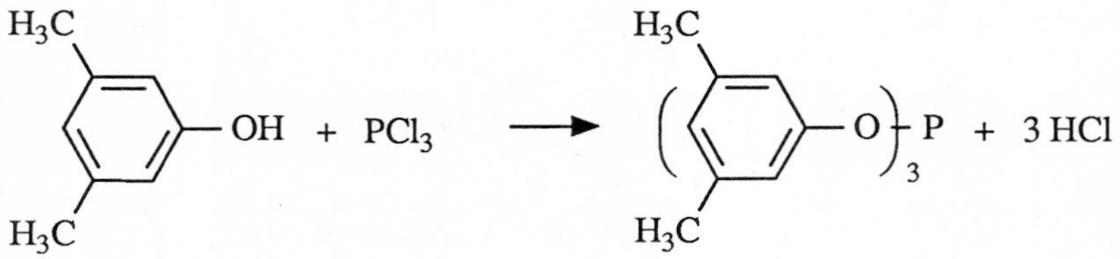

(h) 3,5-dimethylphenol: H_3C, H_3C, OH + PCl_3 ⟶ ((H_3C, H_3C ring)–O)$_3$P + 3 HCl

(i) 3,5-dimethylphenol: H_3C, H_3C, OH + $(CH_3O)_2SO_2$ + NaOH ⟶ (H_3C, H_3C ring) OCH_3

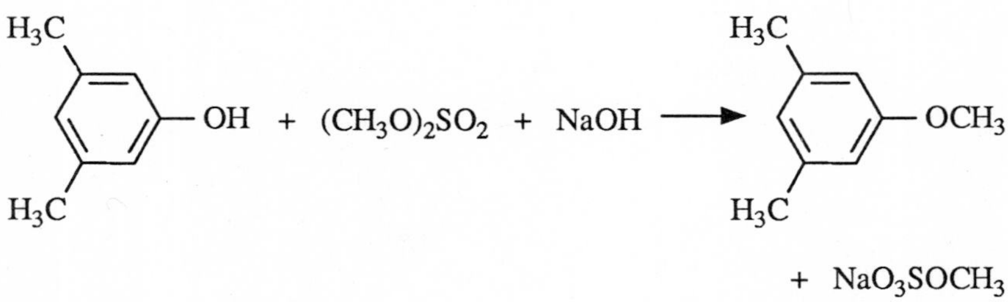

+ NaO_3SOCH_3

12.13 (a) 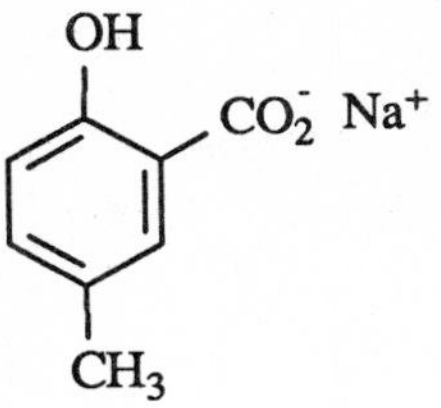 OH, CO_2^- Na^+, CH_3 or O^- 2 Na^+, CO_2^-, CH_3

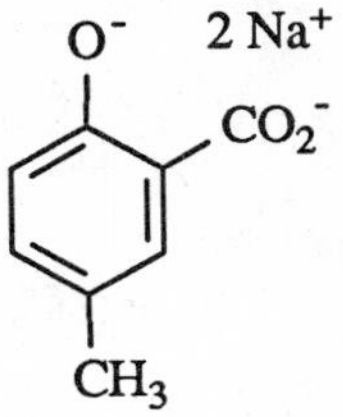

(b) 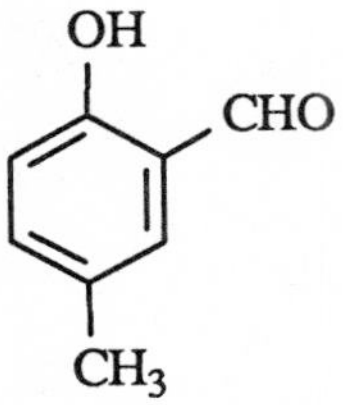 OH, CHO, CH_3 or O^- Na^+, CHO, CH_3

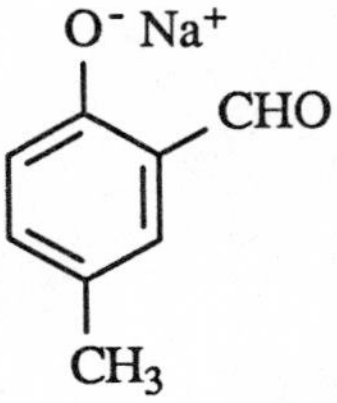

(c) 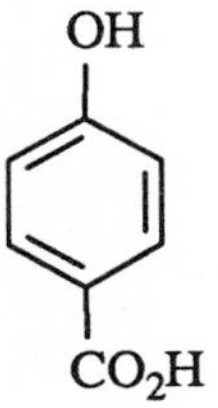 OH, CO_2H

(d) 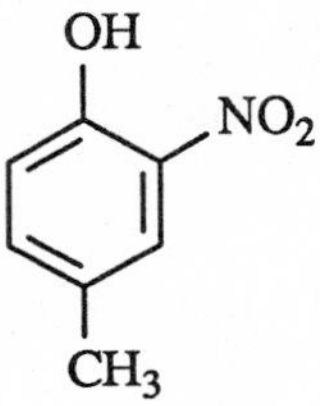 OH, NO_2, CH_3

(e) 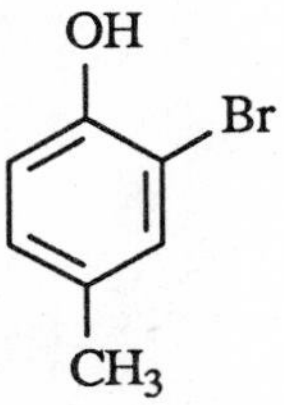 OH, Br, CH_3

12.14 (a) 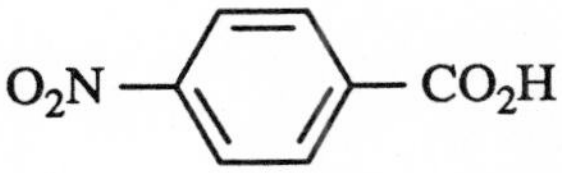 O_2N—⟨benzene⟩—CO_2H (b) 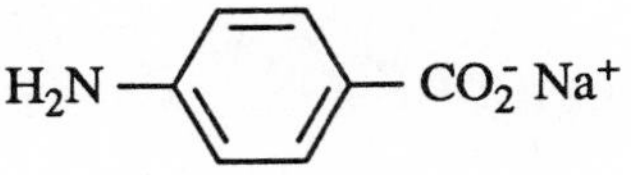 H_2N—⟨benzene⟩—CO_2^- Na^+

(c) 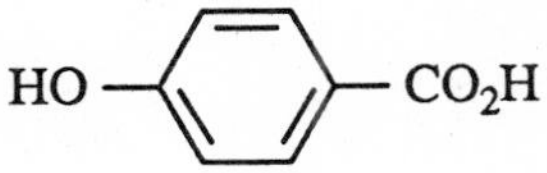 HO—⟨benzene⟩—CO_2H

(d)  CH_3O—⟨benzene⟩—CO_2H + CH_3O—⟨benzene⟩—CO_2CH_3

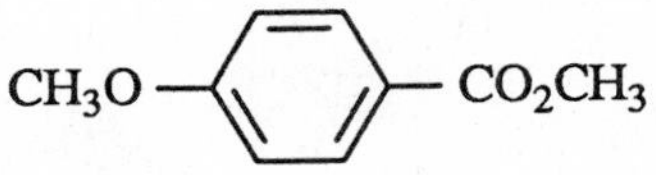

anisic acid

12.15 HO_2C, OH, OH

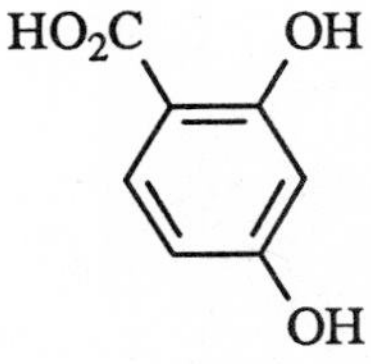

Attack does not occur at the position between the hydroxyl groups because of steric hindrance.

12.16 $C_6H_6 \xrightarrow[H_2SO_4]{HNO_3} C_6H_5NO_2 \xrightarrow[(2)\ OH^-]{(1)\ Fe,\ HCl} C_6H_5NH_2 \xrightarrow[0°]{NaNO_2,\ HCl}$

$C_6H_5N_2^+\ Cl^- \xrightarrow[heat]{H_2O,\ H^+} C_6H_5OH \xrightarrow[heat]{CHCl_3,\ OH^-}$ (benzene ring with CHO and O^- ortho)

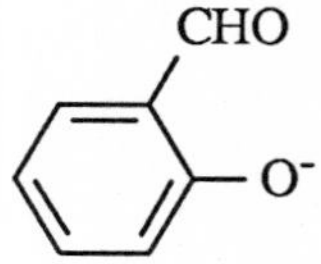

$\xrightarrow[(2)\ H^+]{(1)\ ClCH_2CO_2^-\ Na^+}$ product

12.17 (a) no reaction (b) 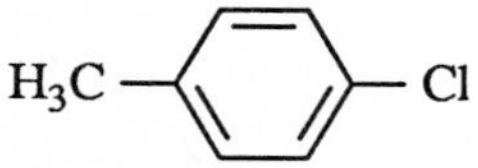 H_3C–(benzene)–Cl

(c) 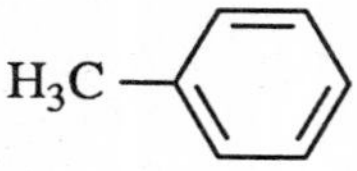 H_3C–(benzene) (d) 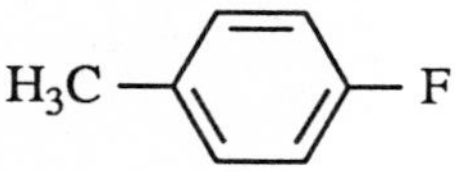 H_3C–(benzene)–F

12.18 (a)  CH_3CH_2–(benzene)–$N_2^+\ Cl^-$ (b) 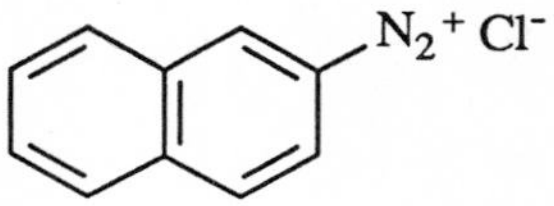 (2-naphthyl)–$N_2^+\ Cl^-$

(c) 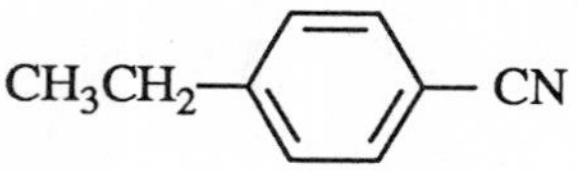 CH_3CH_2–(benzene)–CN

(d) 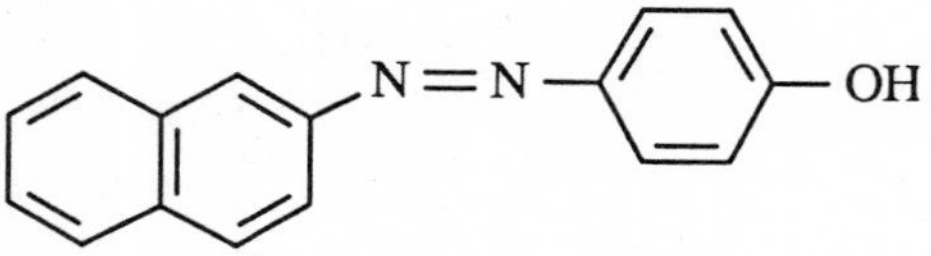 (2-naphthyl)–N=N–(benzene)–OH

12.19 (a) 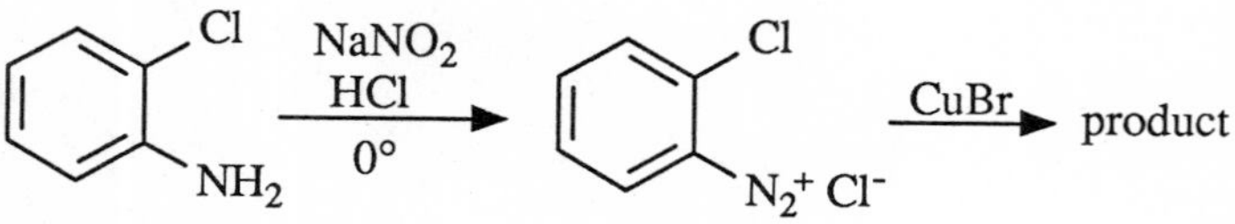 (2-chloroaniline: benzene with Cl and NH_2) $\xrightarrow[0°]{NaNO_2,\ HCl}$ (benzene with Cl and $N_2^+\ Cl^-$) $\xrightarrow{CuBr}$ product

(b) 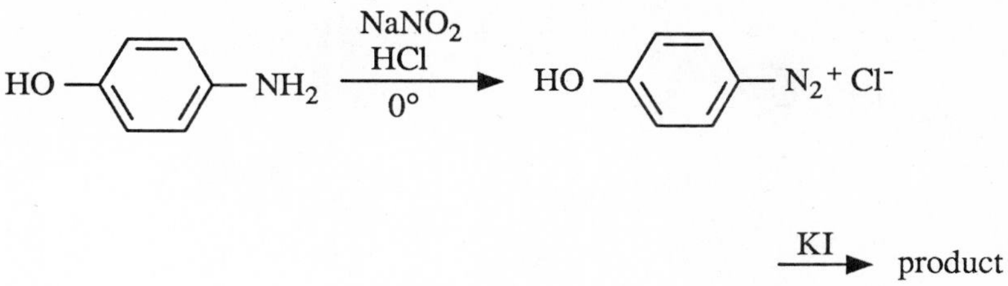

(c) 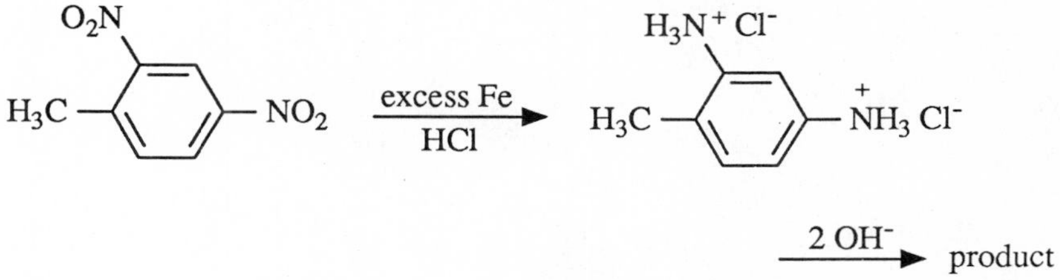

(d) 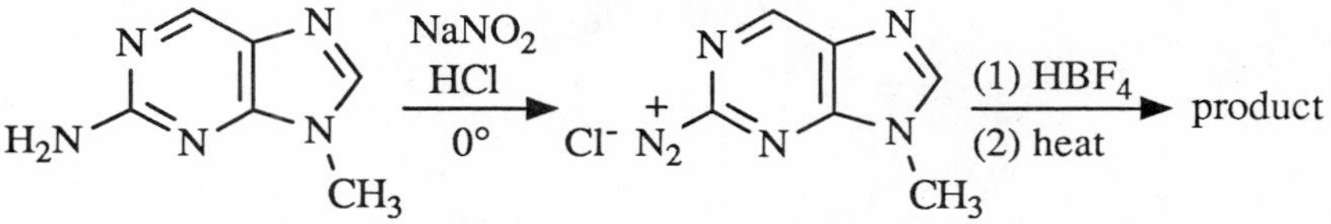

In (d), some of the ring nitrogen atoms may become protonated by the acid.

(e) 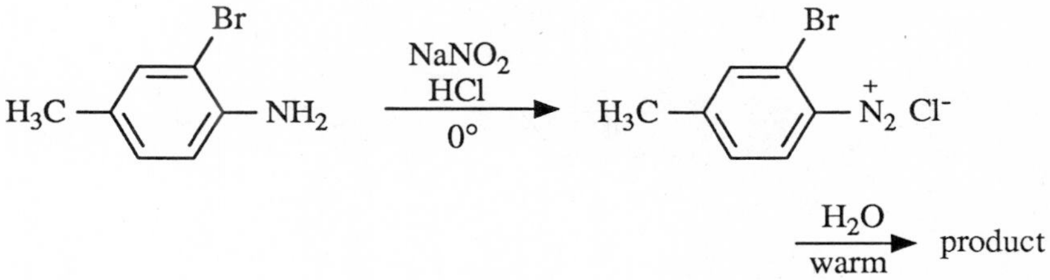

12.20 (a) 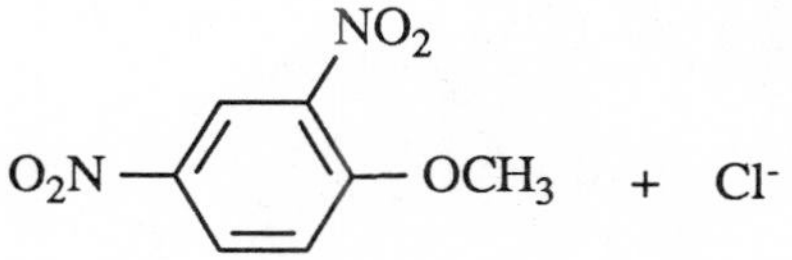

(b) 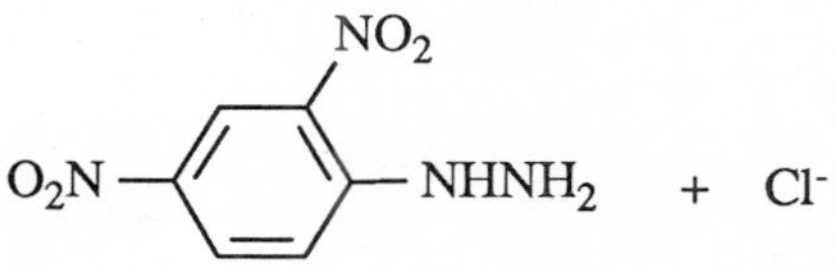

(c) $O_2N-C_6H_3(NO_2)-I$ + Cl^-

(d) $O_2N-C_6H_4-OC_6H_5$ + Cl^-

12.21 $O_2N-C_6H_3(NO_2)-F$ + $H_2N\ddot{}CH_2C(=O)NHCH_2CO_2H$ $\xrightarrow{-H^+}$

$[O_2N-C_6H_3(NO_2)(F)(NHR)^-]$ $\xrightarrow{-F^-}$ $O_2N-C_6H_3(NO_2)-NHCH_2C(=O)NHCH_2CO_2H$

12.22 (a) C_6H_6 + CH_3CH_2Cl $\xrightarrow{AlCl_3}$ $C_6H_5-CH_2CH_3$

(b) C_6H_6 $\xrightarrow[AlCl_3]{CH_3CH_2CH_2C(=O)Cl}$ $C_6H_5-C(=O)CH_2CH_2CH_3$

$\xrightarrow[H_2SO_4]{HNO_3}$ $m\text{-}O_2N-C_6H_4-C(=O)CH_2CH_2CH_3$ $\xrightarrow[KOH,\ heat]{NH_2NH_2}$ product

(c) C_6H_6 $\xrightarrow[FeBr_3]{Br_2}$ C_6H_5-Br $\xrightarrow[ether]{Mg}$ C_6H_5-MgBr

$\xrightarrow[(2)\ H_2O]{(1)\ CH_3CHO}$ $C_6H_5-CH(OH)CH_3$ $\xrightarrow[heat]{H^+}$ $C_6H_5-CH{=}CH_2$

In (c), you might have started with the alpha bromination of $C_6H_5CH_2CH_3$.

(d) $C_6H_5-CH_3 \xrightarrow[FeBr_3]{Br_2}$ Br–C_6H_4–CH_3

12.23 (a) O_2N–$C_6H_3(NO_2)$–Cl + $Na^{+}\ {}^{-}SCH_2-C_6H_5 \longrightarrow$ product

(b) 3,5-$(H_3C)_2C_6H_3$–$NH_2 \xrightarrow[HCl,\ 0°]{NaNO_2}$ 3,5-$(H_3C)_2C_6H_3$–$\overset{+}{N}\equiv N \xrightarrow[heat]{H_2O}$

(c) H_3C–C_6H_4–$NH_2 \xrightarrow[HCl,\ 0°]{NaNO_2}$

H_3C–C_6H_4–$\overset{+}{N}\equiv N \xrightarrow{C_6H_5NH_2}$

(d) H_3C–C_6H_4–$NH_2 \xrightarrow[HCl,\ 0°]{NaNO_2}$

H_3C–C_6H_4–$\overset{+}{N}\equiv N \xrightarrow{KI}$

(e) $C_6H_6 \xrightarrow{HNO_3} C_6H_5NO_2 \xrightarrow[(2)\ NaOH]{(1)\ Fe,\ HCl} C_6H_5NH_2$

$\xrightarrow[HCl,\ 0°]{NaNO_2} C_6H_5\overset{+}{N}\equiv N \xrightarrow[{}^{-}CN]{CuCN}$

(f) H_3C–(2-bromo-1,4-dimethylbenzene: Br)–CH_3 $\xrightarrow[\text{ether}]{\text{Mg}}$ H_3C–(MgBr)–CH_3 $\xrightarrow[\text{(2) } H_2O, H^+]{\text{(1) } CH_3CH_2CHO}$

H_3C–(OH / $CHCH_2CH_3$)–CH_3 $\xrightarrow[\text{heat}]{H^+}$ H_3C–($CH{=}CHCH_3$)–CH_3 $\xrightarrow[\text{Pt}]{H_2}$

12.24 (a) C_6H_5–OH $\xrightarrow[\text{heat, pressure}]{3\ H_2,\ Ni}$ cyclohexyl–OH

(b) C_6H_5–OH $\xrightarrow{HNO_3}$ O_2N–C_6H_4–OH

$\xrightarrow[\text{(2) NaOH}]{\text{(1) Fe, HCl}}$ H_2N–C_6H_4–OH

(c) C_6H_5OH + $CH_3CH_2\overset{O}{\overset{\|}{C}}Cl$ $\xrightarrow{\text{pyridine}}$ $C_6H_5O\overset{O}{\overset{\|}{C}}CH_2CH_3$

(d) $C_6H_5NO_2$ $\xrightarrow[\text{(2) NaOH}]{\text{(1) Fe, HCl}}$ $C_6H_5NH_2$ $\xrightarrow[\text{HCl, 0°}]{NaNO_2}$ $C_6H_5\overset{+}{N}{\equiv}N$

$\xrightarrow[\text{warm}]{H_2O}$ C_6H_5OH $\xrightarrow[\text{(2) } H_2O, H^+]{\text{(1) } CHCl_3,\ {}^-OH,\ 70°}$ (CHO)–C_6H_4–OH

12.25 (a) (Cl)–C_6H_4–CH_2CH_3 + Cl–C_6H_4–CH_2CH_3

(b) $C_6H_5CHBrCH_3$ (c) $C_6H_5CO_2H$

(d) o-$CH_3CH_2CH_2$–C_6H_4–CH_3 + $CH_3CH_2CH_2$–C_6H_4–CH_3 (p)

+ $(CH_3)_2CH$–C_6H_4–CH_3 (p) + o-$CH(CH_3)_2$–C_6H_4–CH_3

(e) o-SO_3H–C_6H_4–CH_3 + HO_3S–C_6H_4–CH_3 (p)

(f) C_6H_{11}–CH_2CH_3

12.26 $C_6H_5CH_3 \xrightarrow{HNO_3}$ o-NO_2–C_6H_4–CH_3 + O_2N–C_6H_4–CH_3 (p)

$\downarrow$ $KMnO_4$, heat (each product)

o-NO_2–C_6H_4–CO_2H and O_2N–C_6H_4–CO_2H (p)

$C_6H_5CH_3 \xrightarrow[\text{heat}]{KMnO_4}$ C_6H_5–CO_2H $\xrightarrow[H_2SO_4]{HNO_3}$ m-O_2N–C_6H_4–CO_2H

12.27 (a) $C_6H_6 \xrightarrow[AlCl_3]{C_6H_{11}\text{–}Cl}$ C_6H_5–C_6H_{11} (cyclohexyl)

(b) $C_6H_6 \xrightarrow[AlCl_3]{CH_3CHClCH_3} C_6H_5CH(CH_3)_2 \xrightarrow{NBS} C_6H_5C(Br)(CH_3)_2 \xrightarrow[ether]{Mg}$

$C_6H_5C(MgBr)(CH_3)_2 \xrightarrow[(2)\ H_2O,\ H^+]{(1)\ HCH(=O)} C_6H_5C(CH_2OH)(CH_3)_2 \xrightarrow{SOCl_2} C_6H_5C(CH_2Cl)(CH_3)_2$

(c) $C_6H_6 \xrightarrow[H_2SO_4]{HNO_3} C_6H_5NO_2 \xrightarrow[(2)\ OH^-]{(1)\ Fe,\ HCl} C_6H_5NH_2 \xrightarrow[HCl,\ 0°]{NaNO_2}$

$C_6H_5N_2^+\ Cl^- \xrightarrow[warm]{H_2O} C_6H_5OH \xrightarrow[(2)\ CH_3I]{(1)\ OH^-} C_6H_5OCH_3$

(d) $C_6H_6 \xrightarrow[AlCl_3]{2\ CH_3I}$ $H_3C-C_6H_4-CH_3$ (para) + *o*-isomer $\xrightarrow[heat]{KMnO_4}$ $HO_2C-C_6H_4-CO_2H$ (para)

(e) $C_6H_5NH_2$ from (c) $\xrightarrow{Br_2}$ 2,4-dibromoaniline (Br–C₆H₃(Br)–NH₂) $\xrightarrow[HCl,\ 0°]{NaNO_2}$

2,4-dibromobenzenediazonium chloride ($Br-C_6H_3(Br)-N_2^+\ Cl^-$) $\xrightarrow{H_3PO_2}$ 1,3-dibromobenzene

or $C_6H_5NO_2$ from (c) $\xrightarrow[FeBr_3]{Br_2}$ m-bromonitrobenzene (NO_2) $\xrightarrow[(2)\ OH^-]{(1)\ Fe,\ HCl}$

m-bromoaniline (NH_2) $\xrightarrow[HCl,\ 0°]{NaNO_2}$ m-bromobenzenediazonium chloride ($N_2^+\ Cl^-$) $\xrightarrow[heat]{CuBr}$

(f) $C_6H_5NO_2$ from (c) $\xrightarrow[H_2SO_4]{SO_3}$ (m-nitrobenzenesulfonic acid: benzene ring with NO_2 and SO_3H) $\xrightarrow[(2)\ H_2O,\ H^+]{(1)\ NaOH,\ fuse}$ (m-nitrophenol: benzene ring with NO_2 and OH)

(g) $C_6H_6 \xrightarrow[FeBr_3]{Br_2} C_6H_5Br \xrightarrow[ether]{Mg} C_6H_5MgBr \xrightarrow[(2)\ H_2O,\ H^+]{(1)\ (CH_3)_2C{=}O} C_6H_5C(OH)(CH_3)_2$

(h) $2\ C_6H_6 + Cl_2CHCH_3 \xrightarrow{AlCl_3} (C_6H_5)_2CHCH_3$

(i) $C_6H_6 \xrightarrow[AlCl_3]{CH_3I} C_6H_5CH_3 \xrightarrow[h\nu]{Br_2} C_6H_5CH_2Br \xrightarrow[CH_3OH]{Na^+\ {}^-OCH_3} C_6H_5CH_2OCH_3$

12.28 $(C_6H_5)_3C^+ + H_2O + HSO_4^-$. The triphenylmethyl cation is highly resonance stabilized because the positive charge is delocalized by the three benzene rings.

$(C_6H_5)_3C{-}\ddot{O}H \underset{}{\overset{H^+}{\rightleftharpoons}} (C_6H_5)_3C{-}\overset{+}{O}H_2 \overset{-\ H_2\ddot{O}:}{\rightleftharpoons}$

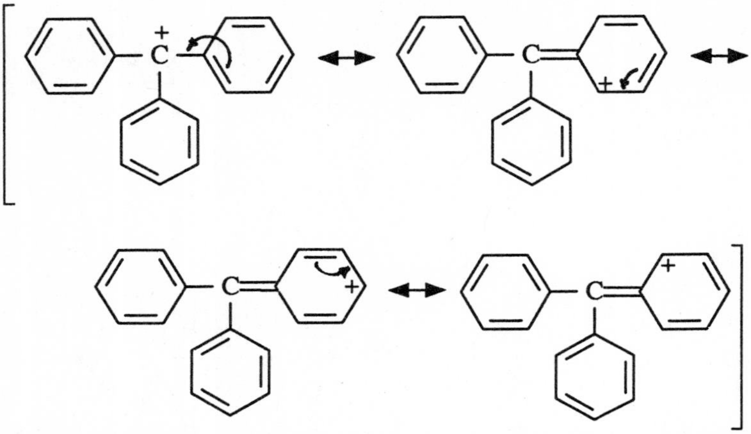

Similar resonance structures (as well as benzene ring resonance structures) can be drawn for the other two rings.

12.29 The two reactions proceed by different mechanisms. Reaction (1) proceeds by an ionic reaction mechanism with a benzylic carbocation as the intermediate. Reaction (2) is a free-radical reaction with a benzylic radical as the intermediate.

(1) $C_6H_5CH{=}CHCH_3 \xrightarrow{H^+} C_6H_5\overset{+}{C}HCH_2CH_3 \xrightarrow{Br^-} C_6H_5CH(Br)CH_2CH_3$

benzylic cation

(2) peroxide + HBr ⟶ peroxide-H + Br·

$C_6H_5CH{=}CHCH_3 \xrightarrow{Br\cdot} C_6H_5\dot{C}HCH(Br)CH_3 \xrightarrow[-\,Br\cdot]{HBr} C_6H_5CH_2CH(Br)CH_3$

benzylic radical

12.30 (a) The unshared electrons on the nitrogen of aniline are delocalized and are less available for donation than those of cyclohexylamine. The positive charge in the product anilinium ion, by contrast, is not delocalized. (The benzene ring resonance structures are not shown here.)

aniline: $C_6H_5\ddot{N}H_2$ ↔ $\overset{+}{N}H_2$ ↔ $\overset{+}{N}H_2$ ↔ $\overset{+}{N}H_2$

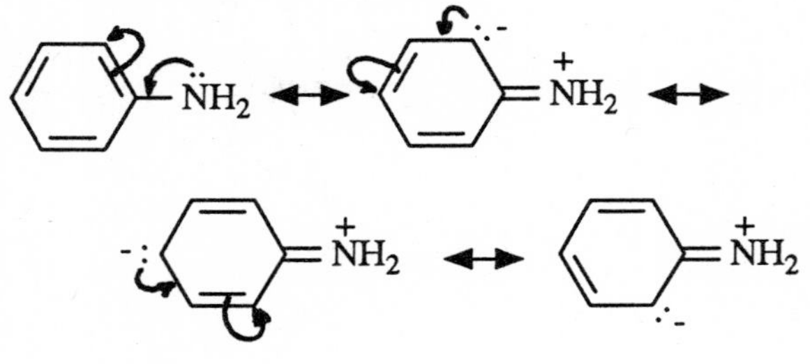

anilinium ion: $C_6H_5\overset{+}{N}H_3$

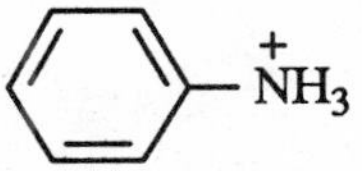

For these reasons, the equilibrium is pushed to the amine side of the equation to a greater extent than it would be for cyclohexylamine.

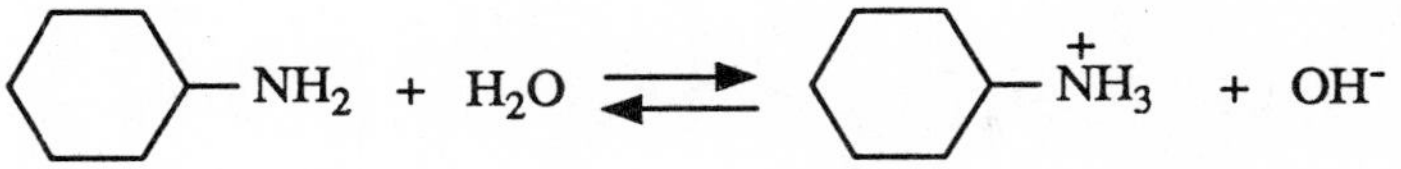

$$C_6H_{11}NH_2 + H_2O \rightleftharpoons C_6H_{11}\overset{+}{N}H_3 + OH^-$$

more basic

$$C_6H_5NH_2 + H_2O \rightleftharpoons C_6H_5\overset{+}{N}H_3 + OH^-$$

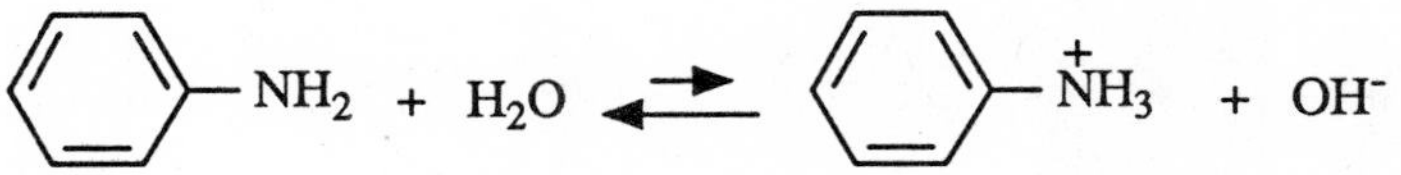

less basic

(b) Less basic, because the nitro group helps delocalize the unshared electrons by resonance and by the inductive effect, causing them to be less available for bonding.

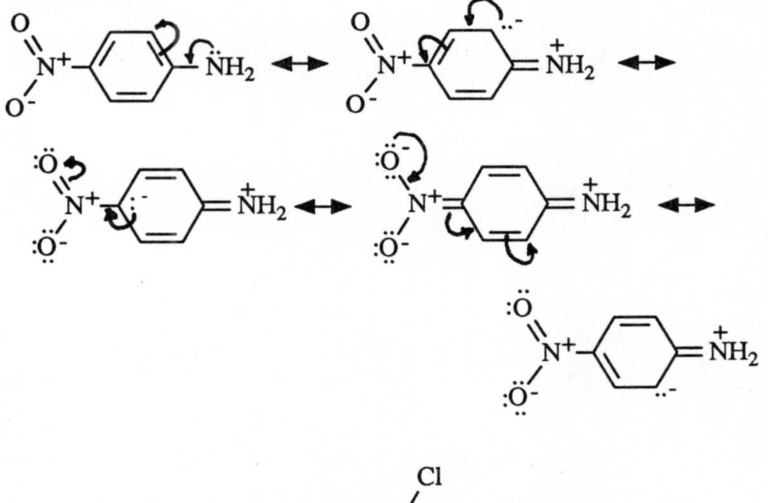

12.31 (a) 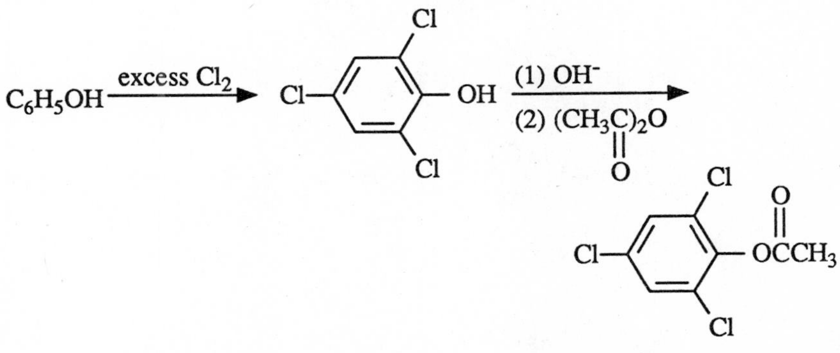 $C_6H_5OH \xrightarrow{\text{excess } Cl_2}$ 2,4,6-trichlorophenol $\xrightarrow[(2)\ (CH_3C(=O))_2O]{(1)\ OH^-}$ 2,4,6-trichlorophenyl acetate ($OCCH_3$ with C=O)

(b) $C_6H_5CH_3 \xrightarrow[h\nu]{Br_2} C_6H_5CH_2Br \xrightarrow[AlBr_3]{C_6H_6} (C_6H_5)_2CH_2 \xrightarrow[h\nu]{Br_2} (C_6H_5)_2CHBr$

or $2\ C_6H_6 \xrightarrow[AlCl_3]{CH_2Cl_2} (C_6H_5)_2CH_2 \xrightarrow[h\nu]{Br_2} (C_6H_5)_2CHBr$

(c) $C_6H_6 \xrightarrow[AlCl_3]{CH_3CH_2Cl} C_6H_5CH_2CH_3 \xrightarrow[H_2SO_4]{HNO_3} O_2N-C_6H_4-CH_2CH_3$

$\xrightarrow[h\nu]{Br_2} O_2N-C_6H_4-CH(Br)CH_3 \xrightarrow{OH^-} O_2N-C_6H_4-CH(OH)CH_3$

$\xrightarrow{H_2CrO_4} O_2N-C_6H_4-C(=O)CH_3$

(d) $C_6H_6 \xrightarrow[FeBr_3]{Br_2} C_6H_5Br \xrightarrow[ether]{Mg} C_6H_5MgBr \xrightarrow[(2)\ H_2O,\ H^+]{(1)\ \text{cyclohexanone}}$

1-phenylcyclohexanol (OH, C_6H_5) $\xrightarrow[-\ H_2O]{H^+}$ 1-phenylcyclohexene ($-C_6H_5$)

(e) $C_6H_6 \xrightarrow[H_2SO_4]{HNO_3} C_6H_5NO_2 \xrightarrow[FeBr_3]{Br_2}$ m-bromonitrobenzene (Br, NO_2) $\xrightarrow[(2)\ OH^-]{(1)\ Fe,\ HCl}$

m-bromoaniline (Br, NH_2) $\xrightarrow[0°]{NaNO_2,\ HCl}$ m-bromobenzenediazonium chloride (Br, $N_2^+\ Cl^-$)

$C_6H_5NO_2 \xrightarrow[(2)\ OH^-]{(1)\ Fe,\ HCl} C_6H_5NH_2 \longrightarrow$ product (with the diazonium salt)

12.32

$NaNO_2$, HCl, 0°; CuBr; (1) Fe, HCl (2) OH^-; $NaNO_2$, HCl, 0°; KI

12.33

KNO_3, H_2SO_4; NH_3, 140°

12.34

H_3C—C₆H₄—Cl $\xrightarrow{\text{NaOH, heat}}$ benzyne $\xrightarrow{H_2O}$ H_3C—C₆H₄—OH + H_3C—C₆H₄—OH

benzyne

12.35 The C-Cl bond of chlorobenzene has double-bond character:

12.36 [reaction scheme: 2-chloropyrimidine + $(CH_3)_2NH$ → resonance-stabilized intermediate (three resonance forms with $\overset{+}{N}H(CH_3)_2$ and Cl on ring carbon) → ($-Cl^-$, $-H^+$) → 2-(dimethylamino)pyrimidine, $N(CH_3)_2$]

lost as $(CH_3)_2\overset{+}{N}H_2$

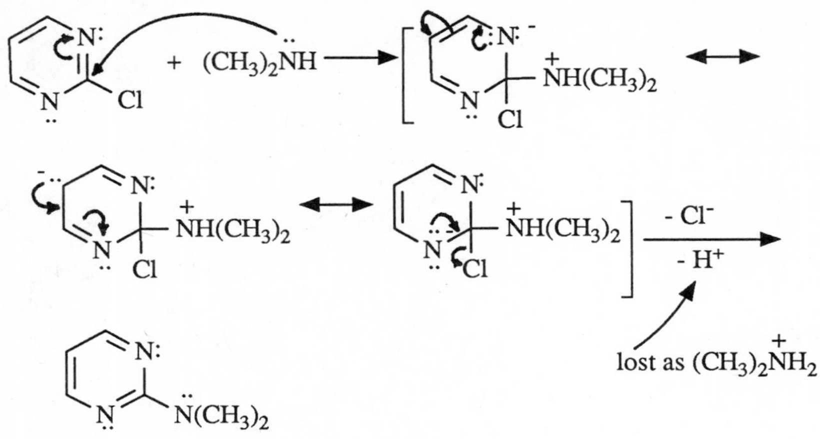

The negative charge in the resonance-stabilized intermediate is delocalized by two electronegative nitrogen atoms.

12.37 (a) 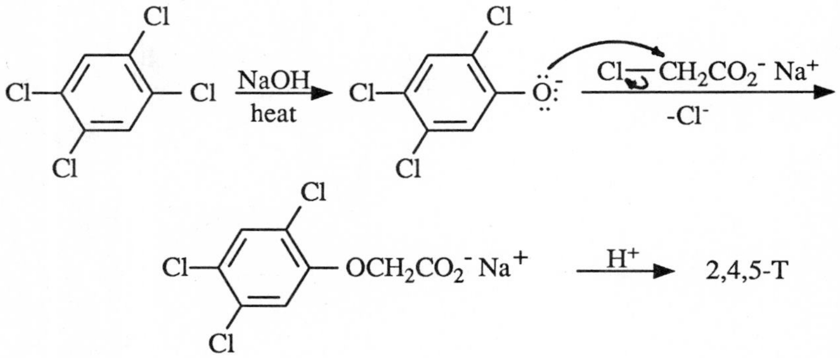 [reaction scheme: 1,2,4,5-tetrachlorobenzene → (NaOH, heat) → 2,4,5-trichlorophenoxide → ($Cl-CH_2CO_2^-\ Na^+$, $-Cl^-$) → 2,4,5-trichlorophenoxy $OCH_2CO_2^-\ Na^+$ → (H^+) → 2,4,5-T]

(b)

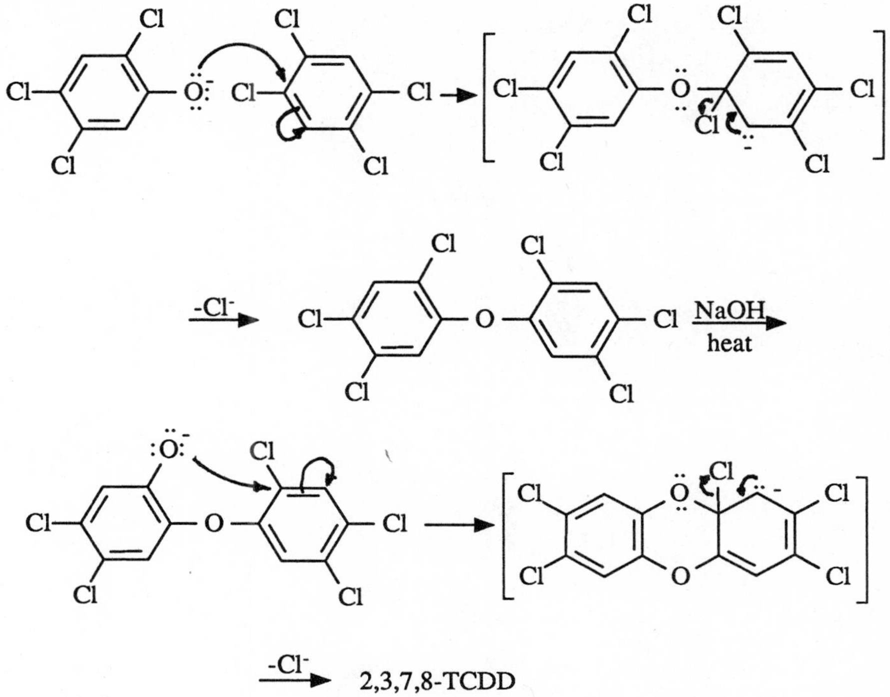

Cl Cl Cl Cl Cl Cl Cl Cl

$\xrightarrow{-Cl^-}$ Cl Cl O Cl Cl Cl Cl $\xrightarrow[\text{heat}]{\text{NaOH}}$

Cl O Cl Cl Cl Cl Cl Cl O O Cl Cl Cl

$\xrightarrow{-Cl^-}$ 2,3,7,8-TCDD

12.38 $C_6H_5Cl \xrightarrow[H_2SO_4]{HNO_3}$ O_2N–C_6H_4–Cl $\xrightarrow[(2)\ OH^-]{(1)\ Fe,\ HCl}$ H_2N–C_6H_4–Cl

o and *p* *o* and *p*

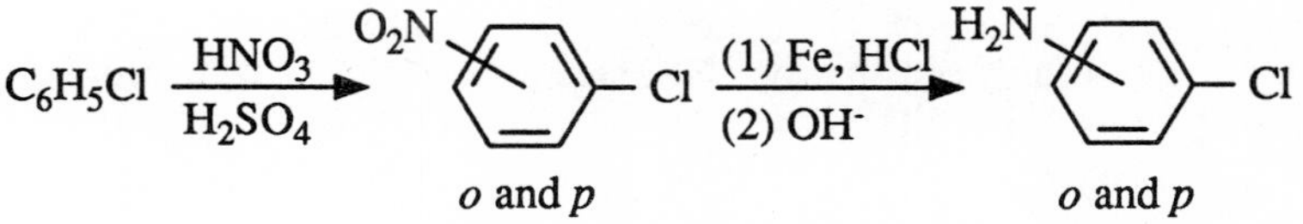

The spectrum is of a *p*-disubstituted benzene (note the pair of doublets in the aromatic region); therefore, the structures are as follows:

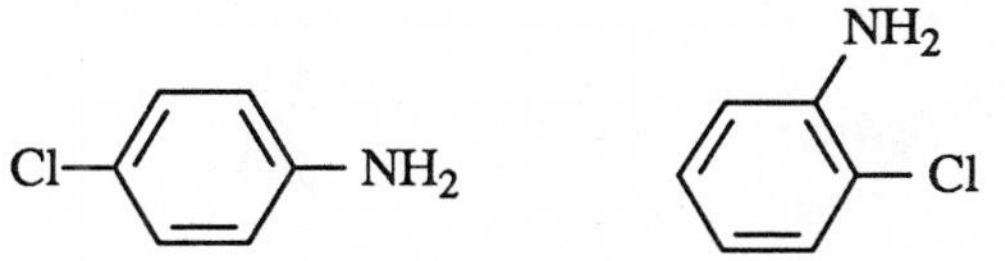

Cl–C_6H_4–NH_2 NH_2 Cl

spectrum shown other component

12.39 H_3C–C_6H_4–OH ⟶ H_3C–C_6H_4–OCH_3

upfield singlet downfield singlet

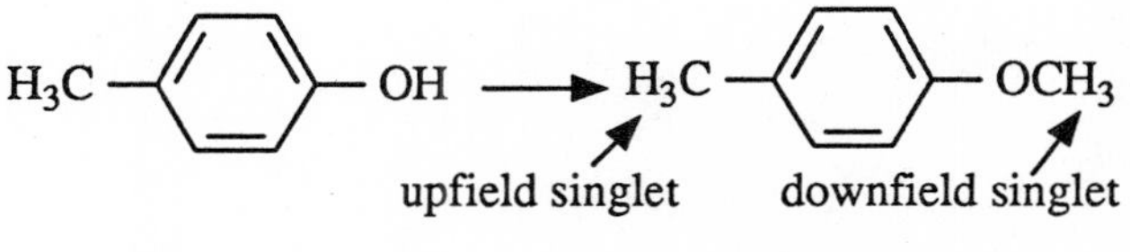

The two sets of doublets in the aromatic region of the spectrum suggest the *p*-substituted isomer.

12.40

C_6H_5OH $\xrightarrow{HNO_3}$ HO–C_6H_4–NO_2 (A) $\xrightarrow[(2)\ CH_3I]{(1)\ NaOH}$ CH_3O–C_6H_4–NO_2 (C) $\xrightarrow[(2)\ base]{(1)\ Fe,\ HCl}$ CH_3O–C_6H_4–NH_2 (E)

C_6H_5OH $\xrightarrow{HNO_3}$ o-$HOC_6H_4NO_2$ (B) $\xrightarrow[(2)\ CH_3I]{(1)\ NaOH}$ o-$CH_3OC_6H_4NO_2$ (D) $\xrightarrow[(2)\ base]{(1)\ Fe,\ HCl}$ o-$CH_3OC_6H_4NH_2$ (F)

The infrared spectra and chemical behavior suggest that A and B are nitrophenols. Compounds C and D are the methyl ethers of these phenols, and compounds E and F are the corresponding aryl amines.

The infrared spectra of A and C show the principal peak in the aromatic C-H bending region at about 860 cm^{-1} (11.6 μm), as we would expect for *p*-disubstituted benzenes. By contrast, B and D show their principal peaks in this region at about 735 cm^{-1} (13.6 μm), as we would expect for *o*-disubstituted benzenes. In the nmr spectra, the -N$\underline{H}_2$ absorption of F is farther downfield than that of E because of the effect of the nearby electronegative oxygen atom.

Chapter 13

Aldehydes and Ketones

13.31 (a) $BrCH_2C(=O)CH_3$ (b) $CH_3C(=O)\text{-}C(=O)CH_3$

(c) $CH_3C(=O)CH_2CH_2\text{-}C_6H_4\text{-}OH$ (para)

(d) $(CH_3)_2C{=}CHCH_2CH_2(CH_3)C{=}C(H)CH{=}O$

13.32 (a) cyclooctanone (b) 1,4-cyclohexanedione

(c) 2-methyl-3-hexanone (d) hydroxypropanone or hydroxyacetone

(e) phenylethanal or phenylacetaldehyde

(f) 4-bromobutanal or γ-bromobutyraldehyde

13.33 (a) $CH_3C(=O)CH_2CHO$ or any other structure with keto and aldehyde groups 1,3 to each other.

(b) $H_2C{=}CHC(=O)CH_3$ or any other structure with a carbon-carbon double bond and a keto group in conjugation.

(c) $BrCH_2CHO$ or any other aldehyde with a bromine on the α carbon (carbon adjacent to the aldehyde carbon).

(d) $CH_3C(=O)CH_2CH_2OH$ or any other 1,3-hydroxyketone.

13.34 (a) $Br_2CHCH(=O) + H_2O \rightleftharpoons Br_2CHCH(OH)OH$

(b) $Br_2CHCH_2CH(=O) + H_2O \rightleftharpoons Br_2CHCH_2CH(OH)OH$

The hydrate in (a) would be more stable than the hydrate in (b). In (a), the bromine atoms are closer to the carbonyl group and, therefore, their electron-withdrawing inductive effect is greater.

13.35 (a) $CH_3CH_2CH(=O) + CH_3OH \underset{}{\overset{H^+}{\rightleftharpoons}} CH_3CH_2CH(OH)OCH_3 \underset{}{\overset{CH_3OH,\ H^+}{\rightleftharpoons}} CH_3CH_2CH(OCH_3)OCH_3 + H_2O$

(b) $CH_3CH(OH)CH_2CH_2C(=O)CH_3 \overset{H^+}{\rightleftharpoons}$

H_3C–(five-membered ring with O)–$C(OH)CH_3$ $\underset{}{\overset{CH_3OH,\ H^+}{\rightleftharpoons}}$ H_3C–(five-membered ring with O)–$C(OCH_3)CH_3$ + H_2O

All stereoisomers are formed.

In (b), the cyclic hemiketal would be formed in preference to the open-chain hemiketal because of the relative stability of the five-membered ring and because the equilibrium leading to an open-chain hemiketal does not favor that compound.

$CH_3CH(OH)CH_2CH_2C(OH)(OCH_3)CH_3 \overset{H^+}{\rightleftharpoons} CH_3CH(OH)CH_2CH_2C(=O)CH_3 + CH_3OH$

not stable

$\overset{H^+}{\rightleftharpoons}$ H_3C–(five-membered ring with O)–$C(OH)CH_3$

stable

Stereoisomers of the cyclic compounds with the methyl groups *cis* or *trans* to each other would be formed. Because these reactions are reversible, the relative percentages of the *cis* and *trans* isomers would depend on their relative stabilities.

racemic (H, O, OH; H_3C, CH_3) *cis* methyls

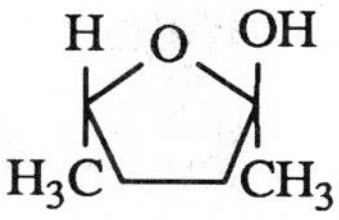

racemic (H, O, CH_3; H_3C, OH) *trans* methyls

(c) 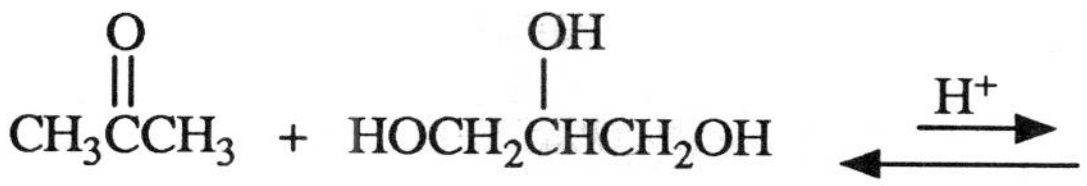 $CH_3C(=O)CH_3$ + $HOCH_2CH(OH)CH_2OH$ $\xrightarrow{H^+}$ (equilibrium)

$(H_3C)_2C(OH)-O-CH(CH_2OH)CH_2OH$ or $(H_3C)_2C(OH)-O-CH_2CH(OH)CH_2OH$ $\xrightarrow{H^+, -H_2O}$ (equilibrium)

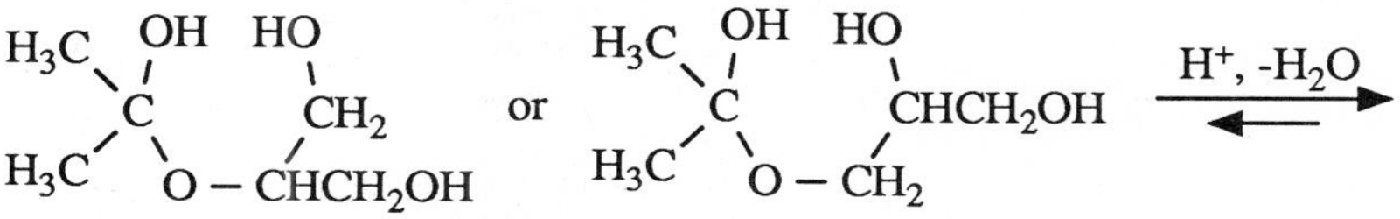

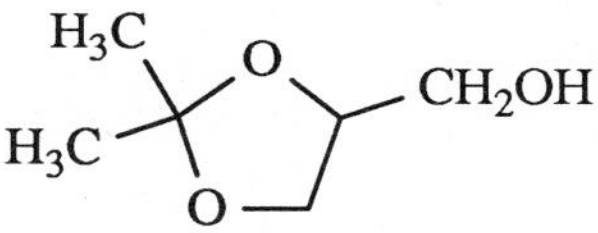

five-membered ring ketal (H_3C, H_3C, O, O, CH_2OH)

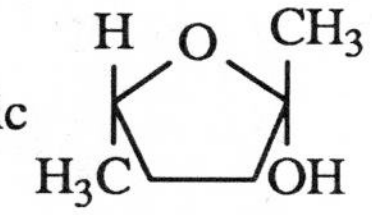

or $(H_3C)_2C(OH)-O-CH_2CH(OH)CH_2OH$ $\xrightarrow{H^+, -H_2O}$ (equilibrium) six-membered ring ketal (H_3C, H_3C, O, O, OH)

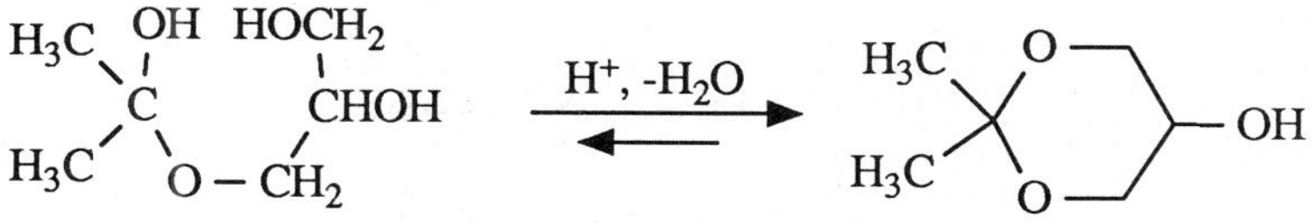

13.36 (a) 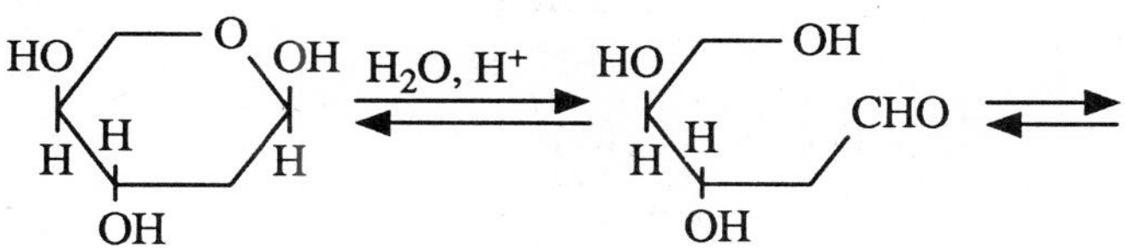 (HO, O, OH, H, H, H, OH) $\xrightarrow{H_2O, H^+}$ (equilibrium) (HO, OH, CHO, H, H, OH) ⇌

(HO, O, H, H, H, OH, OH) + ($HOCH_2$, H, O, OH, H, H, OH) + ($HOCH_2$, H, O, H, H, OH, OH)

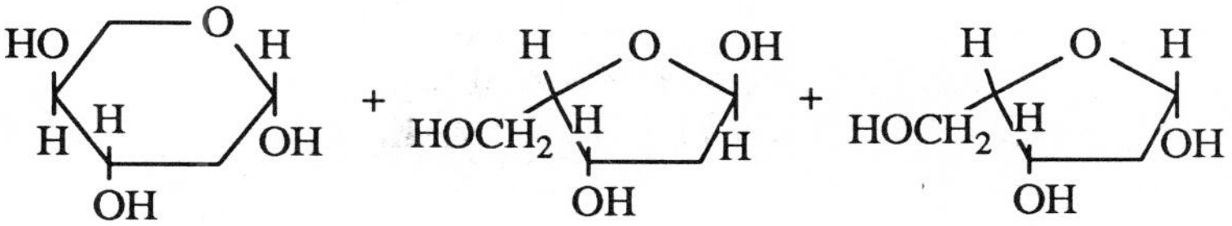

(b) H_3C-CH (cyclic acetal: O–CH_2–CH_2–O) $\overset{H_2O,\ H^+}{\rightleftharpoons}$ $CH_3\overset{OH}{C}HOCH_2CH_2OH$ $\rightleftharpoons$

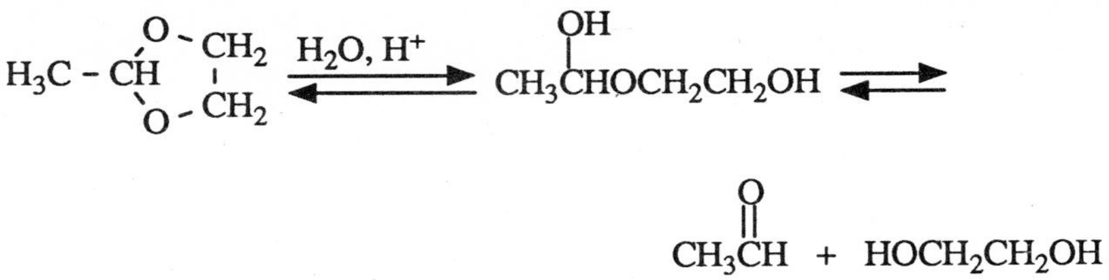

$CH_3\overset{O}{\overset{\|}{C}}H + HOCH_2CH_2OH$

(c) HO, H, H, OH (pyranose ring)–O–(cyclohexyl) $\overset{H_2O,\ H^+}{\rightleftharpoons}$ HO, O, OH, H, H, H, OH (pyranose ring)

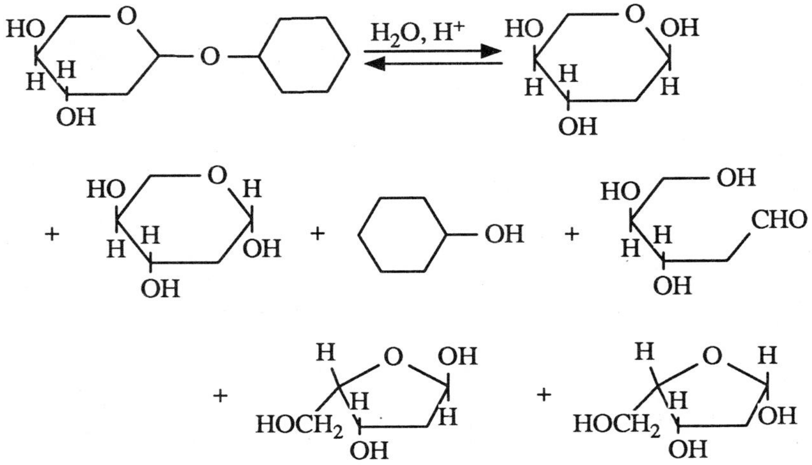

+ HO, O, H, H, H, OH, OH (pyranose ring) + (cyclohexyl)–OH + HO, OH, CHO, H, H, OH (open chain)

+ H, O, OH, H, H, $HOCH_2$, OH (furanose ring) + H, O, H, H, OH, $HOCH_2$, OH (furanose ring)

13.37 $C_6H_5CH{=}\ddot{O}{:}$ $\overset{H^+}{\rightleftharpoons}$ $C_6H_5CH{=}\overset{+}{O}H$ + $H\ddot{O}CH_2C(CH_2OH)_3$ $\rightleftharpoons$ $C_6H_5CH(\ddot{O}H)-\overset{+}{O}(H)CH_2C(CH_2OH)_3$

$\overset{-H^+}{\rightleftharpoons}$ $C_6H_5CH(\ddot{O}H)$–:Ö–CH_2–C$(CH_2OH)_2$–CH_2OH $\overset{H^+}{\rightleftharpoons}$ $C_6H_5CH(\overset{+}{O}H_2)$–ÖCH_2–C$(CH_2OH)_2$–$CH_2OH$

$\overset{-H_2\ddot{O}:}{\rightleftharpoons}$ $C_6H_5CH{=}\overset{+}{O}-CH_2$–C$(CH_2OH)_2$–$H\ddot{O}CH_2$ $\rightleftharpoons$

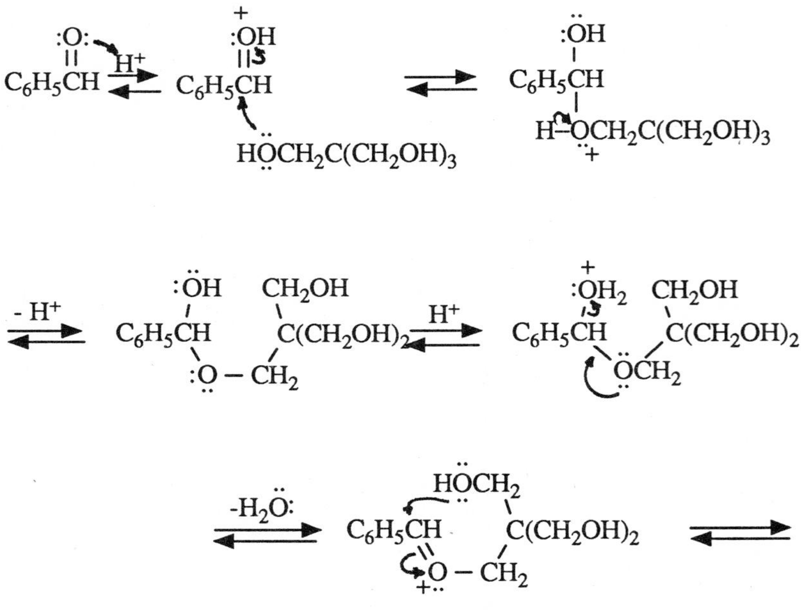

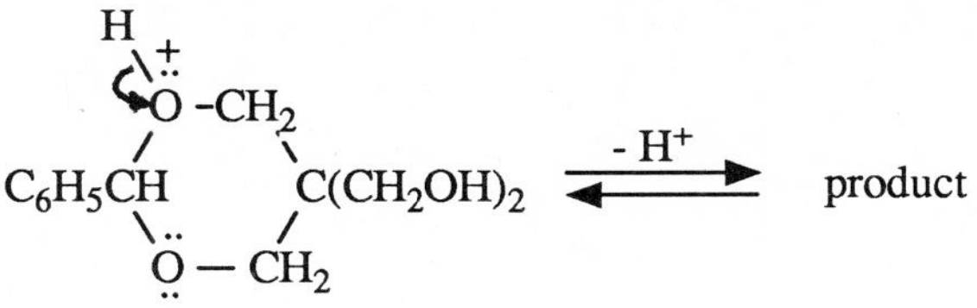

$$C_6H_5CH(\overset{+}{O}H\text{-}CH_2)(\ddot{O}\text{-}CH_2)C(CH_2OH)_2 \xrightleftharpoons{-H^+} \text{product}$$

13.38 (a) 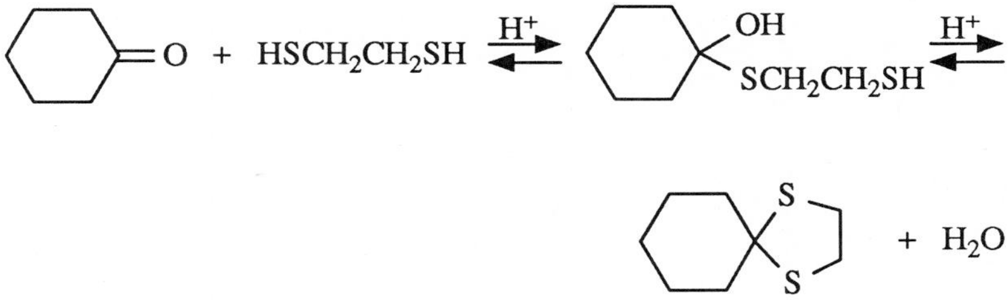 cyclohexanone + $HSCH_2CH_2SH \xrightleftharpoons{H^+}$ 1-hydroxy-1-(SCH_2CH_2SH)cyclohexane $\xrightleftharpoons{H^+}$ cyclohexane spiro-1,3-dithiolane (S, S) + H_2O

(b) 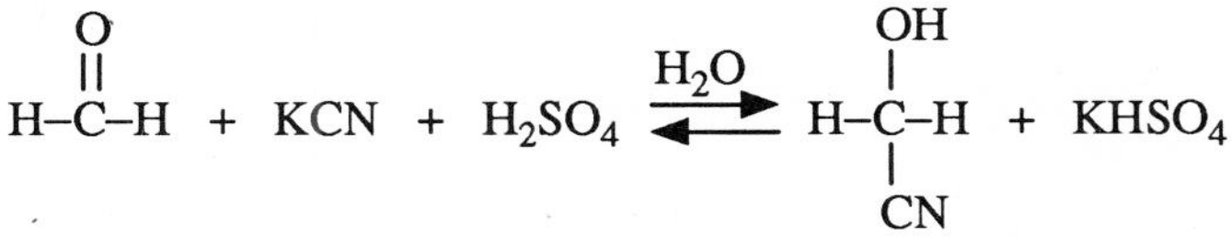 $H\text{-}\overset{O}{\overset{||}{C}}\text{-}H + KCN + H_2SO_4 \xrightleftharpoons{H_2O} H\text{-}\underset{CN}{\underset{|}{\overset{OH}{\overset{|}{C}}}}\text{-}H + KHSO_4$

In (b), KCN and H_2SO_4 generate HCN.

(c) 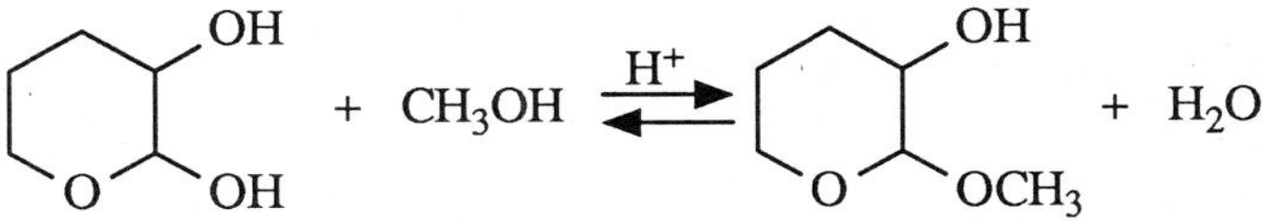 tetrahydropyran-2,3-diol (OH, O, OH) + $CH_3OH \xrightleftharpoons{H^+}$ 2-methoxytetrahydropyran-3-ol (OH, O, OCH_3) + H_2O

In (c), the –OH group that reacts is part of the hemiacetal group.

13.39 (a) 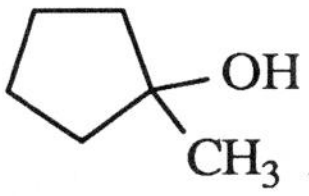 1-methylcyclopentanol (OH, CH_3)

(b) 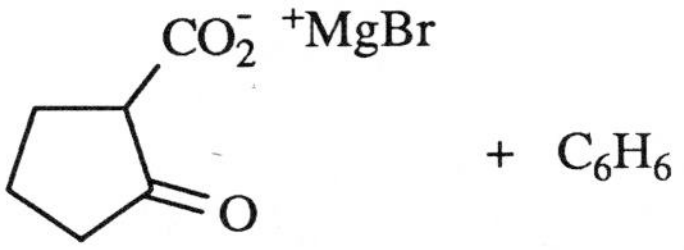 2-oxocyclopentanecarboxylate (CO_2^- ^+MgBr, O) + C_6H_6

(c) 1-naphthalenemethanol (CH_2OH)

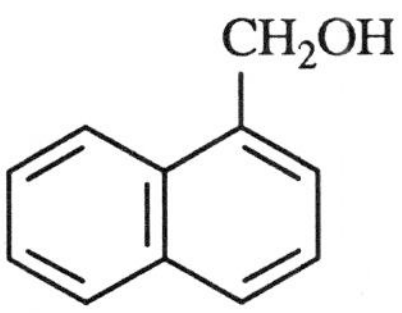

In (b), the starting carbonyl compound contains an acidic proton that reacts with the Grignard reagent at a faster rate than does the keto group.

13.40 (a) $CH_3I \xrightarrow[\text{ether}]{\text{Mg}} CH_3MgI \xrightarrow[\text{(2) } H_2O,\ H^+]{\text{(1) } (CH_3CH_2)_2C{=}O} CH_3CH_2C(OH)(CH_3)CH_2CH_3$

(b) CH_3MgI from (a) $\xrightarrow[\text{(2) } H_2O,\ H^+]{\text{(1) HCHO}} CH_3CH_2OH$

(c) CH_3MgI from (a) $\xrightarrow[\text{(2) } H_2O,\ H^+]{\text{(1) } CH_3CH_2CH_2CHO} CH_3CH_2CH_2CH(OH)CH_3$

racemic

13.41 (a) (1) $C_6H_5CH_2Br \xrightarrow[\text{ether}]{\text{Mg}} C_6H_5CH_2MgBr \xrightarrow[\text{(2) } H_2O,\ H^+]{\text{(1) } CH_2\text{-}CH_2 \text{ (epoxide, O bridging)}} C_6H_5CH_2CH_2CH_2OH$

(2) $C_6H_5CH_2CH_2Br \xrightarrow[\text{ether}]{\text{Mg}} C_6H_5CH_2CH_2MgBr \xrightarrow[\text{(2) } H_2O,\ H^+]{\text{(1) HCHO}} C_6H_5CH_2CH_2CH_2OH$

(b) (1) $C_6H_5Br \xrightarrow[\text{ether}]{\text{Mg}} C_6H_5MgBr \xrightarrow[\text{(2) } H_2O,\ H^+]{\text{(1) } H_3C\text{-}C(=O)\text{-}C_6H_{11}}$ product

(2) $CH_3I \xrightarrow[\text{ether}]{\text{Mg}} CH_3MgI \xrightarrow[\text{(2) } H_2O,\ H^+]{\text{(1) } C_6H_5C(=O)\text{-}C_6H_{11}}$ product

(3) $C_6H_{11}\text{-}Br \xrightarrow[\text{ether}]{\text{Mg}} C_6H_{11}\text{-}MgBr \xrightarrow[\text{(2) } H_2O,\ H^+]{\text{(1) } CH_3C(=O)\text{-}C_6H_5}$ product

In (b), the use of NH_4^+ instead of strong acid to hydrolyze the magnesium alkoxide would minimize dehydration of the tertiary alcohol.

13.42 (a) o-nitrobenzaldehyde + $C_6H_5NH_2$ $\xrightleftharpoons{H^+}$ o-$O_2NC_6H_4CH{=}NC_6H_5$ + H_2O

a 1° amine — an imine

(b) $(CH_3)_2CHCHO$ + pyrrolidine $\xrightleftharpoons{H^+}$ $(CH_3)_2C{=}CH-N(CH_2)_4$ + H_2O

a 2° amine — an enamine

(c) C_6H_5CHO + $(CH_3)_3CNH_2$ $\xrightleftharpoons{H^+}$ $C_6H_5CH{=}NC(CH_3)_3$ + H_2O

a 1° amine — an imine

(d) $CH_3(CH_2)_5CHO + H_2NOH{\cdot}HCl \xrightarrow{Na_2CO_3} CH_3(CH_2)_5CH{=}NOH + H_2O$

an oxime

In (d), the sodium carbonate reacts with the hydroxylamine hydrochloride. The resultant hydroxylamine can then act as a nucleophile.

$$2\ H_3\overset{+}{N}OH\ Cl^- + Na_2CO_3 \longrightarrow 2\ H_2\ddot{N}OH + 2\ NaCl + H_2O + CO_2$$

13.43 (a) $CH_3CH_2CH_2CHO + HN(CH_2CH_3)_2 \xrightleftharpoons{H^+}$

(b) cyclopentanone + pyrrolidine $\xrightleftharpoons{H^+}$

(c) CH_3CH_2CHO + $C_6H_5NH_2$ $\xrightleftharpoons{-H_2O}$

(d) CH_3CH_2CHO + $H_2NNHC_6H_5$ $\xrightleftharpoons{H^+}$

The products of (a) and (b) are enamines because *secondary* amines were used. The product of (c) is an imine because a primary amine was used.

13.44 (b) < (a) < (c). Compound (b) contains the most hindered carbonyl group, and compound (c) contains the least hindered carbonyl group.

13.45 (a) 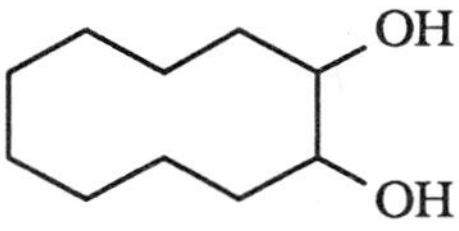 [cyclododecane-1,2-diol structure: OH, OH] (probably a mixture of *cis* and *trans* isomers)

(b) 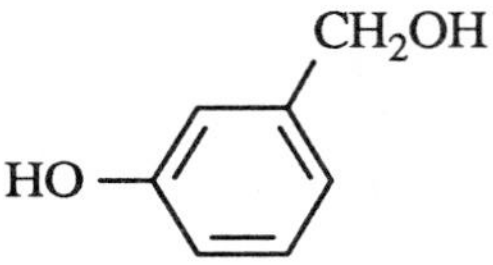 HO—C_6H_4—CH_2OH (meta)

(c) [bicyclo[3.3.0]octane structure] by the Wolff-Kishner reaction

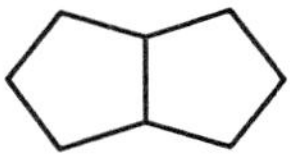

(d) 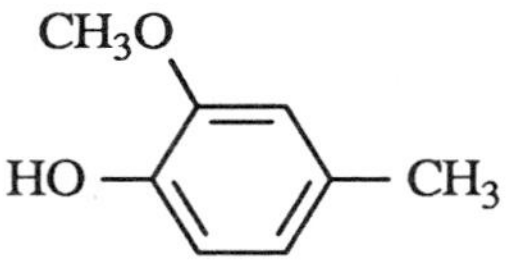 HO—$C_6H_3(OCH_3)$—CH_3 by Clemmenson reduction

(e) $(CH_3)_2CHNH_2$ by reductive amination

(f) $(CH_3)_2CHNHCH_2CH_2OH$, also reductive amination

(g) 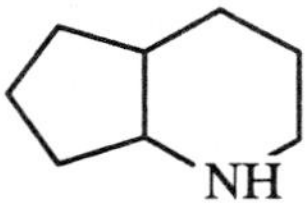 [bicyclic amine structure: NH] by way of a cyclic imine, [bicyclic imine structure: N]

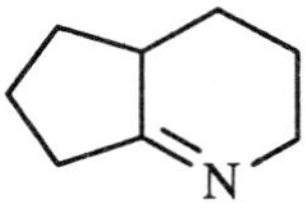

13.46 H, C_6H_5 (ketone) + $\bar{H}$–BH_3 $\xrightarrow{-BH_3}$ H, C_6H_5, $-\ddot{O}:^-$, H $\xrightarrow{H_2O,\ H^+}$ H, C_6H_5, $\ddot{O}H$, H

attack from less hindered side

-OH group is *cis* to the bridge

13.47 Determine the structure of the required Wittig reagent by first circling the portion of the product that has replaced the carbonyl oxygen.

Example: ketone $\longrightarrow$ product (=CHCH=CH_2 circled)

$(C_6H_5)_3P$=CHCH=CH_2 (CHCH=CH_2 circled)

Wittig reagent

(a) CH_2=$CHCH_2Cl$ $\xrightarrow[\text{(2) } CH_3(CH_2)_3Li \text{ (strong base)}]{\text{(1) } (C_6H_5)_3P}$ $(C_6H_5)_3P$=CHCH=CH_2 $\xrightarrow{\text{cyclohexanone}}$

(b) CH_3I $\xrightarrow[\text{(2) } CH_3(CH_2)_3Li]{\text{(1) } (C_6H_5)_3P}$ $(C_6H_5)_3P$=CH_2 $\xrightarrow{\text{CHO}}$

(c) $CH_3OCH_2I \xrightarrow[(2)\ CH_3(CH_2)_3Li]{(1)\ (C_6H_5)_3P} (C_6H_5)_3P{=}CHOCH_3$

$\xrightarrow{CH_3O-C_6H_4-C(=O)-CH_3}$

(d) $BrCF_2H \xrightarrow[(2)\ CH_3(CH_2)_3Li]{(1)\ (C_6H_5)_3P} (C_6H_5)_3P{=}CF_2 \xrightarrow{(CH_3)_2C{=}O}$

13.48 (a) no reaction in acidic solution

(b) $C_6H_{11}-CO_2^-$ or $C_6H_{11}-CO_2H$ after acidification

13.49 (a) and (c). Compound (c) is a hemiacetal, which is in equilibrium with the aldehyde in alkaline solution. Compound (d) is an acetal, which is not in equilibrium with the aldehyde under the alkaline conditions of the Tollens test.

13.50 $HOCH_2CH_2CHO + HOCH_2CH_2OH \overset{H^+}{\rightleftharpoons} HOCH_2CH_2CH(OCH_2CH_2O)$ (cyclic acetal)

$\xrightarrow[OH^-]{KMnO_4} {}^-O_2CCH_2CH(OCH_2CH_2O) \xrightarrow{H^+} HO_2CCH_2CHO$

Alkaline or neutral conditions must be used for the oxidation to prevent the acetal from reverting to the aldehyde.

13.51 (b) < (c) < (a). Butanal is the least acidic because its anion can be stabilized by only one carbonyl group. 2,4-Pentanedione is the most acidic because its anion is stabilized by two carbonyl groups. (See Section 13.9.)

13.52 (a) $\rightleftharpoons$ $^-{:}CH_2CHO + H_2O$

(b) $\rightleftharpoons$ [cyclopentane-1,3-dione anion: O=C–$\ddot{C}^-$H–C=O] + HCO_3^-

(c) $\rightleftharpoons$ $(C_6H_5)_2\ddot{C}^-CO_2CH_2CH_3$ Na^+ + $HOCH_2CH_3$

(d) $\rightleftharpoons$ $CH_3CD_2\overset{O}{\overset{||}{C}}CD_3$ + H_2O (many intermediate steps)

(e) $\rightleftharpoons$ (*S*)-$CH_3CH_2\overset{CH_3}{\overset{|}{C}}H\ddot{C}^-HCHO$ + H_2O

(f) $\rightleftharpoons$ [$CH_3CH_2CH_2\ddot{C}^-(CH_3)CHO\!:$ $\leftrightarrow$ $CH_3CH_2CH_2C(CH_3)\text{=CH-}\ddot{O}\!:^-$] + H_2O

achiral

$\rightleftharpoons$ (*R*)(*S*)-$CH_3CH_2CH_2CH(CH_3)CHO$ + OH^-

13.53 (a) $H_2\ddot{C}^- - N^+(=\ddot{O}\!:)(-\ddot{O}\!:^-)$ $\leftrightarrow$ $H_2C = N^+(-\ddot{O}\!:^-)(-\ddot{O}\!:^-)$

(b) $CH_3\overset{:\ddot{O}}{\overset{||}{C}}-\ddot{C}^-HC\equiv N\!:$ $\leftrightarrow$ $CH_3\overset{:\ddot{O}:^-}{\overset{|}{C}}\text{=CH-}C\equiv N\!:$ $\leftrightarrow$ $CH_3\overset{:\ddot{O}}{\overset{||}{C}}CH\text{=C=}\ddot{N}\!:^-$

(c) $C_6H_5-\ddot{C}^-H-\overset{\ddot{O}:}{\overset{||}{C}}H$ $\leftrightarrow$ $C_6H_5-CH\text{=}\overset{:\ddot{O}:^-}{\overset{|}{C}}H$ $\leftrightarrow$ [ring carbanion (ortho)]=$\overset{\ddot{O}:}{\overset{||}{C}}HCH$

$\leftrightarrow$ [ring carbanion (para)]=$\overset{O}{\overset{||}{C}}HCH$ $\leftrightarrow$ [ring carbanion (ortho)]=$\overset{O}{\overset{||}{C}}HCH$

13.54 (a)

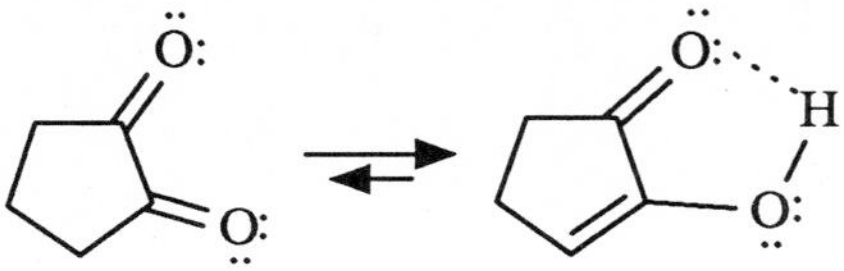

(b) $C_6H_5CCH_3 \rightleftharpoons C_6H_5C{=}CH_2$

(c)

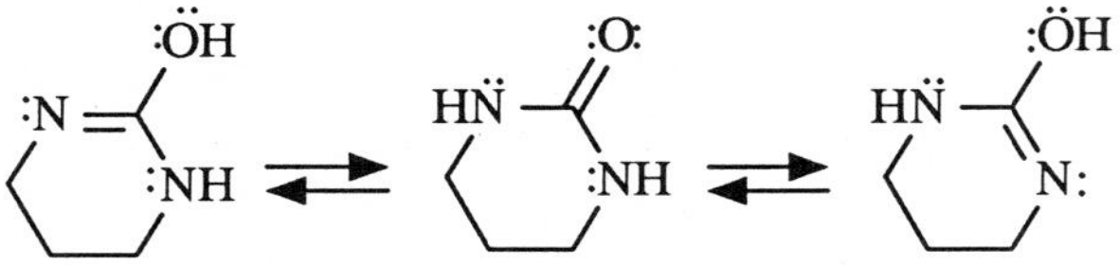

(d)

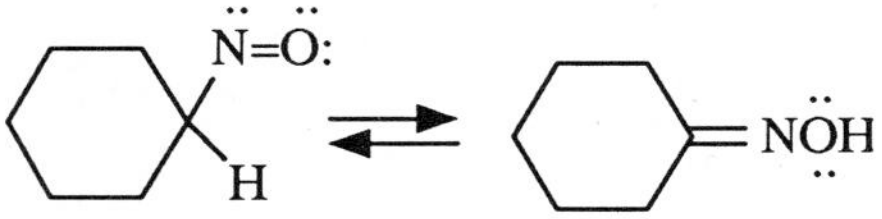

(e)

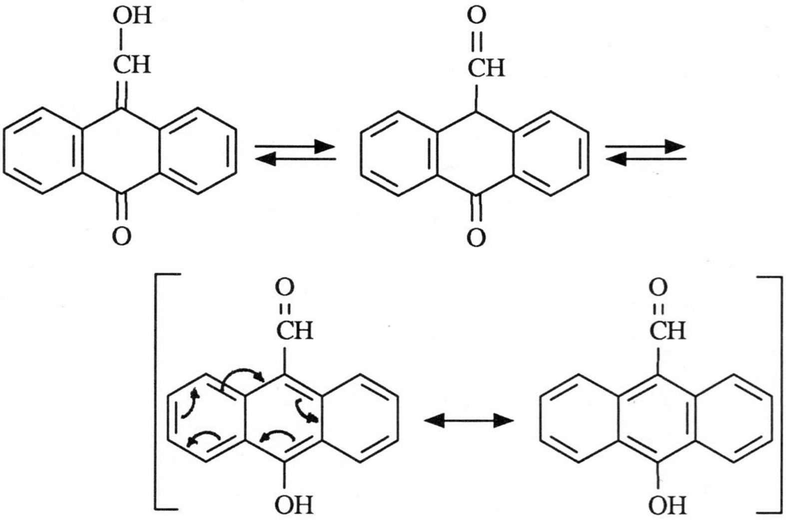

In (e), only pertinent benzene ring resonance structures are shown.

13.55 (a) $C_6H_5\overset{\overset{\displaystyle O}{\|}}{C}CH_2Cl$

(b) 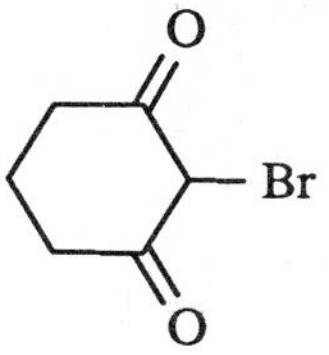 2-bromo-1,3-cyclohexanedione (ring structure: cyclohexane with C=O at positions 1 and 3, Br at position 2)

13.56 (a) $C_6H_{11}\overset{\overset{\displaystyle O}{\|}}{C}CH_3 + I_2/NaOH \longrightarrow C_6H_{11}\overset{\overset{\displaystyle O}{\|}}{C}O^- Na^+ + CHI_3\downarrow$

A yellow precipitate of CHI_3 would be observed.

(b) no reaction

(c) $C_6H_{11}\overset{\overset{\displaystyle OH}{|}}{C}HCH_3 \xrightarrow{I_2} C_6H_{11}\overset{\overset{\displaystyle O}{\|}}{C}CH_3 \xrightarrow[NaOH]{I_2} C_6H_{11}\overset{\overset{\displaystyle O}{\|}}{C}O^- Na^+ + CHI_3\downarrow$

A yellow precipitate would be observed.

13.57 (a) $BrCH_2CH_2\overset{\overset{\displaystyle O}{\|}}{C}CH_3$

(b) $CH_3CHClCH_2CH_2\overset{\overset{\displaystyle O}{\|}}{C}CH_3$

In (b), the double bond is not in conjugation, and therefore Markovnikov's rule applies.

(c) racemic *cis* and *trans* (cycloheptanone ring with O, CH_3 on the adjacent carbon, and OCH_3 on the next carbon)

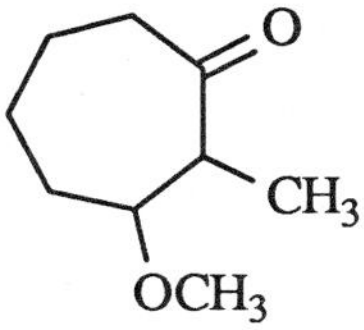

In (c), 1,4-addition would occur because CH_3OH is a weak base and because the ketone is somewhat hindered.

(d) $(CH_3)_3CCH(CO_2CH_2CH_3)_2$

13.58 $CH_2{=}C(CH_3){-}MgBr$ + (cyclohexenone) $\longrightarrow$ $CH_2{=}CCH_3$ ... $\ddot{O}:^- \ \overset{+}{M}gBr$

$\xrightarrow{H_2O,\ H^+}$ CH_3, $CH_2{=}C$, H, H^+, $\ddot{O}{-}H$ $\rightleftharpoons$ $CH_2{=}CCH_3$ ... $=\ddot{O}:$

an enol

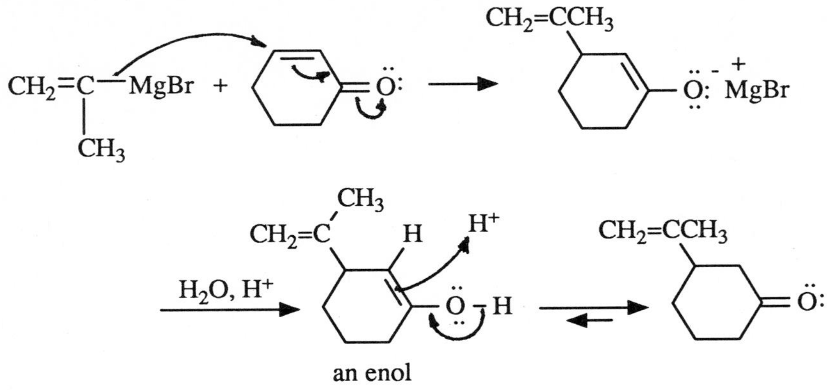

13.59 (a) 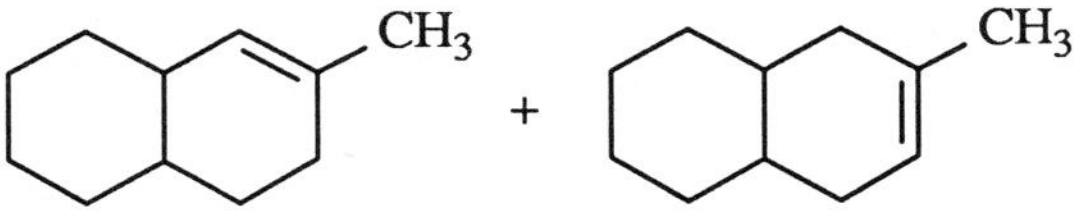 CH_3 + CH_3 (the most substituted alkenes)

(b) 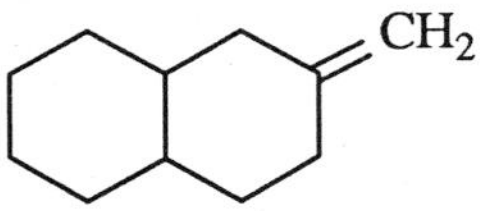 CH_2 + $(C_6H_5)_3PO$

(c) 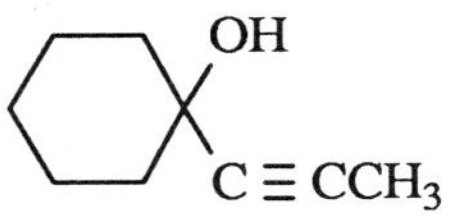 OH, $C\equiv CCH_3$

(d) 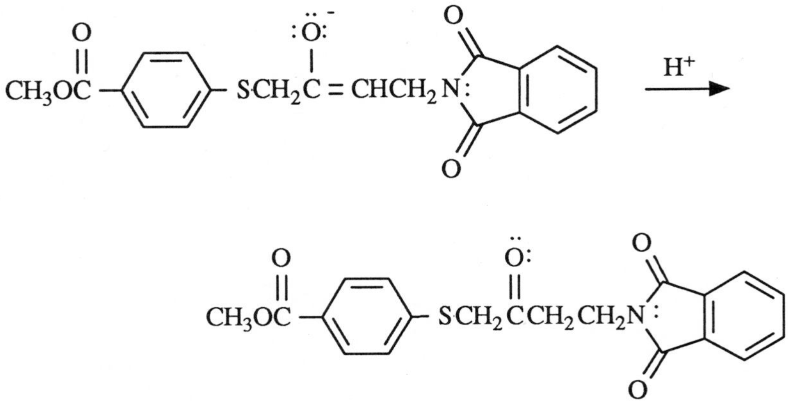 $CH_3OC(=O)$–C_6H_4–$SCH_2C(\ddot{O}:^-){=}CHCH_2N:$ (phthalimide) $\xrightarrow{H^+}$

$CH_3OC(=O)$–C_6H_4–$SCH_2C(=\ddot{O}:)CH_2CH_2N:$ (phthalimide)

(e) [bicyclic ketone structure] + 2 CH_3OH

(f) $CH_3CH_2CH(OH)-CH_2CH(OH)-C_6H_5$

(g) $:\ddot{Br}-\ddot{Cl}: \longrightarrow :\ddot{Br}^{+} + :\ddot{Cl}:^{-}$

$H(H_3C)C=C(H)COCH_3$ + $:\ddot{Cl}:^{-}$ $\longrightarrow$ [$CH_3CH(Cl)CH=C(O^{-})OCH_3$ $\longleftrightarrow$ $CH_3CH(Cl)\ddot{C}^{-}HC(=O)OCH_3$] + $:\ddot{Br}:^{+}$ $\longrightarrow$ $CH_3CH(Cl)-CH(Br)CO_2CH_3$

(h) an acetal $\underset{}{\overset{H^+,\ -CH_3OH}{\rightleftharpoons}}$ [4-methyl-3,4-dihydro-2H-pyran-2-ol] $\rightleftharpoons$

a hemiacetal

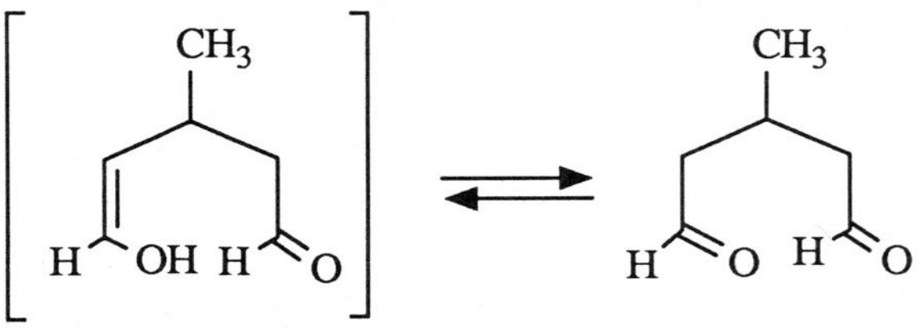

an enol-aldehyde

13.60 (a) 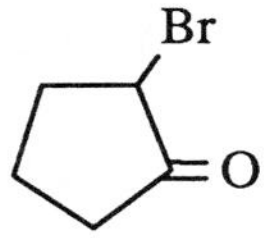 2-bromocyclopentanone (b) 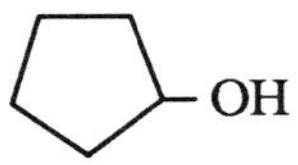 cyclopentanol

(c)  $C_5H_8{=}NNHC_6H_5$

(d) 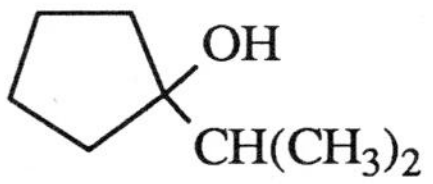 1-$CH(CH_3)_2$-cyclopentanol (OH) + cyclopentylidene$=C(CH_3)_2$

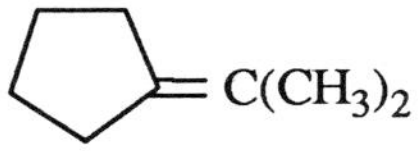

(e) 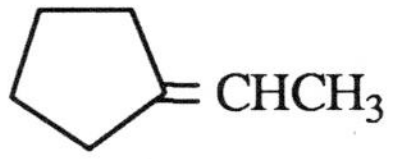 cyclopentylidene$=CHCH_3$ (f) 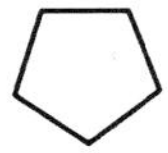 cyclopentane

13.61 (a) $CH_3\overset{O}{\overset{\|}{C}}CH{=}CH_2 \xrightarrow{HCl} CH_3\overset{O}{\overset{\|}{C}}CH_2CH_2Cl \xrightarrow[-CHCl_3]{Cl_2,\ OH^-}$

$^-O\overset{O}{\overset{\|}{C}}CH_2CH_2Cl \xrightarrow{H^+} HO_2CCH_2CH_2Cl$

(b) $CH_3CH_2CH_2CHO \xrightarrow[(2)\ H_2O,\ H^+]{(1)\ NaBH_4} CH_3CH_2CH_2CH_2OH \xrightarrow{HBr}$

$CH_3CH_2CH_2CH_2Br \xrightarrow[(2)\ CH_3(CH_2)_3Li]{(1)\ (C_6H_5)_3P} CH_3CH_2CH_2CH{=}P(C_6H_5)_3$

$$\xrightarrow{CH_3CH_2CH_2CHO} CH_3CH_2CH_2CH{=}CHCH_2CH_2CH_3$$

(c) $$CH_3CH_2CH_2CH_2Br \text{ from (b)} \xrightarrow[\text{ether}]{Mg} CH_3CH_2CH_2CH_2MgBr$$

$$\xrightarrow[(2)\ H_2O,\ H^+]{(1)\ CH_3CH_2\overset{\overset{\displaystyle O}{\|}}{C}CH_3} CH_3CH_2\underset{\underset{\displaystyle CH_3}{|}}{\overset{\overset{\displaystyle OH}{|}}{C}}CH_2CH_2CH_2CH_3$$

(d) $$CH_3CH_2CH_2CH_2OH \text{ from (b)} \xrightarrow{K} CH_3CH_2CH_2CH_2O^- K^+$$

$$\xrightarrow{CH_3CH_2CH_2CH_2Br \text{ from (b)}} (CH_3CH_2CH_2CH_2)_2O$$

(e) $$CH_3CH_2CH_2CHO \xrightarrow{NH_3,\ H_2,\ Pt} CH_3CH_2CH_2CH_2NH_2$$

(f) $$CH_2{=}CHCH_2CHO \xrightarrow[\text{or other mild ox. agent}]{Ag(NH_3)_2^+} CH_2{=}CHCH_2CO_2^-$$

$$\xrightarrow{H^+} CH_2{=}CHCH_2CO_2H$$

(g) $$CH_2{=}CHCH_2CHO \xrightarrow[(2)\ H_2O,\ H^+]{(1)\ NaBH_4} CH_2{=}CHCH_2CH_2OH$$

13.62 (a) $CH_2{=}CHC(=O)CH_2CH_3 \xrightarrow{NH_3} H_2NCH_2CH_2C(=O)CH_2CH_3 \xrightarrow{CH_3CH_2Br}$ product

or $CH_3CH_2Br \xrightarrow{NH_3} CH_3CH_2NH_2 \xrightarrow{CH_2{=}CHC(=O)CH_2CH_3}$ product

(b) 1-cyclohexenyl–$C(=O)CH_3$ + HCN $\xrightarrow{^{-}CN}$ product

In (b), the product would probably be a mixture of *cis* and *trans* isomers (racemic).

(c) 2,2-diphenylcyclohexanone (C_6H_5, C_6H_5, O) $\xrightarrow[^{-}CN]{HCN}$ cyanohydrin (C_6H_5, C_6H_5, HO, CN) $\xrightarrow[-H_2O]{H^+}$ product

(d) $CH_2{=}CHC(=O)H \xrightarrow{HI} ICH_2CH_2C(=O)H \xrightarrow{HOCH_2CH_2OH,\ H^+}$ product

13.63 $RCHO \xrightarrow[H_2,\ Pt]{NH_3} RCH_2NH_2 \xrightarrow{RCHO} RCH_2N{=}CHR$

a 1° amine — *an imine*

$\xrightarrow{H_2,\ Pt} RCH_2NHCH_2R$

a 2° amine

The 1° amine product of the reductive amination can react with RCHO to yield an imine. Reduction of this imine yields a 2° amine.

13.64

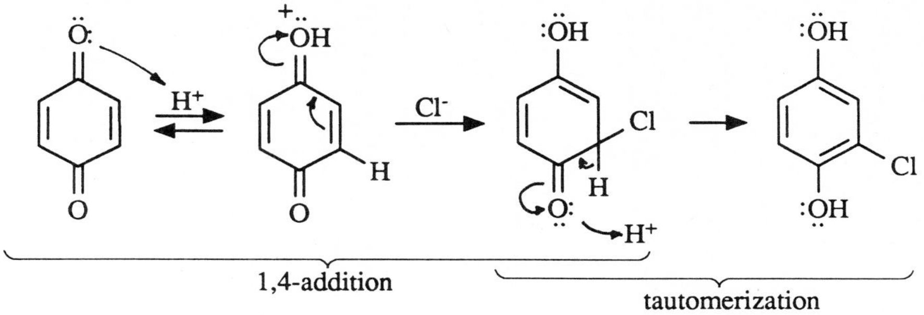

1,4-addition

tautomerization

13.65 $LiAlH_4 + 3\ CH_3CH_2OH \longrightarrow LiAlH(OCH_2CH_3)_3 + 3\ H_2$

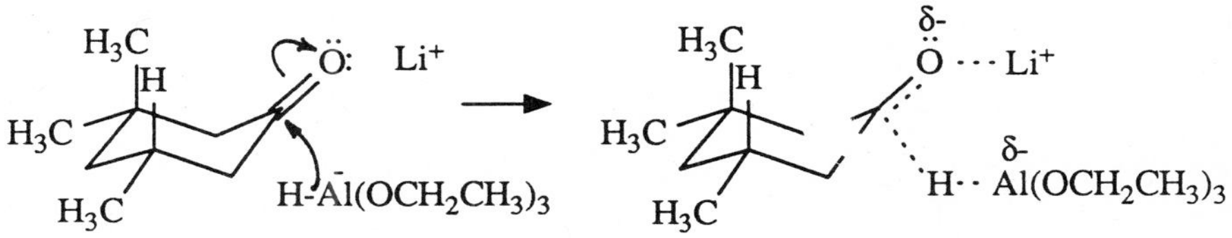

Bulky aluminum hydride ion attacks less hindered side of ring.

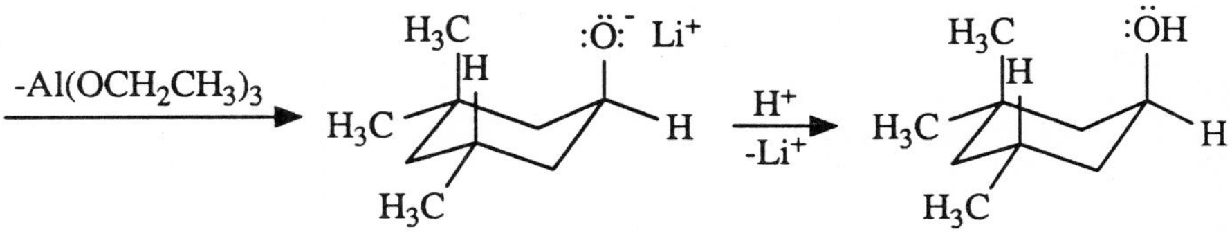

The -OH group is *trans* to the 5-methyl group.

intramolecular hydrogen bond

13.66 (chemical structures of the (Z) and (E) isomers of $CH_3CH_2OC(=O)CH=C(CH_3)NHCH_2C_6H_5$)

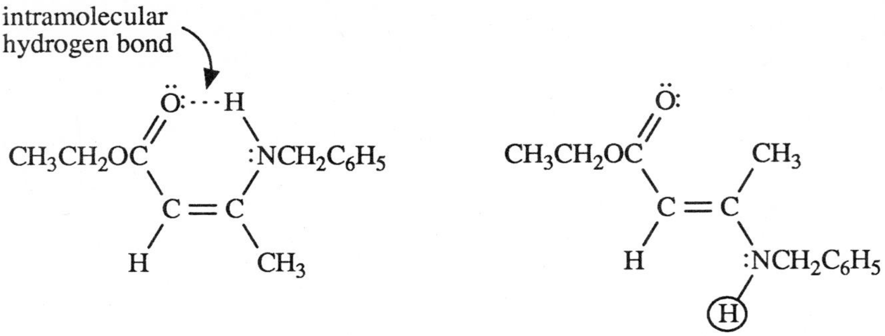

(Z) isomer

(E) isomer

with circled H available for hydrogen bonding

(a) The (Z) isomer undergoes intramolecular hydrogen bonding and thus forms fewer and weaker intermolecular attractions. This isomer would have a lower melting point than the (E) isomer, which can undergo strong intermolecular hydrogen bonding.

(b) Any type of tautomerization would convert the rigid carbon-carbon double bond to a single bond, which can rotate. For example,

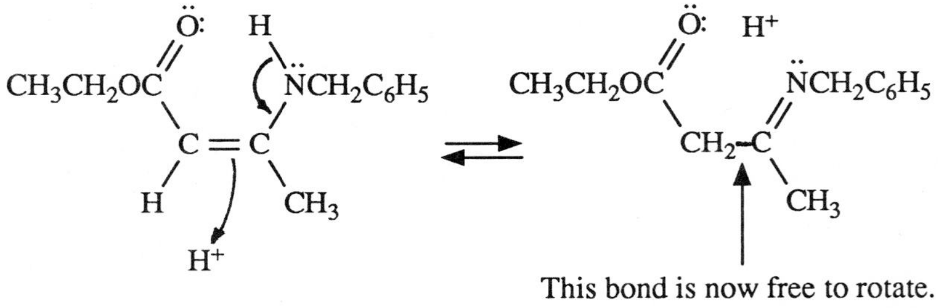

$CH_3CH_2OC(=O)CH=C(CH_3)NHCH_2C_6H_5 \rightleftarrows CH_3CH_2OC(=O)CH_2{-}C(CH_3)=NCH_2C_6H_5 + H^+$

This bond is now free to rotate.

13.67 $CH_3\overset{O}{\overset{\|}{C}}\overset{*}{C}H(C_6H_5)CH_2CH_3$ (3S) $\xrightarrow[(2)\ H_2O]{(1)\ LiAlH_4}$ (2R,3S) 75% + (2S,3S) 25%

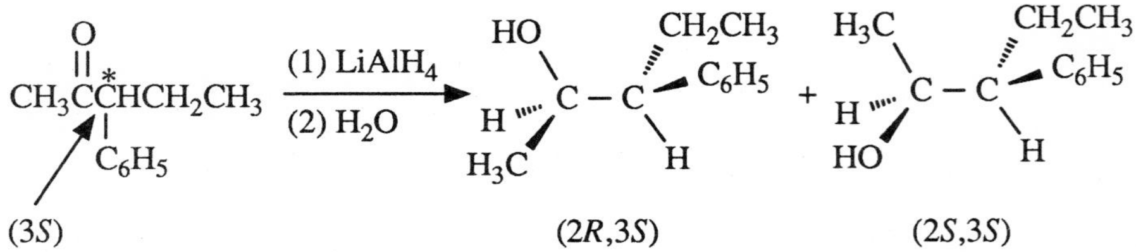

(a) The starting ketone must be the (3S) enantiomer because that chiral carbon is not affected by the reaction.

(b) The transition state leading to the (2R) diastereomer is less hindered and of lower energy than the transition state leading to the (2*S*) diastereomer. Therefore, the (2*R*,3*S*) alcohol is formed at a faster rate.

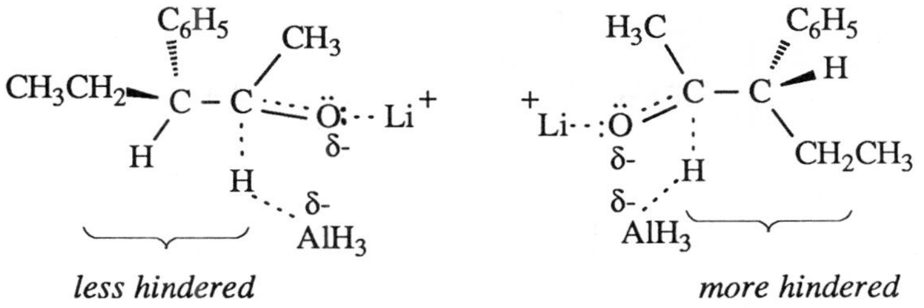

less hindered *more hindered*

The bulky phenyl group is shown in the rear so that the *least hindered transition state* can be drawn in each case. (Any other spatial arrangement of the atoms would show greater hindrance.)

13.68 $CH_3CH{=}O + {}^{-}:OCH_3 \rightleftharpoons CH_3CH(O^-)(OCH_3) \xrightarrow{CH_3OH} CH_3CH(OH)(OCH_3) + CH_3O^-$

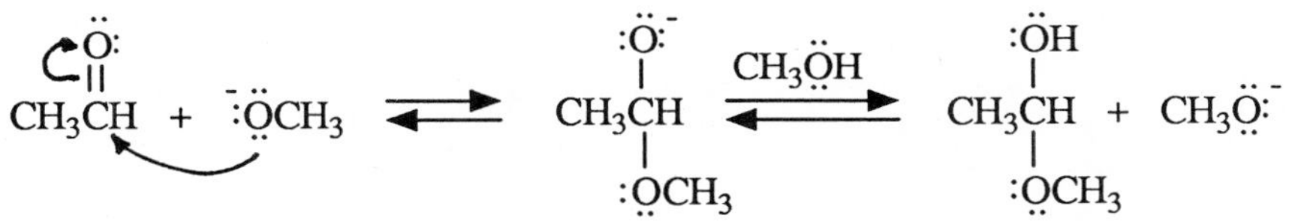

13.69 (a) 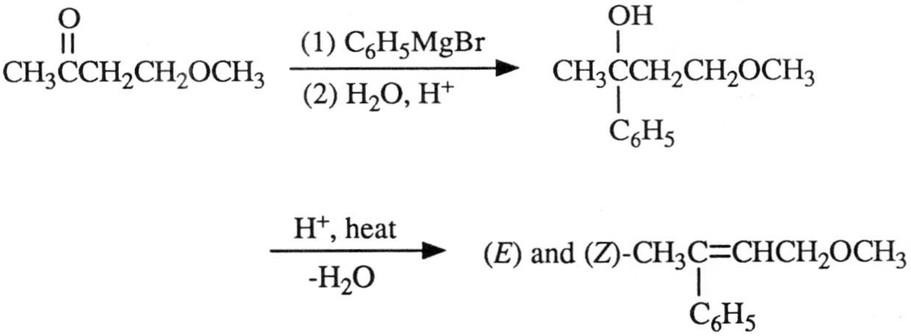 $CH_3COCH_2CH_2OCH_3 \xrightarrow[(2)\ H_2O,\ H^+]{(1)\ C_6H_5MgBr} CH_3C(OH)(C_6H_5)CH_2CH_2OCH_3$

$\xrightarrow[-H_2O]{H^+,\ heat}$ (*E*) and (*Z*)-$CH_3C(C_6H_5){=}CHCH_2OCH_3$

(b) 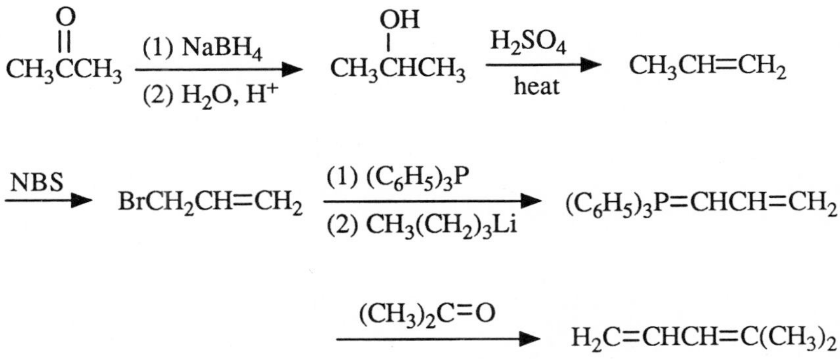 $CH_3COCH_3 \xrightarrow[(2)\ H_2O,\ H^+]{(1)\ NaBH_4} CH_3CH(OH)CH_3 \xrightarrow[heat]{H_2SO_4} CH_3CH{=}CH_2$

$\xrightarrow{NBS} BrCH_2CH{=}CH_2 \xrightarrow[(2)\ CH_3(CH_2)_3Li]{(1)\ (C_6H_5)_3P} (C_6H_5)_3P{=}CHCH{=}CH_2$

$\xrightarrow{(CH_3)_2C=O} H_2C{=}CHCH{=}C(CH_3)_2$

(c) $HCCH_2CH$ (with C=O at each end: $\overset{O}{\overset{\|}{H C}}CH_2\overset{O}{\overset{\|}{C}}H$) $\xrightarrow[(2)\ H_2O,\ H^+]{(1)\ 2\ CH_3CH_2MgBr}$ $CH_3CH_2\underset{}{\overset{OH}{\overset{|}{C}}}HCH_2\overset{OH}{\overset{|}{C}}HCH_2CH_3$

(d) cyclohexanone $\xrightarrow{Cl_2,\ H^+}$ 2-chlorocyclohexanone (Cl) $\xrightarrow[heat]{KOH,\ CH_3CH_2OH}$

2-cyclohexenone $\xrightarrow[1,4\text{-addition}]{H_2O,\ H^+}$ 3-hydroxycyclohexanone (HO) $\xrightarrow{H_2CrO_4}$ 1,3-cyclohexanedione

(e) cyclopentanol (OH) $\xrightarrow{H_2CrO_4}$ cyclopentanone (=O)

cyclopentanol $\xrightarrow{HBr}$ bromocyclopentane (Br) $\xrightarrow[ether]{Mg}$ cyclopentylmagnesium bromide (MgBr)

$\xrightarrow[(2)\ H_2O,\ H^+]{(1)\ \text{cyclopentanone}}$ product

(f) 2-methyl-2-($CH_2CH_2CO_2CH_3$)cyclopentanone (CH_3, =O) $\xrightarrow{Br_2,\ H^+}$ 5-bromo derivative (CH_3, $CH_2CH_2CO_2CH_3$, =O, Br)

$\xrightarrow[heat]{K^+\ {}^-OC(CH_3)_3}$ 5-methyl-5-($CH_2CH_2CO_2CH_3$)-2-cyclopentenone (CH_3, =O)

The ketone contains only one position with alpha hydrogens; therefore, bromination occurs here. Acidic conditions are preferable to alkaline conditions (see Section 13.11). A sterically hindered base should be used in the elimination to minimize formation of the substitution product.

(g) [bicyclic ketone: CH_3 (angular), O, CH_3O, CH_3] $\xrightarrow[\text{(2) } H_2O,\ NH_4^+]{\text{(1) } CH_2{=}CHCH(CH_3)MgBr}$

In (g), a principal by-product might be the following compound, which could arise from an allylic rearrangement of the Grignard reagent.

CH_3 OH $CH_2CH{=}CHCH_3$ CH_3O CH_3

Ammonium chloride was used in the hydrolysis instead of a strong acid to minimize dehydration of the tertiary alcohol. (The complete stereochemistry of the reactant or product was not specified in the literature.)

(h) 2 [o-phenylenediamine: NH_2, NH_2] + [tetrachloroterephthalaldehyde: HC(=O), CH(=O), Cl, Cl, Cl, Cl] $\xrightarrow{\text{buffered solution}}$

(i) $H_2C{=}$[cyclohexane]${=}O$ + $(C_6H_5)_3P{=}CHCH{=}CH_2$ $\longrightarrow$

(j) $CH_3CH{=}CHCH(=O)$ $\xrightarrow{Cl_2,\ H_2O}$ $\left[CH_3\overset{\delta+}{CH}{-}CHCH(=O) \text{ bridged by } \overset{\delta+}{Cl} \right]$ $\longrightarrow$

$CH_3CH(OH){-}CH(Cl)CH(=O)$ $\xrightarrow[-H_2O]{H^+,\ \text{heat}}$ $CH_3CH{=}C(Cl){-}CH(=O)$ $\xrightarrow{Cl_2}$

(k) $CH_3(CH_2)_6CH{=}O \xrightarrow[(2)\ H_2O,\ H^+]{(1)\ CH_2{=}CHMgBr} CH_3(CH_2)_6CH(OH){-}CH{=}CH_2$

$\xrightarrow[cold]{KMnO_4,\ OH^-} CH_3(CH_2)_6CH(OH){-}CH(OH){-}CH_2OH$

or $CH_3(CH_2)_6CH{=}O \xrightarrow[(2)\ H_2O,\ NH_4^+]{(1)\ HC{\equiv}CMgBr} CH_3(CH_2)_6CH(OH){-}C{\equiv}CH \xrightarrow[Pd]{H_2}$

$CH_3(CH_2)_6CH(OH){-}CH{=}CH_2 \xrightarrow{C_6H_5CO_3H} CH_3(CH_2)_6CH(OH)CH{-}CH_2$ (epoxide, O bridging the last two carbons)

$\xrightarrow[(2)\ H^+]{(1)\ KOH,\ H_2O}$

(l) [tricyclic alcohol: OH, H] $\xrightarrow{H_2CrO_4}$ [tricyclic ketone: =O] $\xrightarrow{(C_6H_5)_3P{=}CH_2}$

[tricyclic alkene: =CH₂] $\xrightarrow{BH_3}$ (tricyclic [H] CH_2)$_3$B $\xrightarrow{H_2O_2,\ OH^-}$

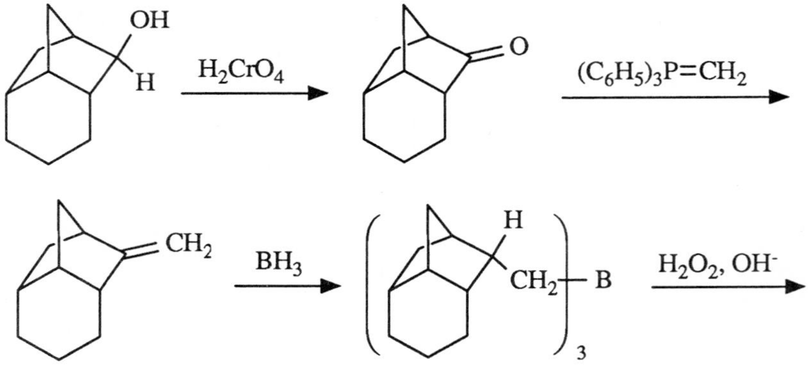

In (l), BH_3 attacks the less hindered side of the alkene molecule. Which side is the less hindered one is not readily apparent in the formulas. A molecular model of the alkene shows that the upper side of the molecule, as we have drawn the formula, in indeed less hindered.

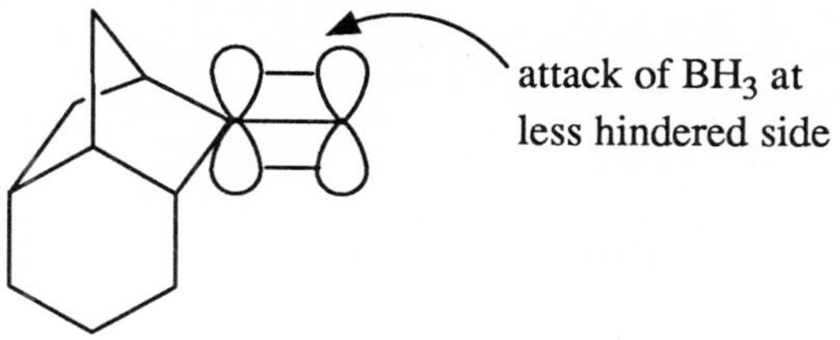

attack of BH_3 at less hindered side

13.70 CH_3O, CH_3O, OCH_3, OCH_3 cyclobutenone + 4 H_2O ⟶ squaric acid + 4 CH_3OH

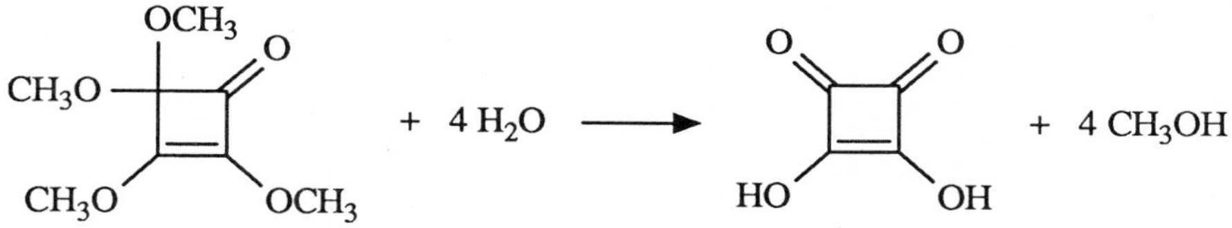

squaric acid

(a) To determine the tautomeric structures, progress around the ring keeping in mind that a carbonyl group requires an alpha hydrogen to undergo tautomerization.

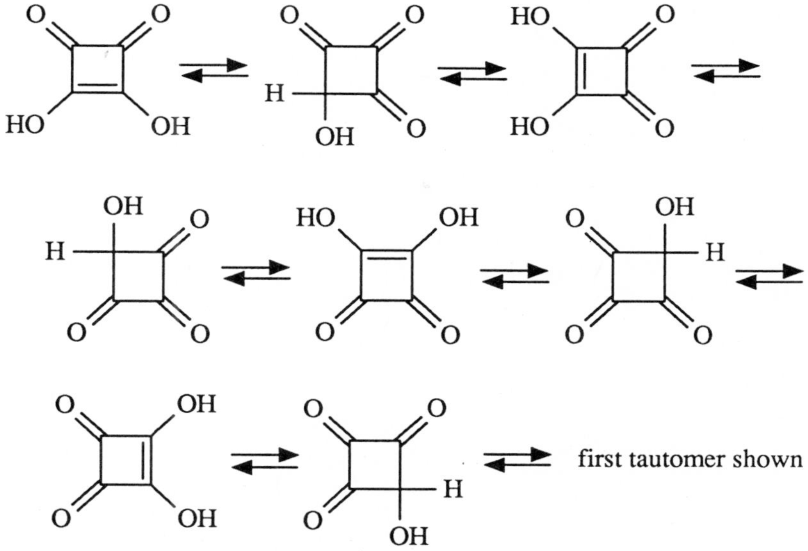

⇌ first tautomer shown

(b) There are two principal tautomeric structures for squaric acid--one contains C-H and O-H bonds, while the other contains no C-H bonds. Proton nmr spectroscopy could be used to estimate the relative quantities of each.

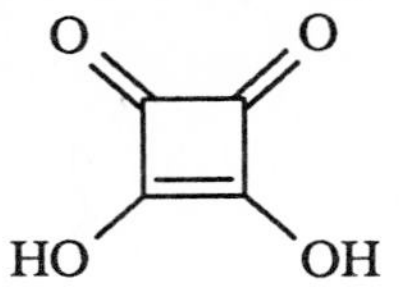

HO, OH (squaric acid structure) — one singlet

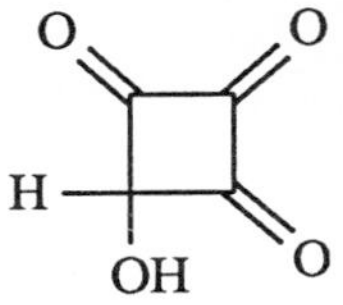

H, OH (tautomer structure) — two singlets

13.71 One of the intermediate oxidation products is a dione, which can undergo enolization and subsequent oxidation at the resulting carbon-carbon double bond.

$$CH_3CH_2CH_2C{\equiv}CCH_2CH_2CH_3 \xrightarrow{KMnO_4} CH_3CH_2CH_2\overset{O}{\overset{\|}{C}}-\overset{O}{\overset{\|}{C}}CH_2CH_2CH_3$$

$$\rightleftharpoons CH_3CH_2CH{=}\overset{HO}{\overset{|}{C}}-\overset{O}{\overset{\|}{C}}CH_2CH_2CH_3 \xrightarrow{KMnO_4}$$

$$CH_3CH_2\overset{O}{\overset{\|}{C}}OH + CO_2 + HO\overset{O}{\overset{\|}{C}}CH_2CH_2CH_3$$

13.72

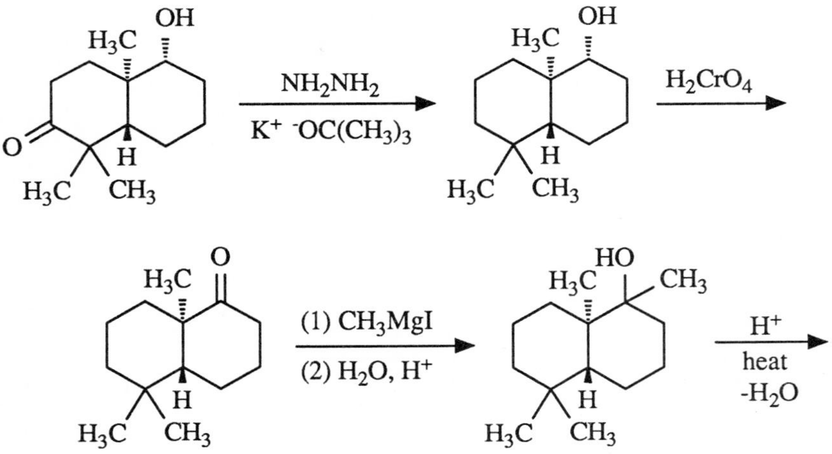

$\xrightarrow[K^+\ ^-OC(CH_3)_3]{NH_2NH_2}$ $\xrightarrow{H_2CrO_4}$

$\xrightarrow[(2)\ H_2O,\ H^+]{(1)\ CH_3MgI}$ $\xrightarrow[heat,\ -H_2O]{H^+}$

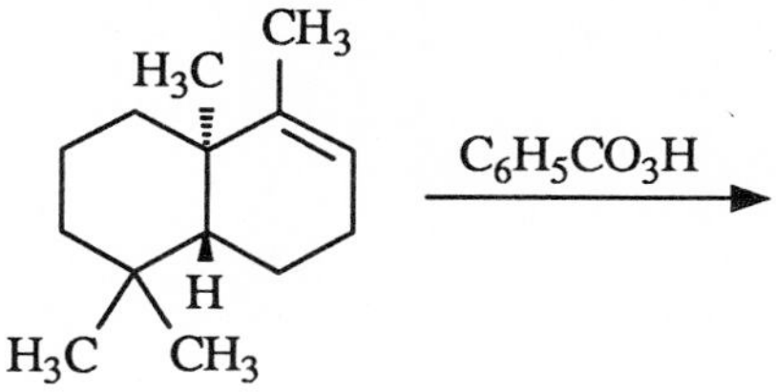

$$\xrightarrow{C_6H_5CO_3H}$$

13.73 $OHCCH_2CH_2CH_2CHO \xrightarrow{H_2O,\ H^+}$ (hydrated) $CH_2(CH_2CHO)(CH_2CH(OH)OH) \longrightarrow$ $\longrightarrow$

$$\longrightarrow$$ 2,6-dihydroxytetrahydropyran (OH, :O:, OH)

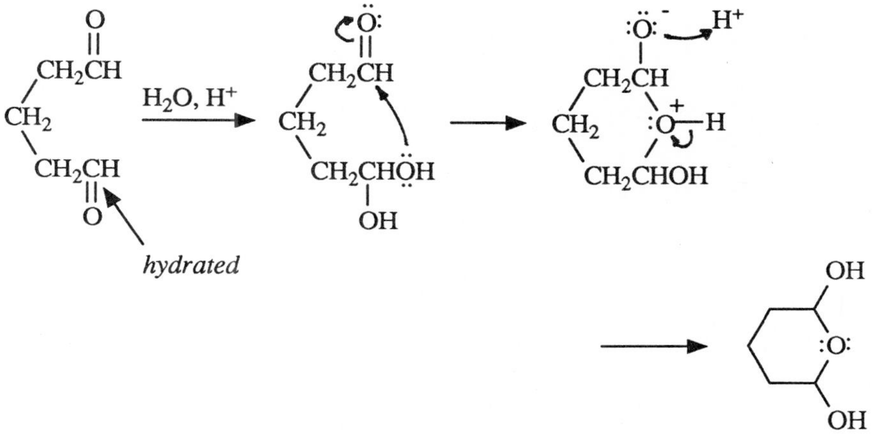

13.74 $HCH{=}O \overset{H^+}{\rightleftharpoons} [H_2C{=}\overset{+}{O}H \longleftrightarrow H_2\overset{+}{C}{-}OH]$ or

$HCH{=}O \overset{ZnCl_2}{\rightleftharpoons} [H_2C{=}\overset{+}{O}{-}\overset{-}{Zn}Cl_2 \longleftrightarrow H_2\overset{+}{C}{-}O\overset{-}{Zn}Cl_2]$

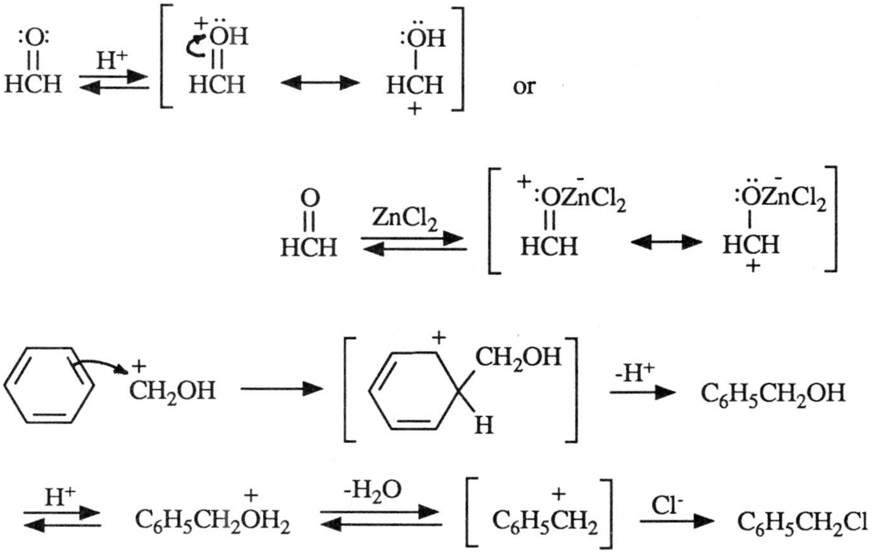

$C_6H_6 + \overset{+}{C}H_2OH \longrightarrow [\ \ ] \xrightarrow{-H^+} C_6H_5CH_2OH$

$\overset{H^+}{\rightleftharpoons} C_6H_5CH_2\overset{+}{O}H_2 \overset{-H_2O}{\rightleftharpoons} [C_6H_5\overset{+}{C}H_2] \xrightarrow{Cl^-} C_6H_5CH_2Cl$

13.75 (Reaction scheme: 2-oxocyclohexyl CH(OH)CO_2CH_3 $\xrightarrow[HCl]{Zn}$ cyclohexyl CH(OH)CO_2CH_3 $\xrightarrow[(2)\ H^+]{(1)\ NaOH}$; (R)-mandelic acid, C_6H_5CH(OH)CO_2H $\xrightarrow{[H]}$ (R)-cyclohexyl CH(OH)CO_2H)

O OH CO_2CH_3 Zn HCl OH CO_2CH_3 (1) NaOH (2) H^+

(R) OH H CO_2H [H] (R) OH H CO_2H

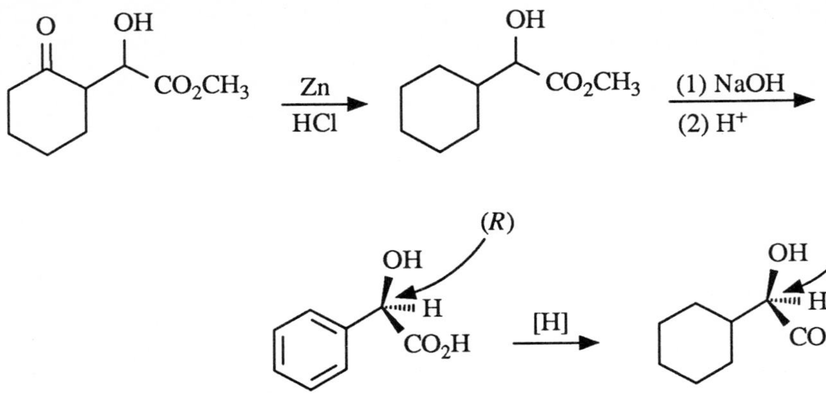

(1) It has one chiral carbon, the alpha carbon.

(2) It is (*R*), the same configuration as the chiral carbon in (*R*)-mandelic acid.

13.76 Civetone, $C_{17}H_{30}O$ (C_nH_{2n-4}), must contain three double bonds or their equivalent. One of these is in a ketone carbonyl group, as determined from the formula and the spectra. Let us consider the chemical properties:

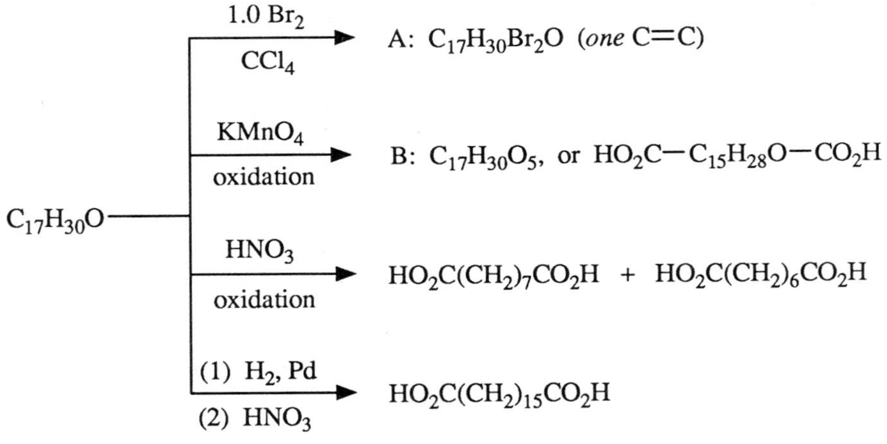

$C_{17}H_{30}O$ —

- $\xrightarrow[CCl_4]{1.0\ Br_2}$ A: $C_{17}H_{30}Br_2O$ (*one* C=C)
- $\xrightarrow[\text{oxidation}]{KMnO_4}$ B: $C_{17}H_{30}O_5$, or $HO_2C{-}C_{15}H_{28}O{-}CO_2H$
- $\xrightarrow[\text{oxidation}]{HNO_3}$ $HO_2C(CH_2)_7CO_2H$ + $HO_2C(CH_2)_6CO_2H$
- $\xrightarrow[(2)\ HNO_3]{(1)\ H_2,\ Pd}$ $HO_2C(CH_2)_{15}CO_2H$

From the preceding information, civetone contains one carbon-carbon double bond. Because it also contains a carbonyl group, it contains one ring. From the permanganate oxidation,

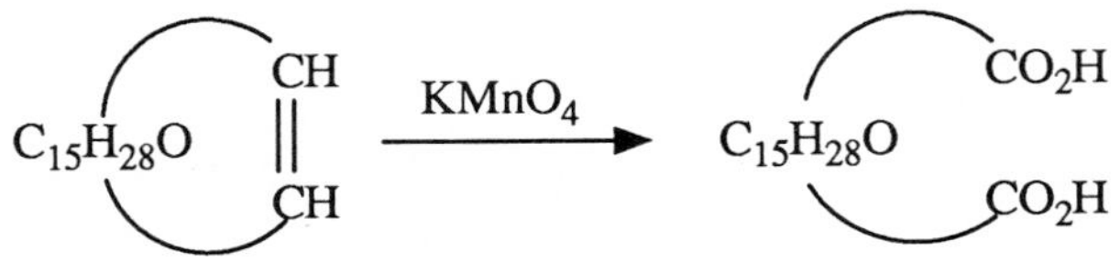

$C_{15}H_{28}O$ (ring) CH=CH $\xrightarrow{KMnO_4}$ $C_{15}H_{28}O$ (CO_2H)(CO_2H)

From the nitric acid oxidation,

$$O{=}C\langle(CH_2)_7CH{=}CH(CH_2)_7\rangle \xrightarrow{HNO_3} HO\overset{O}{\overset{\|}{C}}(CH_2)_7\overset{O}{\overset{\|}{C}}OH \text{ or } HO\overset{O}{\overset{\|}{C}}(CH_2)_6\overset{O}{\overset{\|}{C}}OH$$

From the reduction and subsequent oxidation, we gain confirming evidence for the preceding structure of civetone.

$$O{=}C\langle(CH_2)_7CH{=}CH(CH_2)_7\rangle \xrightarrow[Pd]{H_2} O{=}C\langle(CH_2)_7CH_2{-}CH_2(CH_2)_7\rangle \xrightarrow{HNO_3} HO\overset{O}{\overset{\|}{C}}(CH_2)_6CH_2{-}CH_2(CH_2)_7\overset{O}{\overset{\|}{C}}OH$$

The structures of civetone, compound A, and compound B can be summarized in the following diagram:

$$\underset{\text{civetone}}{O{=}C\langle(CH_2)_7CH{=}CH(CH_2)_7\rangle} \xrightarrow{Br_2} \underset{\text{A}}{O{=}C\langle(CH_2)_7CHBr{-}CHBr(CH_2)_7\rangle}$$

$$\text{civetone} \xrightarrow{KMnO_4} \underset{\text{B}}{O{=}C[(CH_2)_7CO_2H]_2}$$

13.77 (structure: 3-methylcyclopentadecanone ring drawing; O, CH_3)

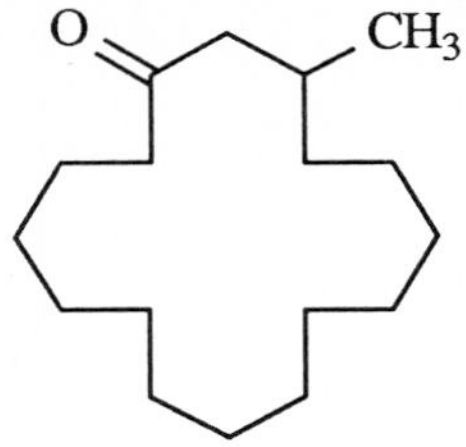

13.78 A: $CH_3\overset{O}{\overset{\|}{C}}(CH_2)_5CH{=}CHCO_2H$ B: $CH_3\overset{O}{\overset{\|}{C}}(CH_2)_5CO_2H$

13.79 (a) $(CH_3)_2C{=}CHCH_2CH{=}CH_2 \xrightarrow[(2)\ H_2O_2,\ OH^-]{(1)\ 9\text{-BBN}}$

$\underset{A}{(CH_3)_2C{=}CHCH_2CH_2CH_2OH} \xrightarrow{\text{pyridine}\cdot CrO_3}$

$\underset{B}{(CH_3)_2C{=}CHCH_2CH_2\overset{O}{\overset{\|}{C}}H} \xrightarrow[(2)\ H_2O,\ NH_4^+]{(1)\ CH_3MgI}$

$(CH_3)_2C{=}CHCH_2CH_2\overset{OH}{\overset{|}{C}}HCH_3$

sulcatol

(b) $(CH_3)_2C{=}CHCH_2CH_2\overset{O}{\overset{\|}{C}}CH_3 \xrightarrow[C_2H_5OH]{NaBH_4} (CH_3)_2C{=}CHCH_2CH_2\overset{OH}{\overset{|}{C}}HCH_3$

the ant pheromone

13.80 $HOCH_2CH_2CH_2CH_2OH$ + O=(cyclohexane-1,4-dione)=O $\overset{H^+}{\rightleftarrows}$

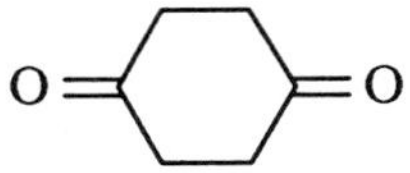

A B

Of the two possible products, A and B, A must be the ketal-ketone and not the diketal.

triplet, δ2.0
infrared carbonyl absorption (1710 cm^{-1})
triplet, δ2.5
multiplet, δ3.8
multiplet, δ1.7

Compound B would not show infrared carbonyl absorption and would show three principal proton nmr signals, not four.

13.81

OCH_3 quartet at δ48.2
singlet at δ95.9
H doublet at δ71.6

CH_3OH / KOH

$C_9H_{13}OI$ A: $C_{10}H_{16}O_2$

Although the formation of the ketal may be concerted, the following stepwise mechanism shows how it can be formed.

$$CH_3\ddot{O}{-}H + {}^{-}:\ddot{O}H \rightleftharpoons CH_3\ddot{O}:^{-} + H_2\ddot{O}:$$

$CH_3\ddot{O}:^{-}$ ⇌ $:\ddot{O}CH_3$ → OCH_3 + $:\ddot{I}:^{-}$

The spectral data in this problem cannot all be interpreted readily. Note that the infrared spectrum of compound A does not contain carbonyl absorption, but that the ^{13}C spectrum does show $-CH_3$ absorption.

13.82 I (d) II (c) III (b)

13.83 A: $C_6H_5CH{=}CHCHO$ B: $C_6H_5CH{=}CHCH_2OH$

The problem does not give sufficient data for determining whether A and B are (*E*) or (*Z*).

Chapter 14

Carboxylic Acids

14.20 (a) 2,2-dimethylpropanoic acid

(b) 2,3-dibromopentanoic acid

(c) 2,5-dihydroxybenzoic acid

(d) *p*-chlorophenoxyethanoic acid (*p*-chlorophenoxyacetic acid)

(e) calcium ethanedioate (calcium oxalate)

(f) sodium *p*-chlorobenzoate

14.21 (a) $CH_2ICH_2CH_2CO_2H$ (b) $HCO_2^- K^+$

(c) $CO_2^- Na^+$ $CO_2^- Na^+$ (d) $C_6H_5CO_2^- Na^+$

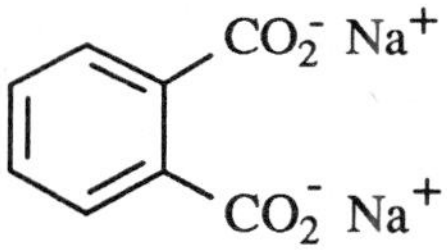

(e) H_3C CO_2H

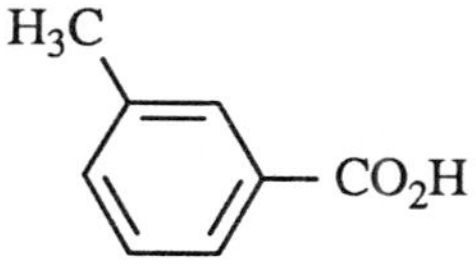

14.22 (a) $CH_3CH_2\overset{O}{\overset{\|}{C}}—$ (b) $CH_3CH_2CH_2\overset{O}{\overset{\|}{C}}—$

(c) O_2N $\overset{O}{\overset{\|}{C}}—$

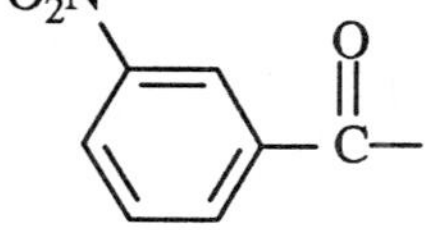

14.23 (a) $H_2\ddot{O}:\cdots\cdots H_2O$

$CH_3CH_2C(=\ddot{O}:\cdots\cdots HO)(OH\cdots\cdots :\ddot{O})CCH_2CH_3$ (cyclic dimer: each C=O hydrogen-bonded to the OH of the other molecule)

$CH_3CH_2C(=\ddot{O}:\cdots\cdots H{-}\ddot{O}{-}H)(OH\cdots)$ (the C=O oxygen and the OH both hydrogen-bonded to one water molecule, $:\ddot{O}—H$)

(b) $OH\cdots:\ddot{O}:\cdots\cdots H$ — CH_3CHC … $:\ddot{O}—H$ … OH (lactic acid: C=O and OH both hydrogen-bonded to one water molecule; the 2-OH also hydrogen-bonded to the C=O oxygen)

HO, $\ddot{O}:\cdots\cdots HO$, OH — CH_3CHC … $CCHCH_3$ … $OH\cdots\cdots :\ddot{O}$ (cyclic dimer of lactic acid)

$CH_3CHOH\cdots\cdots :\ddot{O}{=}C(OH)CHCH_3$, with CO_2H below the first carbon and OH on the second carbon

$H{-}O{-}H\cdots\cdots :\ddot{O}H\cdots\cdots :\ddot{O}{-}H$, with H on the first O and H on the last O; the middle OH belongs to CH_3CHCO_2H

(There are other possibilities)

14.24 (a) $CH_3CH_2CH_2Br \xrightarrow{KCN} CH_3CH_2CH_2CN \xrightarrow[\text{heat}]{H_2O,\ H^+} CH_3CH_2CH_2CO_2H$

or $CH_3CH_2CH_2Br \xrightarrow{Mg,\ ether} CH_3CH_2CH_2MgBr \xrightarrow{CO_2} CH_3CH_2CH_2CO_2MgBr \xrightarrow{H_2O,\ H^+} CH_3CH_2CH_2CO_2H$

(b) $CH_3CH_2CH_2CH_2OH \xrightarrow[\text{heat}]{KMnO_4}$

(c) $CH_3CH_2CH_2CHO \xrightarrow{KMnO_4}$

(d) $CH_3CH_2CH_2CH{=}CHCH_2CH_2CH_3 \xrightarrow[\text{heat}]{KMnO_4}$

In (b), (c), and (d), other oxidizing agents would also be suitable.

(e) $\xrightarrow[\text{heat}]{H_2O,\ H^+}$

(f) $\xrightarrow[\text{heat}]{H_2O,\ H^+}$

In (e) and (f), you might have used alkaline hydrolysis (saponification), followed by acidification.

14.25 (a) Benzene $\xrightarrow[FeBr_3]{Br_2}$ C_6H_5–Br $\xrightarrow[ether]{Mg}$ C_6H_5–MgBr $\xrightarrow[(2)\ H_2O,\ H^+]{(1)\ CO_2}$ C_6H_5–CO_2H

(b) Bromobenzene does not undergo an S_N2 reaction with CN^-. Therefore, the nitrile must be prepared using a benzenediazonium ion.

Benzene $\xrightarrow[H_2SO_4]{HNO_3}$ C_6H_5–NO_2 $\xrightarrow[(2)\ NaOH]{(1)\ Fe,\ HCl}$ C_6H_5–NH_2

$\xrightarrow[H^+]{NaNO_2}$ C_6H_5–$\overset{+}{N}\equiv N$ $\xrightarrow[^-CN]{CuCN}$ C_6H_5–CN

$\xrightarrow[heat]{H_2O,\ H^+}$ C_6H_5–CO_2H

14.26 (a) $pK_a = -\log(1.3 \times 10^{-4}) = 4 - \log 1.3 = 3.9$

(b) $pK_a = -\log(3.65 \times 10^{-5}) = 5 - \log 3.65 = 4.44$

14.27 $K_a = \dfrac{[H^+][A^-]}{[HA]}$ $\qquad pH = -\log[H^+] = 2.50$ $\qquad [H^+] = 10^{-2.50} = 3.20 \times 10^{-3}$

$$K_a = \frac{(10^{-2.50})(10^{-2.50})}{0.010 - (3.20 \times 10^{-3})} = \frac{10^{-5}}{0.00680} = 1.47 \times 10^{-3}$$

14.28 A more electronegative substituent stabilizes the anion relative to the acid to a greater extent; therefore, FCH_2CO_2H is the strongest acid of the three and $BrCH_2CO_2H$ is the weakest.

14.29 (d) < (h) < (e) < (f) < (c) < (b) < (a) < (g). Note the order: alcohol (weakest), water, phenol, carbonic acid (weaker than RCO_2H because HCO_3^- reacts with RCO_2H), carboxylic acids (in order of increasing inductive effect).

14.30 (a) *p*-bromobenzoic acid (b) *m*-bromobenzoic acid

(c) 3,5-dibromobenzoic acid

The reason for these relative acidities is that Br is electron withdrawing by the inductive effect and stabilized the conjugate base relative to the acid. In (c), two Br atoms exert a stronger electron-withdrawing effect than one Br atom.

stabilizes anion by electron-withdrawing inductive effect

Br–C_6H_4–CO_2H ⇌ Br–C_6H_4–CO_2^- + H^+

14.31 *p*-methylphenol < phenol < *p*-nitrophenol.

p-Nitrophenol is a stronger acid than phenol because of resonance stabilization of the anion, which is aided by the nitro group. (See the answer to Problem 12.4 in the text for the resonance structures.) *p*-Methylphenol is a weaker acid than phenol because the methyl group is electron-releasing and destabilizes the anion relative to the conjugate acid.

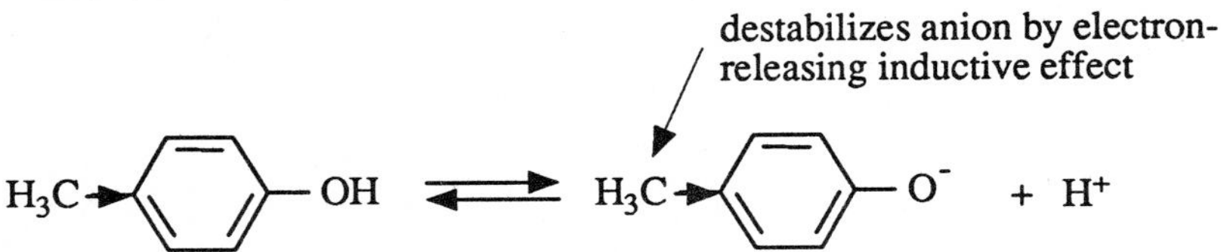

14.32 (b) is the most acidic and (c) is the least acidic.

14.33 In each case, determine which of the parent acids is the weaker acid. The anion of that acid is the stronger base.

(a) $CH_3CH{=}CH^-$ (b) $CH_3CO_2^-$ (c) $ClCH_2CO_2^-$

(d) $(CH_3)_3CO^-$ (e) $ClCH_2CH_2CO_2^-$

14.34 (a) cyclohexane-1,2-di(CO_2^- Na^+) (b) $C_6H_5CH_2O_2CCH_2CH_3$

(c) $CH_3CO_2^-$ + CH_3OH (d) $CH_3CO_2^-$ $CH_3NH_3^+$

(e) CH_3CO_2H + $ClCH_2CO_2^-$ (f) $HO_2CCH_2CO_2^-$ Na^+

(g) no reaction (h) $C_6H_5CO_2CH_3$

(i) $C_6H_5CO_2^-$ Li^+ (j) $C_6H_5O^-$ Li^+

14.35 Mixture $\xrightarrow{NaHCO_3}$ $C_6H_5CO_2^- Na^+$

in Solution A

Remaining mixture $\xrightarrow{NaOH}$ $CH_3CH_2-C_6H_4-O^- Na^+$

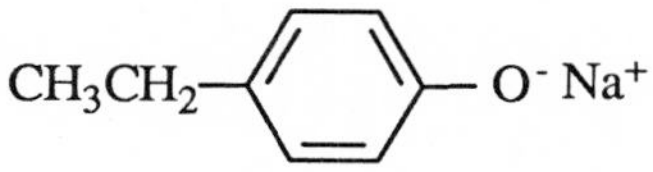

in Solution B

Remaining mixture $\xrightarrow{H_2O}$ no dissolved salts

Solution C

Solution D contains C_6H_5CHO

14.36 (a) Shake with aqueous $NaHCO_3$; the carboxylic acid dissolves and the ester does not.

(b) Shake with aqueous NaOH; *p*-butylphenol dissolves and the ester does not.

(c) Shake with aqueous NaOH; *p*-butylphenol dissolves and the alcohol does not.

14.37 (a) 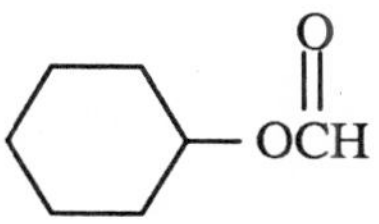 $C_6H_{11}-OCH=O$ (cyclohexyl–O–C(=O)H)

(b) $C_6H_5C(=O)OCH_2CH_2OC(=O)C_6H_5$

(c) $C_6H_5C(=O)OCH_2CH_3$ and $C_6H_5C(={}^{18}O)OCH_2CH_3$

In (c), the intermediate can lose either ^{16}O or ^{18}O:

$$C_6H_5\overset{\overset{\ddot{O}:}{\|}}{C}{}^{18}OH \rightleftharpoons C_6H_5\overset{\overset{+\ddot{O}H}{\|}}{C}{}^{18}OH \rightleftharpoons C_6H_5\underset{\underset{\overset{..}{\underset{+}{H O}}CH_2CH_3}{|}}{\overset{\overset{:\ddot{O}H}{|}}{C}}{}^{18}OH \rightleftharpoons$$

$$CH_3CH_2\ddot{O}H \text{ (attacks carbonyl carbon)}$$

$$C_6H_5\underset{\underset{:\ddot{O}CH_2CH_3}{|}}{\overset{\overset{:\overset{+}{O}H_2}{|}}{C}}{}^{18}\ddot{O}H \quad \text{or} \quad C_6H_5\underset{\underset{:\ddot{O}CH_2CH_3}{|}}{\overset{\overset{:\ddot{O}H}{|}}{C}}{}^{18}\overset{+}{\ddot{O}}H_2$$

$$\updownarrow -H_2O,\ H^+ \qquad\qquad \updownarrow -H_2{}^{18}O,\ H^+$$

$$C_6H_5\underset{\underset{OCH_2CH_3}{|}}{C}={}^{18}O \qquad C_6H_5\underset{\underset{OCH_2CH_3}{|}}{\overset{\overset{O}{\|}}{C}}$$

14.38 (b) < (c) < (a). The order is based upon decreasing steric hindrance.

14.39 (a)

$$CH_3\overset{*}{C}HBrCO_2\overset{*}{C}H(CH_3)CH_2CH_3 \;(R)\;(R) \quad + \quad CH_3\overset{*}{C}HBrCO_2\overset{*}{C}H(CH_3)CH_2CH_3 \;(S)\;(R)$$

(b) These two esters are diastereomers. The enantiomer of the first would be (*S*,*S*), while the enantiomer of the second would be (*R*,*S*).

14.40 (a)

$$H\overset{O}{\overset{\|}{C}}OH + \text{excess } CH_3CH_2OH \overset{H^+}{\rightleftharpoons} H\overset{O}{\overset{\|}{C}}OCH_2CH_3 + H_2O$$

(b)

$$HO\overset{O}{\overset{\|}{C}}-\overset{O}{\overset{\|}{C}}OH + \text{excess } CH_3CH_2OH \overset{H^+}{\rightleftharpoons}$$

$$CH_3CH_2O\overset{O}{\overset{\|}{C}}-\overset{O}{\overset{\|}{C}}OCH_2CH_3 + 2\,H_2O$$

(c) 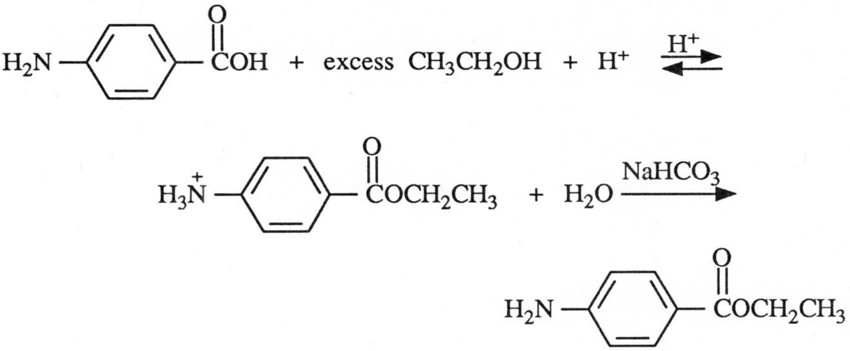 $H_2N-C_6H_4-COOH$ + excess CH_3CH_2OH + H^+ $\overset{H^+}{\rightleftharpoons}$

$H_3\overset{+}{N}-C_6H_4-COOCH_2CH_3$ + H_2O $\xrightarrow{NaHCO_3}$

$H_2N-C_6H_4-COOCH_2CH_3$

14.41 (a) $(CH_3)_3CCH_2OH$ (b) $HOCH_2CH_2CH_2CH_2OH$

(c) 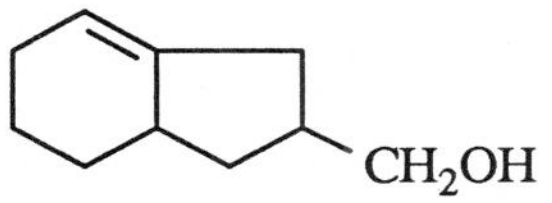 CH_2OH

14.42 *o*-Phthalic acid has a smaller pK_1 (more acidic) because intramolecular hydrogen bonding helps stabilize the anion resulting from the first ionization.

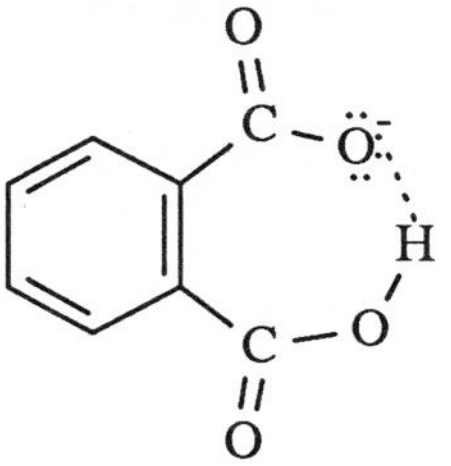

However, the pK_2 of *o*-phthalic acid is larger because of (1) a stabilized reactant, and (2) a dianion that is destabilized by the proximity of two negative charges.

14.43 (a) 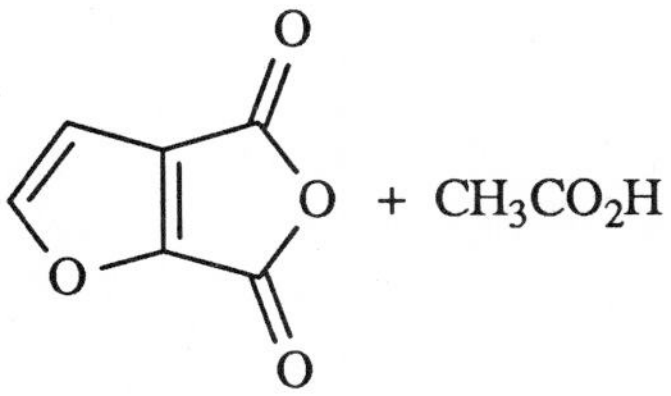 + CH_3CO_2H (b) 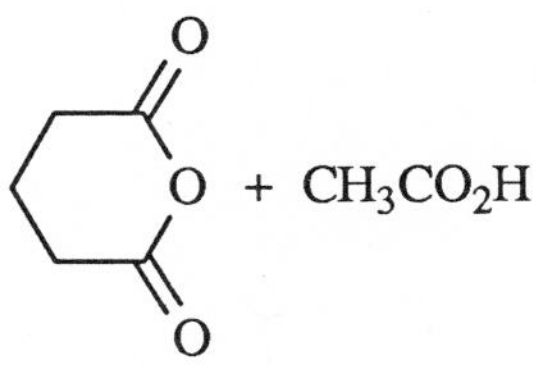 + CH_3CO_2H

14.44 (a) 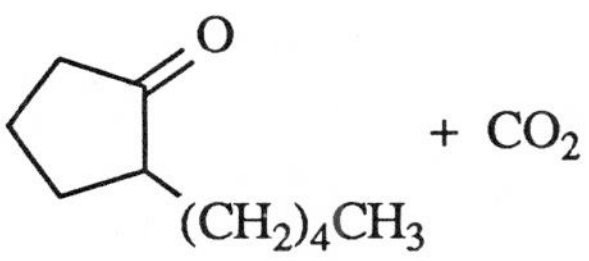 + CO_2 (b) 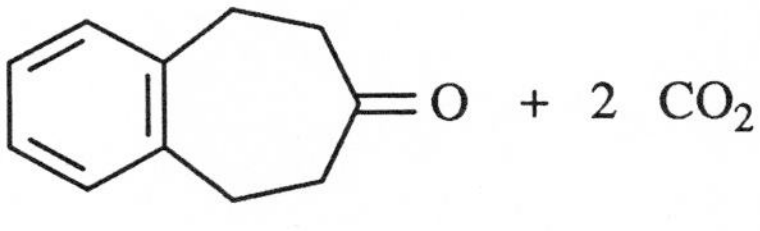 =O + 2 CO_2

$(CH_2)_4CH_3$

In (b), the intermediate would be the following β-keto carboxylic acid:

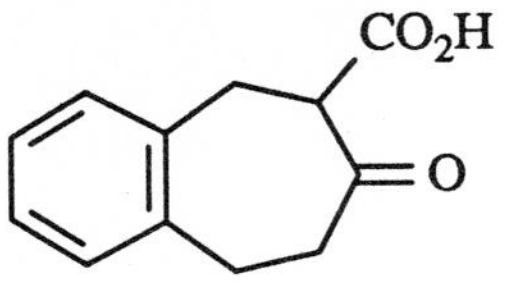

CO_2H

$=O$

14.45 (a) $(HOCH_2)_2C(CO_2CH_2CH_3)_2 \xrightarrow[-CH_3CH_2OH]{HBr,\ H_2O} (HOCH_2)_2C(CO_2H)_2$

$\xrightarrow[-CO_2]{heat} (HOCH_2)_2CHCO_2H \xrightarrow[-H_2O]{H^+} CH_2{=}\underset{\displaystyle CH_2OH}{\underset{|}{C}}CO_2H \xrightarrow{HBr}$ product

(b)

O

$CO_2CH_2CH_3$

$CH_3CH_2O_2C$

O

$\xrightarrow[\text{heat}]{H_2O,\ H^+}$

$- 2\ CH_3CH_2OH$

O

CO_2H

HO_2C

O

$\xrightarrow[-\ 2\ CO_2]{heat}$ product

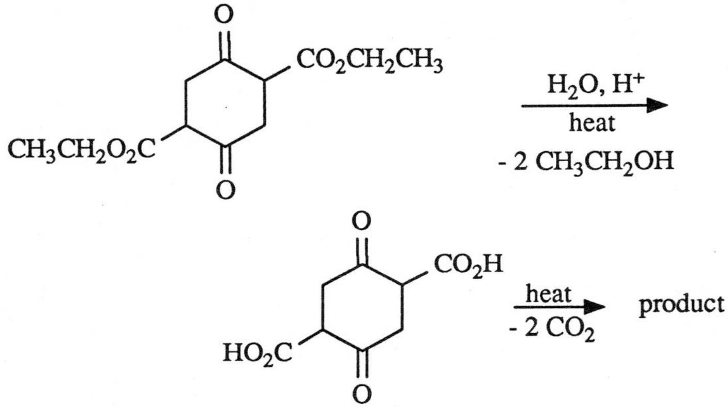

Each ester group undergoes hydrolysis and decarboxylation in separate steps. Consequently, intermediates such as the following would be formed.

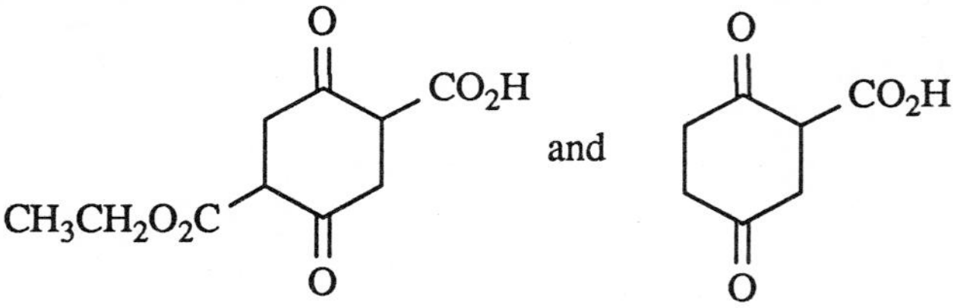

O

CO_2H

$CH_3CH_2O_2C$

O

and

O

CO_2H

O

the β -CO_2H group

CO_2H

14.46 $HO_2CCH_2CH\overset{|}{}\underset{\displaystyle \overset{||}{O}}{C}CO_2H \xrightarrow{-CO_2} HO_2CCH_2CH_2\overset{\displaystyle O}{\overset{||}{C}}CO_2H$

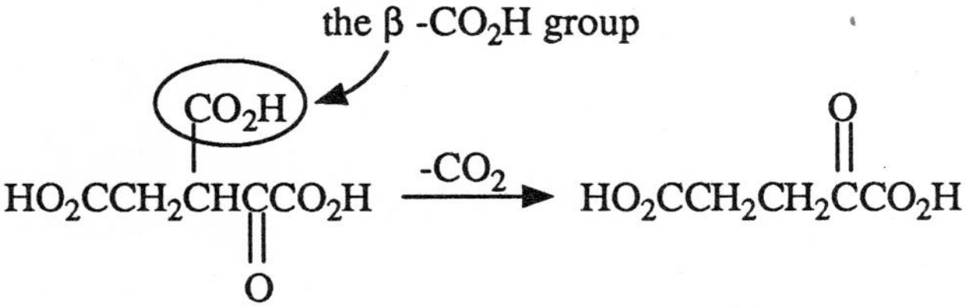

The carboxyl group that is lost is the only one β to a carbonyl group.

14.47 (a) [cyclohexane ring with Cl and CO_2H on adjacent carbons]

cis and *trans* (racemic)

(b) [cyclohexane ring with OH and CO_2H on adjacent carbons]

cis and *trans* (racemic)

(c) $CH_3CH(NHCH_3)CH_2CO_2^-\ H_3\overset{+}{N}CH_3$

(racemic)

14.48 (a) $CH_3CH_2CH_2CH(CO_2^-)_2\ 2\ Na^+ + CO_2 + H_2O$

(b) $CH_3CH_2CH_2CH(CH_2OH)_2$

(c) $CH_3CH_2CH_2CH(CO_2CH_3)_2$

In (c), some decarboxylation might occur. This side reaction would yield $CH_3CH_2CH_2CH_2CO_2H$, which would be converted to $CH_3CH_2CH_2CH_2CO_2CH_3$.

(d) $CH_3CH_2CH_2CH(CO_2^-)_2\ 2\ Na^+$

(e) $CH_3CH_2CH_2CH_2CO_2H$

(f) $CH_3CH_2CH_2CH(CO_2^-)_2 + 2\ CH_4 + 2\ Mg^{2+} + 2\ I^-$

(g) $CH_3CH_2CH_2CH(CO_2^-)_2\ 2\ NH_4^+$

14.49 (a) (1) KCN
(2) dil. HCl, heat
(3) neutralize with NaOH

(b) (1) Mg, diethyl ether
(2) CO_2
(3) dil. HCl

(c) (1) CrO_3, 2 pyridine
(2) HCN
(3) dil. HCl, heat

(d) (1) excess KCN
(2) dil. HCl, heat

(e) (1) HBr, H_2O_2 (or BH_3, followed by $Br_2 + OH^-$)
(2) KCN
(3) dil. HCl, heat

(f) (1) Mg, diethyl ether
(2) CO_2
(3) dil. HCl

(g) dil. HCl, heat

(h) (1) Mg, diethyl ether
(2) $H_2C{-}CH_2$ (ethylene oxide, with O bridging)
(3) H_2O, H^+
(4) H_2CrO_4

14.50 (a) 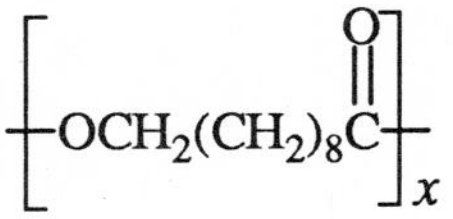

(b) 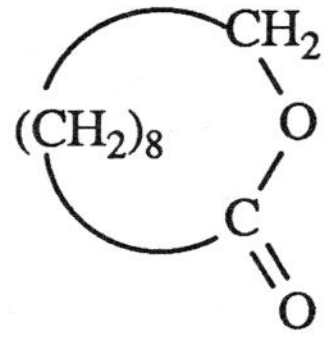

(c) 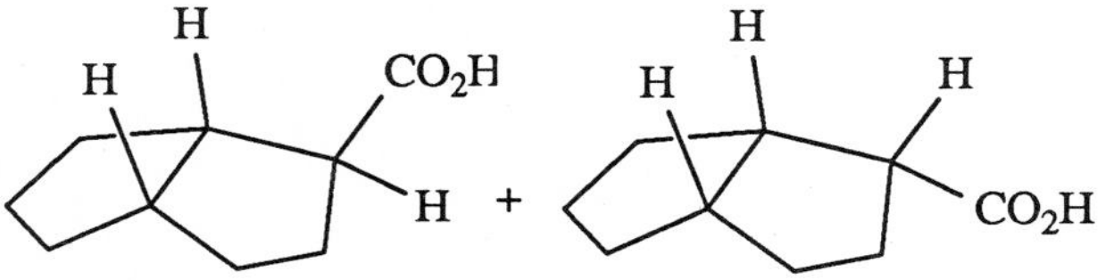

(d) 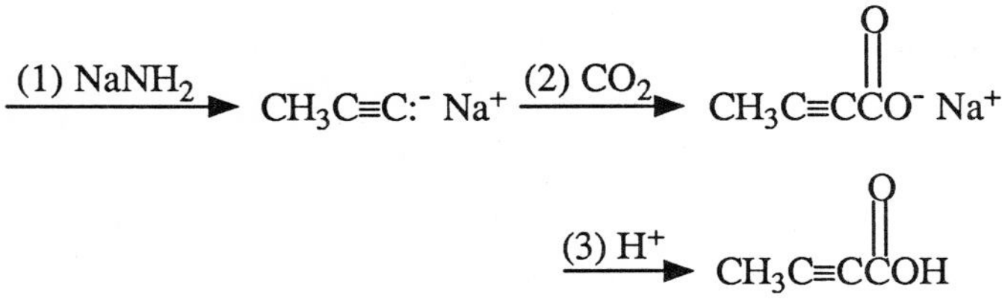

14.51 The oxidation is similar to the oxidation of an alkylbenzene:

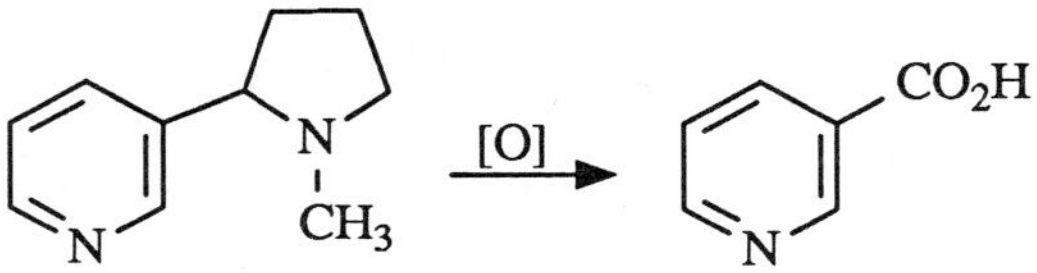

nicotine nicotinic acid

14.52 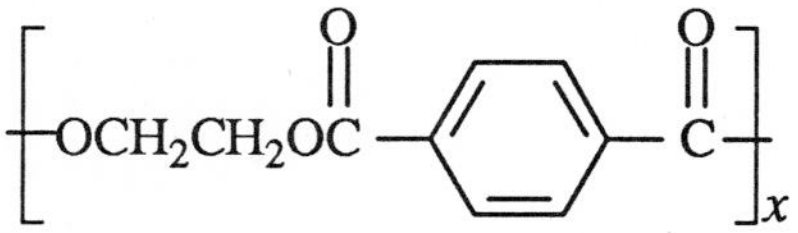

14.53 (a) $^{14}CH_3I \xrightarrow{OH^-} {}^{14}CH_3OH \xrightarrow{CH_3CH_2CO_2H,\ H^+}$

or $^{14}CH_3I \xrightarrow{CH_3CH_2CO_2^-}$

(b) $^{14}CH_3I \xrightarrow{KCN} {}^{14}CH_3CN \xrightarrow[heat]{H_2O,\ H^+}$

(c) $^{14}CH_3I \xrightarrow{OH^-} {}^{14}CH_3OH \xrightarrow{CrO_3 \cdot 2\text{ pyridine}} H^{14}CHO$

$\xrightarrow[\text{(2) } H_2O,\ H^+]{\text{(1) } CH_3MgI} CH_3{}^{14}CH_2OH \xrightarrow{H_2CrO_4}$

14.54 (a) $C_6H_5CO_2H \xrightarrow[FeBr_3]{Br_2}$ $m\text{-}BrC_6H_4CO_2H$ $\xrightarrow[\text{(2) } H_2O,\ H^+]{\text{(1) } LiAlH_4}$

(b) $C_6H_5CO_2H \xrightarrow[\text{(2) } H_2O,\ H^+]{\text{(1) } LiAlH_4} C_6H_5CH_2OH \xrightarrow{HCl} C_6H_5CH_2Cl \xrightarrow{CN^-}$

(c) $C_6H_5CO_2H \xrightarrow[H_2SO_4]{HNO_3}$ $m\text{-}O_2NC_6H_4CO_2H$ $\xrightarrow[\text{(2) } OH^-]{\text{(1) Fe, HCl}}$

$m\text{-}H_2NC_6H_4CO_2H \xrightarrow[\text{(2) } H_2O,\ OH^-]{\text{(1) } CH_3OH,\ H^+,\text{ heat}} m\text{-}H_2NC_6H_4CO_2CH_3$

14.55 $Y{-}C_6H_4{-}OSO_2{-}CH_3 + :\ddot{O}CH_3^- \longrightarrow \left[Y{-}C_6H_4{-}SO_2{-}\overset{\delta-}{O}\cdots CH_3\cdots\overset{\delta-}{O}CH_3 \right]$

S_N2 transition state

→ products

If Y withdraws electron density by the inductive effect, it stabilizes the S_N2 transition state, lowering its energy and thus increasing the rate of reaction. The expected rates (slowest to fastest) would be in the following order, the same order as the ability of the groups to withdraw electron density:

(a) $-CH_3$ < (c) $-H$ < (d) $-Br$ < (b) $-NO_2$

↑ electron-releasing inductive effect (at $-CH_3$)

↑ greatest electron-withdrawing inductive effect (at $-NO_2$)

14.56 Using Ar– to represent the dimethoxyphenyl group:

$$H_3C(H)(Ar)C-C(CO_2CH_2CH_3)_2(NH_2) \xrightarrow{H_2O,\ H^+} H_3C(H)(Ar)C-C(CO_2H)_2(\overset{+}{N}H_3)$$

$$\xrightarrow[-CO_2]{heat}$$ (3R) (2S) + (3R) (2R)

A pair of diastereomers

14.57 (a) $CH_3COH \xrightleftharpoons{H^+} CH_3C(=\overset{+}{O}H)OH$ (+ $H_2{}^{18}O$:) $\rightleftharpoons CH_3C(OH)(OH)(^{18}\overset{+}{O}H_2) \xrightleftharpoons{-H^+} CH_3C(OH)(OH)(^{18}OH) \xrightleftharpoons{H^+}$

$CH_3C(\overset{+}{O}H_2)(OH)(^{18}OH) \xrightleftharpoons{-H_2O} CH_3C(=\overset{+}{O}H)(^{18}OH) \xrightleftharpoons{-H^+} CH_3C(=O)-{}^{18}OH$

↕

$CH_3C(OH)(=^{18}\overset{+}{O}H) \xrightleftharpoons{-H^+} CH_3C(=^{18}O)OH$

(b) CH_2CO_2H (cyclopentenyl) $\xrightleftharpoons{H^+}$ cyclopentyl cation with $CH_2-C(=O)OH$ $\rightleftharpoons$

bicyclic lactone ($\overset{+}{O}H$) $\xrightleftharpoons{-H^+}$ bicyclic lactone

(c) [Mechanism scheme: bromocubyl ketone attacked by :OH⁻; tetrahedral intermediate loses Br^- to give cubanecarboxylic acid (in brackets)]

$\xrightarrow[-H_2O]{OH^-}$ cubane-CO_2^- $\xrightarrow{H_2O,\ H^+}$ cubane-CO_2H

(d) $BrCH_2(CH_2)_4CH_2CO_2H \xrightarrow[-Br^-,\ -H_2O]{2\ OH^-} HOCH_2(CH_2)_4CH_2CO_2^- \xrightarrow{2\ H^+}$

$HOCH_2(CH_2)_4CH_2C(=OH^+)OH$ → [cyclic oxonium tetrahedral intermediate with OH, OH] ⇌ [cyclic intermediate with $:OH_2^+$ and OH]

$\underset{-H_2O}{\rightleftharpoons}$ [protonated eight-membered lactone, $=\overset{+}{O}H$] $\underset{-H^+}{\rightleftharpoons}$ [eight-membered lactone]

(e) $C_6H_5CCl_3 \xrightarrow[-Cl^-]{OH^-} C_6H_5CCl_2(OH) \xrightarrow[-H_2O,\ -Cl^-]{OH^-} C_6H_5CCl(=O) \xrightarrow{OH^-}$

$C_6H_5C(O^-)(OH)Cl \xrightarrow{-Cl^-} C_6H_5COH(=O) \xrightarrow[-H_2O]{OH^-} C_6H_5CO_2^- \xrightarrow{H_2O,\ H^+} C_6H_5CO_2H$

14.58 $HO_2CCH(OH)-CH(OH)CO_2H \xrightarrow[-H_2O]{heat} HO_2CCH{=}C(OH)CO_2H$ (*an enol*) $\longrightarrow HO_2CCH_2C(=O)CO_2H$ (β *to C=O*)

$\xrightarrow{-CO_2} CH_3C(=O)CO_2H$

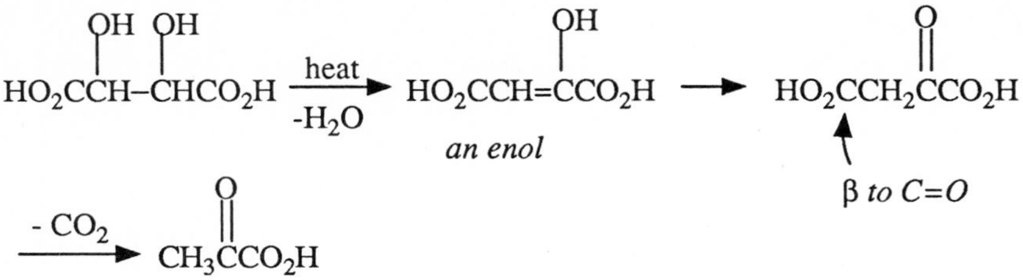

14.59 A carboxylate does not contain a true carbonyl group; it is a composite of two resonance structures:

$RC(\cdots O^{\delta-})(\cdots O^{\delta-})$ not $RC(=\ddot{O}:)(-\ddot{O}:^-)$ or $RC(-\ddot{O}:^-)(=\ddot{O}:)$

resonance structures

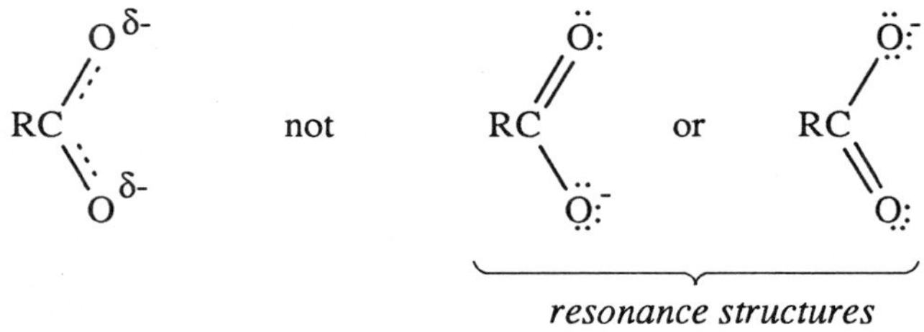

14.60 The general formula for A shows two sites of unsaturation. When A is converted to B, a molecule of water is lost. These reactions suggest that a diacid is converted to an anhydride.

The addition of two molecules of CH_3OH with the loss of two molecules of water suggest the formation of a diester. The $LiAlH_4$ reduction to a compound with *no* sites of unsaturation and with the loss of two oxygen atoms suggests the reduction of a diacid. The following scheme shows the conclusions.

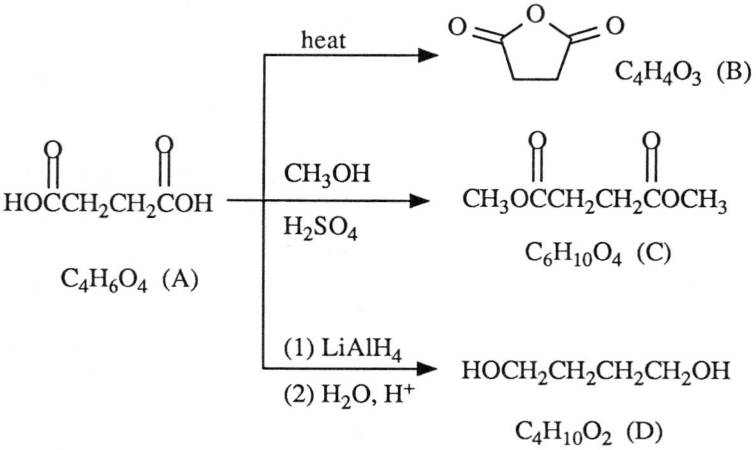

$HOC(=O)CH_2CH_2C(=O)OH$ $C_4H_6O_4$ (A)

- heat → succinic anhydride (cyclic, O=C–O–C=O ring) $C_4H_4O_3$ (B)
- CH_3OH, H_2SO_4 → $CH_3OC(=O)CH_2CH_2C(=O)OCH_3$ $C_6H_{10}O_4$ (C)
- (1) $LiAlH_4$ (2) H_2O, H^+ → $HOCH_2CH_2CH_2CH_2OH$ $C_4H_{10}O_2$ (D)

14.61 (a) $CH_3(CH_2)_4CH_2Br \xrightarrow[\text{ether}]{\text{Mg}} CH_3(CH_2)_4CH_2MgBr \xrightarrow[\text{(2) } H_2O,\ H^+]{\text{(1) } \overset{O}{CH_2CH_2}}$

$CH_3(CH_2)_4CH_2CH_2CH_2OH \xrightarrow{\text{HBr}} CH_3(CH_2)_7Br \xrightarrow{\text{KCN}} CH_3(CH_2)_7CN$

$\xrightarrow[\text{heat}]{H_2O,\ H^+}$

(b) $CH_3CH_2CH_2CH{=}CHCO_2H \xrightarrow[\text{(2) } CN^-]{\text{(1) neutralize}} CH_3CH_2CH_2\overset{CN}{C}HCH_2CO_2H$

$\xrightarrow[\text{heat}]{H_2O,\ H^+} CH_3CH_2CH_2\overset{CO_2H}{C}HCH_2CO_2H \xrightarrow[\text{heat}]{\text{excess } CH_3OH,\ H^+}$

(c) cyclopentadiene + $HO_2C(H)C{=}C(H)CO_2H$ $\xrightarrow{\text{heat}}$

a cis-diacid

14.62 (a) $CH_3CH_2Br \xrightarrow{\text{KCN}} CH_3CH_2CN \xrightarrow[\text{heat}]{H_2O,\ H^+} CH_3CH_2CO_2H$

$CH_3CH_2Br \xrightarrow{OH^-} CH_3CH_2OH \xrightarrow[\text{heat}]{CH_3CH_2CO_2H,\ H^+} CH_3CH_2CO_2CH_2CH_3$

A Grignard reaction could also be used to prepare the carboxylic acid.

(b) cyclopentanol $\xrightarrow{H_2CrO_4}$ cyclopentanone $\xrightarrow{Cl_2,\ H^+}$ 2-chlorocyclopentanone (Cl) $\xrightarrow{\text{KCN}}$

2-cyanocyclopentanone (CN) $\xrightarrow[\text{(2) } H^+]{\text{(1) } H_2O,\ OH^-,\ \text{heat}}$

(c) 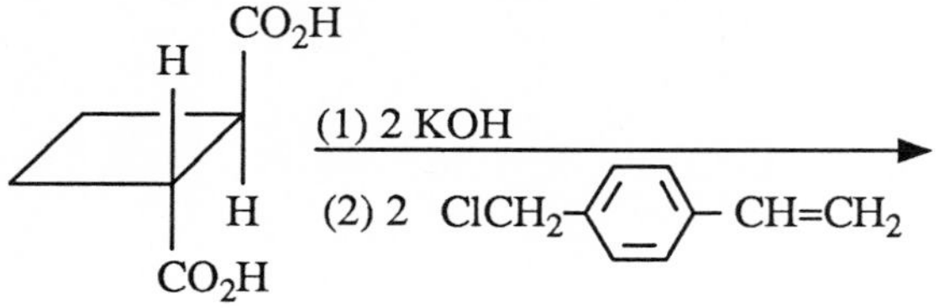 [cyclobutane-1,2-dicarboxylic acid, trans] $\xrightarrow[\text{(2) 2 } ClCH_2\text{-}C_6H_4\text{-}CH{=}CH_2]{\text{(1) 2 KOH}}$

14.63 From the formula ($C_4H_8O_2$) and the infrared spectrum (RCO_2H), we can determine the partial structure $C_3H_7CO_2H$ for compound A. The only possible structures for compound A are $CH_3CH_2CH_2CO_2H$ and $(CH_3)_2CHCO_2H$. Compound B must therefore be either $CH_3CH_2CH_2CH_2OH$ or $(CH_3)_2CHCH_2OH$. The nmr spectrum shows typical isopropyl group absorption: a septet (area 1) for $(CH_3)_2C\underline{H}$- and a doublet (area 6) for $(C\underline{H}_3)_2CH$. Therefore:

$$\underset{\text{A}}{(CH_3)_2CHCO_2H} \xrightarrow[\text{(2) } H_2O,\ H^+]{\text{(1) } LiAlH_4} \underset{\text{B}}{(CH_3)_2CHCH_2OH}$$

14.64 $$\underset{\text{A}}{CH_3\overset{Br}{\overset{|}{C}}HCH_2CO_2H} \xrightarrow[\text{(2) } H^+]{\text{(1) } (CH_3)_3CO^-} \underset{\text{B}}{CH_2{=}CHCH_2CO_2H} + \underset{\text{C}}{CH_3CH{=}CHCO_2H}$$

A could not be $(CH_3)_2CBrCO_2H$ or $CH_3\overset{CH_2Br}{\overset{|}{C}}HCO_2H$ because only one alkenyl acid could result from the elimination reaction. For B, note the $=CH_2$ bending mode at about 920 cm^{-1} (10.8 μm) in the infrared spectrum. The nmr spectrum of C shows CH_3 absorption (the doublet at about δ1.9) and $C\underline{H}{=}C\underline{H}$ downfield (besides the offset $-CO_2H$). From these data, you cannot tell if C is the *cis* or *trans* acid.

Chapter 15

Derivatives of Carboxylic Acids

15.31 (b) < (a) < (c), based primarily upon the difference in electronegativities between the two atoms of the bond.

15.32 (b) < (a) < (c)

15.33 (a) 3-methylbutanoyl chloride

(b) *p*-bromobenzoyl bromide

(c) 3-butenoyl chloride

15.34 (a)

$$\textit{cis}\text{-}CH_3(CH_2)_7CH{=}CH(CH_2)_7CO_2H + SOCl_2 \xrightarrow{\text{heat}} \textit{cis}\text{-}CH_3(CH_2)_7CH{=}CH(CH_2)_7\overset{O}{\overset{\|}{C}}Cl + SO_2 + HCl$$

(b)

$$3\ \textit{cis}\text{-}CH_3(CH_2)_7CH{=}CH(CH_2)_7CO_2H + PCl_3 \xrightarrow{\text{heat}} 3\ \textit{cis}\text{-}CH_3(CH_2)_7CH{=}CH(CH_2)_7\overset{O}{\overset{\|}{C}}Cl + H_3PO_3$$

15.35 (a) $CH_3CH_2\overset{O}{\overset{\|}{C}}OH$ + HCl

(b) $CH_3CH_2\overset{O}{\overset{\|}{C}}O$–(cyclohexyl) + HCl

(c) $CH_3CH_2\overset{O}{\overset{\|}{C}}O$–(p-phenylene)–Br + HCl

(d) $CH_3CH_2\overset{O}{\overset{\|}{C}}$ - N(piperidine ring) + (piperidinium ring, $\overset{+}{N}H_2$) Cl^-

(e) $CH_3CH_2\overset{O}{\overset{\|}{C}}OCH_2CH_3 + CH_3CH_2OH + NaCl$

(f) $CH_3CH_2\overset{O}{\overset{\|}{C}}O^- Na^+ + H_2O + NaCl$

(g) $CH_3CH_2\overset{O}{\overset{\|}{C}}O\overset{O}{\overset{\|}{C}}H + NaCl$

(h) $CH_3CH_2CH_2OH$

(i) $CH_3CH_2\overset{OH}{\overset{|}{C}}(CH_2CH_3)_2$

(j) $CH_3\underset{\underset{Br}{|}}{C}H\overset{O}{\overset{\|}{C}}OH$

15.36 (a) $HO\overset{O}{\overset{\|}{C}}OH \longrightarrow H_2O + CO_2$

(b) $CH_3CH_2O\overset{O}{\overset{\|}{C}}OCH_2CH_3$

(c) $Cl\overset{O}{\overset{\|}{C}}OCH_2CH_3$

(d) $H_2N\overset{O}{\overset{\|}{C}}OCH_2CH_3 + NH_4^+ Cl^-$

In (d), note that the more reactive group is the first to leave; continued treatment with NH_3 would yield $H_2N\overset{\overset{\displaystyle O}{\|}}{C}NH_2$.

15.37 (a) pentanoic propanoic anhydride

(b) *p*-chlorobenzoic phenylethanoic anhydride or *p*-chlorobenzoic phenylacetic anhydride

(c) benzoic methanoic anhydride or benzoic formic anhydride

(d) ethanoic methanoic anhydride or acetic formic anhydride

15.38 (a)

$$CH_3CH_2\overset{\overset{\displaystyle O}{\|}}{C}O^- + Cl\overset{\overset{\displaystyle O}{\|}}{C}CH_2CH_2CH_2CH_3 \longrightarrow \text{product}$$

$$\text{or } CH_3CH_2\overset{\overset{\displaystyle O}{\|}}{C}Cl + {}^-O\overset{\overset{\displaystyle O}{\|}}{C}CH_2CH_2CH_2CH_3 \longrightarrow \text{product}$$

(b)

$$Cl\text{–}C_6H_4\text{–}\overset{\overset{\displaystyle O}{\|}}{C}O^- + Cl\overset{\overset{\displaystyle O}{\|}}{C}CH_2\text{–}C_6H_5 \longrightarrow \text{product}$$

$$\text{or } Cl\text{–}C_6H_4\text{–}\overset{\overset{\displaystyle O}{\|}}{C}Cl + {}^-O\overset{\overset{\displaystyle O}{\|}}{C}CH_2\text{–}C_6H_5 \longrightarrow \text{product}$$

(c)

$$H\overset{\overset{\displaystyle O}{\|}}{C}O^- + Cl\overset{\overset{\displaystyle O}{\|}}{C}C_6H_5 \longrightarrow \text{product}$$

(d)

$$H\overset{\overset{\displaystyle O}{\|}}{C}O^- + Cl\overset{\overset{\displaystyle O}{\|}}{C}CH_3 \longrightarrow \text{product}$$

In (c) and (d), only the method shown is feasible: $HC(=O)Cl$ (HCCl with C=O) is unstable.

15.39 (a) $C_6H_4(CH_2CO_2^-Na^+)(CO_2^-Na^+)$ + H_2O (b) $C_6H_4(CH_2COOH)(COOH)$

(c) $C_6H_4(CH_2COOCH_2CH_3)(COOH)$ + $C_6H_4(CH_2COOH)(COOCH_2CH_3)$

(d) $C_6H_4(CH_2COOCH_2CH_3)(CO_2^-Na^+)$ + $C_6H_4(CH_2CO_2^-Na^+)(COOCH_2CH_3)$

(e) $C_6H_4(CH_2CONHCH_3)(CO_2^-\ H_3\overset{+}{N}CH_3)$ + $C_6H_4(CH_2CO_2^-\ H_3\overset{+}{N}CH_3)(CONHCH_3)$

15.40 $CH_3CH_2O-C_6H_4-NH_2$ + $CH_3C(=O)-OC(=O)CH_3$ + pyridine $\longrightarrow$

$CH_3CH_2O-C_6H_4-NH-\overset{O}{\overset{||}{C}}CH_3 + CH_3\overset{O}{\overset{||}{C}}O^- +$ (pyridinium, $C_5H_5N^+-H$)

15.41 (a) (o-$C_6H_4(CO_2H)_2$) $\xrightarrow[-H_2O]{heat}$ (phthalic anhydride) $\xrightarrow{CH_3OH}$ product

(b) (1,1-cyclohexanediacetic acid, CH_2CO_2H / CH_2CO_2H) $\xrightarrow[-H_2O]{heat}$ (cyclic anhydride)

$\xrightarrow{C_6H_5CH_2OH}$ product

15.42 (a) ethyl phenylethanoate or ethyl phenylacetate

(b) methyl *o*-aminobenzoate

15.43 (a) $C_6H_5-CH_2\overset{O}{\overset{||}{C}}OH$ + excess CH_3CH_2OH $\xrightarrow[heat]{H^+}$

(b) (o-$H_2NC_6H_4\overset{O}{\overset{||}{C}}OH$) + CH_3OH + H^+ $\xrightarrow{heat}$

In (b), excess acid must be used because of the presence of the basic amino group. Finally, the ester-amine salt must be neutralized.

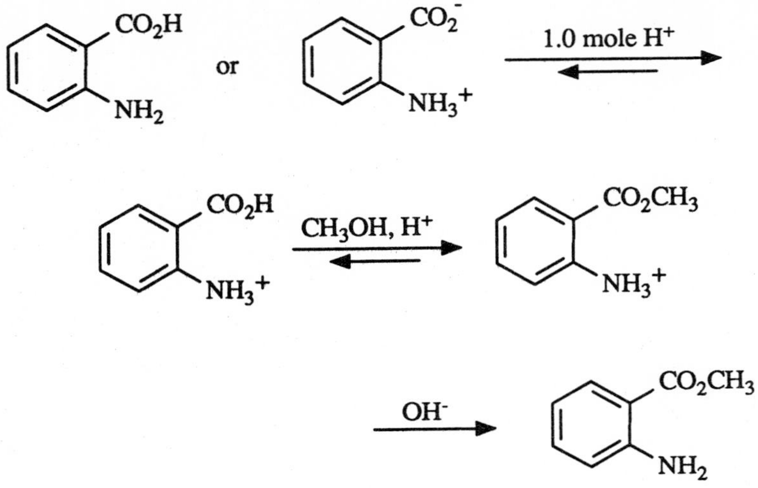

CO_2H, NH_2 or CO_2^-, NH_3^+ $\xrightleftharpoons{\text{1.0 mole } H^+}$

CO_2H, NH_3^+ $\xrightleftharpoons{CH_3OH,\ H^+}$ CO_2CH_3, NH_3^+

$\xrightarrow{OH^-}$ CO_2CH_3, NH_2

15.44 CO_2H, H_3C, CH_3, CH_3 + CH_3CH_2OH $\xrightleftharpoons{H^+}$ no ester

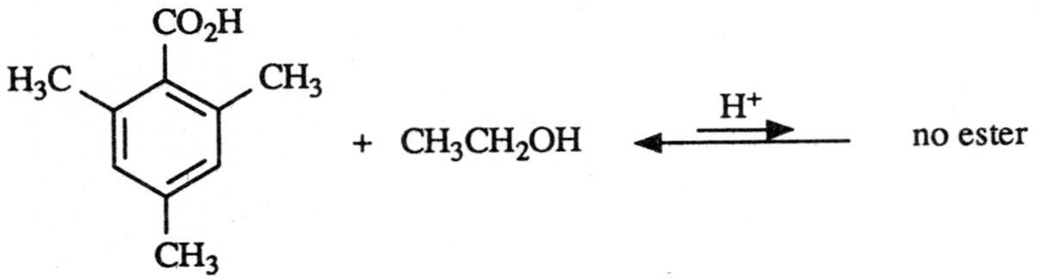

The esterification reaction proceeds by way of a hindered intermediate. In this case, the intermediate would be sufficiently hindered that it cannot be formed.

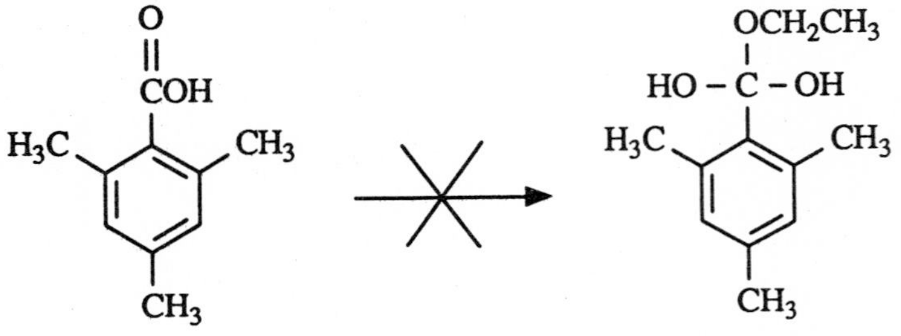

O, COH, H_3C, CH_3, CH_3 ⟶ (crossed arrow) OCH_2CH_3, HO – C – OH, H_3C, CH_3, CH_3

not formed

To synthesize the ester, we must use a more reactive carboxylic acid derivative. For example:

$$\underset{\text{ArCOH}}{\overset{\text{O}}{\|}} \xrightarrow{\text{SOCl}_2} \underset{\text{ArCCl}}{\overset{\text{O}}{\|}} \xrightarrow[\text{pyridine}]{\text{CH}_3\text{CH}_2\text{OH}} \underset{\text{ArCOCH}_2\text{CH}_3}{\overset{\text{O}}{\|}}$$

15.45 (a) $CH_3CH_2CH_2\overset{O}{\overset{\|}{C}}OCH_3 + H_2O \overset{H^+}{\rightleftharpoons} CH_3CH_2CH_2\overset{O}{\overset{\|}{C}}OH + CH_3OH$

(b) $CH_3CH_2CH_2\overset{O}{\overset{\|}{C}}NH_2 + CH_3OH$

(c) $CH_3CH_2CH_2\overset{O}{\overset{\|}{C}}OCH_2CH_3 + {}^-OCH_3$

(d) $CH_3CH_2CH_2\overset{O}{\overset{\|}{C}}O^- Na^+ + CH_3OH$

(e) $CH_3CH_2CH_2C(O^-\,{}^+MgBr)(C_6H_5)-C_6H_5 + CH_3O^-\,{}^+MgBr$

(f) $CH_3CH_2CH_2CH_2OH + CH_3OH$

15.46 Saponification is the reaction of choice because heating in dilute HCl could lead to addition of HCl and H_2O to the carbon-carbon double bond and to allylic rearrangement.

$$CH_3CH{=}CHCH_2O\overset{O}{\overset{\|}{C}}CH_3 \xrightarrow[\text{(2) neutralize with } H^+]{\text{(1) } H_2O,\ OH^-,\ \text{heat}} CH_3CH{=}CHCH_2OH + CH_3CO_2H$$

15.47 (a) $CH_3CO_2CH_2CH_3 \xrightarrow[\text{(2) } H^+]{\text{(1) } H_2O,\ OH^-,\ \text{heat}} CH_3CO_2H + CH_3CH_2OH$

(b) $CH_3CO_2CH_2CH_3 \xrightarrow[\text{(2) } H_2O,\ H^+]{\text{(1) } LiAlH_4} 2\ CH_3CH_2OH$

(c) $CH_3CO_2CH_2CH_3 \xrightarrow[\text{(2) } H_2O,\ H^+]{\text{(1) } 2\ CH_3MgI} (CH_3)_3COH + CH_3CH_2OH$

(d) $CH_3CO_2CH_2CH_3 \xrightarrow[\text{(2) } SOCl_2]{\text{(1) } H_2O,\ H^+,\ \text{heat}} CH_3\overset{O}{\overset{\|}{C}}Cl$

$\xrightarrow[AlCl_3]{C_6H_6} CH_3\overset{O}{\overset{\|}{C}}C_6H_5$

(e) $CH_3CO_2CH_2CH_3 \xrightarrow[\text{heat}]{NaOH,\ H_2O} CH_3CO_2^-\ Na^+ + HOCH_2CH_3$

(f) $CH_3CO_2CH_2CH_3 \xrightarrow{CH_3NH_2} CH_3\overset{O}{\overset{\|}{C}}NHCH_3 + HOCH_2CH_3$

15.48 (a) [cocaine structure: bicyclic ring with NCH_3 bridge, CO_2CH_3, H, $O_2CC_6H_5$, H] + 2 NaOH $\xrightarrow{H_2O}$

cocaine

[structure: bicyclic ring with NCH_3 bridge, $CO_2^-\ Na^+$, H, OH, H] + $C_6H_5CO_2^-\ Na^+$ + CH_3OH $\xrightarrow{\text{excess } H^+}$

saponification products

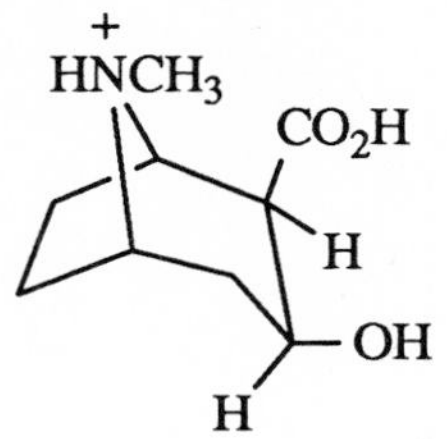

$\overset{+}{H}NCH_3$ … CO_2H … H … OH … H $+ C_6H_5CO_2H + CH_3OH$

products after acidification

(b) nepetalactone (CH_3, CH_3, O) + NaOH $\xrightarrow{H_2O}$ [CH_3, CHOH, $\overset{O}{\overset{\|}{C}}O^- Na^+$, CH_3]

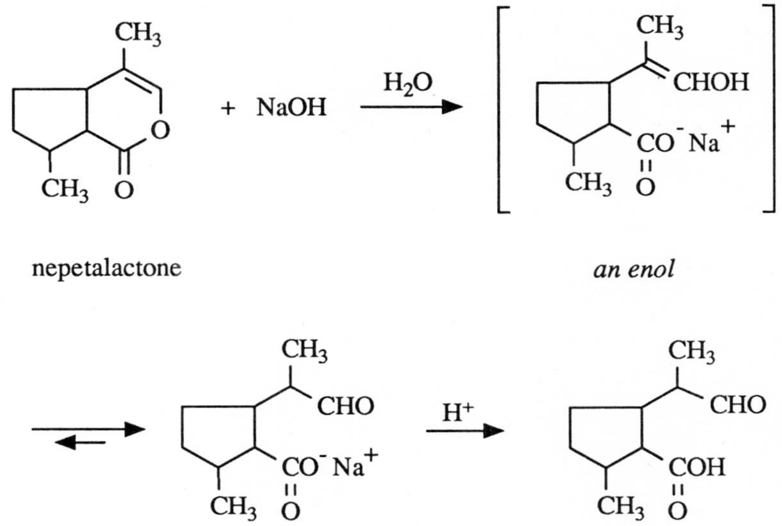

nepetalactone

an enol

$\rightleftarrows$ CH_3, CHO, $\overset{O}{\overset{\|}{C}}O^- Na^+$, CH_3 $\xrightarrow{H^+}$ CH_3, CHO, $\overset{O}{\overset{\|}{C}}OH$, CH_3

saponification product

product after acidification

15.49 (a) $BrCH_2CH_2CH_2\overset{O}{\overset{\|}{C}}OH$

(b) $CH_3\overset{OH}{\overset{|}{C}}(\underset{CH_3}{\underset{|}{C}}{=}CHCH_3)_2 + CH_3CH_2OH$

(c) $ClCH_2\overset{O}{\overset{\|}{C}}NH_2 + CH_3CH_2OH$

(d) $CH_3C(=O)CH_2OH + CH_3OC(=O)CH_3$

(e) $O_2N{-}C_6H_4{-}CO_2H + CH_3OH$

(f) $C_6H_5C(=O)CH_2C(=O)NHC_6H_5 + CH_3CH_2OH$

15.50 (a) octanamide

(b) *N*-phenylethanamide (*N*-phenylacetamide; acetanilide)

(c) *N*-phenyl-2-ethylbutanamide

(d) *N,N*-diethylbenzamide

15.51 (1) $CH_3(CH_2)_4C(=O){-}Cl + 2\ CH_3NH_2 \longrightarrow$

$CH_3(CH_2)_4C(=O)NHCH_3 + CH_3\overset{+}{N}H_3\ Cl^-$

(2) $CH_3(CH_2)_4C(=O){-}OC(=O)(CH_2)_4CH_3 + 2\ CH_3NH_2 \longrightarrow$

$CH_3(CH_2)_4C(=O)NHCH_3 + CH_3(CH_2)_4C(=O)O^-\ CH_3\overset{+}{N}H_3$

(3) $CH_3(CH_2)_4\overset{O}{\overset{||}{C}}-OCH_3 + CH_3NH_2 \longrightarrow$

$CH_3(CH_2)_4\overset{O}{\overset{||}{C}}NHCH_3 + CH_3OH$

15.52 $H_2N-C_6H_3(OCH_2CH_3)-\overset{O}{\overset{||}{C}}OCH_2CH_3 + CH_3\overset{O}{\overset{||}{C}}-O\overset{O}{\overset{||}{C}}CH_3 \longrightarrow$

Ethopabate + $CH_3\overset{O}{\overset{||}{C}}OH$

15.53 (a) (1) $HO-C_6H_4-\overset{+}{N}H_3\ Cl^- + CH_3\overset{O}{\overset{||}{C}}OH$

(2) $Na^+\ {}^-O-C_6H_4-NH_2 + CH_3\overset{O}{\overset{||}{C}}O^-\ Na^+$

(b) (1) $Cl-C_6H_4-\overset{+}{N}H_3\ Cl^- + (CH_3)_2\overset{+}{N}H_2\ Cl^- + \left[HO\overset{O}{\overset{||}{C}}OH\right]$

$\downarrow$

$H_2O + CO_2$

(2) $Cl-C_6H_4-NH_2 + {}^-O\overset{O}{\overset{||}{C}}O^- + (CH_3)_2NH$

15.54 (a) $\xrightarrow[\text{(2) } H_2O,\ H^+]{\text{(1) } LiAlH_4}$ $CH_3CH_2\overset{+}{N}(CH_3)HC_6H_5$ (A) $\xrightarrow{OH^-}$ $CH_3CH_2N(CH_3)C_6H_5$ (B)

A

B

(b) $\xrightarrow[\text{(2) } H_2O,\ H^+]{\text{(1) } LiAlH_4}$ cyclohexene ring bearing two adjacent $CH_2\overset{+}{N}H(CH_3)_2$ groups

C

$\xrightarrow{OH^-}$ cyclohexene ring bearing two adjacent $CH_2N(CH_3)_2$ groups

D

15.55 (a) 2-methylpropanenitrile

(b) *p*-methylphenylethanenitrile or (*p*-methylphenyl)ethanenitrile

15.56 (a) $CH_3CH(C_6H_5)C{\equiv}N$ (b) $CH_3CH(Cl)CH_2C{\equiv}N$

15.57 (a) 1-naphthylamine (NH_2) $\xrightarrow[0°]{NaNO_2,\ HCl}$ 1-naphthalenediazonium chloride ($\overset{+}{N}_2\ Cl^-$) $\xrightarrow[\text{heat}]{CuCN}$

(b) $(CH_3)_2C{=}CHCH_2Cl$ $\xrightarrow[S_N2]{Na^+\ {}^-CN}$

15.58 (a) $HOOC(CH_2)_3COOH$ written as $HO\overset{O}{\overset{\|}{C}}CH_2CH_2CH_2\overset{O}{\overset{\|}{C}}OH + 2\ NH_4^+$

(b) 2-methylbenzoic acid (benzene ring with CH_3 and CO_2H ortho) $+\ NH_4^+$

(c) benzene ring with CH_2CO_2H and CO_2H ortho

(d) $H_2NCH_2CH_2CH_2CH_2CH_2NH_2$

15.59 (a) $CH_3CH_2\underset{CH_3}{\underset{|}{C}}H\overset{O}{\overset{\|}{C}}O\overset{O}{\overset{\|}{C}}-$cyclohexyl

(b) cyclohexyl$-\overset{O}{\overset{\|}{C}}O-$cyclopentyl

(c) cyclohex-3-enyl$-CH_2OH + CH_3CH_2OH$

(d) $CH_3CH_2CH_2\overset{O}{\overset{\|}{C}}Cl$

(e) (R)-$CH_3\overset{*}{C}H(CH_2)_5CH_3$ with $NH\overset{}{C}(=O)CH_3$ on the starred carbon $+$ (R)-$CH_3\overset{*}{C}H(CH_2)_5CH_3$ with ${}^+NH_3$ on the starred carbon, ${}^-O\overset{}{C}(=O)CH_3$

(f) $CH_3CH_2\overset{O}{\overset{\|}{C}}CH(CH_3)_2$

(g) $2\ CH_3CH_2OD + \left[DOC(=O)OD \right] \rightarrow D_2O + CO_2$

(h) $C_6H_5C(=O)-{}^{18}OCH(CH_3)_2 + C_6H_5CO_2H$

(i) $Cl-C_6H_4-NH-C_6H_4-Cl + CH_3CO_2^-$

(j) $CH_3C(=O)-NHCH_2CO_2H + CH_3CO_2H$

(k) $C_5H_5N{:} + CH_3(CH_2)_5CO_2H \rightleftharpoons CH_3(CH_2)_5CO_2^- + C_5H_5NH^+$

$CH_3(CH_2)_5CO_2^- \xrightarrow{CH_3(CH_2)_5C(=O)Cl} CH_3(CH_2)_5C(=O)OC(=O)(CH_2)_5CH_3$

(l) $O_2N-C_6H_4-CH_2OH + CH_3CO_2^-\ Na^+$

(m) $(CH_3CH_2)_3COH + 2\ CH_3CH_2OH$

(n) $C_6H_5CH(Br)CO_2H$ (o) $(CH_3)_3CCBr_2CO_2H$

(p) $CH_2{=}CHCO_2H + HCO_2CH_3$ (transesterification)

(q) $CH_2{=}CHCO_2CH_2CH_2OCH_2CH_3$ + CH_3OH

(r) 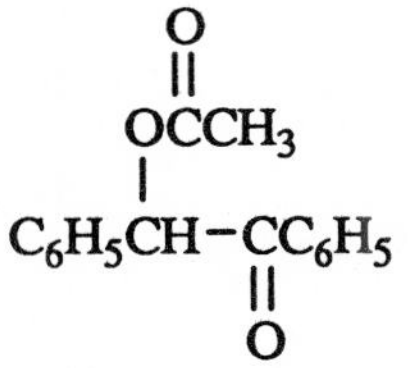 + CH_3CO_2H

(s) 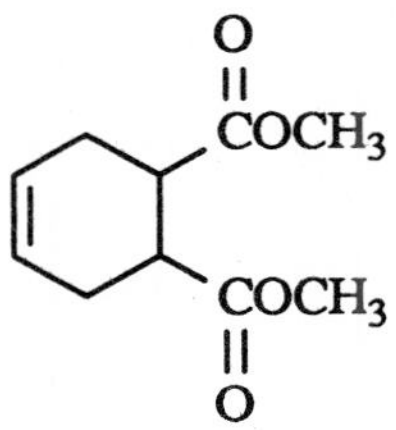

(t) 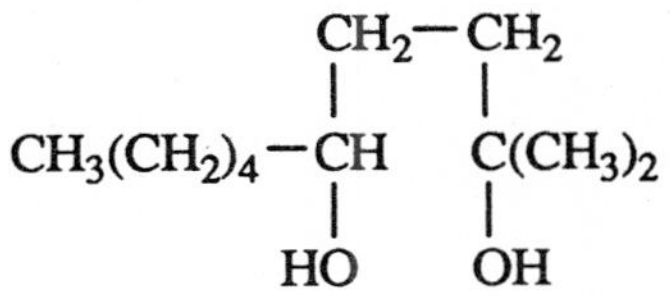

15.60 (a) $HOCH_2CH_2CH_2OH$ + 2 CH_3OH

(b) and (c) $CH_3CH_2CH_2OH$

(d) $CH_3CH_2CH_2NH_2$

(e) $C_6H_5CH_2CH_2NHCH_3$

15.61 (a) 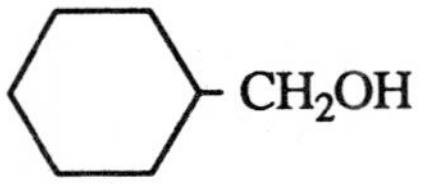 + CH_3OH

(b) 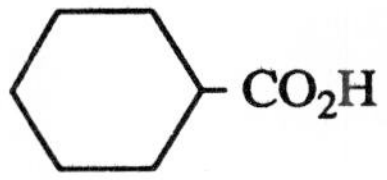

15.62

$$
\begin{array}{l}
\text{CH}_3\text{CH}_2\text{C(OH)(CH}_3\text{)}-\text{CH(OH)CH}_2\text{CH}_2-\text{CH}=\text{C(CH}_2\text{CH}_3\text{)}-\text{CH}_2\text{CH}_2-\text{C(CH}_3\text{)}=\text{CH}-\text{CO}_2^-\ \text{Na}^+
\end{array}
$$

Both the epoxide group and the ester group would react. The stereochemistry of the double bonds would remain unchanged.

15.63 (a) $(CH_3CO)_2O$ + 2 $H_2NCH(CH_3)_2$ ⟶

(b) excess $(CH_3CO)_2O$ + [five-membered ring containing O, with substituents HO, HO, OH, CH_2OH] ⟶

15.64 (a) $CH_3-C_6H_4-OH$ $\xrightarrow{\text{NaOH}}$ $CH_3-C_6H_4-O^-\ Na^+$ $\xrightarrow{(CH_3)_3CCCl\ (C=O)}$ product

(b) 2,6-dimethylphenol ($C_6H_3(CH_3)_2-OH$) $\xrightarrow{\text{LiOH}}$ $C_6H_3(CH_3)_2-O^-\ Li^+$ $\xrightarrow{CH_3CH_2CCl\ (C=O)}$ product

15.65 $HO-\ddot{O}H + :\ddot{O}H^- \rightleftharpoons HO-\ddot{O}:^- + H_2\ddot{O}:$

(m-chlorobenzoyl chloride) $C(=O)Cl$ + $^-:\ddot{O}-OH$ ⟶ tetrahedral intermediate $C(O^-)(Cl)(OOH)$ $\xrightarrow{-Cl^-}$ product

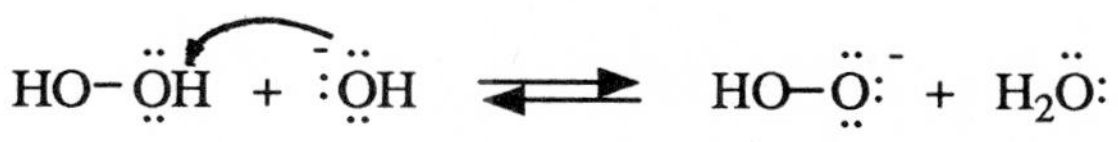

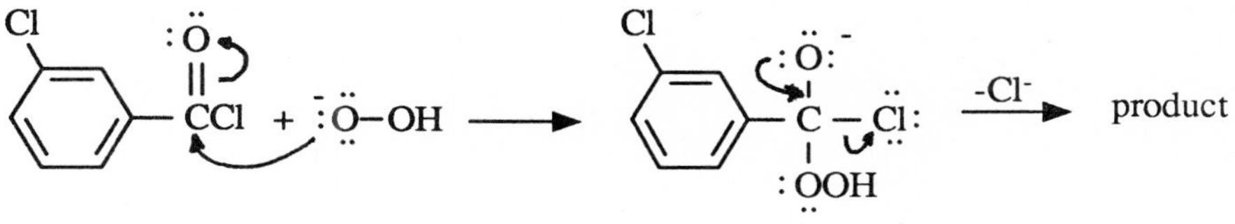

The function of the hydroxide is to convert the hydrogen peroxide to a nucleophile.

15.66 (a) 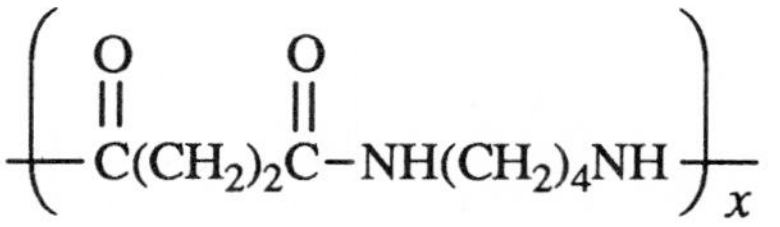 $\left[\overset{O}{\overset{\|}{C}}(CH_2)_2\overset{O}{\overset{\|}{C}}-NH(CH_2)_4NH \right]_x$

(b) $\left[NHCH(CH_3)\overset{O}{\overset{\|}{C}} \right]_x$

(c) $\left[OCH_2-C_6H_{10}-CH_2O\overset{O}{\overset{\|}{C}}-C_6H_4-\overset{O}{\overset{\|}{C}} \right]_x$

15.67 (ethylene carbonate) and $\left[\overset{O}{\overset{\|}{C}}OCH_2CH_2O \right]_x$

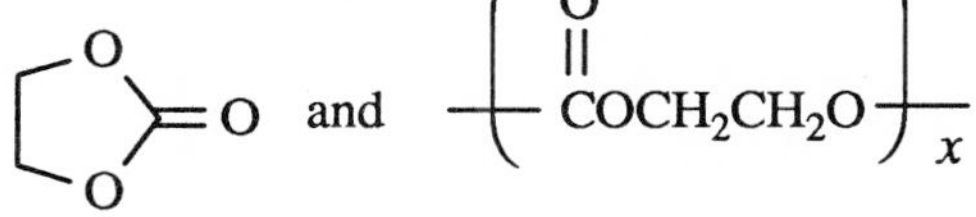

15.68 (a) 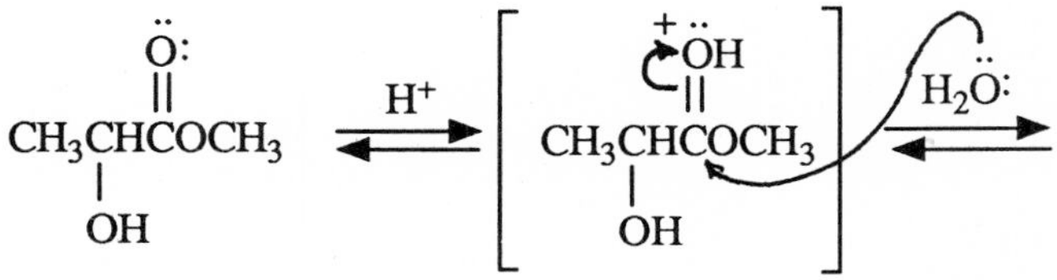 $CH_3CH(OH)\overset{\ddot{O}:}{\overset{\|}{C}}OCH_3 \overset{H^+}{\rightleftharpoons} \left[CH_3CH(OH)\overset{+\ddot{O}H}{\overset{\|}{C}}OCH_3 \right] \overset{H_2\ddot{O}:}{\rightleftharpoons}$

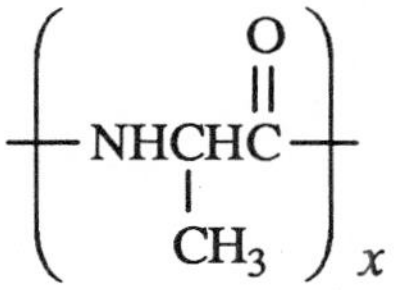

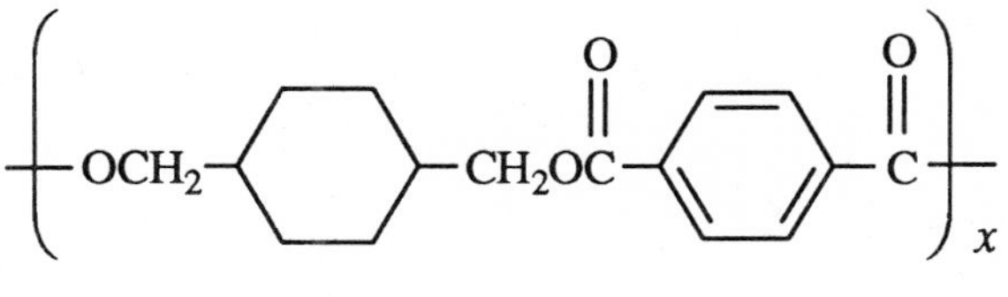

$$\left[\begin{array}{c} \ddot{O}H \\ | \\ CH_3CH-COCH_3 \\ | \quad | \\ HO \quad :\ddot{O}H_2^+ \end{array}\right] \rightleftharpoons \begin{array}{c} :\ddot{O}-H \\ | \quad + \\ CH_3CH-C-\ddot{O}CH_3 \\ | \quad | \quad | \\ HO \quad :\ddot{O}H \quad H \end{array}$$

$$\xrightleftharpoons{-H^+,\ -CH_3\ddot{O}H} \begin{array}{c} \ddot{O}: \\ || \\ CH_3CHCOH \\ | \\ OH \end{array}$$

plus some $CH_2{=}CHCO_2H$ from a dehydration reaction

(b) $$\begin{array}{c} :\ddot{O} \\ || \\ (CH_3)_2CHCOCH_3 \end{array} \xrightarrow{:\ddot{O}H^-} \left[\begin{array}{c} :\ddot{O}:^- \\ | \\ (CH_3)_2CH-C-\ddot{O}CH_3 \\ | \\ :\ddot{O}-H \end{array}\right] \quad :\ddot{O}H^-$$

$$\xrightarrow[-H_2\ddot{O}:]{-CH_3\ddot{O}:^-} \begin{array}{c} :\ddot{O} \\ || \\ (CH_3)_2CHC\ddot{O}:^- \end{array}$$

15.69 (a) $$CH_3(CH_2)_3CH_2Br \xrightarrow[(2)\ H_2O,\ H^+,\ heat]{(1)\ KCN} CH_3(CH_2)_4CO_2H$$

or $$CH_3(CH_2)_3CH_2Br \xrightarrow[(3)\ H_2O,\ H^+]{(1)\ Mg,\ ether;\ (2)\ CO_2} CH_3(CH_2)_4CO_2H$$

(b) $CH_3(CH_2)_3\overset{\overset{O}{\|}}{C}H \xrightarrow{HCN} CH_3(CH_2)_3\overset{\overset{OH}{|}}{C}HCN \xrightarrow[\text{heat}]{H_2O,\ H^+} CH_3(CH_2)_3\overset{\overset{OH}{|}}{C}HCO_2H$

(c) $C_6H_5CH_2Br \xrightarrow{KCN} C_6H_5CH_2CN \xrightarrow{H_2,\ Pt} C_6H_5CH_2CH_2NH_2$

(d) $CH_3CO_2H \xrightarrow{SOCl_2} CH_3\overset{\overset{O}{\|}}{C}Cl \xrightarrow{\text{cyclohexyl-}NH_2} CH_3\overset{\overset{O}{\|}}{C}NH\text{-cyclohexyl}$

(e) $CH_3CH_2CH_2CO_2H \xrightarrow{NaOH} CH_3CH_2CH_2CO_2^-\ Na^+$

$CH_3\overset{\overset{O}{\|}}{C}OH \xrightarrow{SOCl_2} CH_3\overset{\overset{O}{\|}}{C}Cl$

$CH_3CH_2CH_2CO_2^-\ Na^+ + CH_3\overset{\overset{O}{\|}}{C}Cl \longrightarrow CH_3\overset{\overset{O}{\|}}{C}-O\overset{\overset{O}{\|}}{C}CH_2CH_2CH_3$

acetic butanoic anhydride

In (e), you might have used acetate ion and butanoyl chloride.

(f) $HO\overset{\overset{O}{\|}}{C}CH_2\overset{\overset{O}{\|}}{C}OH \xrightarrow{2\ SOCl_2} Cl\overset{\overset{O}{\|}}{C}CH_2\overset{\overset{O}{\|}}{C}Cl \xrightarrow[R_3N:]{2\ (CH_3)_3COH}$

$(CH_3)_3CO\overset{\overset{O}{\|}}{C}CH_2\overset{\overset{O}{\|}}{C}OC(CH_3)_3 + 2\ R_3\overset{+}{N}H\ Cl^-$

In (f), an acid halide must be used because of steric hindrance.

(g) $2\ CH_3CH_2CH_2CH_2MgBr \xrightarrow[(2)\ H_2O,\ H^+]{(1)\ HC(=O)OCH_2CH_3}$

$(CH_3CH_2CH_2CH_2)_2CHOH$

When two R groups of an alcohol are the same, always consider the Grignard reaction of an ester.

(h) $CH_3(CH_2)_8CH_2CH_2CO_2H \xrightarrow[PBr_3\ catalyst]{Br_2} CH_3(CH_2)_8CH_2CHBrCO_2H$

$\xrightarrow[heat]{(CH_3)_3CO^-\ K^+}$ *trans*-$CH_3(CH_2)_8CH{=}CHCO_2^-\ K^+ \xrightarrow{H^+}$ product

Because there is only one set of hydrogens beta to the Br, only one alkene would be formed. Under E2 conditions, the *trans* isomer would predominate. Using the hindered *t*-butoxide ion as the base would minimize the substitution reaction.

(i) $C_6H_5C(=O)OH \underset{}{\overset{CH_3OH,\ H^+}{\rightleftharpoons}} C_6H_5C(=O)OCH_3 \xrightarrow[(2)\ H_2O,\ H^+]{(1)\ 2\ C_6H_5MgBr}$

See Answer 15.69(g).

15.70 (a) $CH_2{=}CHC(=O)NH_2 \xrightarrow{Br_2} CH_2BrCHBrC(=O)NH_2$

A

$\xrightarrow[heat]{H_2O,\ OH^-} CH_2{=}CHCO_2^- + NH_3$

(b) The infrared spectrum would show double NH stretching absorption at about 3125-3570 cm^{-1} (2.8-3.2 μm) plus amide NH (bending) and C=O absorption at about 1515-1670 cm^{-1} (6.0-6.6 μm) and 1650-1700 cm^{-1} (5.88 μm), respectively. Also, $=CH_2$ absorption would be observed at 3135 cm^{-1} (3.2 μm) and 887 cm^{-1} (11.3 μm). C=C absorption would appear in the same general region as the amide I and II bands, about 1620-1680 cm^{-1} (5.95-6.17 μm).

The proton nmr spectrum would show a singlet for NH_2 (area 2) and the characteristic 8-12 peaks (total area 3) around δ5 for $-CH{=}CH_2$.

15.71 CH_3O–(2,6-dimethylphenyl)–Br $\xrightarrow[\text{ether}]{\text{Mg}}$ CH_3O–(2,6-dimethylphenyl)–MgBr

$\xrightarrow{CH_3O\text{–}(2,6\text{-dimethylphenyl})\text{–}COCl}$ CH_3O–(2,6-dimethylphenyl)–C(=O)–(2,6-dimethylphenyl)–OCH_3

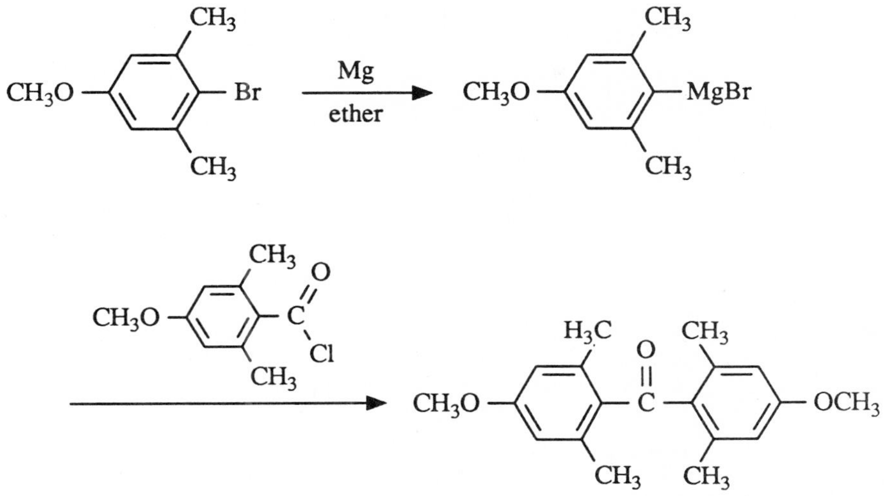

The reaction of a Grignard reagent with an acid halide proceeds stepwise. The product of the first reaction is a ketone, which usually undergoes a second reaction to yield the alkoxide of a tertiary alcohol. In this particular example, the ketone is too hindered to undergo further reaction with the hindered Grignard reagent.

15.72 (a) $:\ddot{Br}-\ddot{Br}:$ + $CH_2{=}CHCH_2CH_2CO_2H$ $\xrightarrow{-Br^-}$ $CH_2(\ddot{Br}^{\delta+})-\overset{\delta+}{C}HCH_2CH_2C(=\ddot{O}:)-\ddot{O}H$

$\longrightarrow$ $:\ddot{Br}CH_2CH$–CH_2 / $:\ddot{O}$ CH_2 / C / || / $:\overset{+}{O}H$ (five-membered ring) $\underset{}{\overset{-H^+}{\rightleftharpoons}}$ product

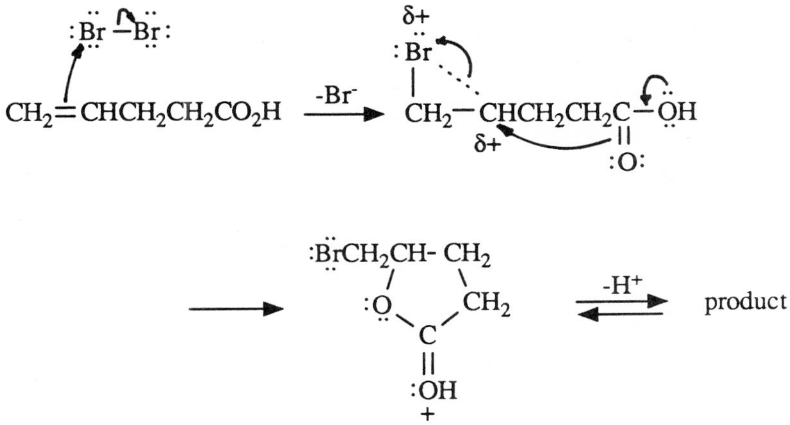

(b)

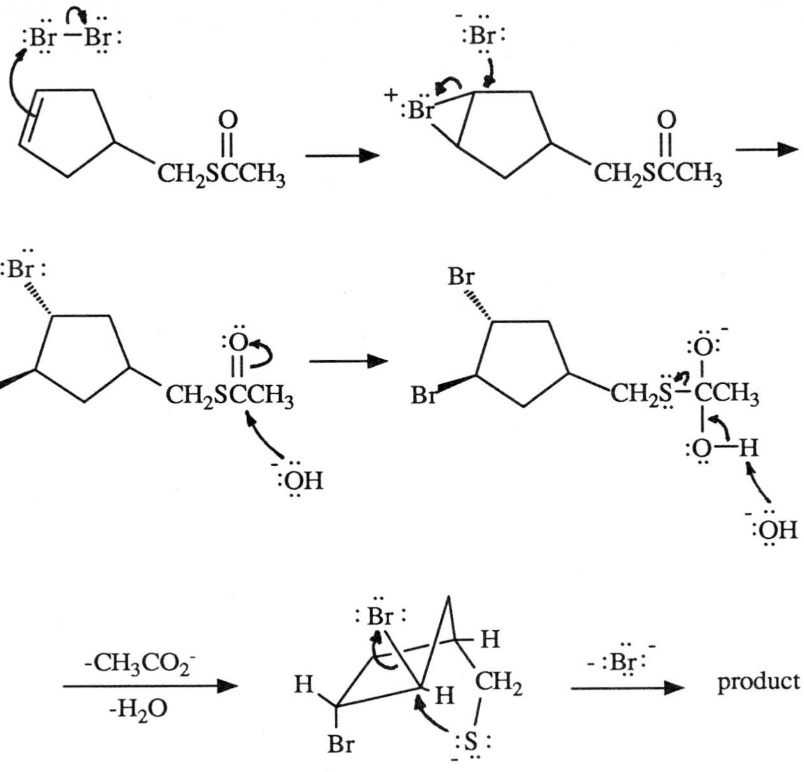

$\xrightarrow{-CH_3CO_2^-,\ -H_2O}$ $\xrightarrow{-:Br:^-}$ product

15.73

$\xrightarrow{-H^+}$ $\xrightarrow{-Cl^-}$

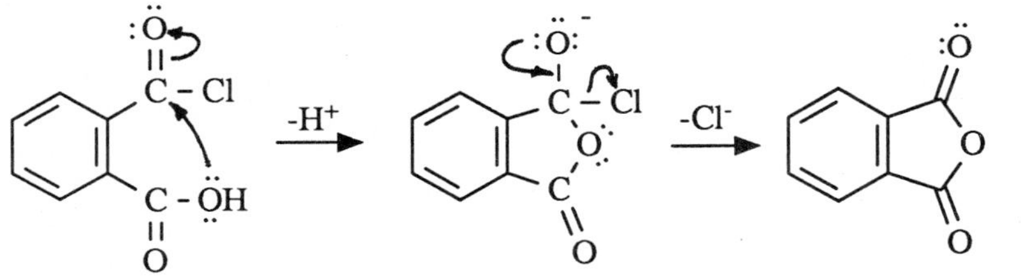

phthalic anhydride

15.74 $HCCH_2CH_2CO_2H$ $\xrightarrow{HCN}$ $HCCH_2CH_2COH$ (CN) $\longrightarrow$

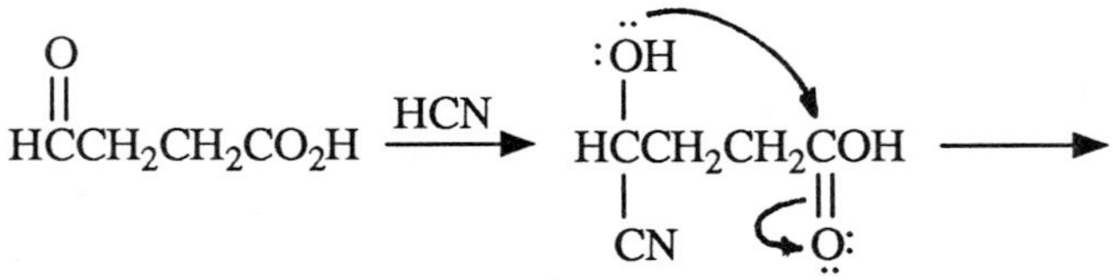

$$:\!CN\!:^- \quad \text{(mechanism: protonated cyclic hemiacetal bearing CN; } -O^-\text{ abstracts H from } H{-}CN)$$

$$\xrightarrow{} \text{(cyclic ring with CN, } \ddot{O}\text{, OH, :OH)} \xrightarrow[-CN^-]{HCN} \text{(CN ring, O, OH, } \ddot{O}H_2^+)$$

$$\xrightarrow{-H_2O} \text{(CN ring, O, C}{=}\ddot{O}^+{-}H\text{)} \xrightarrow{-HCN} \text{(CN-substituted lactone, C=O)}$$

15.75 $C_6H_5NH_2 + CH_3\overset{O}{\overset{\|}{C}}{-}O\overset{O}{\overset{\|}{C}}CH_3 \longrightarrow C_6H_5NH{-}\overset{O}{\overset{\|}{C}}CH_3 \xrightarrow[AlCl_3]{ClCH_2\overset{O}{\overset{\|}{C}}Cl}$

$$ClCH_2\overset{O}{\overset{\|}{C}}{-}C_6H_4{-}NH\overset{O}{\overset{\|}{C}}CH_3 \xrightarrow[(2)\ OH^-]{(1)\ H_2O,\ H^+} ClCH_2\overset{O}{\overset{\|}{C}}{-}C_6H_4{-}NH_2$$

In the Friedel-Crafts reaction, the more reactive acid chloride group reacts faster than does the alpha chlorine.

15.76 $CH_3\overset{O}{\overset{\|}{C}}\underset{CO_2CH_3}{\underset{|}{C}H}CH_2CO_2CH_3 \xrightarrow[-2\ CH_3OH,\ -CO_2]{H_2O,\ H^+,\ heat} CH_3\overset{O}{\overset{\|}{C}}CH_2CH_2CO_2H \xrightarrow{SOCl_2}$

$C_8H_{12}O_5$ A ($C_5H_8O_3$)

$$CH_3\overset{O}{\overset{\|}{C}}CH_2CH_2\overset{O}{\overset{\|}{C}}Cl \rightleftharpoons CH_3\overset{OH}{\overset{|}{C}}{=}CHCH_2\overset{O}{\overset{\|}{C}}Cl \xrightarrow{-HCl} CH_3{-}\text{(unsaturated five-membered lactone)}$$

an enol B ($C_5H_6O_2$)

15.77

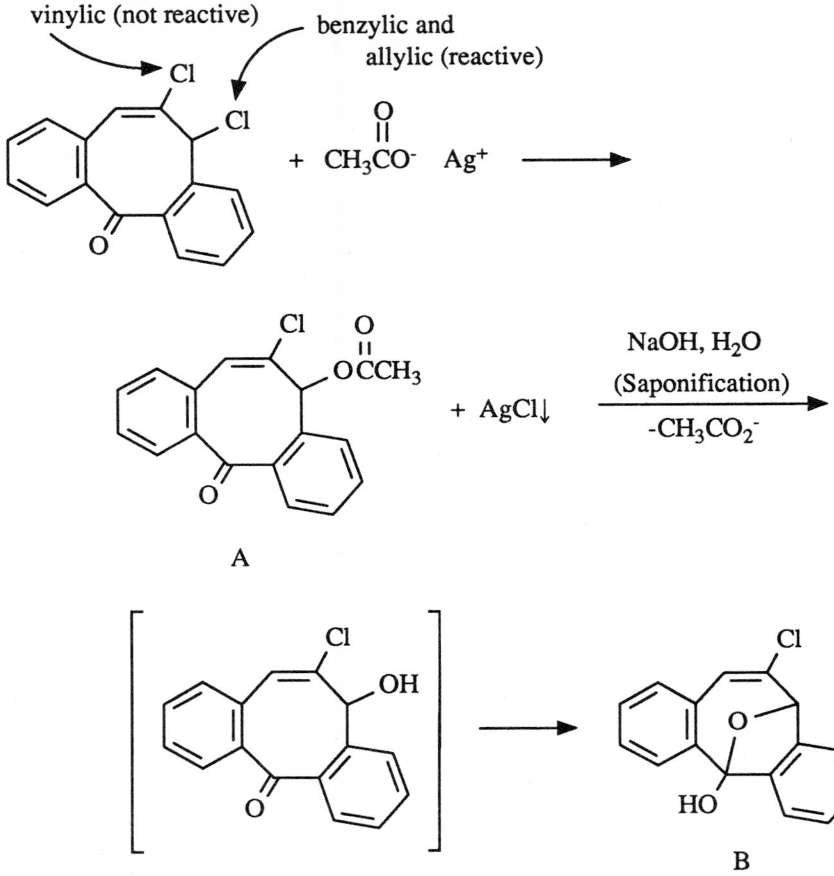

vinylic (not reactive)

benzylic and allylic (reactive)

$+\ CH_3\overset{O}{\overset{||}{C}}O^-\ Ag^+ \longrightarrow$

$+\ AgCl\downarrow \xrightarrow[-CH_3CO_2^-]{NaOH,\ H_2O\ (Saponification)}$

A

B

15.78 (a) $CH_3CH_2CH_2Br \xrightarrow{KCN} CH_3CH_2CH_2CN \xrightarrow[heat]{H_2O,\ H^+}$

$CH_3CH_2CH_2CO_2H \xrightarrow{SOCl_2} CH_3CH_2CH_2\overset{O}{\overset{||}{C}}Cl$

$\xrightarrow[(2)\ H_2O,\ H^+]{(1)\ LiAlH[OC(CH_3)_3]_3} CH_3CH_2CH_2\overset{O}{\overset{||}{C}}H$

or a Grignard reaction could be used to convert 1-bromopropane to butanoic acid

(b) $CH_3CO_2H \xrightarrow[PCl_3]{Cl_2} ClCH_2CO_2H \xrightarrow[\text{heat}]{CH_3CH_2OH,\ H^+}$

$ClCH_2CO_2CH_2CH_3 \xrightarrow{KCN}$ product

(c) cyclohexyl–$CO_2H \xrightarrow{SOCl_2}$ cyclohexyl–$\overset{O}{\overset{\|}{C}}Cl$

$\xrightarrow[-(CH_3)_2\overset{+}{N}H_2\ Cl^-]{2\ (CH_3)_2NH}$ cyclohexyl–$\overset{O}{\overset{\|}{C}}N(CH_3)_2$

(d) (lactone structure) $\xrightarrow[\text{transesterification}]{CH_3CH_2OH,\ HBr}$ $HOCH_2CH_2CH_2\overset{O}{\overset{\|}{C}}OCH_2CH_3$

a lactone

$\xrightarrow[-H_2O]{HBr}$ $BrCH_2CH_2CH_2\overset{O}{\overset{\|}{C}}OCH_2CH_3$

(e) $HO\overset{O}{\overset{\|}{C}}CH_2\overset{O}{\overset{\|}{C}}CH_2\overset{O}{\overset{\|}{C}}OH \xrightarrow[\text{heat}]{(CH_3CO)_2O}$

$\xrightarrow{1\ CH_3OH}$ product

(f) (2-bromocyclohexene-1-carbaldehyde: Br, CHO) $\xrightarrow[(2)\ H^+]{(1)\ Ag^+ \text{ or other mild oxidizing agent}}$ (Br, CO_2H)

$\xrightarrow{SOCl_2}$ 2-bromo-1-cyclohexenecarbonyl chloride (CCl, $\overset{\|}{O}$, Br) $\xrightarrow[-(CH_3)_2\overset{+}{N}H_2\ Cl^-]{2\ (CH_3)_2NH}$ product

(g) $C_6H_6 \xrightarrow[AlCl_3]{CH_3I} C_6H_5CH_3 \xrightarrow[H_2SO_4]{HNO_3} O_2N-C_6H_4-CH_3$

$\xrightarrow[heat]{KMnO_4} O_2N-C_6H_4-CO_2H \xrightarrow[(2)\ OH^-]{(1)\ Fe,\ HCl}$

$H_2N-C_6H_4-CO_2H \xrightarrow[heat]{CH_3CH_2OH,\ H^+}$

$H_3\overset{+}{N}-C_6H_4-CO_2CH_2CH_3 \xrightarrow{OH^-}$

$H_2N-C_6H_4-CO_2CH_2CH_3$

(h) H, CO_2H / H, CO_2H (maleic acid) $\xrightarrow{heat}$ maleic anhydride (O, O, O) $\xrightarrow[heat]{NH_3}$ maleimide (O, NH, O)

$\xrightarrow[heat]{\text{1,3-butadiene}}$ cis-tetrahydrophthalimide (H, O, NH, H, O)

(i) $ClCCH_2CH_2CO_2CH_3$ (C=O) $\xrightarrow[cold]{[(CH_3)_2CHCH_2]CuLi}$ product

The acid chloride could be prepared by the following sequence:

(succinic anhydride) $\xrightarrow{\text{1 } CH_3OH}$ $HO\overset{O}{\overset{\|}{C}}CH_2CH_2CO_2CH_3$ $\xrightarrow{SOCl_2}$

The cuprate could be prepared as follows:

$(CH_3)_2CHCH_2Br \xrightarrow{Li} (CH_3)_2CHCH_2Li \xrightarrow{CuI}$

(j) (cyclohexyl)$-CH_2OH \xrightarrow{H_2CrO_4}$ (cyclohexyl)$-CO_2H \xrightarrow{SOCl_2}$ (cyclohexyl)$-\overset{O}{\overset{\|}{C}}Cl$

(cyclohexyl)$-CH_2OH \xrightarrow{SOCl_2}$ (cyclohexyl)$-CH_2Cl \xrightarrow[\text{(2) CuI}]{\text{(1) Li}}$

$\left(\text{(cyclohexyl)}-CH_2\right)_2CuLi \xrightarrow{\text{(cyclohexyl)}-\overset{O}{\overset{\|}{C}}Cl}$ product

(k) (cyclobutane-1,1-diyl)$(CO_2CH_2CH_3)_2 \xrightarrow[\text{(2) } H_2O,\ H^+]{\text{(1) } LiAlH_4}$ (cyclobutane-1,1-diyl)$(CH_2OH)_2 \xrightarrow{\text{2 TsCl}}$

(cyclobutane-1,1-diyl)$(CH_2OTs)_2 \xrightarrow{NaCN}$ (cyclobutane-1,1-diyl)$(CH_2CN)_2 \xrightarrow{H_2O,\ H^+}$

(cyclobutane-1,1-diyl)$(CH_2CO_2H)_2 \xrightarrow{CH_3CH_2OH,\ H^+}$

In the second reaction, you may have used HBr instead of tosyl chloride.

15.79 (a)

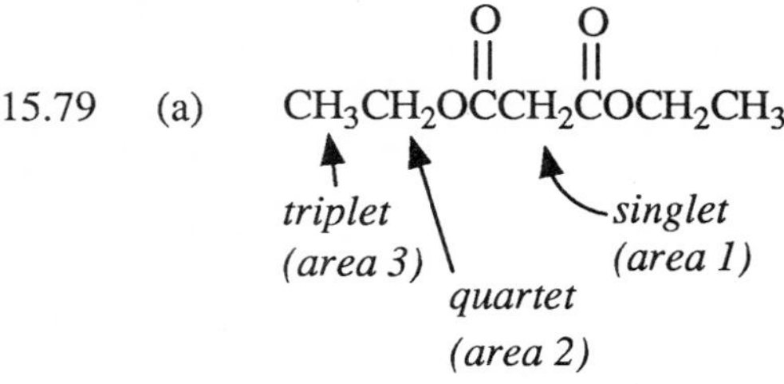

(b)

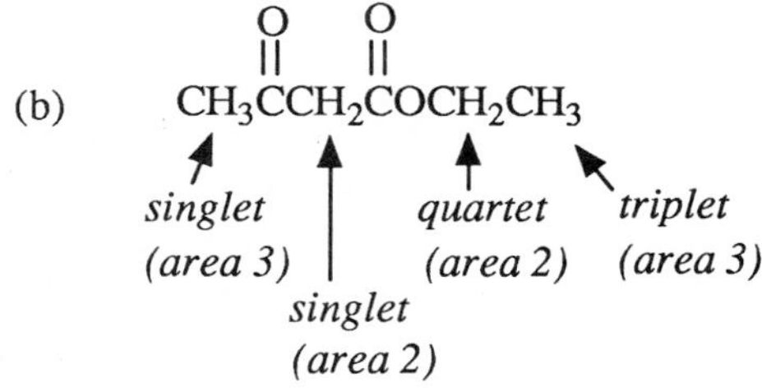

(c)

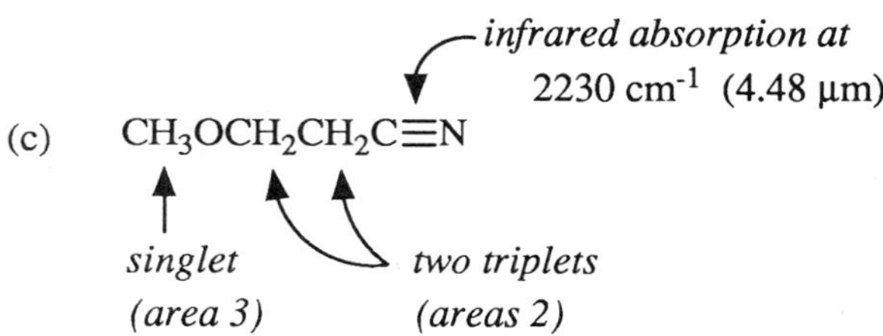

15.80 Free rotation around the amide C-N bond is restricted because of the resonance delocalization of the nitrogen's unshared pair of electrons.

$$\mathrm{RC(=\ddot{O}\!:)\!-\!\ddot{N}R_2} \longleftrightarrow \mathrm{RC(\!-\!:\ddot{O}\!:^-)\!=\!\overset{+}{N}R_2}$$

Therefore, one of the two ethyl groups in the amide is *cis* to the oxygen and the other one is *trans*. The carbons of these two ethyl groups are nonequivalent and, consequently, give separate signals in the ^{13}C nmr spectrum.

^{-}O — $C=N^{+}$ — CH_2CH_3 ; H ; H

cis

^{-}O — $C=N^{+}$ — H ; H ; CH_2CH_3

trans

15.81 *A:* $CH_3\overset{O}{\overset{\|}{C}}OCH{=}CH_2 \xrightarrow[\text{heat}]{H_2O,\ H^+} CH_3CO_2H + [HOCH{=}CH_2] \rightarrow CH_3CHO$

singlet (area 3) — CH_3; *multiplet (area 1)* — CH; *multiplet (area 2)* — CH_2

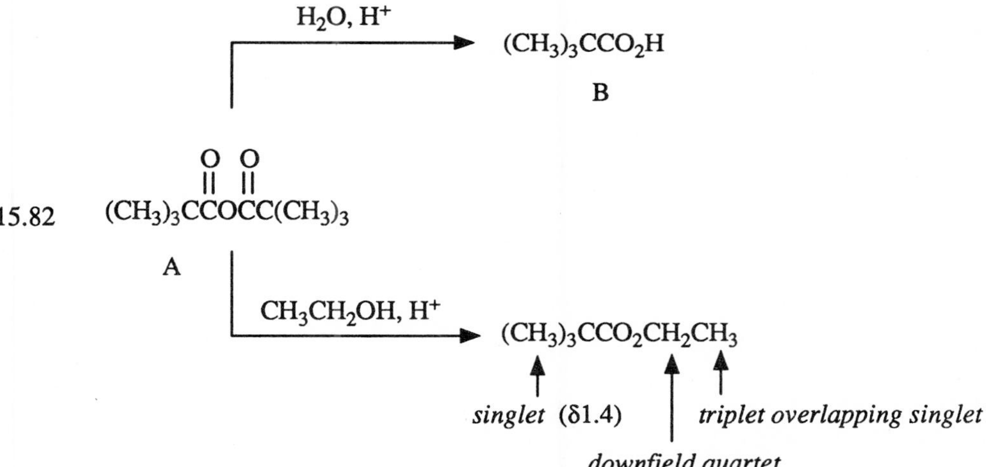

Although the nmr spectrum of C is not easy to analyze, the chemistry implies an ethyl ester.

Chapter 16

Enolates and Carbanions: Building Blocks for Organic Synthesis

16.26 (a) $C_6H_5C(CH_3)\text{(H)}CO_2C_2H_5$ (b) $CH_3C(=O)C\text{(H}_2\text{)}CN$

(c) $OHCC(CH_3)\text{(H)}CH{=}CHCHO$

16.27 (a) $C_2H_5O_2CCH_2CN + C_2H_5\ddot{\underset{..}{O}}:^- \rightleftharpoons C_2H_5O_2C\overset{-}{\underset{..}{C}}HCN + C_2H_5\ddot{\underset{..}{O}}H$

(b) 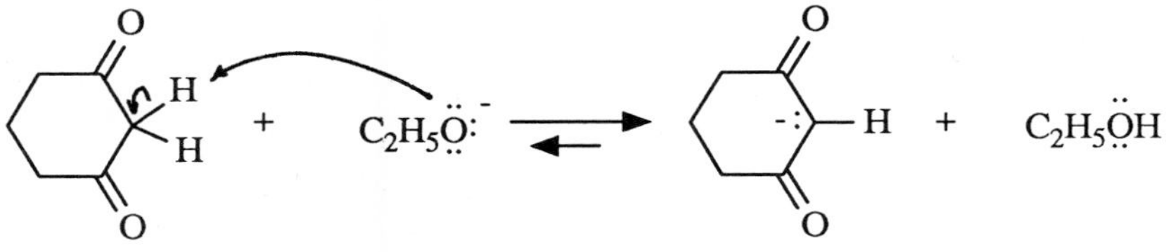 cyclohexane-1,3-dione (CH₂ between carbonyls) + $C_2H_5\ddot{\underset{..}{O}}:^- \rightleftharpoons$ carbanion of cyclohexane-1,3-dione ($-\!:$ at C–H) + $C_2H_5\ddot{\underset{..}{O}}H$

(c) 2-cyanocyclohexanone (CN, H) + $C_2H_5\ddot{\underset{..}{O}}:^- \rightleftharpoons$ carbanion ($-\!:$) of 2-cyanocyclohexanone (CN) + $C_2H_5\ddot{\underset{..}{O}}H$

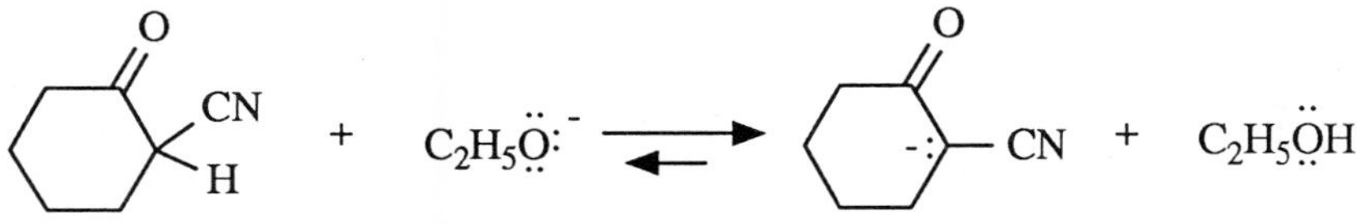

16.28 (a) 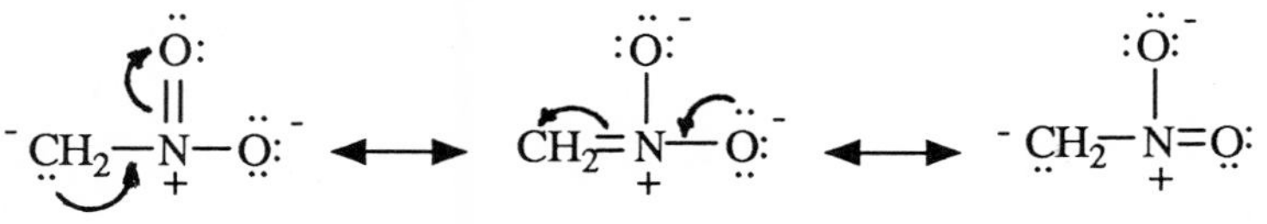 $^-\underset{..}{C}H_2-\overset{+}{N}(=\ddot{O}:)-\ddot{\underset{..}{O}}:^- \longleftrightarrow CH_2{=}\overset{+}{N}(-\ddot{\underset{..}{O}}:^-)-\ddot{\underset{..}{O}}:^- \longleftrightarrow {}^-\underset{..}{C}H_2-\overset{+}{N}(-\ddot{\underset{..}{O}}:^-){=}\ddot{O}:$

(b) $^{-}\ddot{C}H_2-CH=CH-\overset{O}{\overset{\|}{C}}OC_2H_5 \longleftrightarrow CH_2=CH\overset{-}{\ddot{C}}H-\overset{:\ddot{O}}{\overset{\|}{C}}OC_2H_5 \longleftrightarrow$

$$CH_2=CHCH=\overset{:\ddot{O}:^-}{\overset{|}{C}}OC_2H_5$$

(c) $CH_3\overset{:\ddot{O}}{\overset{\|}{C}}-\overset{-}{\ddot{C}}HCN \longleftrightarrow CH_3\overset{:\ddot{O}:^-}{\overset{|}{C}}=CH-C\equiv N:$

$$\longleftrightarrow CH_3\overset{O}{\overset{\|}{C}}CH=C=\ddot{N}:^-$$

16.29 (b) < (a) < (e) < (d) < (c) < (f). The compounds with hydrogens alpha to only one carbonyl group, (b) and (a), are less acidic than ethanol (e). Those with hydrogens alpha to two carbonyl groups are more acidic. None of the other compounds is as acidic as a carboxylic acid (f). Also note that a hydrogen alpha to an aldehyde group is slightly more acidic than one alpha to an ester group.

16.30 (a) $\rightleftharpoons CH_3CO_2C_2H_5 + CH_3\overset{O}{\overset{\|}{C}}\overset{-}{\ddot{C}}HCO_2C_2H_5$

(b) $\rightleftharpoons CH_3\overset{O}{\overset{\|}{C}}\overset{-}{\ddot{C}}HNO_2 + H_2O$

(c) $\rightleftharpoons$ (cyclohexyl carbanion, $^{-}$:)$-CO_2C_2H_5$ + $CH_3\overset{O}{\overset{\|}{C}}CH_2\overset{O}{\overset{\|}{C}}CH_3$

(d) $\rightleftharpoons CH_3\overset{O}{\overset{\|}{C}}\overset{-}{\ddot{C}}H\overset{O}{\overset{\|}{C}}H$

16.31 (a) 2,2-dimethylcyclopentane-1,3-dione: $O=$ ring $=O$ with H_3C and CH_3 on the carbon between the two C=O groups

(b) $CH_3\overset{O}{\overset{\|}{C}}CHCO_2C_2H_5$ with a cyclohexyl group on the CH

(c) cyclohexyl$-CH_2\overset{O}{\overset{\|}{C}}CH_3$ from the decarboxylation of $CH_3\overset{O}{\overset{\|}{C}}CHCO_2H$ with a cyclohexyl group on the CH

(d) $CH_3\overset{O}{\overset{\|}{C}}CHCO_2C_2H_5$ with $CH_2\underset{O}{\underset{\|}{C}}CH_3$ on the CH

(e) $CH_3\overset{O}{\overset{\|}{C}}CH_2CH_2\overset{O}{\overset{\|}{C}}CH_3$

(f) O_2N–(benzene ring, NO_2 ortho to the side chain)–$\underset{CN}{\underset{|}{C}}HCO_2C_2H_5$ by nucleophilic aromatic substitution

16.32 (a) $CH_2(CO_2CH_3)_2$ $\xrightarrow[\text{(2) cyclopentyl}-CH_2Br]{\text{(1) } NaOCH_3}$

cyclopentyl$-CH_2CH(CO_2CH_3)_2$ $\xrightarrow[\text{heat}]{H_2O,\ H^+}$

(b) $CH_2(CO_2CH_3)_2 \xrightarrow[(2)\ CH_3CHBrCO_2C_2H_5]{(1)\ NaOCH_3}$

$$C_2H_5O_2C\underset{\displaystyle}{\overset{\displaystyle CH_3}{\overset{|}{C}}}HCH(CO_2CH_3)_2 \xrightarrow[\text{heat}]{H_2O,\ H^+}$$

(c) $CH_3\overset{O}{\overset{\|}{C}}CH_2CO_2CH_3 \xrightarrow[(2)\ C_6H_5CH_2Br]{(1)\ NaOCH_3} CH_3\overset{O}{\overset{\|}{C}}\underset{\underset{CH_2C_6H_5}{|}}{C}HCO_2CH_3 \xrightarrow[\text{heat}]{H_2O,\ H^+}$

(d) $CH_2(CO_2CH_3)_2 \xrightarrow[(2)\ (CH_3)_2C{=}CHCH_2Br]{(1)\ NaOCH_3}$

$$(CH_3)_2C{=}CHCH_2CH(CO_2CH_3)_2 \xrightarrow[(2)\ HC{\equiv}CCH_2Br]{(1)\ NaOCH_3}$$

16.33 (2-bromophenyl)$-CH_3 \xrightarrow{NBS}$ (2-bromophenyl)$-CH_2Br$

$$CH_2(CO_2CH_2CH_3)_2 \xrightarrow{NaOC_2H_5} {}^{-}CH(CO_2CH_2CH_3)_2$$

$$\xrightarrow[(2)\ H_2O,\ H^+]{(1)\ \text{(2-bromophenyl)}-CH_2Br} \text{(2-bromophenyl)}-CH_2CH(CO_2H)_2$$

$$\xrightarrow[-CO_2]{\text{heat}} \text{(2-bromophenyl)}-CH_2CH_2CO_2H$$

16.34 (a) 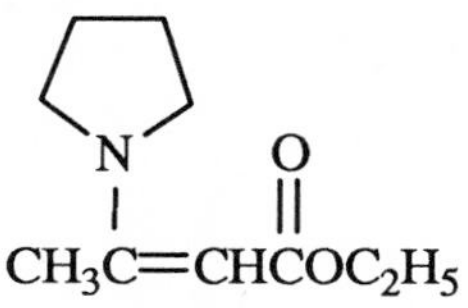 $CH_3C(N(CH_2)_4)=CHC(=O)OC_2H_5$ (b) 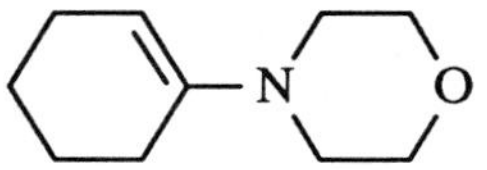 1-morpholino-cyclohexene

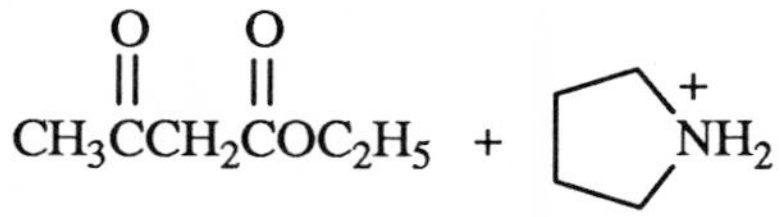

16.35 (a) 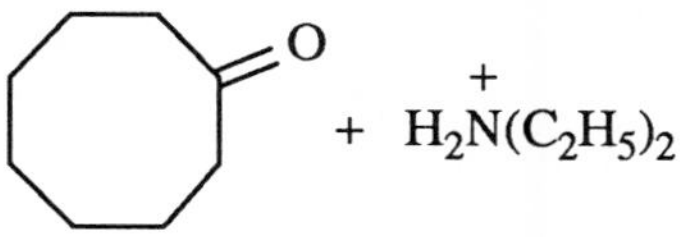 cyclooctanone + $H_2\overset{+}{N}(C_2H_5)_2$

(b) $CH_3\overset{O}{\overset{\|}{C}}CH_2\overset{O}{\overset{\|}{C}}OC_2H_5$ + pyrrolidinium ($\overset{+}{N}H_2$)

(c) $(CH_3)_2CH\overset{O}{\overset{\|}{C}}H$ + pyrrolidinium ($\overset{+}{N}H_2$)

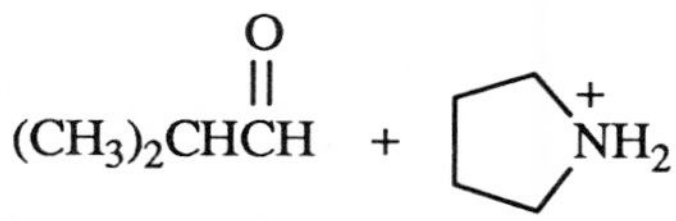

(d) 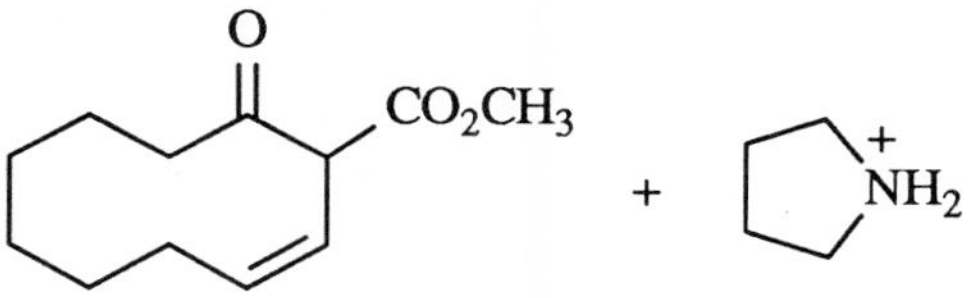 2-(CO_2CH_3)-cycloalkenone + pyrrolidinium ($\overset{+}{N}H_2$)

In (b) and (d), the ester group could also be hydrolyzed, depending on conditions.

16.36 cyclopentanone (=Ö:) + H–:N(piperidine) ⇌ tetrahedral intermediate with :Ö:⁻ and $\overset{+}{N}$–H (piperidinium), H^+

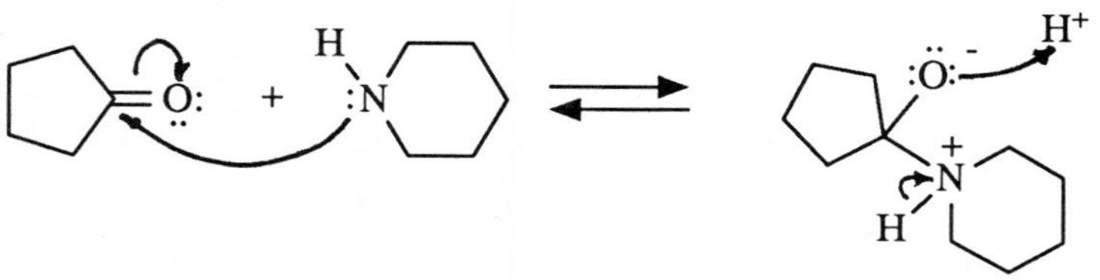

H^+

$-H^+$

$-H_2O$

C_6H_5C-Cl

C_6H_5CCl

$-Cl^-$

CC_6H_5

16.37 (a)

HN

CH_3 CH_2Cl

CH_2 CH_3

H_2O, H^+

(b)

HN

O_2N NO_2 Cl

$-Cl^-$

one resonance structure of the enamine

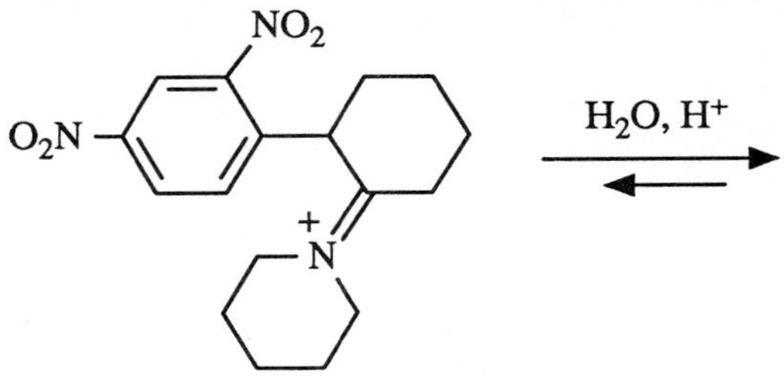

$\xrightarrow[\longleftarrow]{H_2O,\ H^+}$

(c) $\xrightarrow{HN}$ $\xrightarrow{ClC(=O)(CH_2)_8C(=O)Cl}$

$\xrightarrow[\longleftarrow]{H_2O,\ H^+}$

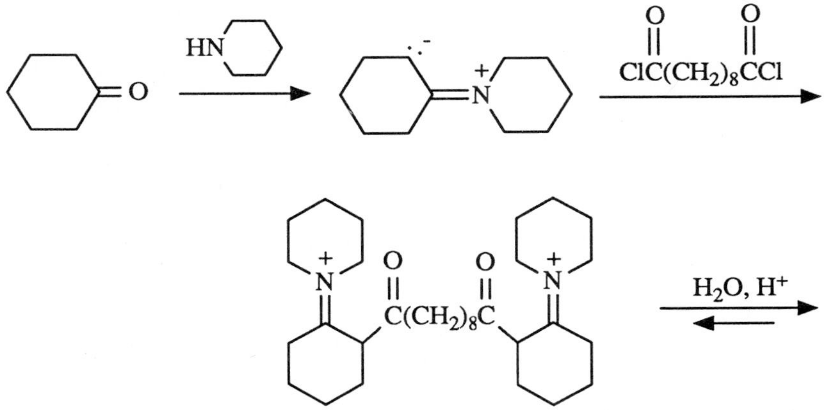

16.38 (a) $:\overset{-}{C}H_2-\overset{\overset{..}{:O:}}{\overset{||}{C}}CH_3 \longleftrightarrow CH_2=\overset{\overset{..}{:O:^-}}{\overset{|}{C}}CH_3$ Formaldehyde does not contain an acidic hydrogen.

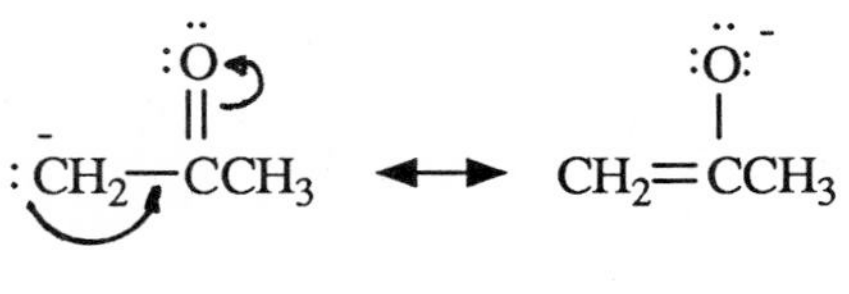

(b) HCHO

(c) $CH_3\overset{O}{\overset{||}{C}}\overset{-}{\underset{..}{C}}H_2 + H\overset{\overset{..}{:O:}}{\overset{||}{C}}H \rightleftharpoons CH_3\overset{O}{\overset{||}{C}}CH_2\overset{\overset{..}{:O:^-}}{\overset{|}{C}}H_2 \xrightarrow[\longleftarrow]{H-\ddot{O}H}$

$CH_3\overset{O}{\overset{||}{C}}CH_2\overset{:\ddot{O}H}{\overset{|}{C}}H_2 + {}^-:\ddot{O}H$

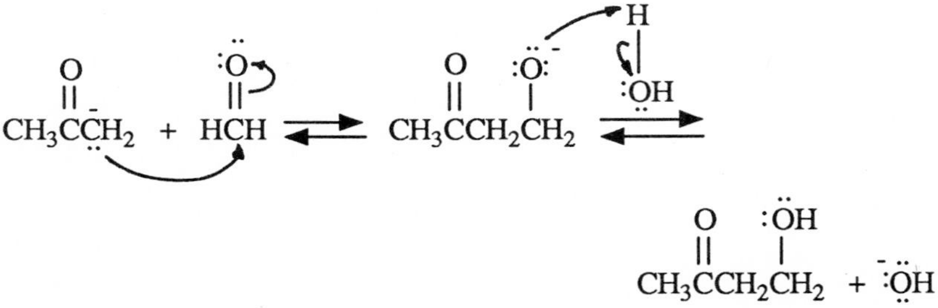

16.39 (a) 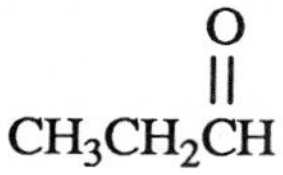 $CH_3CH_2CH{=}O$ (b) $C_6H_5CH_2CH{=}O$

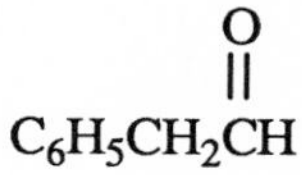

(c) $C_6H_5CH{=}O$ + $C_6H_5CH_2CH{=}O$

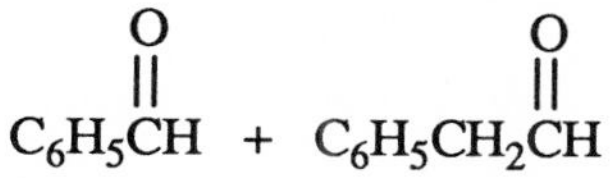

(d) 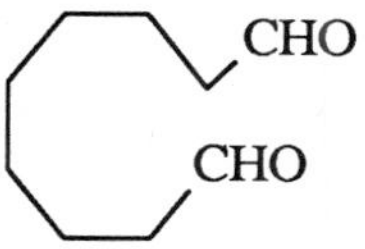 CHO CHO or $HC(=O)(CH_2)_7CH(=O)$

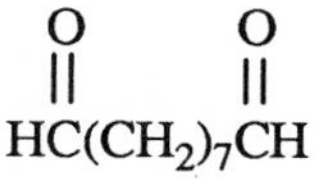

16.40 (a) 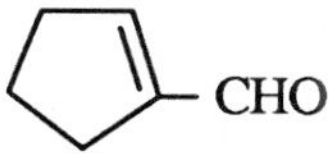 CHO (b) 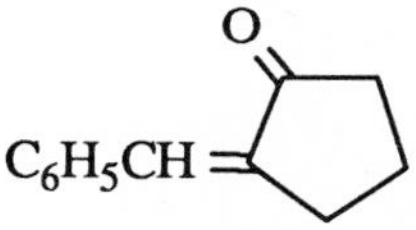 $C_6H_5CH=$

(E) and (Z)

(c) 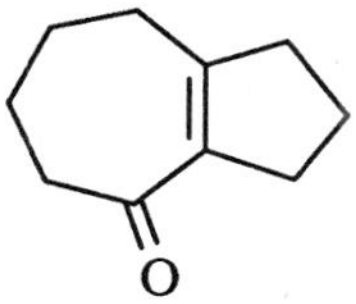 O (d) 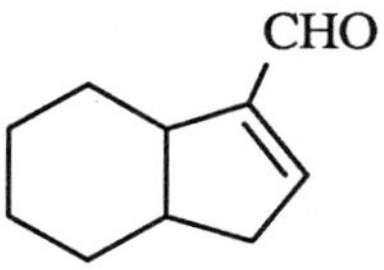 CHO

The mechanism for (c) follows:

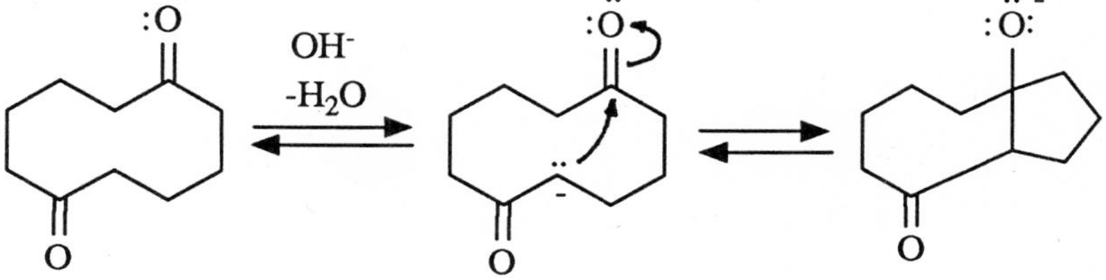

:Ö: OH⁻ −H_2O :Ö: :Ö:⁻ O O O

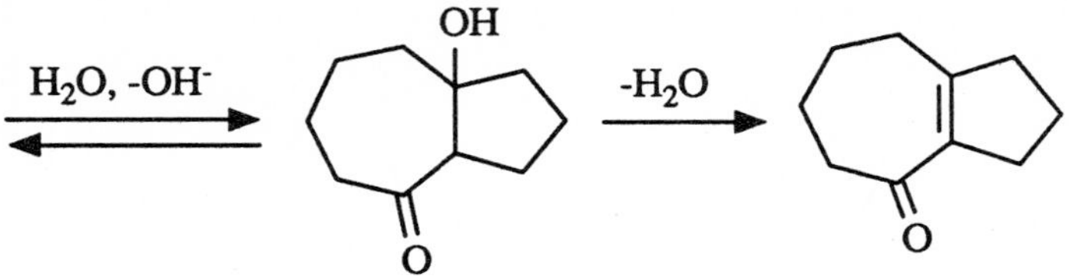

$\xrightleftharpoons{H_2O,\ -OH^-}$ OH ... O $\xrightarrow{-H_2O}$ O

16.41 (a) 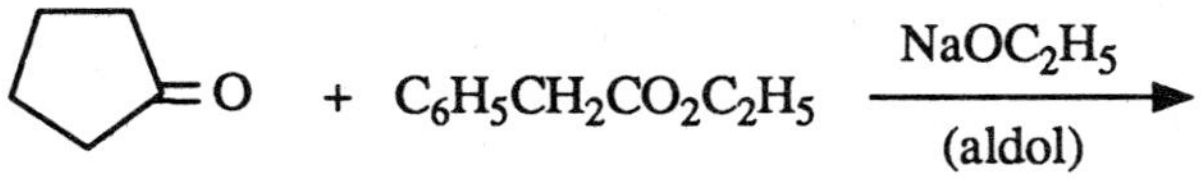 =O + $C_6H_5CH_2CO_2C_2H_5$ $\xrightarrow[\text{(aldol)}]{NaOC_2H_5}$

(b) 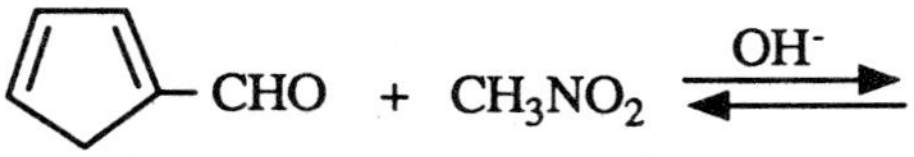 –CHO + CH_3NO_2 $\xrightleftharpoons{OH^-}$

(c) C_6H_5CHO + $CH_3\overset{O}{\overset{\|}{C}}CH_3$ $\xrightleftharpoons{OH^-}$

(d) $CH_3\overset{O}{\overset{\|}{C}}CH_3$ $\xrightleftharpoons{C_6H_5CHO,\ OH^-}$ $C_6H_5CH{=}CH\overset{O}{\overset{\|}{C}}CH_3$ $\xrightleftharpoons{C_6H_5CHO,\ OH^-}$

16.42 (a) $CH_3(CH_2)_3\overset{O}{\overset{\|}{C}}(CH_2)_3CH_3$

(b) 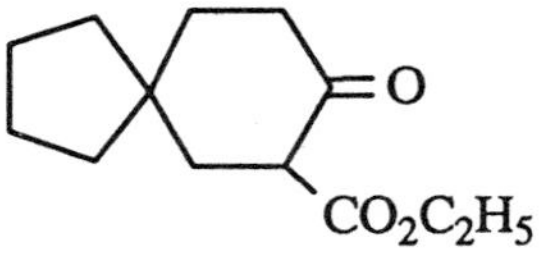 =O, $CO_2C_2H_5$

(c) $C_6H_5CH(CO_2C_2H_5)_2$

(d) 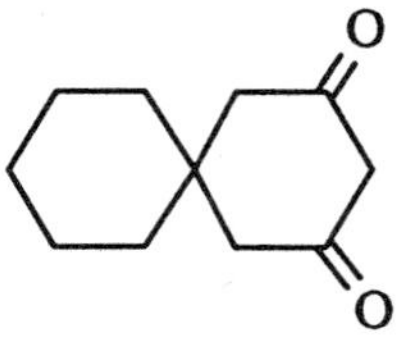 O, O

(e) 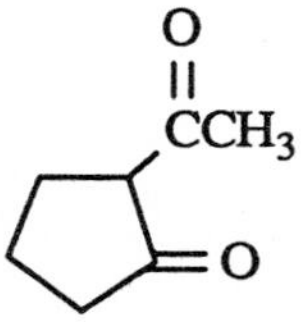 $\overset{O}{\overset{\|}{C}}CH_3$, =O

(f) 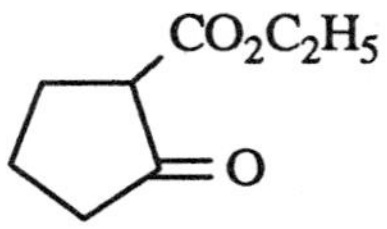 $CO_2C_2H_5$, =O

16.43 (a) $(CH_3)_3CCO_2C_2H_5$ + $CH_3CH_2CO_2C_2H_5$ $\xrightarrow[(2)\ H^+]{(1)\ NaOC_2H_5}$

(b) [benzene ring with two $CO_2C_2H_5$ groups] + $CH_3CO_2C_2H_5$ $\xrightarrow[(2)\ H^+]{(1)\ NaOC_2H_5}$

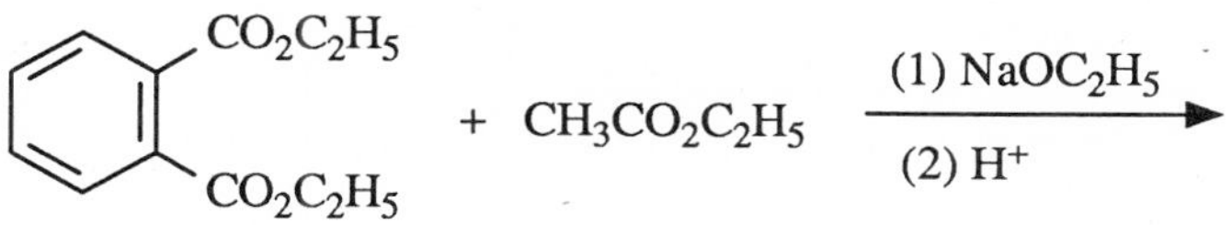

(c) [chain diester: $CO_2C_2H_5$, $CO_2C_2H_5$, CH_3] $\xrightarrow[(2)\ H^+]{(1)\ NaOC_2H_5}$

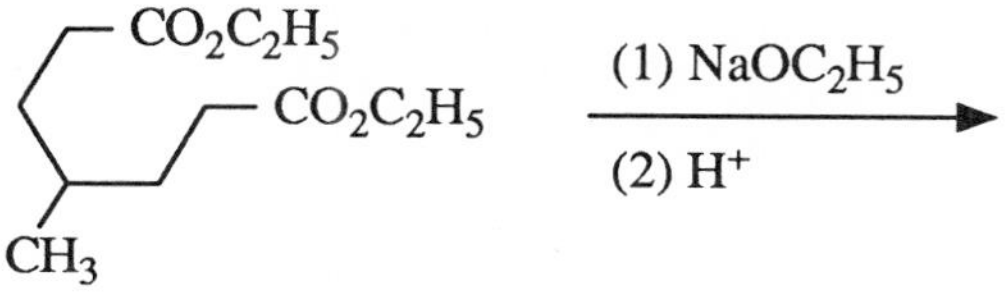

(d) $C_6H_5CO_2C_2H_5$ + $CH(CN)CO_2C_2H_5$ [$CH_2CO_2C_2H_5$ with CN] $\xrightarrow[(2)\ H^+]{(1)\ NaOC_2H_5}$

(e) [cyclohexane with $CO_2C_2H_5$ and $CH_2CH_2CH_2CO_2C_2H_5$] (1) $NaOC_2H_5$ (2) H_2O, OH^-, heat (3) H^+ →

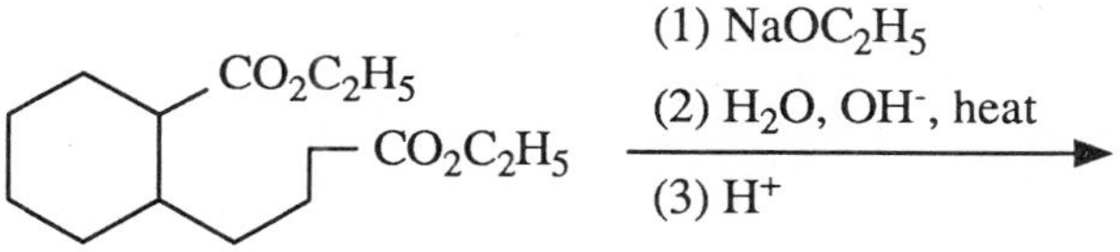

The other possible ester condensation product is not formed because the equilibrium reaction favors the more stable enolate ion.

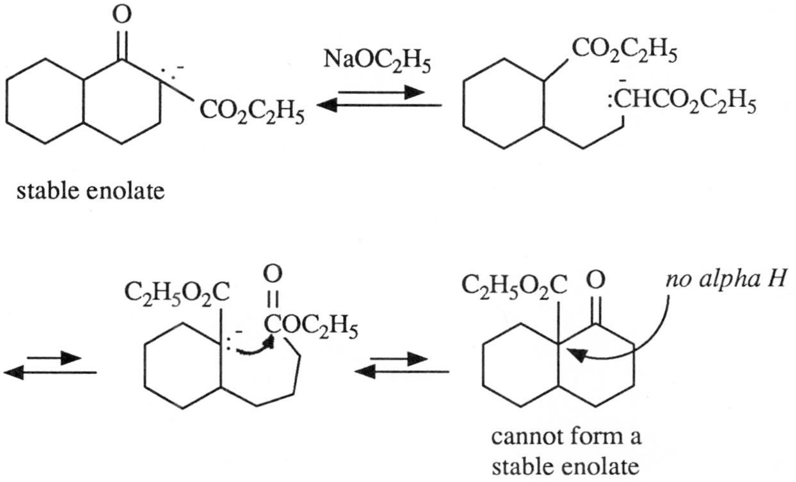

O

$NaOC_2H_5$

$CO_2C_2H_5$

$CO_2C_2H_5$

$:\bar{C}HCO_2C_2H_5$

stable enolate

$C_2H_5O_2C$ O COC_2H_5

$C_2H_5O_2C$ O

no alpha H

cannot form a stable enolate

(Not all the steps in the reversible reaction mechanism are shown here.)

(f) $C_6H_5CH_2CN \xrightarrow[(3)\ H^+]{(1)\ NaOC_2H_5,\ (2)\ CH_3CO_2C_2H_5}$ or $C_6H_5CH(Br)C(=O)CH_3 \xrightarrow{NaCN}$

16.44 (a) $CH_2CH_2C(=O)CH_2CH_3$ / $CH(CO_2C_2H_5)_2$ (b) $CH_2CH_2C(=O)CH_2CH_3$ / $C(CO_2C_2H_5)_2$ / $CH_2CH_2C(=O)CH_2CH_3$

(c) $(CH_3\overset{O}{\overset{\|}{C}})_2CH$ attached to a cyclohexanone ring (=O) (d) $CH_3CHCH_2CO_2C_2H_5$ / $C_6H_5CHCO_2C_2H_5$

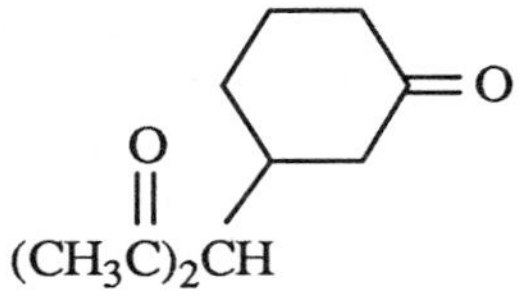

16.45 (a) $CH_3C(=O)CH_2CO_2C_2H_5 + CH_2{=}CHCN \xrightarrow[(2)\ H^+]{(1)\ NaOC_2H_5}$

(b) 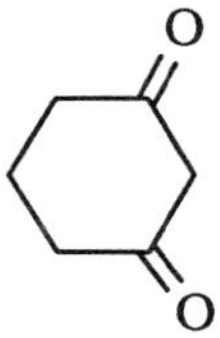 1,3-cyclohexanedione (O, O) $+ CH_2{=}CHCO_2C_2H_5 \xrightarrow[(2)\ H^+]{(1)\ NaOC_2H_5}$

(c) $CH_3CH(Cl)NO_2 + CH_2{=}CHCN \xrightarrow[(2)\ H^+]{(1)\ NaOC_2H_5}$

(d) 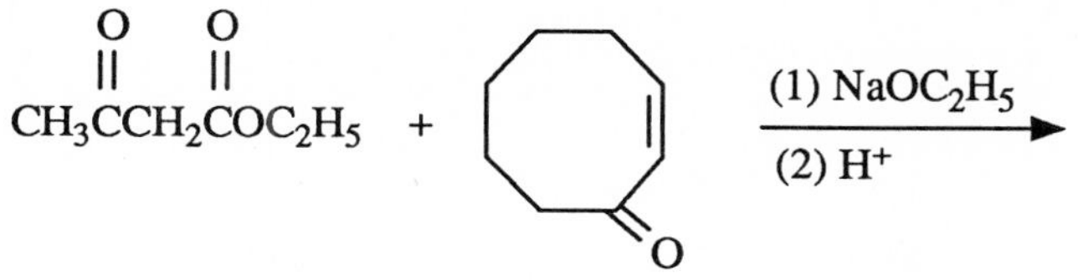 $CH_3\overset{O}{\overset{||}{C}}CH_2\overset{O}{\overset{||}{C}}OC_2H_5$ + (cyclooctenone) $\xrightarrow[(2)\ H^+]{(1)\ NaOC_2H_5}$

16.46 (a) 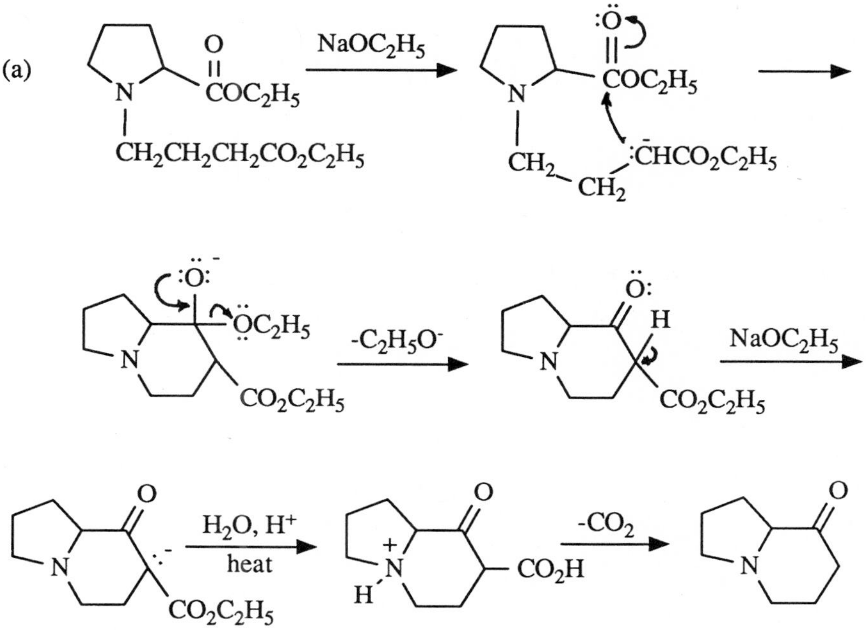 N-substituted 2-carbethoxypyrrolidine with $CH_2CH_2CH_2CO_2C_2H_5$ $\xrightarrow{NaOC_2H_5}$ $\ddot{C}HCO_2C_2H_5$ enolate attacks $\overset{O}{\overset{||}{C}}OC_2H_5$ $\longrightarrow$

tetrahedral intermediate ($:\ddot{O}:^-$, $\ddot{O}C_2H_5$, $CO_2C_2H_5$) $\xrightarrow{-C_2H_5O^-}$ β-keto ester (H, $CO_2C_2H_5$) $\xrightarrow{NaOC_2H_5}$

enolate ($CO_2C_2H_5$) $\xrightarrow[heat]{H_2O,\ H^+}$ ($\overset{+}{N}$–H, CO_2H) $\xrightarrow{-CO_2}$ ketone

a β-keto acid

To see why this ring closure product is formed in preference to the other possibility, see Answer 16.43(e). Neutralization would free the amine.

(b) (*E*) and (*Z*) $CH_3CH_2CH{=}\overset{CN}{\overset{|}{C}}CO_2C_2H_5$

In (b), alanine was used as an acidic buffering agent.

(c) Lithium diisopropylamide is a very strong base.

$R_2\ddot{N}:^-$

H O

H_3C

OC_2H_5

$-R_2\ddot{N}H$

H_3C

O

OC_2H_5

$BrCH_2CH_2CH_2Cl$

S_N2

H_3C O

$ClCH_2CH_2CH_2$

OC_2H_5

In the final step, the bromide ion is displaced instead of the chloride ion because Br^- is a better leaving group. Another product that you might have predicted is the following one:

O

H_3C

OC_2H_5

$CH_2CH_2CH_2Cl$

(d)

O

O O

$CH_2CCH_2COC_2H_5$

most acidic

K_2CO_3

$:\ddot{O}$

CH_2

C=O

$^-:CH$

$CO_2C_2H_5$

1,4-addition

$^-:\ddot{O}:$

$C_2H_5O_2C$

O

H^+ and tautomerization

O

$C_2H_5O_2C$

O

16.47 [cyclohexanone + pyrrolidine (N–H) → enamine ↔ iminium resonance structures]

$\xrightarrow[\text{1,4-addition}]{CH_2{=}CHCCH_3\ (\text{C=O})}$ [iminium with $CH_2CH{=}CCH_3$ bearing $:\ddot{O}:^-$] $\xrightarrow[\text{tautomerization}]{H_2O,\ H^+ \text{ and}}$

an enolate

[iminium with $CH_2CH_2CCH_3$ (C=O)] $\xrightarrow{\text{hydrolysis}}$ [2-($CH_2CH_2CCH_3$)cyclohexanone] + [pyrrolidinium, $\overset{+}{N}H_2$]

16.48 (a) The key to solving this synthesis problem is to identify the $C_6H_5CH{=}$ group in the product. This group is likely to arise from benzaldehyde, C_6H_5CHO, and suggests that an aldol condensation or related reaction should be used.

$$C_6H_5CH(=O) + C_6H_5CH_2COC_2H_5(=O) \underset{}{\overset{NaOC_2H_5}{\rightleftharpoons}} C_6H_5CH(OH)-CH(C_6H_5)COC_2H_5(=O)$$

$$\xrightarrow{-H_2O} C_6H_5CH{=}C(C_6H_5)COC_2H_5(=O)$$

(b) The product is an α-substituted cyclohexanone--a type of product that can be obtained by an enamine synthesis.

[cyclohexanone] $\xrightarrow{\text{HN (piperidine)}}$ [1-piperidinocyclohexene, N]

$\xrightarrow{CH_3CH{=}CHCH_2Cl}$ [cyclohexylidene piperidinium ion with $CH_2CH{=}CHCH_3$ substituent, $=\overset{+}{N}$] $\xrightarrow{H_2O,\ H^+}$

Other amines could have been used to form the enamine.

(c) $CH_2(CO_2CH_3)_2 \xrightarrow[(2)\ (CH_3)_2CHBr]{(1)\ NaOCH_3}$

Elimination, yielding propene, might be a side reaction.

(d) $C_6H_5CHO + CH_2(CN)_2 \xrightarrow[(2)\ H^+]{(1)\ NaOC_2H_5 \text{ or } NaOH}$

(e) [cyclohexanone] $=O + HCOC_2H_5$ (with C=O) $\xrightarrow[(2)\ H^+]{(1)\ NaOC_2H_5}$

(f) $C_6H_5CO_2C_2H_5 + CH_3CO_2C_2H_5 \xrightarrow{NaOC_2H_5}$

$C_6H_5\overset{O}{\overset{\|}{C}}{-}CH_2CO_2C_2H_5 \xrightarrow[-CH_3CH_2OH]{C_6H_5NH_2}$

In alkaline solution, imine formation should not be a problem.

(g) $C_2H_5O_2C(CH_2)_5CO_2C_2H_5 \xrightarrow{NaOC_2H_5}$ [cyclohexanone carbanion with $CO_2C_2H_5$]

$\xrightarrow[-Br^-]{Br(CH_2)_6CO_2C_2H_5}$ [2-cyclohexanone bearing $CO_2C_2H_5$ and $(CH_2)_6CO_2C_2H_5$] $\xrightarrow[heat]{H_2O,\ H^+}$

$$\text{(2-oxocyclohexane with } CO_2H \text{ and } (CH_2)_6CO_2H \text{ on C-1)} \xrightarrow{-CO_2}$$

(h) $$C_2H_5O_2C{-}CO_2C_2H_5 + CH_3CH_2CH_2CO_2C_2H_5 \xrightarrow[(2)\ H^+]{(1)\ NaOC_2H_5}$$

(i) $$\text{(cyclohex-3-enone)}{=}O + CH_2(CO_2C_2H_5)_2 \xrightarrow[(2)\ H^+]{(1)\ NaOC_2H_5}$$

16.49 *Attack by ^-OH:*

$$HC(=O){-}H + {}^-{:}\ddot{O}H \longrightarrow HC(H)(O^-)(OH)$$

Hydride transfer:

$$HC(O^-)(H)(OH) + H_2C{=}O \longrightarrow \left[HC(=O)(OH) + CH_3O^- \right] \longrightarrow HC(=O)(O^-) + CH_3OH$$

16.50 (a) The product is a substituted acetic acid; therefore, a malonic ester synthesis can be used.

$$CH_2(CO_2C_2H_5)_2 \xrightarrow{NaOC_2H_5} {}^-{:}CH(CO_2C_2H_5)_2 \xrightarrow{CH_3(CH_2)_6Br}$$

$$CH_3(CH_2)_6CH(CO_2C_2H_5)_2 \xrightarrow[heat]{H_2O,\ H^+} CH_3(CH_2)_6CH(CO_2H)CO_2H \xrightarrow{-CO_2}$$

(b) A β-diketone can be obtained from an ester condensation.

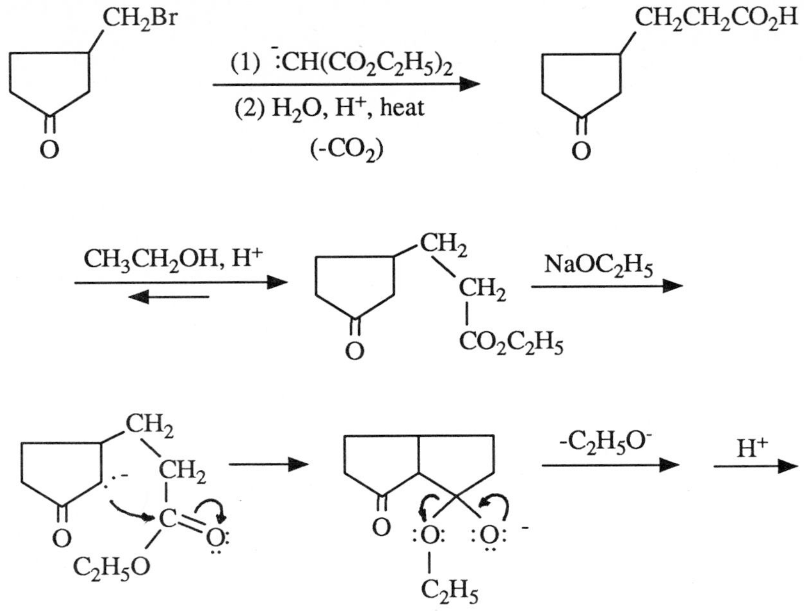

CH_2Br (cyclopentanone) —(1) $^{-}$:$CH(CO_2C_2H_5)_2$; (2) H_2O, H^+, heat ($-CO_2$)→ $CH_2CH_2CO_2H$ (cyclopentanone)

CH_3CH_2OH, H^+ ⇌ (cyclopentanone)-$CH_2CH_2CO_2C_2H_5$ —$NaOC_2H_5$→

(enolate)-$CH_2CH_2C(=O)OC_2H_5$ → (bicyclic alkoxide, :O: C_2H_5, :O:$^-$) —$-C_2H_5O^-$→ —H^+→

(c) Dialkylation of a compound with somewhat acidic hydrogens, followed by a Wolff-Kishner reduction of the carbonyl group, yields the product.

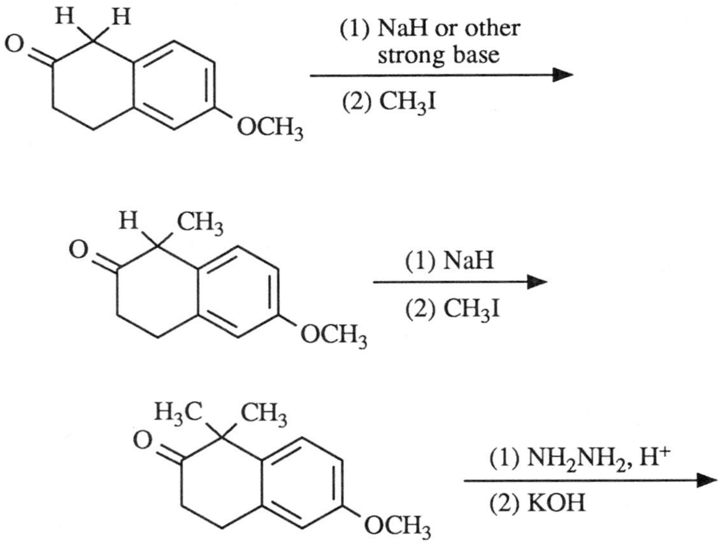

H H, O, OCH_3 —(1) NaH or other strong base; (2) CH_3I→

H CH_3, O, OCH_3 —(1) NaH; (2) CH_3I→

H_3C CH_3, O, OCH_3 —(1) NH_2NH_2, H^+; (2) KOH→

(d) A crossed Claisen condensation between a formate ester (no acidic hydrogens) and the given reactant (acidic benzylic hydrogens) will yield the product.

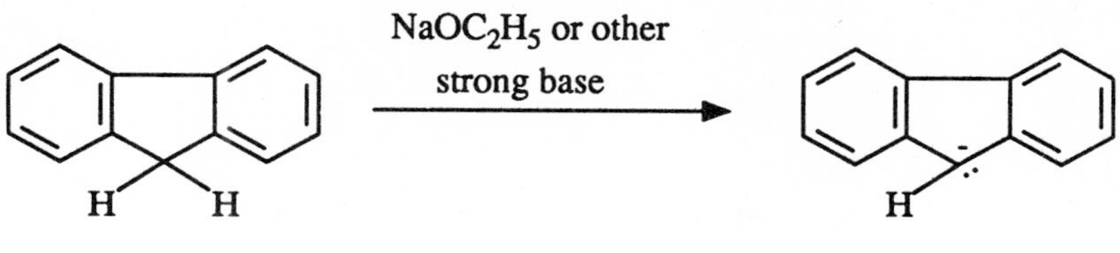

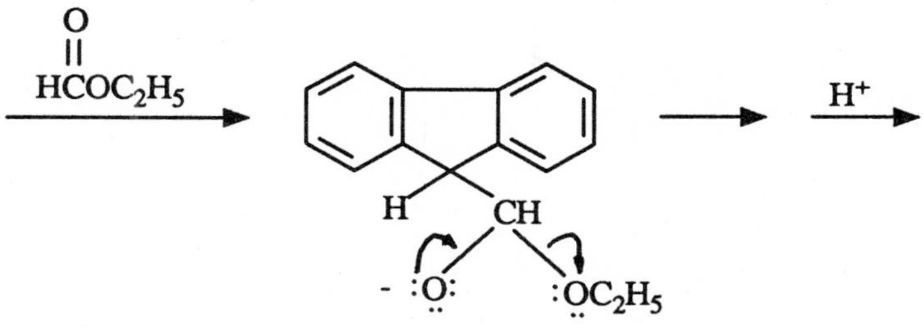

NaOC$_2$H$_5$ or other strong base

$\overset{O}{\overset{||}{HCOC_2H_5}}$

H$^+$

16.51 (a) An α,β-unsaturated ketone can be obtained by an aldol condensation.

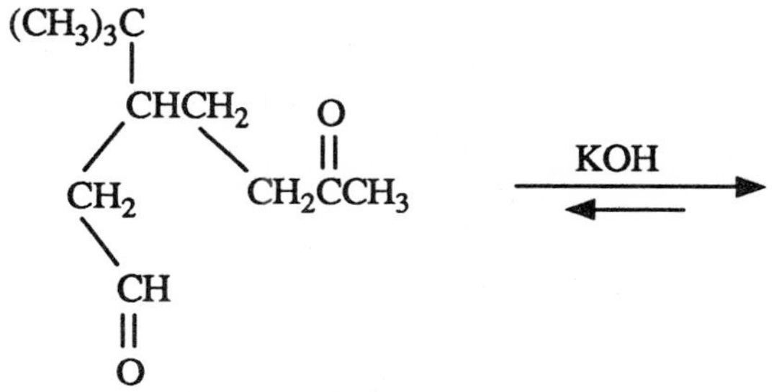

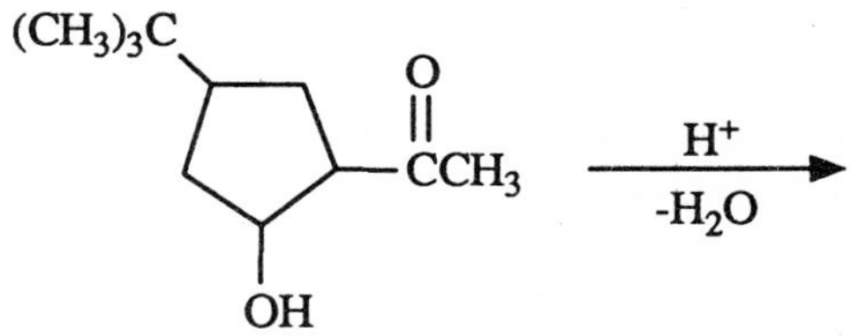

KOH

H$^+$
-H$_2$O

(b) 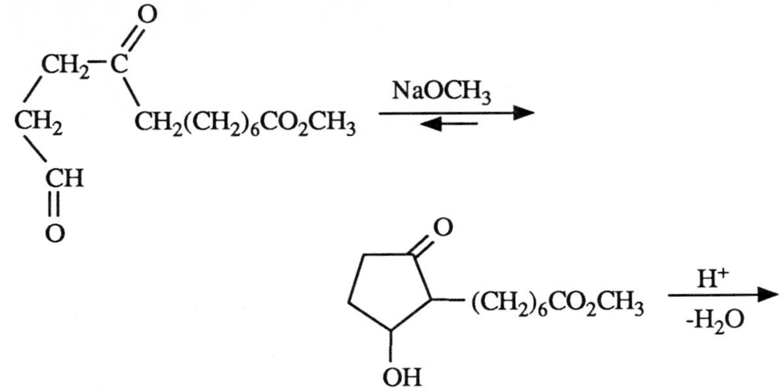 CH_2–CH_2–CH_2–C(=O)–$CH_2(CH_2)_6CO_2CH_3$ with CH=O terminal $\xrightarrow{NaOCH_3}$ (reversible)

2-[(CH$_2$)$_6$CO$_2$CH$_3$]-3-hydroxycyclopentanone $\xrightarrow[-H_2O]{H^+}$

A three-membered ring or a seven-membered ring would be less likely to be formed than the five-membered ring.

(c) 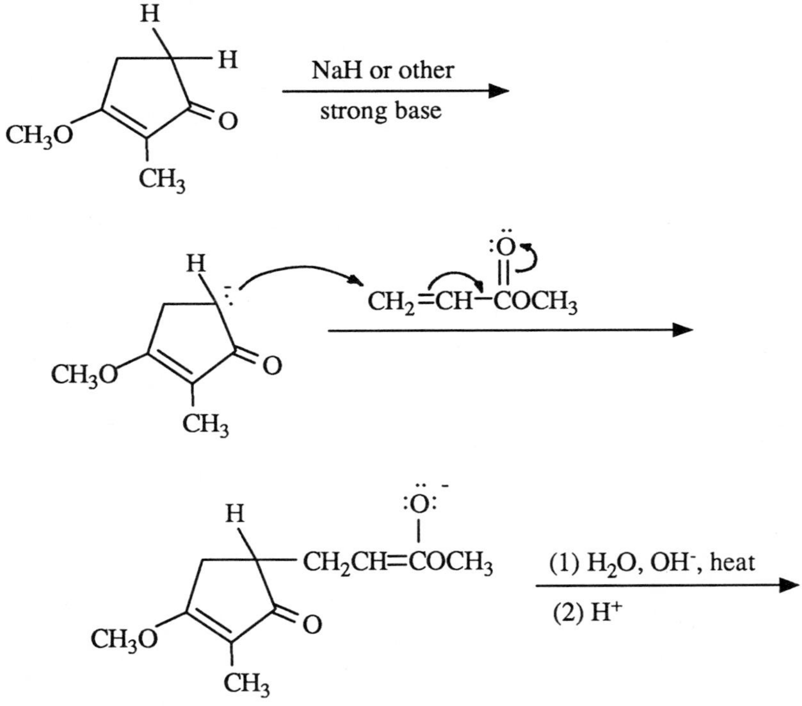 3-methoxy-2-methylcyclopent-2-enone (CH_3O, CH_3, =O, H, H) $\xrightarrow[\text{strong base}]{\text{NaH or other}}$

Carbanion (H, :) of the cyclopentenone (CH_3O, CH_3, =O) + CH_2=CH–C(=O)OCH$_3$ ⟶

H–C(ring)–CH_2CH=C(–$\ddot{O}$:$^-$)OCH$_3$ (ring: CH_3O, CH_3, =O) $\xrightarrow[\text{(2) } H^+]{\text{(1) } H_2O, OH^-, \text{heat}}$

an enolate of an ester

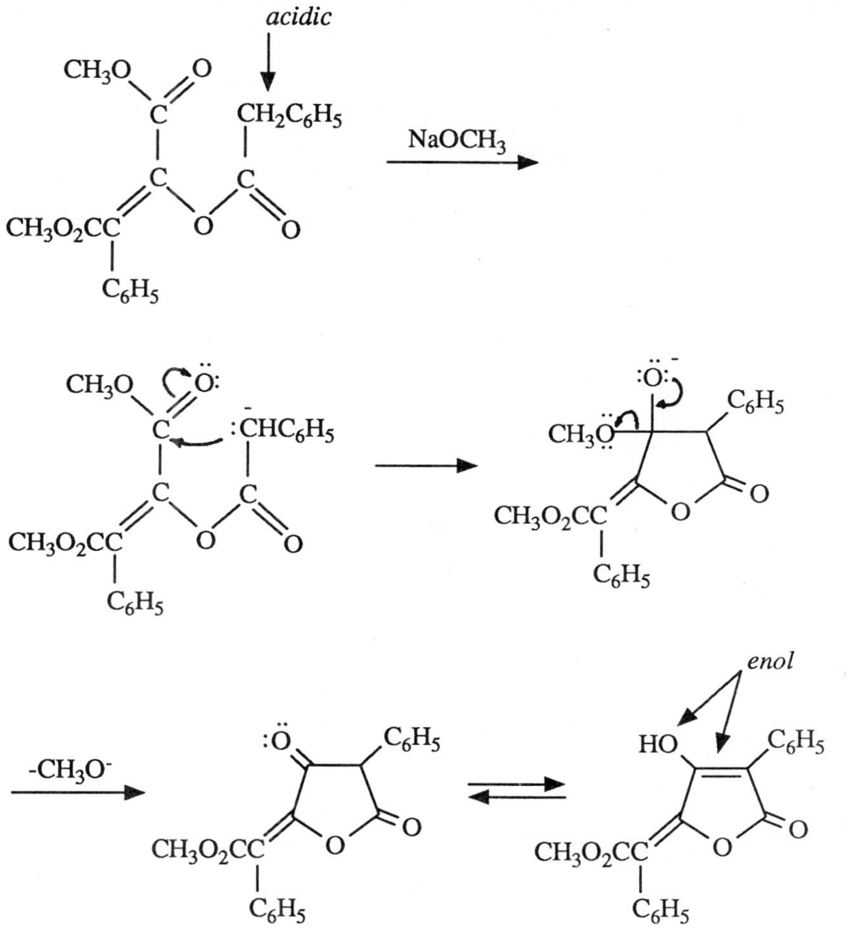

(d) [reaction scheme: CH_3O_2C ester/lactone with $CH_2C_6H_5$ (acidic) treated with $NaOCH_3$]

acidic

$NaOCH_3$

$-CH_3O^-$

enol

Note that vulpinic acid is a stable enol. Its stability arises from extensive conjugation of the enol C=C bond.

16.52 (a) 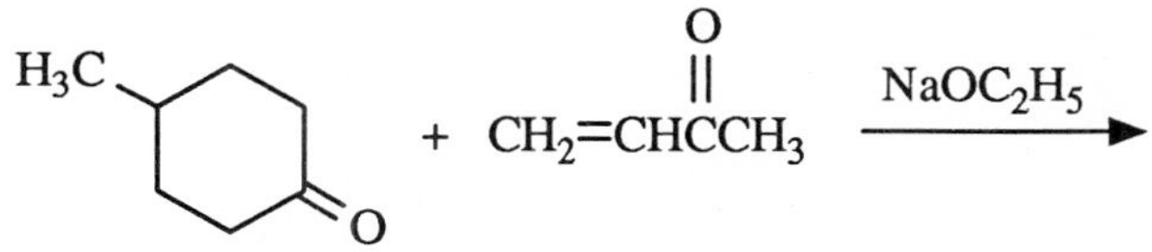 [4-methylcyclohexanone] + $CH_2{=}CHCCH_3$ (with C=O) $\xrightarrow{NaOC_2H_5}$

(b) 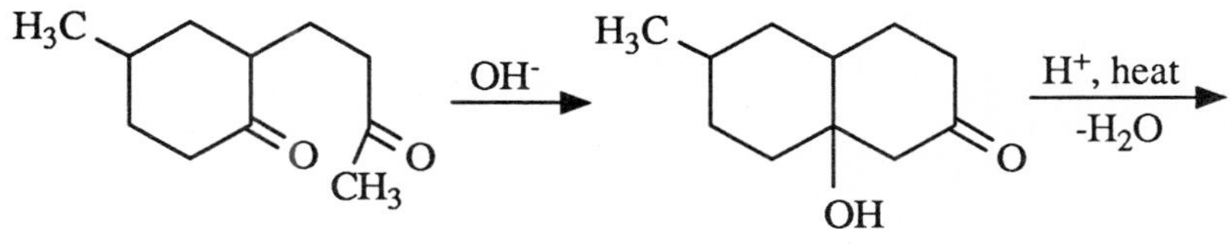 $\xrightarrow{OH^-}$ $\xrightarrow[-H_2O]{H^+,\ heat}$

16.53 (a) $CH_3CO_2H + CH_3CH_2OH \underset{}{\overset{\text{dil. HCl, heat}}{\rightleftharpoons}} CH_3CO_2C_2H_5 + H_2O$

(b) $C_2H_5OH \xrightarrow{\text{Na}} C_2H_5ONa \xrightarrow[\text{(2) } H^+]{\text{(1) 2 } CH_3CO_2C_2H_5} CH_3\overset{O}{\overset{\|}{C}}CH_2CO_2C_2H_5$

(c) $CH_3CH_2OH \xrightarrow{PBr_3} CH_3CH_2Br \xrightarrow[NaOC_2H_5]{CH_3\overset{O}{\overset{\|}{C}}CH_2CO_2C_2H_5}$

$CH_3\overset{O}{\overset{\|}{C}}\underset{CH_2CH_3}{\underset{|}{C}}HCO_2C_2H_5 \xrightarrow[\text{heat}]{\text{dil. HCl}} CH_3\overset{O}{\overset{\|}{C}}CH_2CH_2CH_3$

(d) $CH_2(CO_2C_2H_5)_2 \xrightarrow[\text{(2) } NaOC_2H_5,\ C_2H_5Br]{\text{(1) } NaOC_2H_5,\ C_2H_5Br} (CH_3CH_2)_2C(CO_2C_2H_5)_2$

$\xrightarrow[\text{heat}]{\text{dil. HCl}} (CH_3CH_2)_2CHCO_2H$

(e) $CH_3CO_2H \xrightarrow[Br_2]{PBr_3} BrCH_2\overset{O}{\overset{\|}{C}}Br \xrightarrow{CH_3CH_2OH} BrCH_2CO_2C_2H_5$

16.54 (a) $C_6H_5CH_2\overset{O}{\overset{\|}{C}}CH_3 + CH_2{=}CH\overset{O}{\overset{\|}{C}}CH_3 \xrightarrow{NaOC_2H_5}$

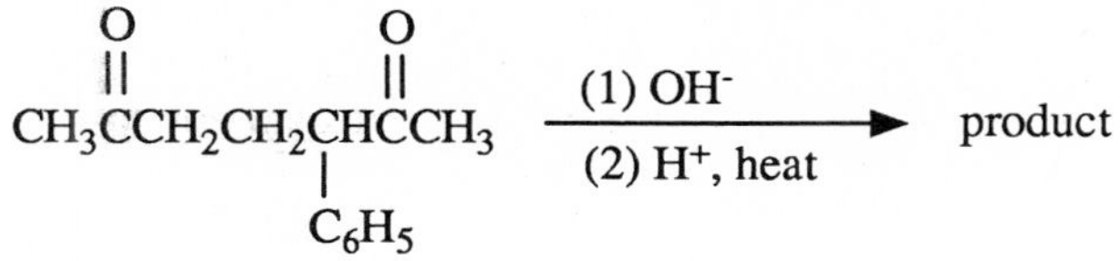
(1) OH⁻
(2) H⁺, heat
product

Loss of the most acidic proton in the last step would not lead to cyclization because a four-membered ring would result.

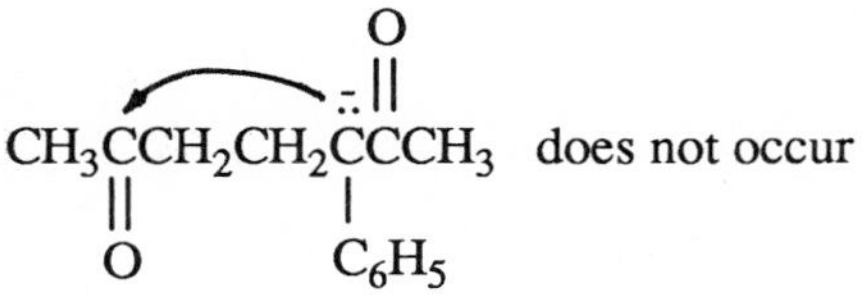
does not occur

(b) 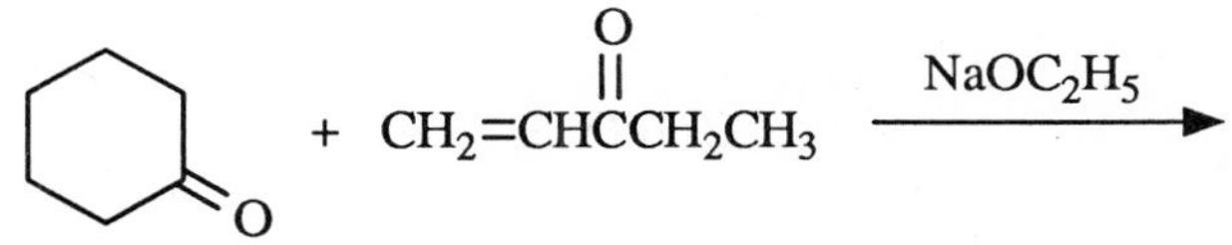
NaOC2H5

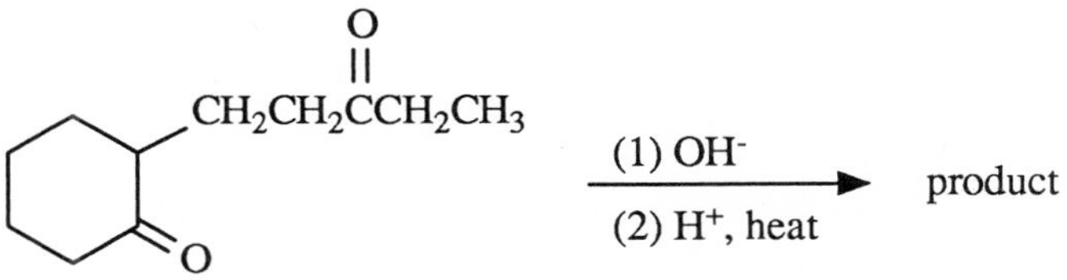
(1) OH⁻
(2) H⁺, heat
product

16.55 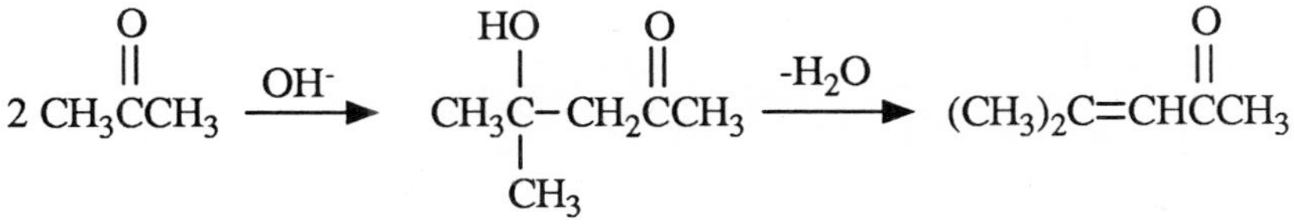
OH⁻
-H2O

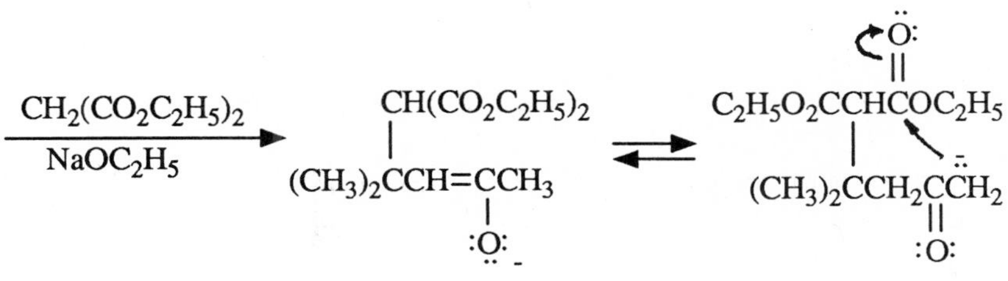
CH2(CO2C2H5)2
NaOC2H5

an enolate

$\rightleftharpoons$ [cyclic tetrahedral intermediate: ring bearing $C_2H_5O_2C$, two H_3C, ring C=O, and $C(:\ddot{O}:^-)(\ddot{O}C_2H_5)$] $\xrightleftharpoons{-C_2H_5O^-}$ [2-carbethoxy-3,3-dimethylcyclohexane-1,5-dione: $C_2H_5O_2C$, H_3C, H_3C, O, O]

$\xrightarrow[\text{heat}]{H_2O,\ H^+}$ [5,5-dimethylcyclohexane-1,3-dione: H_3C, H_3C, O, O]

16.56 $CH_2(CO_2C_2H_5)_2 \xrightarrow[(2)\ Br(CH_2)_nBr]{(1)\ NaOC_2H_5} Br(CH_2)_nCH(CO_2C_2H_5)_2$

$\xrightarrow{NaOC_2H_5}$ [ring: $(CH_2)_n$ — $C(CO_2C_2H_5)_2$] $\xrightarrow[\text{heat}]{H_2O,\ H^+}$ [ring: $(CH_2)_n$ — $CHCO_2H$]

where n = 4 or 5 and $(CH_2)_n$ may contain alkyl substituents.

16.57 $R'\overset{O}{\overset{\|}{C}}CHRCO_2C_2H_5 \xrightarrow[(2)\ ClCH_2CO_2C_2H_5]{(1)\ NaOC_2H_5} R'\overset{O}{\overset{\|}{C}}\underset{CH_2CO_2C_2H_5}{\underset{|}{C}}RCO_2C_2H_5$

(by an ester condensation)

$\xrightarrow[\text{heat}]{H_2O,\ H^+} R'\overset{O}{\overset{\|}{C}}CHRCH_2CO_2H$

16.58 (a) $CH_2(CO_2C_2H_5)_2 \xrightarrow[(2)\ Br(CH_2)_4Br]{(1)\ NaOC_2H_5}$ [CH_2CH_2 / CH_2 — $CH(CO_2C_2H_5)_2$ / CH_2Br]

$\xrightarrow{NaOC_2H_5}$ (cyclopentane-1,1-$(CO_2C_2H_5)_2$) $\xrightarrow[\text{heat}]{H_2O,\ H^+}$ (cyclopentane)$-CO_2H$

(b) $CH_2(CO_2C_2H_5)_2 \xrightarrow[(2)\ Br(CH_2)_4Br]{(1)\ NaOC_2H_5} Br(CH_2)_4CH(CO_2C_2H_5)_2$

$\xrightarrow[CH_2(CO_2C_2H_5)_2]{NaOC_2H_5} (C_2H_5O_2C)_2CH(CH_2)_4CH(CO_2C_2H_5)_2$

$\xrightarrow[\text{heat}]{H_2O,\ H^+} HO_2C(CH_2)_6CO_2H$

We would use a dilute solution in (a) and an excess of malonic ester in (b).

16.59 (a) $H\text{-}CH_2\overset{O}{\overset{\|}{C}}H \underset{}{\overset{CO_3^{2-}}{\rightleftharpoons}} {}^{-}:CH_2\overset{O}{\overset{\|}{C}}H$ (attacks $\overset{:\ddot{O}}{\overset{\|}{HCH}}$) $\rightleftharpoons {}^{-}:\ddot{\underset{..}{O}}CH_2\text{-}CH_2\overset{O}{\overset{\|}{C}}H \overset{H_2O}{\rightleftharpoons}$

$HOCH_2CH_2\overset{O}{\overset{\|}{C}}H \overset{CO_3^{2-}}{\rightleftharpoons} HOCH_2\overset{..}{C}H\overset{O}{\overset{\|}{C}}H \overset{\overset{:\ddot{O}}{\overset{\|}{HCH}}}{\rightleftharpoons} HOCH_2\overset{\overset{:\ddot{O}:^{-}}{|}}{\overset{HCH}{\overset{|}{C}}}HCHO$

$\overset{H_2O}{\rightleftharpoons} HOCH_2\overset{\overset{CH_2OH}{|}}{C}HCHO \overset{CO_3^{2-}}{\rightleftharpoons} (HOCH_2)_2\overset{..}{C}CHO \overset{\overset{:\ddot{O}}{\overset{\|}{HCH}}}{\rightleftharpoons}$

$(HOCH_2)_2\overset{\overset{:\ddot{O}:^{-}}{|}}{\overset{HCH}{\overset{|}{C}}}CHO \overset{H_2O}{\rightleftharpoons}$ product

(b)

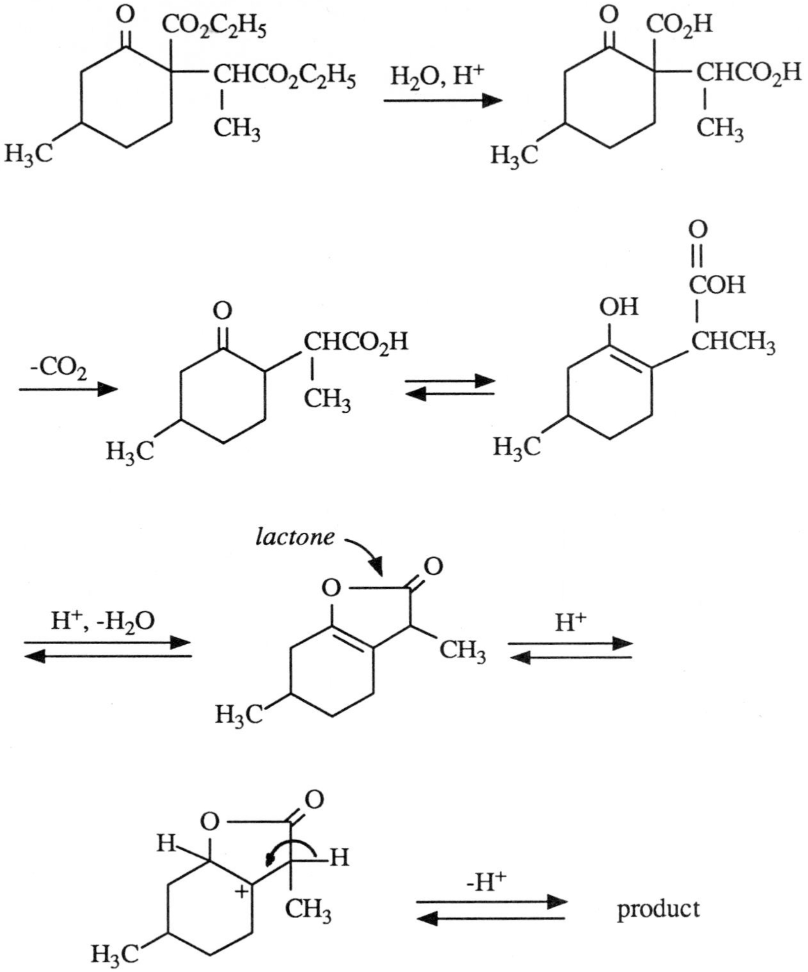

O
CO2C2H5
CHCO2C2H5
CH3
H3C
H2O, H+
O
CO2H
CHCO2H
CH3
H3C
-CO2
O
CHCO2H
CH3
H3C
O
COH
OH
CHCH3
H3C
lactone
O
O
H+, -H2O
CH3
H+
H3C
O
O
H
H
+
CH3
H3C
-H+
product

Protonation of the other sp^2 carbon would lead to a less stable carbon-carbon double bond.

(c)

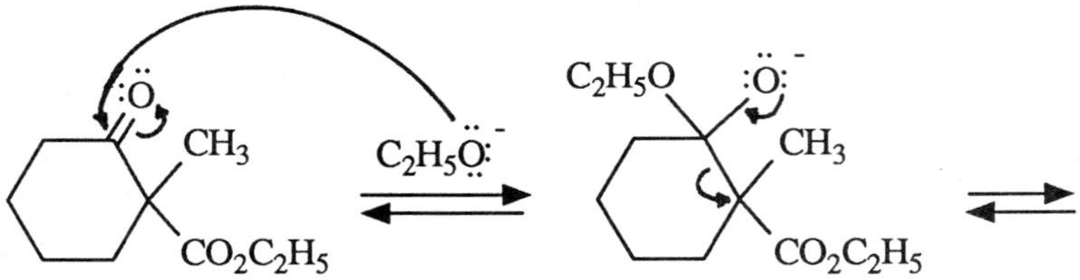

:O:
CH3
CO2C2H5
C2H5O:
C2H5O
:O:
CH3
CO2C2H5

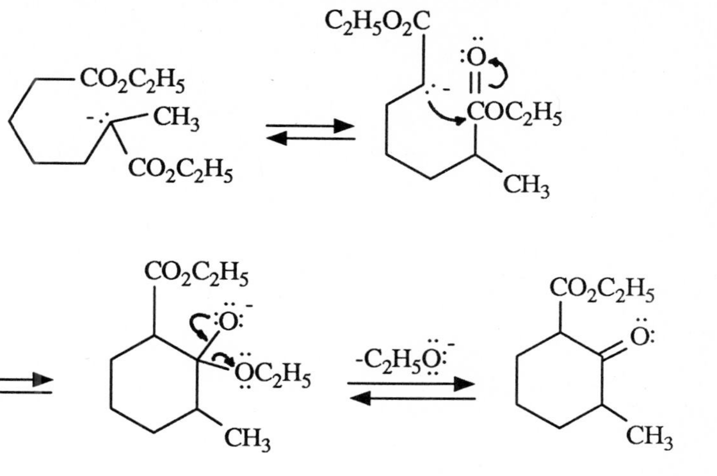

The first steps in this rearrangement are a reverse ester condensation. Then a forward ester condensation yields the rearranged product.

16.60

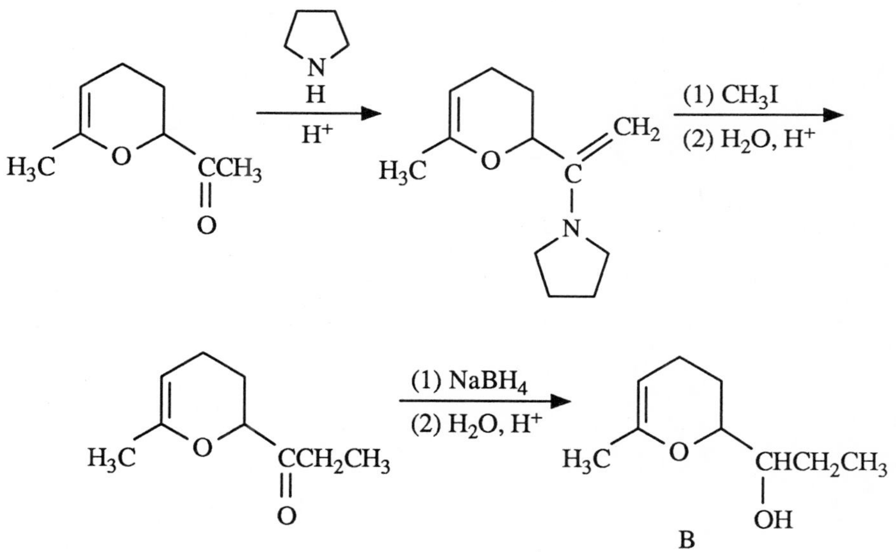

(1) $NaBH_4$ (2) H_2O, H^+

B

H^+

H H_3C O H $HOCHCH_2CH_3$ ⇌ H H_3C H + O H $H\ddot{O}CHCH_2CH_3$ ⇌

B

H_3C O H — $\ddot{O}^+$-$CHCH_2CH_3$ $\xrightarrow{-H^+}$ brevicomin

16.61 H_2NCCH_2CN (C=O) $\xrightarrow{OH^-}$ $H_2NC\ddot{C}HCN$ (C=O, carbanion) + $C_2H_5OCCH_2{}^{14}CCH_3$ (two C=O) $\xrightarrow{aldol}$

$C_2H_5OCCH_2{}^{14}CCH_3$ (C=O, $:\ddot{O}:^-$) with $H_2NCCHCN$ (C=O) $\underset{-CH_3O^-}{\xrightarrow{CH_3OH,}}$ $C_2H_5OCCH_2{}^{14}CCH_3$ (C=O, $:\ddot{O}H$) with $H_2NCCHCN$ (C=O) $\xrightarrow{-H_2O}$

CH_3 ^{14}C HC CHCN $:\ddot{O}$=C C=O C_2H_5O $:NH_2$ ⟶ CH_3 CN $^-:\ddot{O}$ $\overset{+}{N}$ H_2 O C_2H_5O $\xrightarrow{OH^-, -H_2O}$

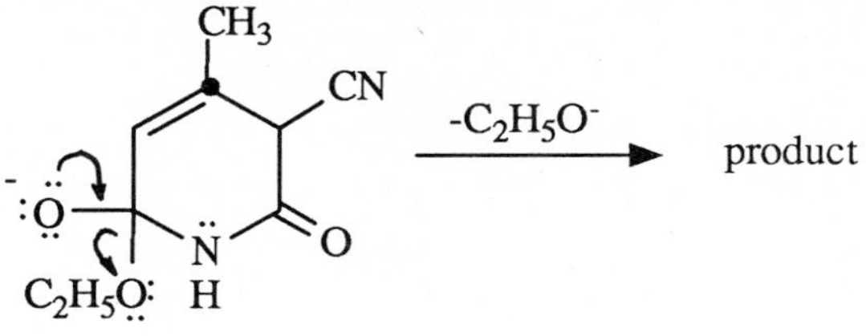

CH$_3$

CN

$-C_2H_5O^-$ → product

$^-$:Ö: — C_2H_5Ö: — N̈ — H — O

16.62 I: $CH_3CH_2CH(CO_2C_2H_5)CCO_2C_2H_5$ (with C=O) II: $CH_3CH_2CH_2CCO_2H$ (with C=O)

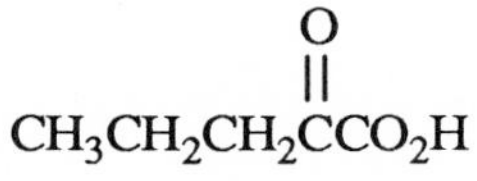

16.63 $C_2H_5O_2C(CH_2)_3CO_2C_2H_5$ $\xrightarrow{NaOC_2H_5}$ $C_2H_5O_2C$—$\overset{-}{HC}$:—CH_2—CH_2—$C_2H_5O_2C$ + :Ö=COC_2H_5—$CO_2C_2H_5$

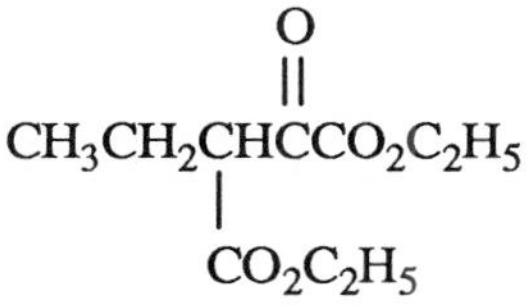

→ $C_2H_5O_2C$—HC—C(—:Ö:$^-$)(—ÖC_2H_5)(—$CO_2C_2H_5$), CH_2, CH_2, $C_2H_5O_2C$ $\xrightarrow{-C_2H_5O^-}$ $C_2H_5O_2C$—CH—C(=Ö:)—$CO_2C_2H_5$, CH_2, CH_2, $CO_2C_2H_5$

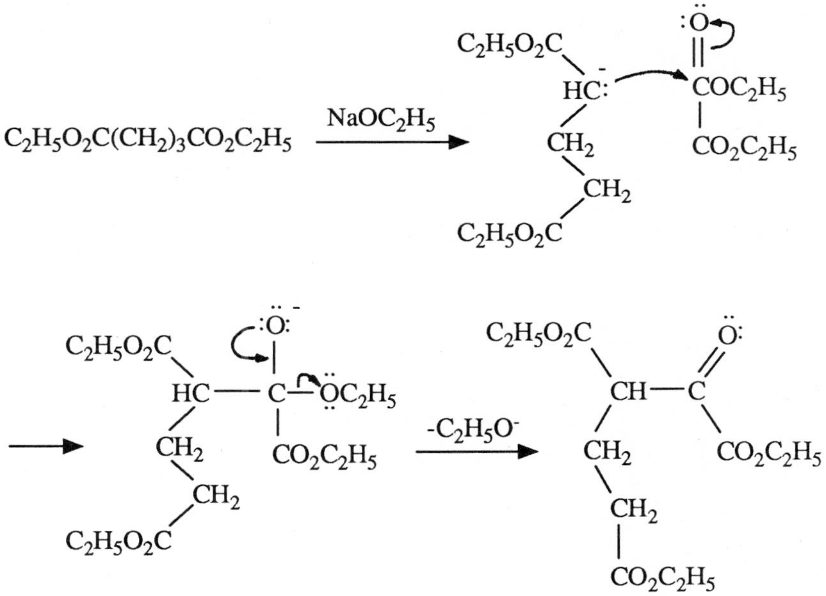

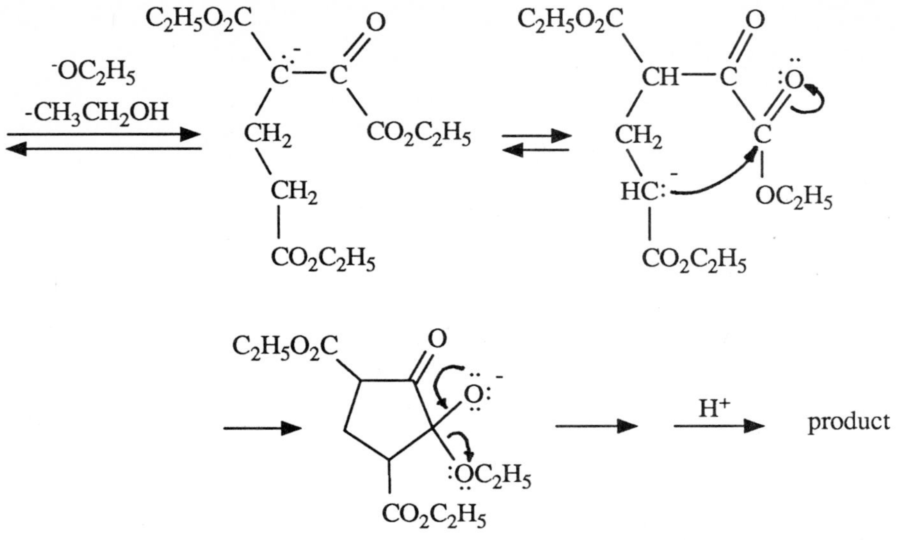

$^{-}OC_2H_5$, $-CH_3CH_2OH$

$C_2H_5O_2C$ ... O ... CH_2 ... $CO_2C_2H_5$ ... CH_2 ... $CO_2C_2H_5$

$C_2H_5O_2C$ ... CH—C ... O ... CH_2 ... C ... HC: ... OC_2H_5 ... $CO_2C_2H_5$

$C_2H_5O_2C$ ... O ... O: ... $:OC_2H_5$ ... $CO_2C_2H_5$ → H^+ → product

16.64 I, $(C_2H_5O_2C)_2CHCH_2CH_2CH(CO_2C_2H_5)_2$

II, $C_2H_5O_2C$ ... $CO_2C_2H_5$ ... $C_2H_5O_2C$ ... $CO_2C_2H_5$

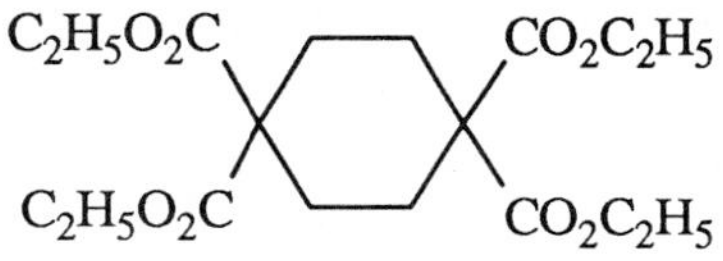

III, *cis* and *trans* HO_2C— —CO_2H

16.65 CH_3 ... O: ... $C(CH_3)_2$ ... HO: → 1,4-addition → CH_3 ... O: ... HO—$C(CH_3)_2$ *an enolate* ⇌ H_2O, $-OH^-$ ⇌ CH_3 ... O ... HO—$C(CH_3)_2$

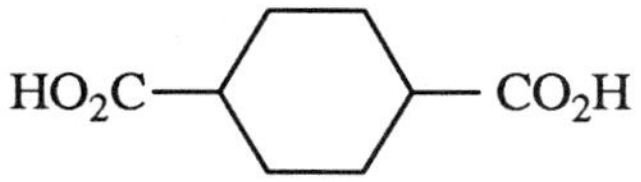

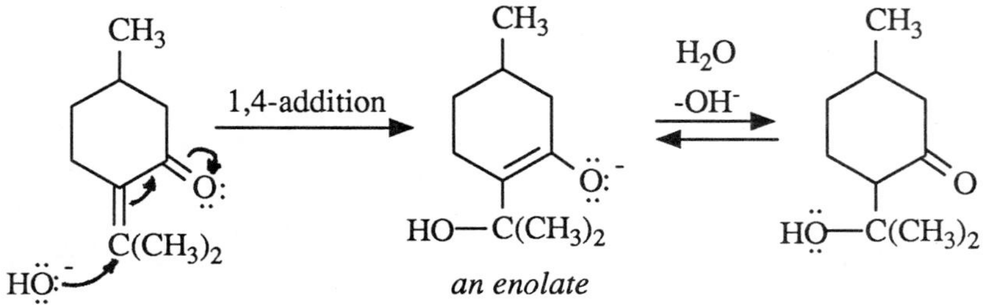

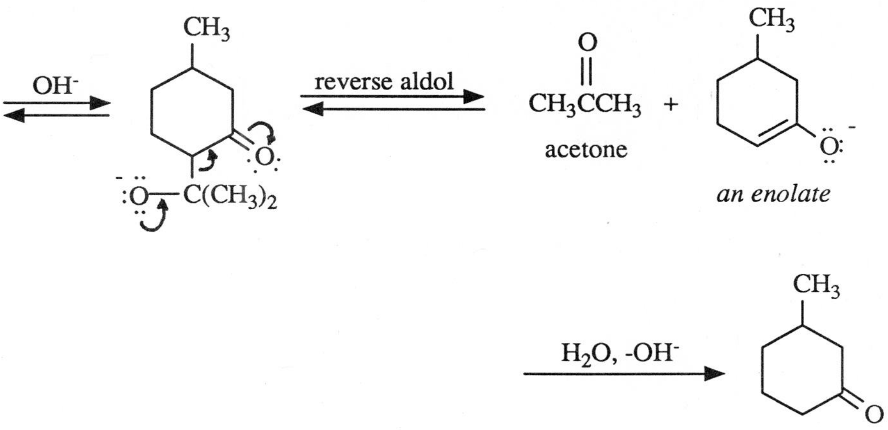

$\xrightarrow{OH^-}$ (reverse aldol) $\rightleftharpoons$ CH_3CCH_3 (acetone) + an enolate

$\xrightarrow{H_2O,\ -OH^-}$

16.66 (a) + $Br-CH_2C(CO_2C_2H_5){=}CHCO_2C_2H_5$ $\xrightarrow{-Br^-}$

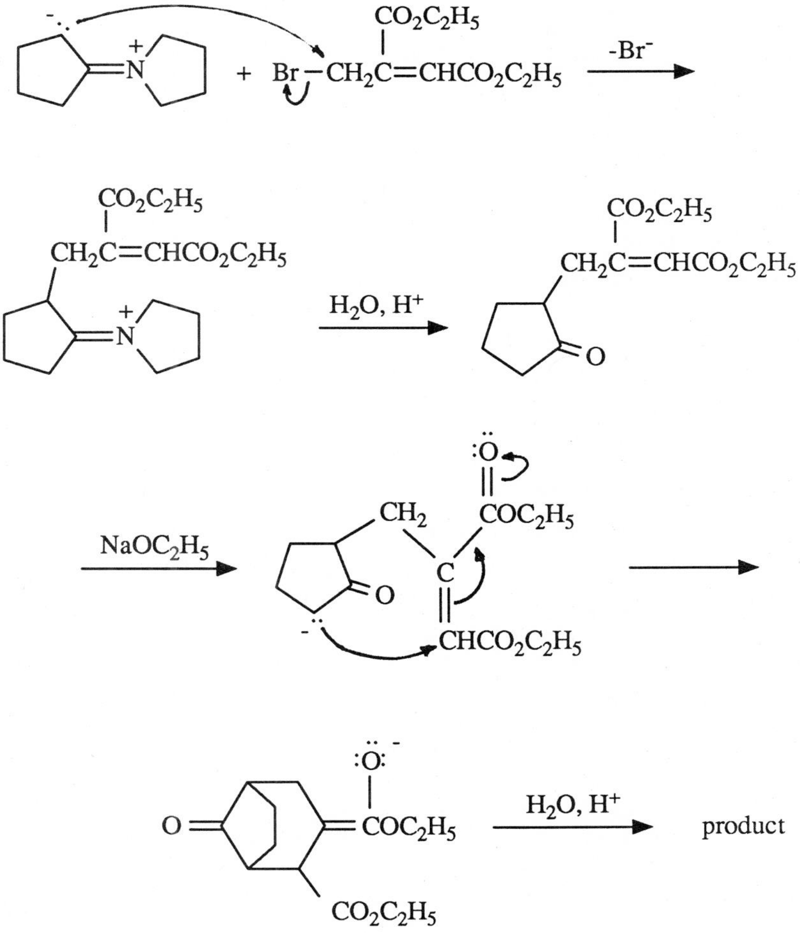

$\xrightarrow{H_2O,\ H^+}$

$\xrightarrow{NaOC_2H_5}$ $\longrightarrow$

$\xrightarrow{H_2O,\ H^+}$ product

(b) [cyclohexylidene morpholinium enamine] + $CH_2{=}CH{-}CCH_3$ (with C=O) —1,4-addition→

[adduct: $CH_2CH{=}C(O^-)CH_3$ on cyclohexylidene morpholinium] —H_2O, H^+→ [2-($CH_2CH_2CCH_3$, C=O)cyclohexanone] —OH^-, aldol→

[enolate $CH_2CH_2C(O)CH_2^-$ on cyclohexanone] → [bicyclic ketone with O^-] —H_2O, $-OH^-$→

[bicyclic ketone with HO] —H^+, heat, $-H_2O$→ product

(c) HCH (C=O) + $HN(CH_3)_2$ ⇌ [$^-O{-}CH_2NH^+(CH_3)_2$] ⇌

$HOCH_2N(CH_3)_2$ —H^+→ $H_2O^+{-}CH_2{-}N(CH_3)_2$

—$-H_2O$→ $CH_2{=}N^+(CH_3)_2$

an iminium ion

$$C_6H_5\overset{O}{\overset{\|}{C}}CH_3 \rightleftharpoons C_6H_5\overset{OH}{\overset{|}{C}}{=}CH_2 \xrightarrow{CH_2{=}\overset{+}{N}(CH_3)_2}$$

an enol

$$C_6H_5\overset{:\ddot{O}-H}{\overset{|}{\underset{+}{C}}}CH_2CH_2\ddot{N}(CH_3)_2 \xrightarrow{-H^+} \text{product}$$

16.67 (a) 2 cyclopentanone $\xrightarrow[\text{aldol}]{OH^-}$ 2-(1-hydroxycyclopentyl)cyclopentanone (O, HO) $\xrightarrow[\text{heat}]{H^+}$

(b) benzene-1,2-dicarbaldehyde (CHO, CHO) + $CH_3CH_2\overset{O}{\overset{\|}{C}}CH_2CH_3 \xrightarrow[-H_2O]{OH^-}$

$C_6H_4(CHO)-CH{=}\overset{H_3C}{\overset{|}{C}}-\overset{O}{\overset{\|}{C}}CH_2CH_3 \xrightarrow{OH^-}$

ring: CH=C(CH3)–C=O–CH(CH3)–CH(OH) fused to benzene $\xrightarrow[-H_2O]{H^+,\ heat}$ product

(c) dilute $CH_3\overset{O}{\overset{\|}{C}}CH_2CO_2C_2H_5 \xrightarrow[(2)\ Br(CH_2)_4Br]{(1)\ NaOC_2H_5}$

$$CH_3\overset{O}{\overset{\|}{C}}CHCO_2C_2H_5 \text{ with } CH_2CH_2CH_2CH_2Br \text{ chain on CH} \xrightarrow{NaOC_2H_5}$$

(d) $$C_2H_5O_2CCH_2CH_2\overset{C_2H_5O_2C}{\overset{|}{C}H}\overset{CH_3}{\overset{|}{C}H}CO_2C_2H_5 \xrightarrow{NaOC_2H_5}$$

$$C_2H_5O_2C\overset{-}{\underset{\cdot\cdot}{C}}HCH_2\overset{C_2H_5O_2C}{\overset{|}{C}H}\overset{CH_3}{\overset{|}{C}H}CO_2C_2H_5 \xrightarrow[(2)\ H^+]{(1)\text{ ester condensation}}$$

The starting triester can be prepared by the following sequence:

$$C_2H_5O_2CCH_2\overset{CH_3}{\overset{|}{C}H}CO_2C_2H_5 \xrightarrow{NaOC_2H_5}$$

$$C_2H_5O_2C\overset{-}{\underset{\cdot\cdot}{C}}H\overset{CH_3}{\overset{|}{C}H}CO_2C_2H_5 \xrightarrow[(2)\ H^+]{(1)\ CH_2{=}CHCO_2C_2H_5}$$

(e) cyclopentyl–Br $\xrightarrow[\text{ether}]{Mg}$ cyclopentyl–MgBr $\xrightarrow[(2)\ H_2O,\ H^+]{(1)\ \overset{O}{CH_2{-}CH_2}}$

cyclopentyl–CH_2CH_2OH $\xrightarrow[\text{pyridine}]{CrO_3}$ cyclopentyl–$CH_2\overset{O}{\overset{\|}{C}}H$

$$\xrightarrow[\text{aldol}]{HC(=O)-CO_2^- Na^+, OH^-} \text{cyclopentyl}-CH(CH{=}O)-CH(OH)-CO_2^- \xrightarrow{H^+} \text{product}$$

(f) $$CH_3C(=O)-CH_2-C(=O)OC_2H_5 \xrightarrow{NaOC_2H_5}$$

$$CH_3C(=O)-\overset{-}{\underset{..}{C}}H-C(=O)OC_2H_5 \xrightarrow{CH_3C(=O)-CH(Cl)-C(=O)OC_2H_5} \text{product}$$

(g) $$CH_2(CO_2C_2H_5)_2 \xrightarrow{NaOC_2H_5} {}^{-}{:}CH(CO_2C_2H_5)_2$$

$$\xrightarrow[\text{Michael}]{CH_3CH{=}CHCO_2C_2H_5} CH_3-CH(CH(CO_2C_2H_5)_2)(CH_2CO_2C_2H_5)$$

$$\xrightarrow[-CO_2]{H_2O, H^+, \text{heat}} CH_3-CH(CH_2CO_2H)_2 \xrightarrow[\text{heat}]{(CH_3CO)_2O} \text{product}$$

(h) $$C_2H_5O_2C-CH_2-C(=O)CH_3 \xrightarrow{NaOC_2H_5} C_2H_5O_2C-HC{:}^{-}(-C(=O)CH_3) \;\curvearrowright\; CH_2Br-C(=O)-CH_2CH_3 \xrightarrow{-Br^-}$$

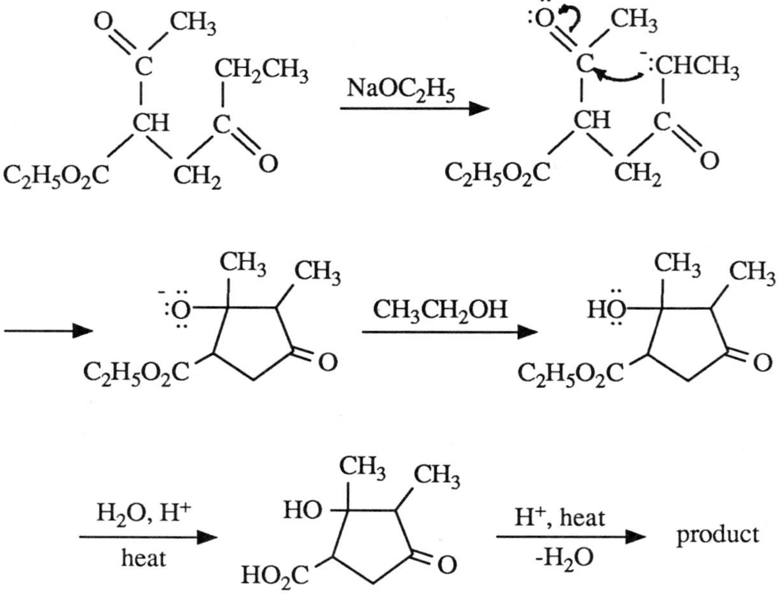

$NaOC_2H_5$ →

→

CH_3CH_2OH →

H_2O, H^+, heat →

H^+, heat, $-H_2O$ → product

(i)

CH_3CH_2OH, H^+, heat →

(1) $NaOC_2H_5$
(2) $CH_2{=}CHCCH_2CH_3$ (C=O)
→ $CH_3CH_2CCH_2CH_2$ (C=O) … $CO_2C_2H_5$

H_2O, H^+, heat, $-CO_2$ → $CH_3CH_2CCH_2CH_2$ (C=O) …

$NaOC_2H_5$ →

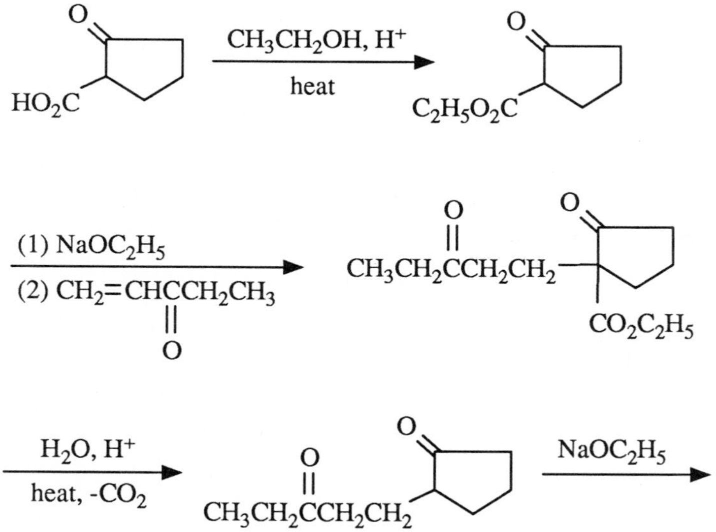

H_3C, $HC:^-$, O, $O{=}C$, CH_2CH_2 → aldol → H_3C, OH, O

$\xrightarrow[\text{heat}]{H^+}$ H_3C, O $\xrightarrow{H_2,\ Pt}$

(j) + O, O $\xrightarrow{\text{heat}}$ O, O $\xrightarrow{H_2,\ Ni}$

O, O $\xrightarrow[(2)\ CH_3CH_2CH_2Br]{(1)\ NaOC_2H_5}$

(k) $CH_3CH_2\overset{O}{\overset{\|}{C}}CH_2\overset{O}{\overset{\|}{C}}OC_2H_5$ $\xrightarrow{NaOC_2H_5}$ $CH_3CH_2\overset{O}{\overset{\|}{C}}\overset{-}{\underset{\cdot\cdot}{C}}H\overset{O}{\overset{\|}{C}}OC_2H_5$

$\xrightarrow[\text{1,4-addition}]{\text{(2-cyclohexenone, O)}}$ O, CH, $C_2H_5O_2C$, CH_2CH_3, $C{=}O$ $\xrightarrow{NaOC_2H_5}$

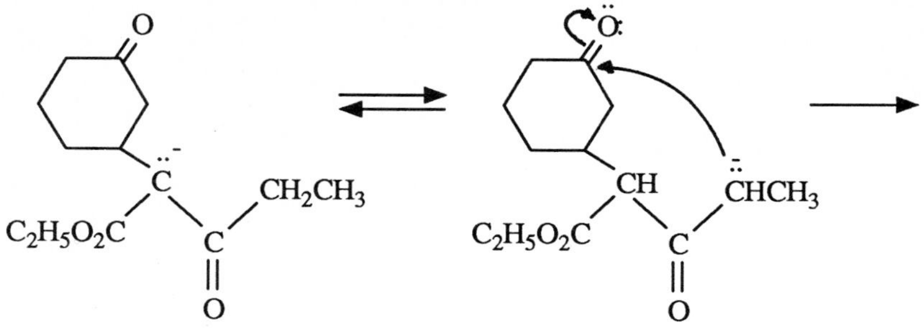

$C_2H_5O_2C$ — C: (carbanion) — C(=O) — CH_2CH_3 ⇌ $C_2H_5O_2C$ — CH — C(=O) — :CHCH₃ ⟶

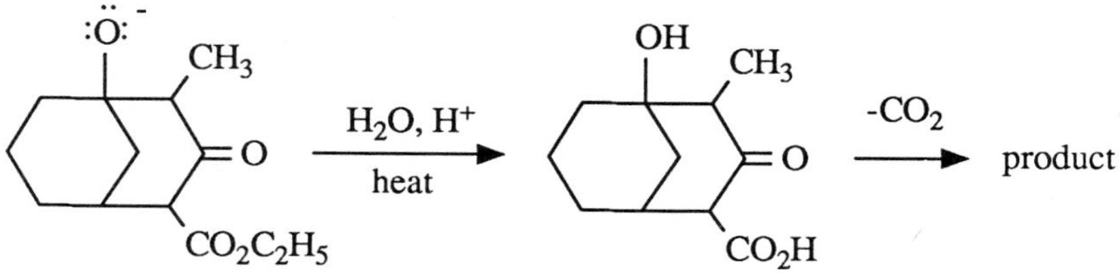

H_2O, H^+, heat ⟶ $-CO_2$ ⟶ product

$CO_2C_2H_5$; CO_2H ; CH_3 ; OH

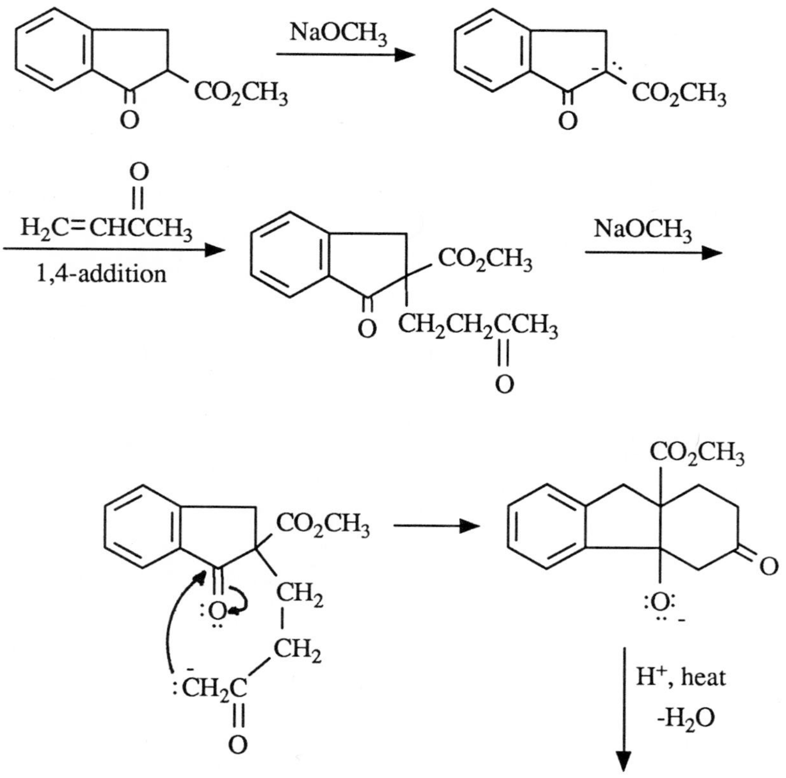

(l) $\xrightarrow{NaOCH_3}$ CO_2CH_3

$H_2C{=}CHCCH_3$ (1,4-addition) ⟶ CO_2CH_3, $CH_2CH_2CCH_3$ $\xrightarrow{NaOCH_3}$

CO_2CH_3, CH_2, CH_2, :$\bar{C}H_2C$ ⟶ CO_2CH_3, :O:$^-$

H^+, heat, $-H_2O$

product

(m) 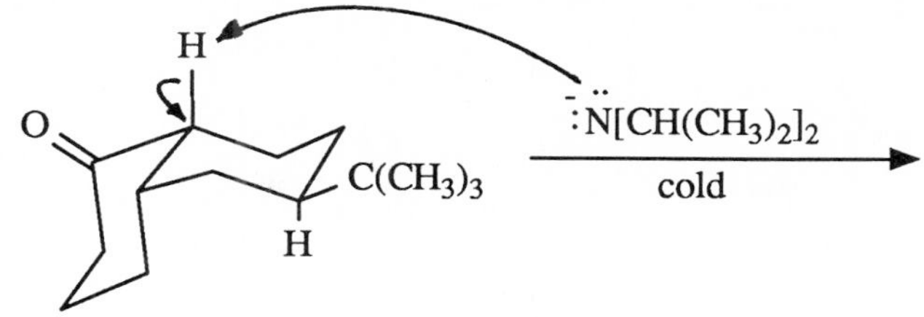
H
O
C(CH$_3$)$_3$
H
:N̈[CH(CH$_3$)$_2$]$_2$
cold

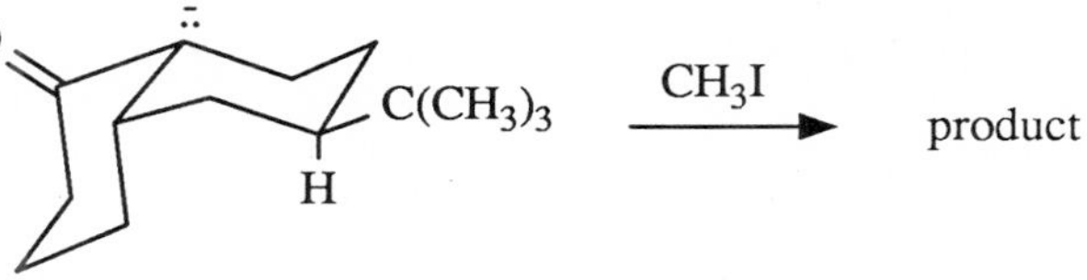
O
C(CH$_3$)$_3$
H
CH$_3$I
product

16.68 $CH_3\overset{O}{\overset{\|}{C}}CH_2CH_2\overset{O}{\overset{\|}{C}}CH_2CH_2CH{=}CHCH_2CH_3$ (*cis*) $\xrightarrow{Al_2O_3}$ *cis*-jasmone

MW 182   MW 164

C=O, but no OH

The molecular weights of the starting dione and jasmone differ by 18, suggesting an aldol addition reaction followed by loss of water. The following structure for jasmone is the only one to fit the nmr data:

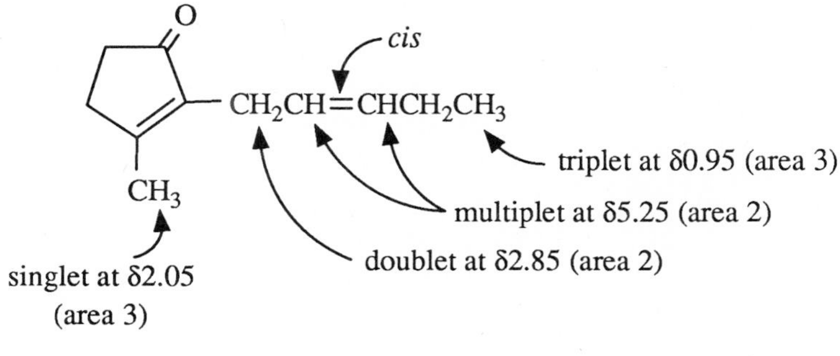
O
cis
CH$_2$CH=CHCH$_2$CH$_3$
CH$_3$
triplet at δ0.95 (area 3)
multiplet at δ5.25 (area 2)
doublet at δ2.85 (area 2)
singlet at δ2.05
(area 3)

16.69 $CH_3\overset{O}{\overset{\|}{C}}-\underset{CH_3}{\overset{CH_3}{C}}-CO_2C_2H_5$ $\xrightarrow{H\overset{O}{\overset{\|}{C}}OC_2H_5,\ Na}$ $H\overset{O}{\overset{\|}{C}}-CH_2\overset{O}{\overset{\|}{C}}-\underset{CH_3}{\overset{CH_3}{C}}-CO_2C_2H_5$

keto form of A

$\rightleftarrows$ $HOCH{=}CH\overset{O}{\overset{\|}{C}}-\underset{CH_3}{\overset{CH_3}{C}}-CO_2C_2H_5$ $\xrightarrow{H_2,\ Ni}$ $HOCH_2CH_2\overset{O}{\overset{\|}{C}}-\underset{CH_3}{\overset{CH_3}{C}}-CO_2C_2H_5$

enol form of A

B

$\xrightarrow[-H_2O\ \text{or}\ C_2H_5OH]{H^+,\ \text{heat}}$ $CH_2{=}CH\overset{O}{\overset{\|}{C}}-\underset{CH_3}{\overset{CH_3}{C}}-CO_2C_2H_5$ + (cyclic lactone D: six-membered ring with C=O, C(CH_3)(CH_3), ring O, and lactone C=O)

C

D

For compound C, the two methyl groups give rise to the singlet (area 6); the ethyl group gives rise to the triplet and the quartet; the three vinyl protons give rise to the downfield doublets. For compound D, the two methyl groups give rise to the singlet; the alpha protons give rise to the triplet at δ2.71; the other two protons give rise to the other triplet, farther downfield because of the neighboring oxygen atom.

Chapter 17

Amines

17.19 (a) 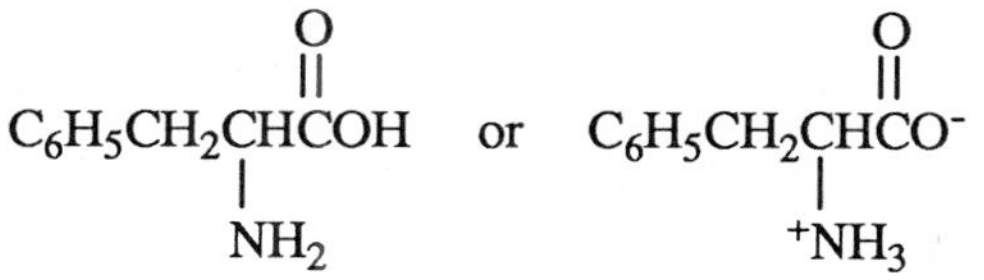 $C_6H_5CH_2CH(NH_2)COOH$ or $C_6H_5CH_2CH(^+NH_3)COO^-$

(because the amino group and the carboxyl group undergo an acid-base reaction)

(b) 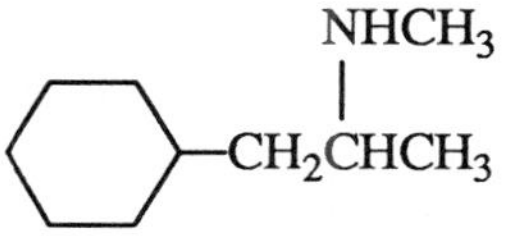 cyclohexyl–$CH_2CH(NHCH_3)CH_3$ (c) $C_6H_5N(CH_3)_2$

(d) 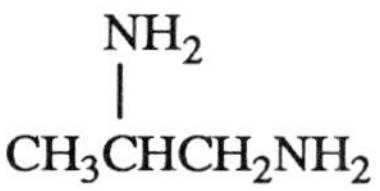 $CH_3CH(NH_2)CH_2NH_2$

17.20 (a) *N*-ethyl-*N*-methylbenzylamine

(b) *cis*-1,2-cyclohexanediamine

(c) *N,N*-diethyldiphenylmethanamine; *N,N*-diethyldiphenylmethylamine; or *N,N*-diethyl(diphenylmethyl)amine

(d) 4-aminopentanal

17.21 (a) one chiral C: one pair of enantiomers

(b) no isolable stereoisomers

(c) achiral (The N does not have four different attached groups.)

(d) one chiral N and one chiral C: four stereoisomers representing two pairs of enantiomers. (The diastereomeric pairs are also geometric isomers.)

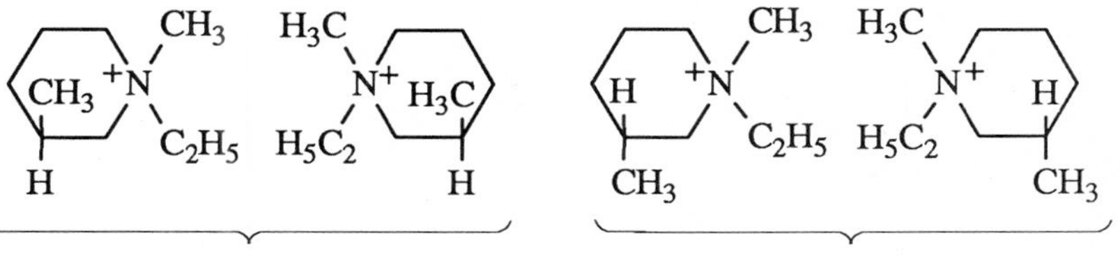

enantiomers *enantiomers*

(e) one chiral N: one pair of enantiomers

(f) one double bond: one pair of geometric isomers, or diastereomers:

H, OH, C=N, C_6H_5 H, C=N, C_6H_5, OH

17.22 (a) Cyclohexylamine forms stronger hydrogen bonds with water than does cyclohexanol. (The partially positive H of H_2O is more strongly held by the more basic N atom.)

(b) Dimethylamine forms hydrogen bonds in the pure state, while trimethylamine does not.

(c) The branching around the nitrogen of dimethylamine diminishes its ability to hydrogen bond effectively because of steric hindrance.

17.23 (a), (b), (c), (d), and (e). Compound (f) has no unshared valence electrons, which are necessary for an amine to act as a nucleophile.

17.24 (a) cyclopentyl–CH_2Br → (1) potassium phthalimide (N^- K^+); (2) H_2O, OH^-, heat

Reaction of the halide with NH_3 would not be as satisfactory as the preceding sequence because of overalkylation.

or cyclopentyl–Br → (Na^+ ^-CN) cyclopentyl–CN → (1) $LiAlH_4$; (2) H_2O, H^+; (3) neutralize

(b) cyclohexyl–OH → (H_2CrO_4) cyclohexanone (=O) → (HCN)

1-cyanocyclohexanol (cyclohexane ring bearing CN and OH on the same carbon) $\xrightarrow[\text{(2) } H_2O,\ H^+ \text{ (3) neutralize}]{\text{(1) } LiAlH_4}$

(c) $BrCH(CO_2C_2H_5)_2$ $\xrightarrow{\text{potassium phthalimide } (N^-\ K^+)}$

N-phthalimido $NCH(CO_2C_2H_5)_2$ $\xrightarrow[\text{(2) } BrCH_2CO_2C_2H_5]{\text{(1) } NaOC_2H_5}$

N-phthalimido $NC(CO_2C_2H_5)_2$ with $CH_2CO_2C_2H_5$ on the same carbon $\xrightarrow[\text{heat}]{H_2O,\ H^+}$

(d) $(CH_3)_2CHOH$ $\xrightarrow{H_2CrO_4}$ $(CH_3)_2C{=}O$ $\xrightarrow[\text{heat, pressure}]{H_2,\ Ni,\ NH_3}$

Starting with an isopropyl halide would be less satisfactory because of elimination side reactions.

(e) cyclohexanol (cyclohexyl–OH) $\xrightarrow{H_2CrO_4}$

cyclohexanone (cyclohexane ring $=O$) $\xrightarrow[\text{heat, pressure}]{H_2,\ Ni,\ H_2NCH_2CH_2CH_3}$

(f) $CH_3CH_2CH_2CH_2OH \xrightarrow{CrO_3 \cdot 2\ pyridine} CH_3CH_2CH_2CHO$

$\xrightarrow[\text{heat, pressure}]{H_2,\ Ni,\ (CH_3)_2NH}$

or $CH_3CH_2CH_2CH_2Br$ $\xrightarrow{\text{potassium phthalimide } (N^-\ K^+)}$

$CH_3CH_2CH_2CH_2—N$(phthalimide) $\xrightarrow[\text{heat}]{H_2O,\ OH^-}$

$CH_3CH_2CH_2CH_2NH_2 \xrightarrow[H_2,\ Ni]{2\ H_2C{=}O}$ product

As the final step, reductive amination would be preferable to treatment with CH_3I because the latter might lead to overalkylation.

17.25 (a) $CH_3(CH_2)_3CH_2OH \xrightarrow{HBr} CH_3(CH_2)_3CH_2Br$

$\xrightarrow[\text{(2) } H_2O,\ OH^-,\ \text{heat}]{\text{(1) potassium phthalimide } (N^-\ K^+)}$ $CH_3(CH_2)_3CH_2NH_2$

(b) $CH_3(CH_2)_3CH_2Br$ from (a) $\xrightarrow{KCN} CH_3(CH_2)_4CN$

$$\xrightarrow[(2)\ H_2O,\ H^+]{(1)\ LiAlH_4} CH_3(CH_2)_4CH_2NH_2$$

(c) $$CH_3(CH_2)_3CH_2OH \xrightarrow{H_2CrO_4} CH_3(CH_2)_3CO_2H \xrightarrow[(2)\ NH_3]{(1)\ SOCl_2}$$

$$CH_3(CH_2)_3\overset{\overset{\displaystyle O}{\|}}{C}NH_2 \xrightarrow{Br_2,\ OH^-} CH_3(CH_2)_3NH_2$$

17.26 (a) $$C_6H_6 \xrightarrow[H_2SO_4]{HNO_3} C_6H_5NO_2 \xrightarrow[HCl]{Fe}$$

$$C_6H_5\overset{+}{N}H_3\ Cl^- \xrightarrow{OH^-} C_6H_5NH_2$$

(b) $$C_6H_5\overset{\overset{\displaystyle O}{\|}}{C}NH_2 \xrightarrow{Br_2,\ OH^-} C_6H_5NH_2$$

(c) $$C_6H_5NH_2 \xrightarrow{(CH_3\overset{\overset{\displaystyle O}{\|}}{C})_2O} C_6H_5NH\overset{\overset{\displaystyle O}{\|}}{C}CH_3 + CH_3CO_2H$$

(d) $$(R)\text{-}CH_3\overset{\overset{\displaystyle OH}{|}}{C}HCH_2CH_3 \xrightarrow{TsCl} (R)\text{-}CH_3\overset{\overset{\displaystyle OTs}{|}}{C}HCH_2CH_3$$

$$\xrightarrow{\text{phthalimide procedure}} (S)\text{-}CH_3\overset{\overset{\displaystyle NH_2}{|}}{C}HCH_2CH_3$$

(e) $$C_6H_5CH_3 \xrightarrow{KMnO_4} C_6H_5CO_2H \xrightarrow{SOCl_2} C_6H_5\overset{\overset{\displaystyle O}{\|}}{C}Cl \xrightarrow{NH_3}$$

$$C_6H_5\overset{O}{\overset{\|}{C}}NH_2 \xrightarrow[\text{(2) } H_2O,\ H^+ \quad \text{(3) neutralize}]{\text{(1) } LiAlH_4} C_6H_5CH_2NH_2$$

(f) $$CH_3CO_2H \xrightarrow[\text{heat}]{CH_3CH_2OH,\ H^+} CH_3CO_2C_2H_5 \xrightarrow{NH_3} CH_3\overset{O}{\overset{\|}{C}}NH_2$$

17.27 (a) Aniline, $C_6H_5NH_2$, is more basic because the bromine of *p*-bromoaniline is electron-withdrawing and decreases the electron density on the nitrogen atom.

(b) Tetramethylammonium hydroxide, $(CH_3)_4N^+\ OH^-$, is more basic because it is more ionic, comparable to NaOH.

(c) *p*-Nitroaniline is more basic because it has only one electron-withdrawing $-NO_2$ group.

(d) Ethylamine is more basic because ethanolamine has an electron-withdrawing -OH group and can also form an internal hydrogen bond, both of which decrease its basicity.

(e) *p*-Toluidine is more basic because the $-CH_3$ group is electron-releasing, while the $-CCl_3$ group is electron-withdrawing.

17.28 In each case, the stronger base holds the proton.

(a) 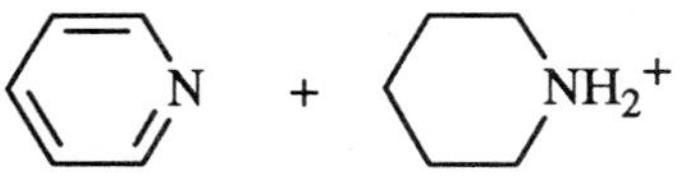 pyridine (N) + piperidinium (NH_2^+)

(b) 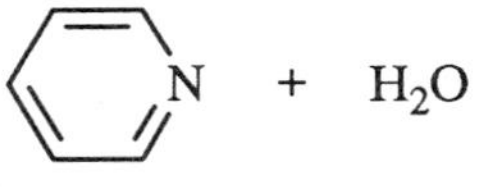 pyridine (N) + H_2O

(c) $C_6H_5NH_2$ + $(CH_3)_3NH^+\ Cl^-$

(d) no appreciable reaction

(e) 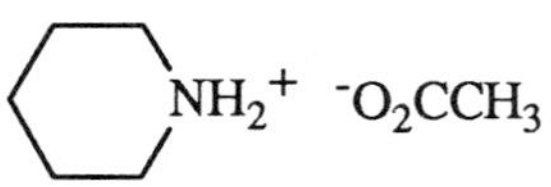 piperidinium (NH_2^+) $^-O_2CCH_3$

17.29 (a) (3) < (1) < (2)

(b) (2) < (3) < (1)

(c) (1) < (2) < (3)

(d) (2) < (1)

17.30 (a) *N,N*-dimethylcyclohexanammonium bromide; *N,N*-dimethylcyclohexylammonium bromide; or *N,N*-dimethylcyclohexylamine hydrobromide

(b) *p*-bromoanilinium bromide or *p*-bromoaniline hydrobromide

(c) *N,N,N*-trimethyl-2-chloroethanammonium chloride or *N,N,N*-trimethyl-2-chloroethylammonium chloride

17.31 (a) $\text{(pyrrolidine)}NCC_6H_5$ (with C=O) + $\text{(pyrrolidine)}\overset{+}{N}H_2\ Cl^-$

(b) $\text{(pyrrolidine)}NCCH_3$ (with C=O) + $CH_3CO_2^-$ $\text{(pyrrolidine)}\overset{+}{N}H_2$

(c) $\text{(pyrrolidine)}\overset{+}{N}(CH_3)_2\ I^-$

(d) $\text{(pyrrolidine)}N-C(=O)-C_6H_4-CO_2^-$ + $\text{(pyrrolidine)}\overset{+}{N}H_2$

(e) $\text{(pyrrolidine)}NSO_2C_6H_5$ + $\text{(pyrrolidine)}\overset{+}{N}H_2\ Cl^-$

(f) $\text{(pyrrolidine)}NCH_2CH_3$ (g) $\text{(pyrrolidine)}NNO$

(h) $\text{(pyrrolidine)}\overset{+}{N}H_2\ Cl^-$ (i) $\text{(pyrrolidine)}NC(CH_3)=CH_2$

17.32 (a) Dissolve in diethyl ether. Wash with dilute acid to remove the amine. Wash with dilute $NaHCO_3$ to remove the acid. The alcohol remains in the ether.

(b) Washing with dilute acid removes the amine.

17.33 (a) C_6H_5 ... N ... CH_3 (Note the conjugation.)

(b) $N(CH_3)_2$ + $N(CH_3)_2$

17.34 Using R to represent the $CH_3(CH_2)_6$– group,

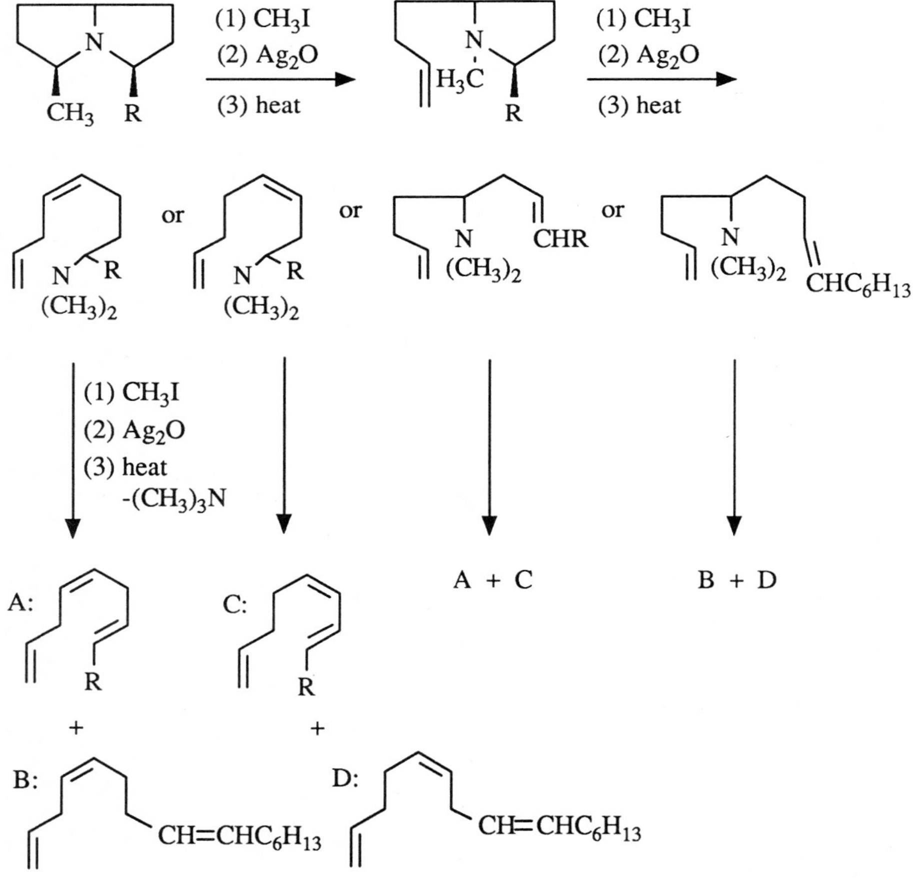

17.35 N—H

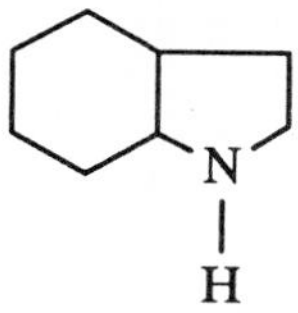

17.36 (a) $(CH_3CH_2)_2NCH_2CH_2OH + NaCl + H_2O$

(b) $CH_3NH{-}CH_2CO_2^- \xrightarrow[(2)\ H^+]{(1)\ ClCH_2CO_2^-} CH_3N(CH_2CO_2H)_2$

or $CH_3\overset{+}{N}HCH_2CO_2^-$ with CH_2CO_2H on the N

(c) 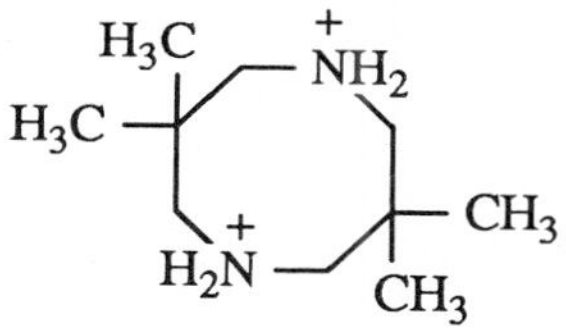 H_3C, H_3C, $\overset{+}{N}H_2$, CH_3, $H_2\overset{+}{N}$, CH_3

(d) $CH_3N(NO)CO_2C_2H_5$ A 2° amine yields an *N*-nitrosoamine.

(e) $C_6H_5CH_2CH(N{=}CHC_6H_5)CO_2CH_3$ A 1° amine yields an imine.

(f) 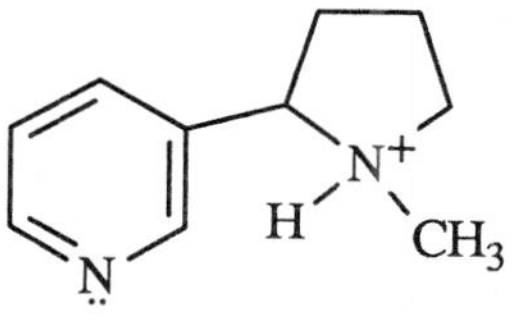 N^+, H, CH_3, N

(g) N-pyrrolidinyl–$\overset{CH_3}{|}$$CHCO_2CH_2CH_3$ + pyrrolidinium ($\overset{+}{N}H_2$) Br^-

17.37 (a) $(CH_3CH_2)_2\overset{+}{N}HCH_2\overset{O}{\overset{||}{C}}NH$–(2,6-dimethylphenyl) HSO_4^-

(b) $(CH_3CH_2)_2\overset{+}{N}HCH_2CO_2H\ Cl^-$ + $H_3\overset{+}{N}$–(2,6-dimethylphenyl) Cl^-

(c) $(CH_3CH_2)_2NCH_2CO_2^-\ Na^+$ + H_2N–(2,6-dimethylphenyl)

Hindrance around the amide group would make the reactions leading to the products shown in (b) and (c) very slow.

17.38 (a) Treat with a cold aqueous solution of $NaNO_2$ + HCl. Aniline forms a diazonium salt, while *n*-hexylamine gives off nitrogen gas.

(b) Treat with cold dilute HCl. *n*-Octylamine dissolves while octanamide does not.

(c) Treat with dilute NaOH:

$$(CH_3CH_2)_3NH^+\ Cl^- \xrightarrow{OH^-} (CH_3CH_2)_3N + H_2O + Cl^-$$

amine smell

$$(CH_3CH_2)_4N^+\ Cl^- \xrightarrow{OH^-} \text{no reaction}$$ *(odorless)*

17.39 (a) Phthalimide $\xrightarrow[\text{(2) } BrCH(CO_2C_2H_5)_2]{\text{(1) } OH^-}$ N-phthalimidyl $NCH(CO_2C_2H_5)_2$

$\xrightarrow[\text{(2) } CH_3CH_2CHBrCH_3]{\text{(1) } NaOC_2H_5}$ phthalimidyl $N-C(CO_2C_2H_5)_2$ bearing $CH_3CHCH_2CH_3$ $\xrightarrow[\text{heat}]{H_2O,\ H^+}$

phthalic acid ($C_6H_4(CO_2H)_2$) + $CH_3CH_2CH(CH_3)CH(\overset{+}{N}H_3)CO_2H$ + CO_2

$\xrightarrow{\text{neutralize}}$ $CH_3CH_2CH(CH_3)CH(NH_2)CO_2H$

mixture of stereoisomers

(b) $C_6H_5CH_3 \xrightarrow{NBS} C_6H_5CH_2Br \xrightarrow{KCN} C_6H_5CH_2CN$

$\xrightarrow[\text{(2) } H_2O,\ H^+ \text{ (3) neutralize}]{\text{(1) } LiAlH_4}$ $C_6H_5CH_2CH_2NH_2$

(c) $C_6H_5CO_2H \xrightarrow{SOCl_2} C_6H_5\overset{O}{\overset{\|}{C}}Cl$

$CH_3CH_2\overset{O}{\overset{\|}{C}}CH_2CH_3 \xrightarrow[H^+,\ -H_2O]{\text{pyrrolidine (N–H)}}$ $CH_3CH{=}C(NC_4H_8)CH_2CH_3$ (pyrrolidine enamine) $\longrightarrow$

$$\underset{\text{(iminium: } C_6H_5C(=O)-CH(CH_3)-C(=N^+\text{pyrrolidinium})CH_2CH_3)}{CH_3CH{-}C(C(=O)C_6H_5)=N^+C_4H_8{-}CH_2CH_3} \xrightarrow{H_2O,\ H^+} C_6H_5\overset{O}{\overset{\|}{C}}\underset{CH_3}{\underset{|}{C}}H\overset{O}{\overset{\|}{C}}CH_2CH_3$$

(d) $$C_6H_5CO_2H \xrightarrow{SOCl_2} C_6H_5\overset{O}{\overset{\|}{C}}Cl \xrightarrow{NH_3} C_6H_5\overset{O}{\overset{\|}{C}}NH_2$$

$$\xrightarrow{Br_2,\ OH^-} C_6H_5NH_2$$

(e) $$C_6H_5CH{=}CH_2 \xrightarrow{C_6H_5CO_3H} C_6H_5\overset{\ \ O}{CH{-}CH_2}\ \text{(epoxide)}$$

$$\xrightarrow{(CH_3)_2NH} C_6H_5\overset{OH}{\overset{|}{C}}HCH_2N(CH_3)_2$$

The amine attacks the less hindered epoxide carbon.

(f) $$C_6H_5CH_3 \xrightarrow[AlCl_3]{CH_3I} H_3C{-}C_6H_4{-}CH_3\ (para)$$

$$\xrightarrow[\text{(2) NBS or } Br_2,\ h\nu]{\text{(1) separate from } o \text{ isomer}} BrCH_2{-}C_6H_4{-}CH_2Br$$

$$\xrightarrow{\text{excess } NH_3}$$

In the final step, you may have used potassium phthalimide instead of NH_3.

(g) 3,5-$(CH_3O)_2C_6H_3$–CN $\xrightarrow[\text{mild}]{H_2O,\ H^+}$ 3,5-$(CH_3O)_2C_6H_3$–$\overset{O}{\overset{\|}{C}}NH_2$

$\xrightarrow{Br_2,\ OH^-}$ 3,5-$(CH_3O)_2C_6H_3$–NH_2

17.40 (a) HO–C_6H_4–NO_2 $\xrightarrow[(2)\ CH_3CH_2Br]{(1)\ OH^-}$ C_2H_5O–C_6H_4–NO_2

$\xrightarrow[(2)\ OH^-]{(1)\ Fe,\ HCl}$ C_2H_5O–C_6H_4–NH_2

$\xrightarrow{(CH_3\overset{O}{\overset{\|}{C}})_2O}$ C_2H_5O–C_6H_4–$NH\overset{O}{\overset{\|}{C}}CH_3$

(b) HO_3S–C_6H_4–NH_2 $\xrightarrow[0°]{NaNO_2,\ HCl}$ HO_3S–C_6H_4–$N_2^+\ Cl^-$

$\xrightarrow{C_6H_5N(CH_3)_2}$ HO_3S–C_6H_4–N=N–C_6H_4–$N(CH_3)_2$

(c) C_6H_{11}–Br $\xrightarrow[(2)\ CO_2;\ (3)\ H_2O,\ H^+]{(1)\ Mg,\ ether}$ C_6H_{11}–CO_2H $\xrightarrow{SOCl_2}$

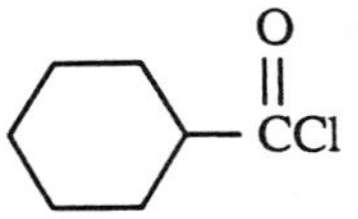

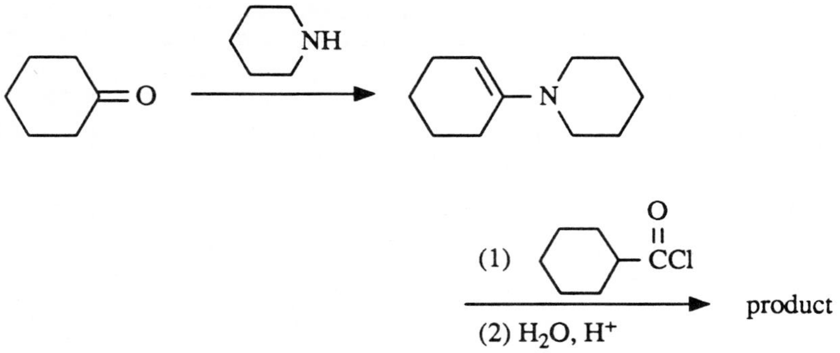

$$\text{Cyclohexyl–}\overset{O}{\overset{\|}{C}}Cl$$

$$\text{Cyclohexanone} \xrightarrow{\text{piperidine (NH)}} \text{1-(cyclohex-1-enyl)piperidine}$$

$$\xrightarrow[\text{(2) } H_2O,\ H^+]{\text{(1) Cyclohexyl–}\overset{O}{\overset{\|}{C}}Cl} \text{product}$$

17.41 (a) 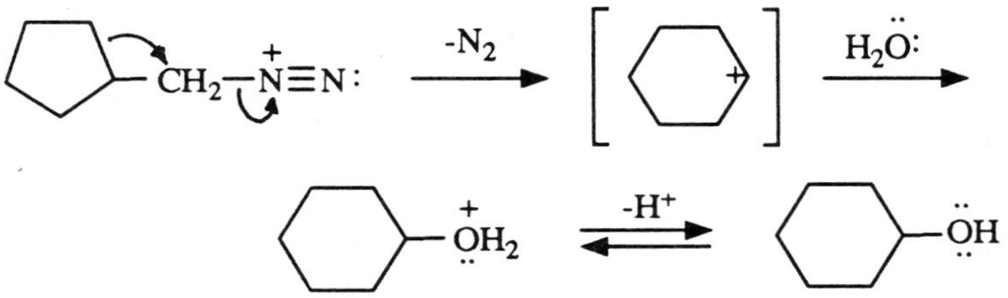 $\text{Cyclopentyl–}CH_2\text{–}\overset{+}{N}\equiv N: \xrightarrow{-N_2} [\text{cyclohexyl cation } (+)] \xrightarrow{H_2\ddot{O}:}$

$$\text{Cyclohexyl–}\overset{+}{O}H_2 \underset{}{\overset{-H^+}{\rightleftharpoons}} \text{Cyclohexyl–}\ddot{O}H$$

(b) 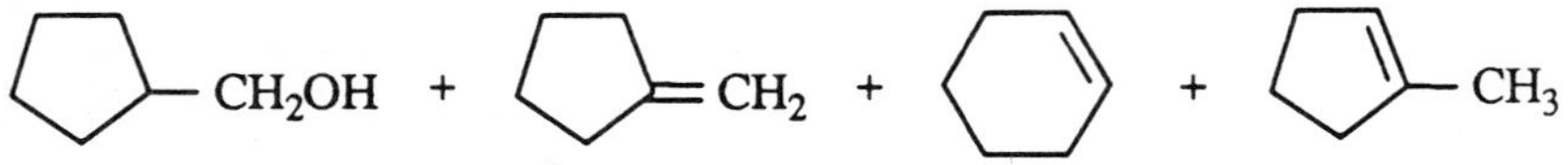 $\text{Cyclopentyl–}CH_2OH + \text{Cyclopentylidene}{=}CH_2 + \text{cyclohexene} + \text{1-methylcyclopentene (–}CH_3\text{)}$

Another possibility for the second alcohol is 1-methylcyclopentanol (CH_3, OH on the same ring carbon)

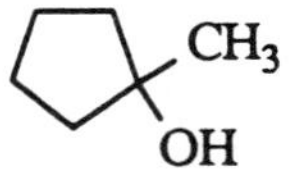

17.42 $CH_3CH_2CH_2\overset{O}{\overset{\|}{C}}NH_2 \xrightarrow{Br_2,\ OH^-} CH_3CH_2CH_2\ddot{N}{=}C{=}\ddot{O}: \xrightarrow{CH_3\ddot{O}H}$

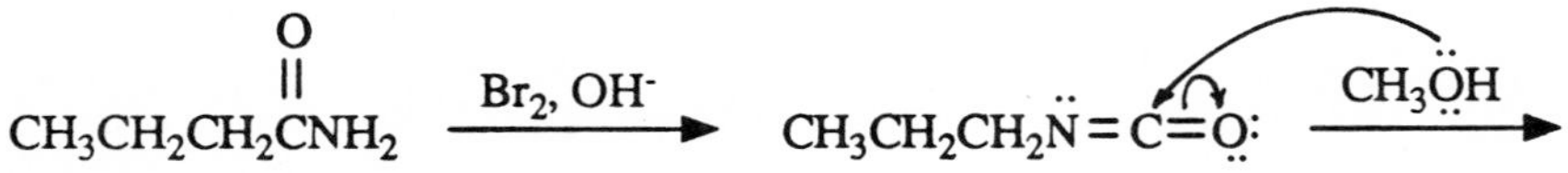

$$CH_3CH_2CH_2\ddot{N}{=}C(\ddot{O}^-)(H\ddot{O}^+CH_3) \xrightarrow{-H^+} CH_3CH_2CH_2\ddot{N}{=}C(\ddot{O}^-)OCH_3 \quad (CH_3\ddot{O}{-}H)$$

$$\xrightarrow{-CH_3\ddot{O}^-} CH_3CH_2CH_2\ddot{N}HC(=O)OCH_3$$

17.43 2-(CH_2CH_2CN)cyclohexanone $\xrightarrow[\text{catalyst}]{H_2}$ 2-($CH_2CH_2CH_2NH_2$)cyclohexanone

$\xrightarrow{-H_2O}$ octahydroquinoline (C=N) $\xrightarrow{H_2}$ decahydroquinoline (N–H)

17.44 $H_3C(C_6H_5)C(H)-C(H)(CH_3)\overset{+}{N}(CH_3)_2$ with $:\ddot{O}^-$ $\rightarrow$ $C_6H_5(H_3C)C=C(H)CH_3$ + $HO\ddot{N}(CH_3)_2$

17.45 $C_6H_5CH_2CH(\overset{+}{N}H_2H)C(=O)\ddot{O}{-}H$ + $^-:\ddot{O}H$ $\rightarrow$ $C_6H_5CH_2CH(\ddot{N}H_2)C(=O)\ddot{O}:^-$ + $C_6H_5C(=\ddot{O})-\ddot{Cl}:$ $\longrightarrow$

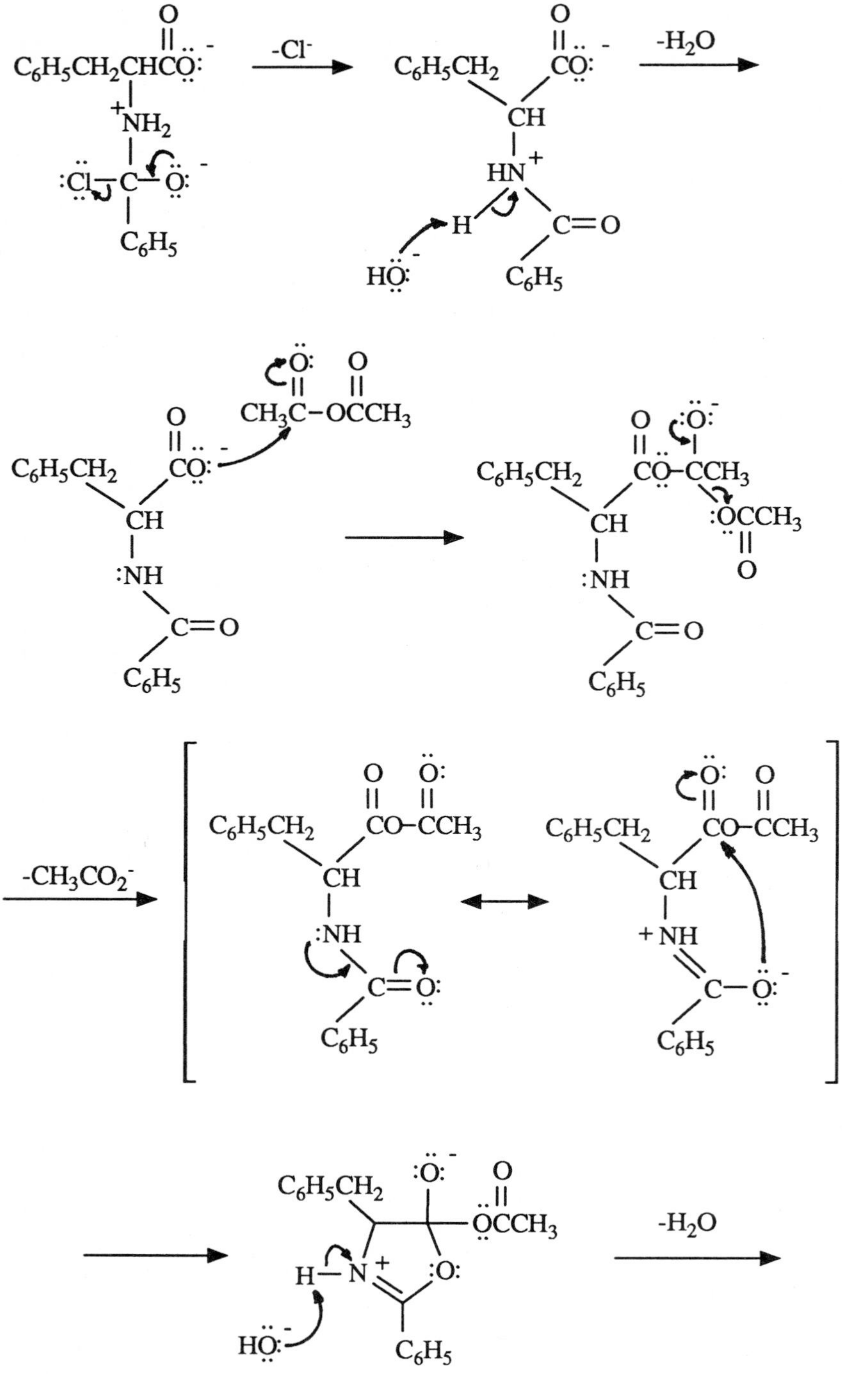

$C_6H_5CH_2CHCO^-$ (O) / $^+NH_2$ / $Cl-C-O^-$ / C_6H_5 $\xrightarrow{-Cl^-}$ $C_6H_5CH_2$ / CO^- (O) / CH / HN^+ / H / C=O / C_6H_5 / HO^- $\xrightarrow{-H_2O}$

$C_6H_5CH_2$ / CO^- (O) / CH / :NH / C=O / C_6H_5 + $CH_3C-OCCH_3$ (O, O) $\longrightarrow$ $C_6H_5CH_2$ / $CO-CCH_3$ (O, O^-) / CH / :NH / C=O / C_6H_5 / :OCCH_3 (O)

$\xrightarrow{-CH_3CO_2^-}$ [$C_6H_5CH_2$ / $CO-CCH_3$ (O, O:) / CH / :NH / C=O: / C_6H_5 $\longleftrightarrow$ $C_6H_5CH_2$ / $CO-CCH_3$ (O:, O) / CH / + NH / $C-O^-$ / C_6H_5]

$\longrightarrow$ $C_6H_5CH_2$ / $:O^-$ / $OCCH_3$ (O) / $H-N^+$ / :O: / C_6H_5 / HO^- $\xrightarrow{-H_2O}$

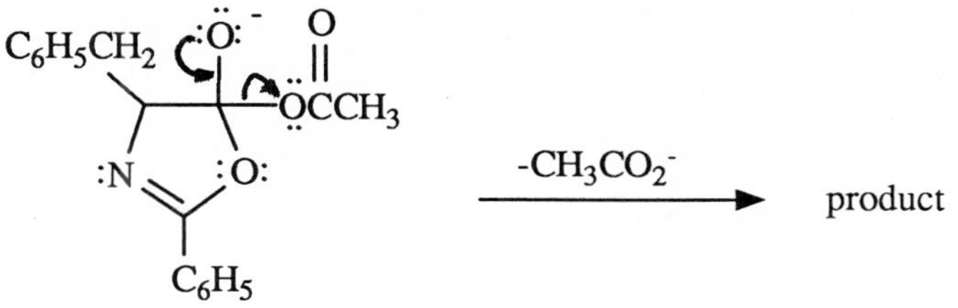

$C_6H_5CH_2$ ... $\ddot{O}CCH_3$ ... :N ... :O: ... C_6H_5 $\xrightarrow{-CH_3CO_2^-}$ product

17.46 $C_6H_5\overset{O}{\overset{||}{C}}CH_2CH_2CH_2Cl \xrightarrow[H^+]{HOCH_2CH_2OH}$ $C_6H_5-C(OCH_2CH_2O)-CH_2CH_2CH_2Cl$

A

$\xrightarrow{\text{potassium phthalimide } (N^- K^+)}$ $C_6H_5-C(OCH_2CH_2O)-CH_2CH_2CH_2-N(\text{phthalimide})$

B

$\xrightarrow[(2)\ HCl,\ heat]{(1)\ KOH,\ CH_3CH_2OH,\ heat}$ $\left[C_6H_5\overset{O}{\overset{||}{C}}CH_2CH_2CH_2NH_2\right]$

$\longrightarrow$ C_6H_5–(1-pyrroline, N)

C ($C_{10}H_{11}N$)

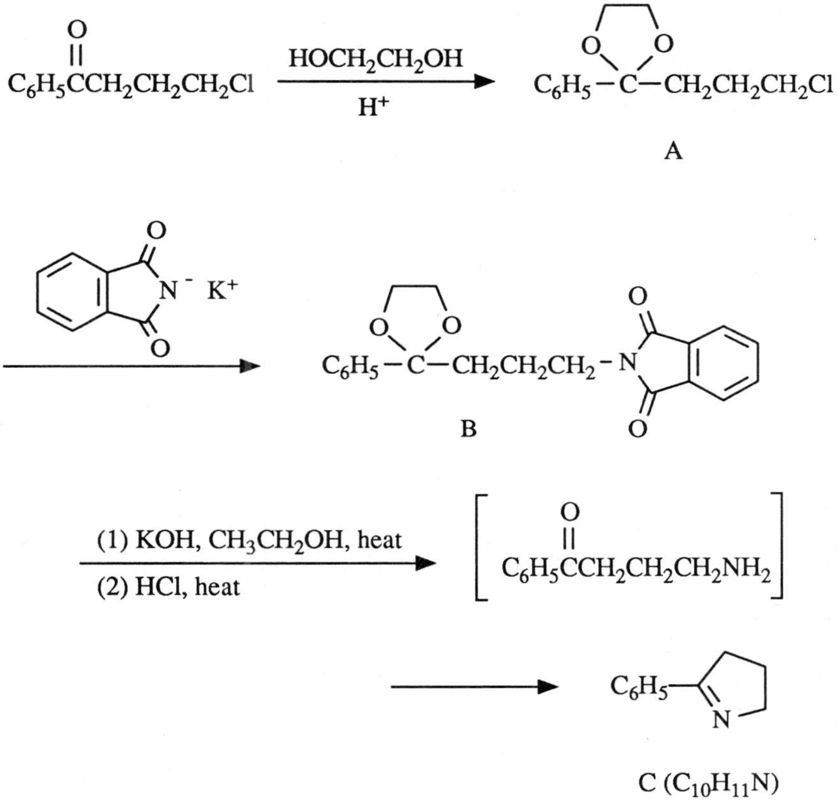

17.47 $CH_3(CH_2)_3Br \xrightarrow[(2)\ OH^-]{(1)\ NH_3}$ $CH_3(CH_2)_3NH_2$ + $[CH_3(CH_2)_3]_2NH$

A B

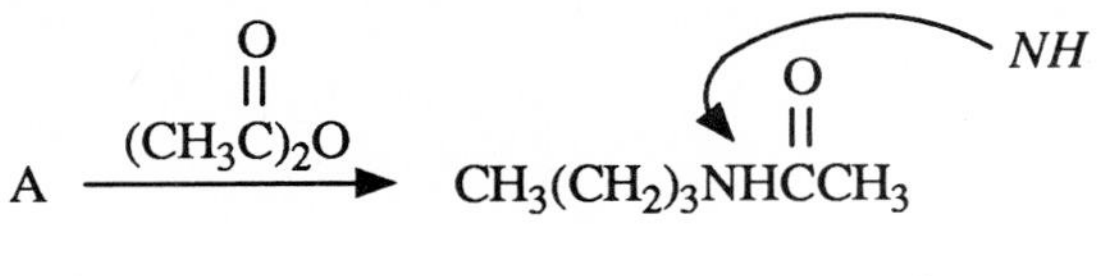

$$A \xrightarrow{(CH_3\overset{O}{\overset{\|}{C}})_2O} CH_3(CH_2)_3NH\overset{O}{\overset{\|}{C}}CH_3$$

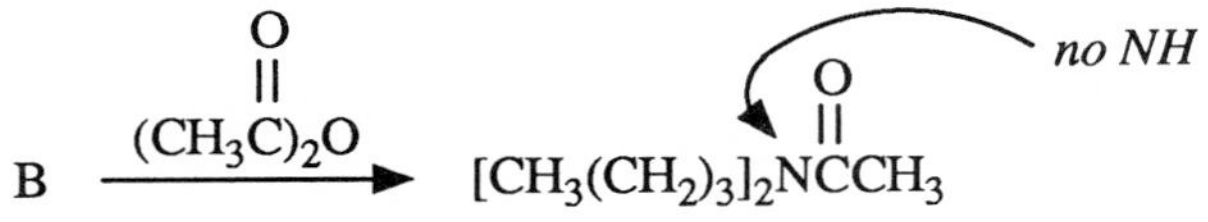

C (*NH*)

$$B \xrightarrow{(CH_3\overset{O}{\overset{\|}{C}})_2O} [CH_3(CH_2)_3]_2N\overset{O}{\overset{\|}{C}}CH_3$$

D (*no NH*)

17.48 (a) $C_6H_5CH_2CH_2NH_2$ and (b) $C_6H_5\underset{}{\overset{CH_3}{\overset{|}{C}}}HNH_2$

Besides nmr absorption from aryl protons, (a) would exhibit two triplets and a singlet (area ratio 1 : 1 : 1). However, (b) would show a quartet, a doublet, and a singlet (area ratio 1 : 3 : 2).

17.49 A: 2-CH_3-C_6H_4-$N(CH_3)\overset{O}{\overset{\|}{C}}CH_3$ B: 2-CH_3-C_6H_4-$NHCH_3$

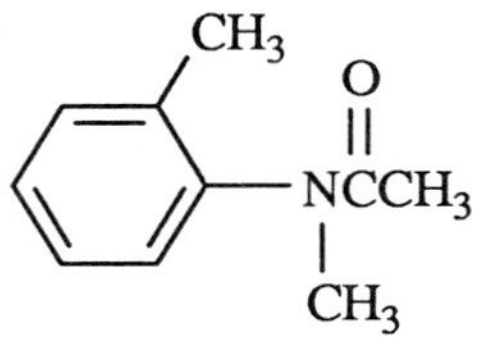

C: 2-CH_3-cyclohexyl-$NHCH_3$

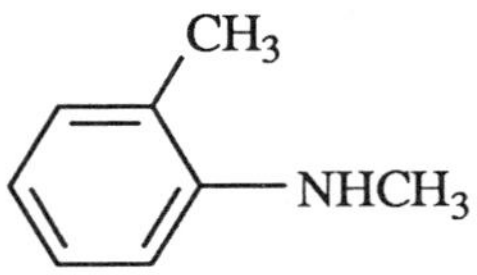

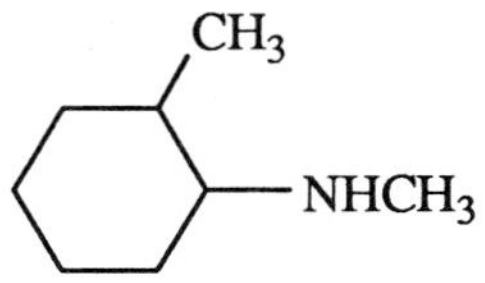

Chapter 18

Polycyclic and Heterocyclic Aromatic Compounds

18.10 (a) 1,8-naphthalenediamine

(b) 5-amino-2-naphthalenesulfonic acid or 5-aminonaphthalene-2-sulfonic acid

(c) 2-(6-methoxy-2-naphthyl)propanoic acid

(d) 5-hydroxynaphthoquinone

18.11 (a) CO_2H OH (b) CH_2CO_2

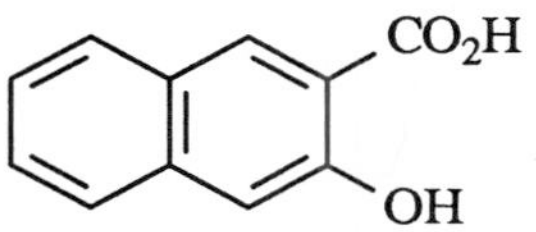

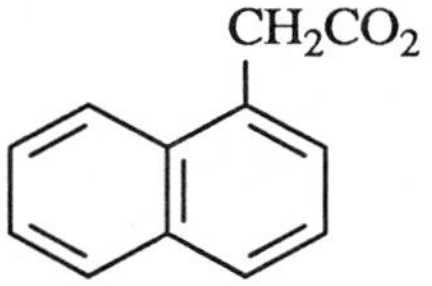

(c) SO_3^- Na^+ $(CH_3)_3C$ $C(CH_3)_3$

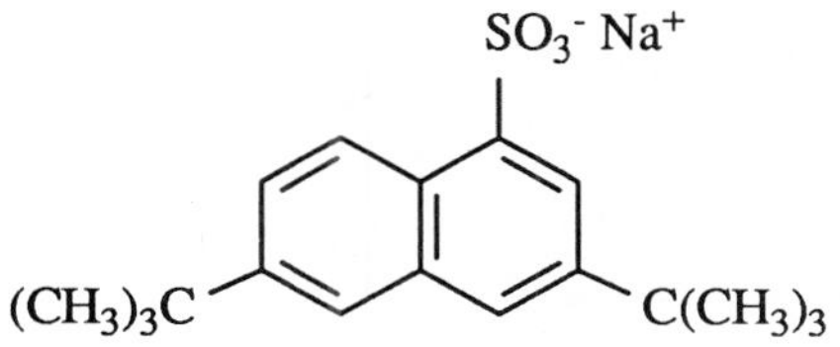

18.12 (a) 14, aromatic (b) 12 (c) 14, aromatic

18.13 (a) The product of this oxidation would depend on how vigorously the reaction is carried out.

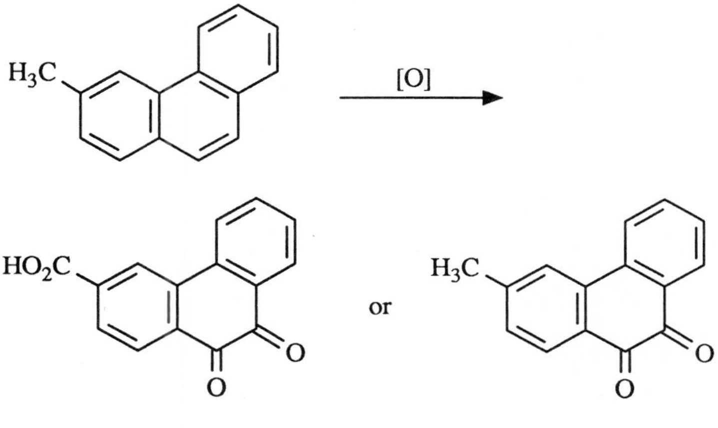

H_3C [O] →

HO_2C O O or H_3C O O

a quinone

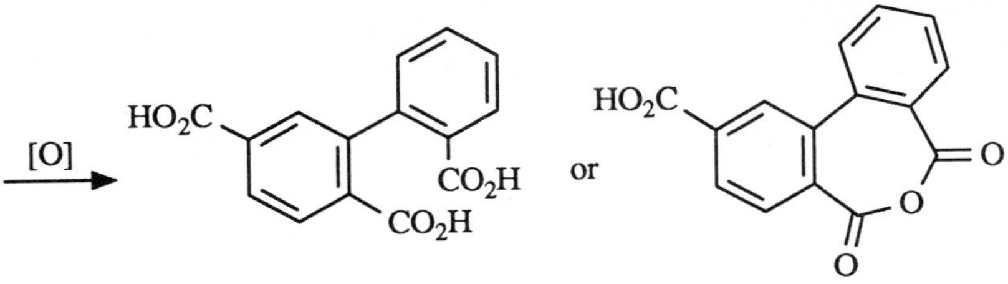

[O] →

HO_2C CO_2H CO_2H or HO_2C O O O

a tricarboxylic acid *an anhydride*

(b)

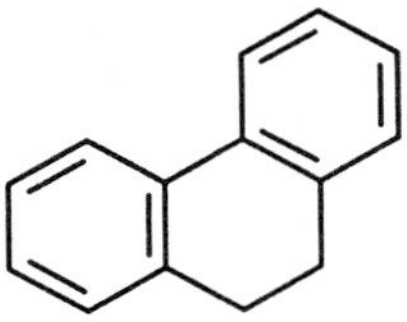

(c) 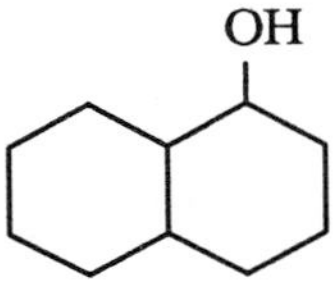 OH or, showing stereochemistry: H H H OH

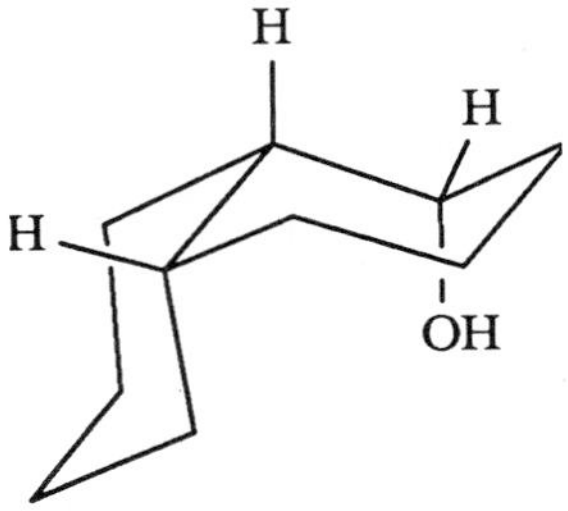

18.14 (a) 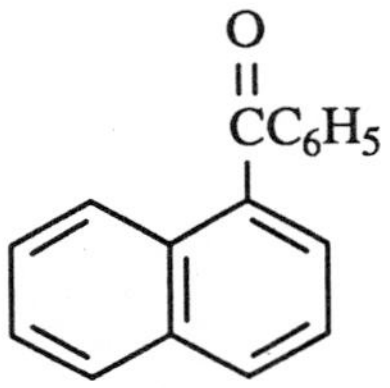 O || CC_6H_5 (b) no reaction

(c) 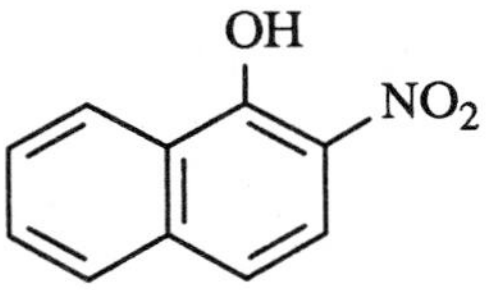 OH NO_2 + OH NO_2

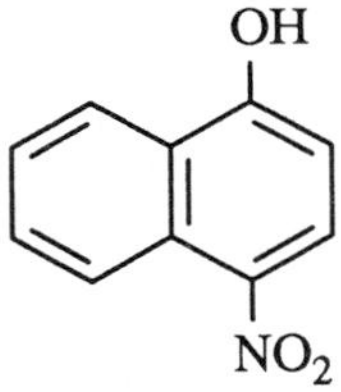

(d) 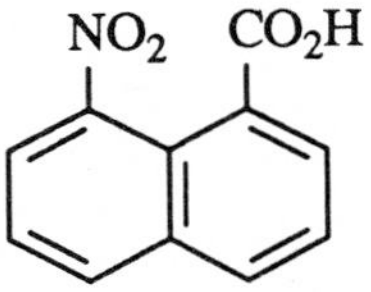 NO_2 CO_2H + CO_2H NO_2

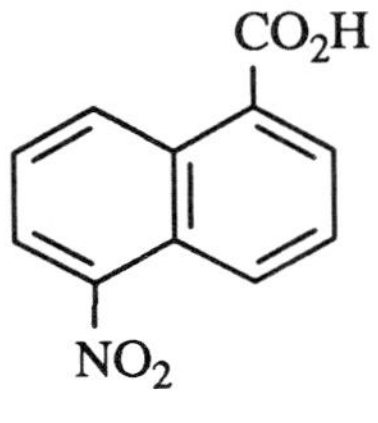

(e) $\xrightarrow{(1)\ CH_3\overset{O}{\overset{||}{C}}Cl}$ [2-naphthyl $NH\overset{O}{\overset{||}{C}}CH_3$]

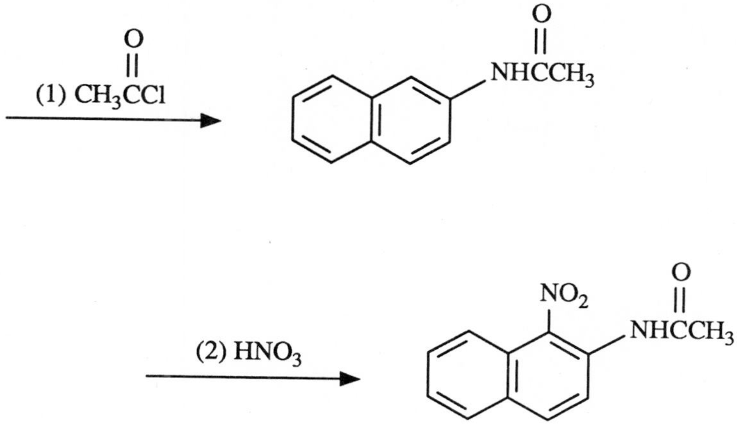

$\xrightarrow{(2)\ HNO_3}$ [1-NO_2, 2-$NH\overset{O}{\overset{||}{C}}CH_3$ naphthalene]

18.15 [1-ethylnaphthalene, CH_2CH_3] $\xrightarrow{NO_2^+}$ [resonance structures of the arenium ion with CH_2CH_3, H, NO_2 and + charge] $\longleftrightarrow$

CH_2CH_3, +, H, NO_2 $\longleftrightarrow$ CH_2CH_3, +, H, NO_2 $\longleftrightarrow$

CH_2CH_3, +, H, NO_2 $\longleftrightarrow$ CH_2CH_3, +, H, NO_2 $\longleftrightarrow$

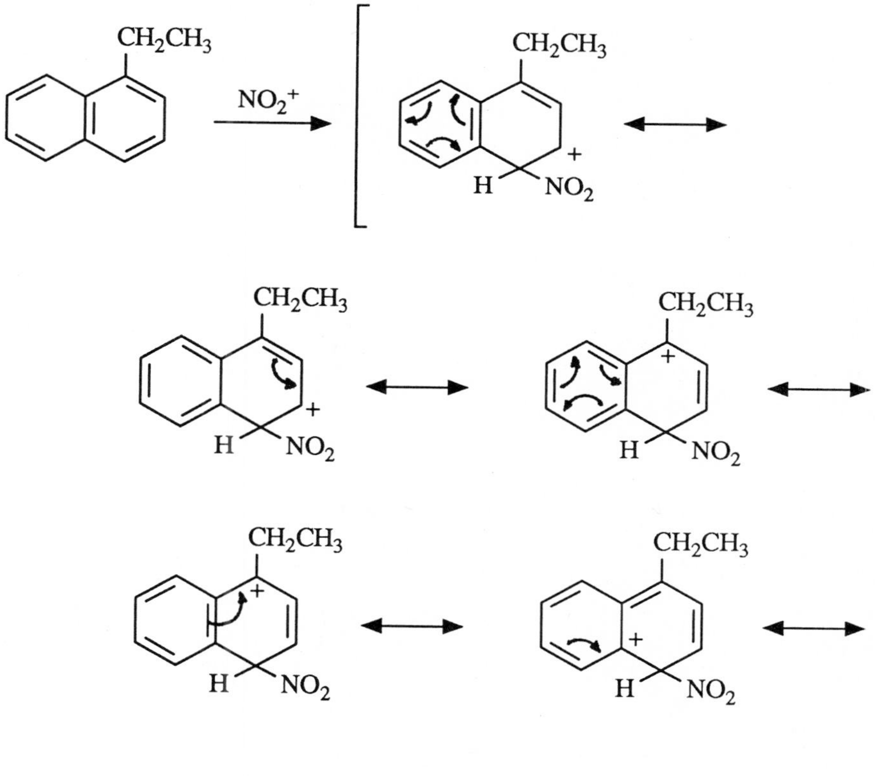

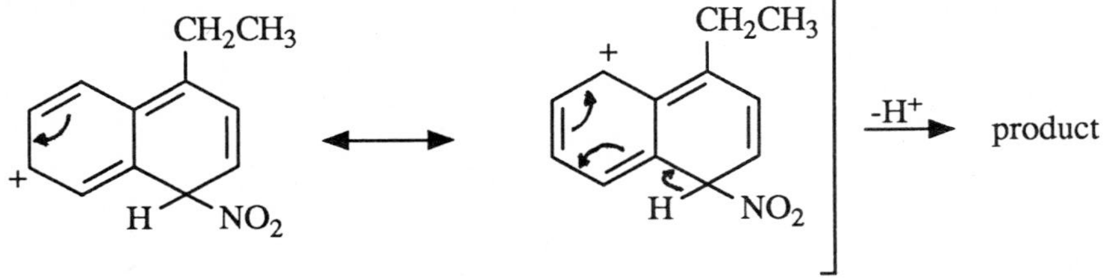

The first four resonance structures each contain a benzenoid ring. These four resonance structures are major contributors.

18.16 (a)

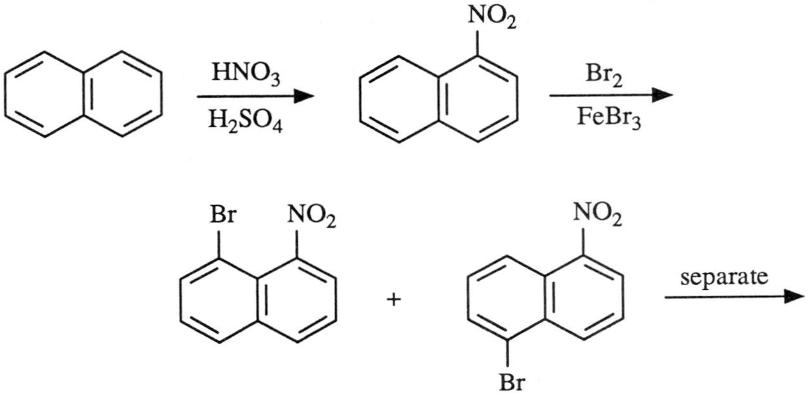

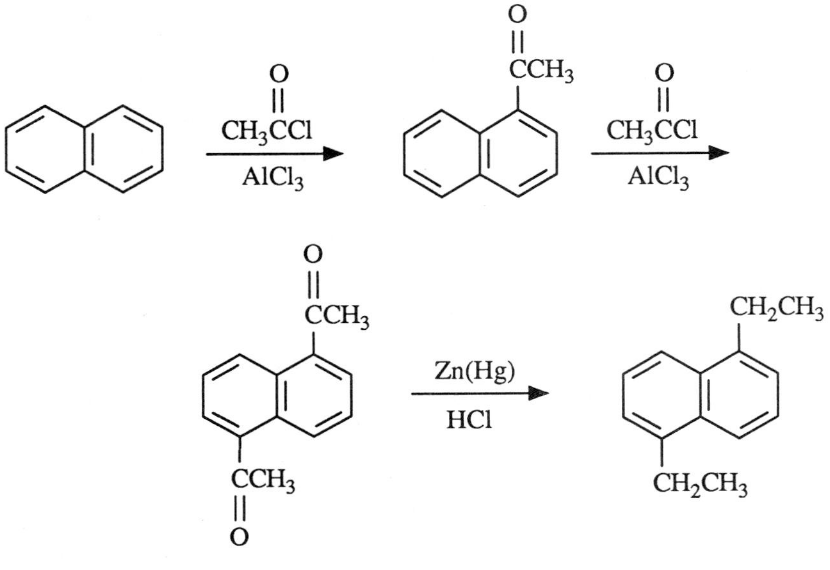

In (b), we would not expect the 1,8-diacetylnaphthalene to be a major by-product because of steric hindrance. A direct alkylation using 2

$CH_3CH_2Cl + AlCl_3$ could not be used in (b) because both R groups would substitute on the same ring.

18.17 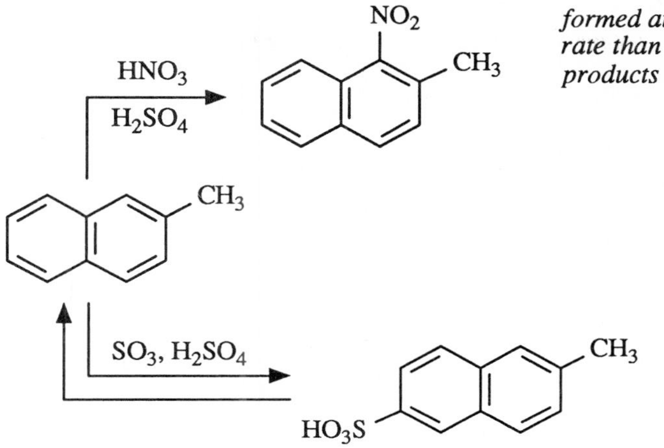

formed at a faster rate than other substitution products (kinetic control)

most stable product (equilibrium control)

18.18 (a) 2-methoxyfuran

(b) *N*-methyl-3-methylpyrrole or 1,3-dimethylpyrrole

(c) 3-hydroxy-5-hydroxymethyl-2-methyl-4-pyridinecarboxylic acid or 3-hydroxy-5-hydroxymethyl-2-methylpyridine-4-carboxylic acid

(d) 5,7-dichloro-8-hydroxyquinoline

18.19 (a) 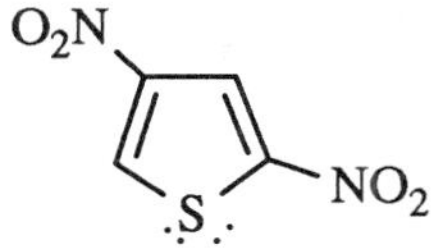

(b) 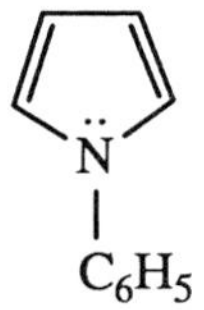

(c) 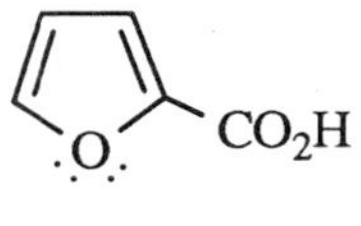

(d) 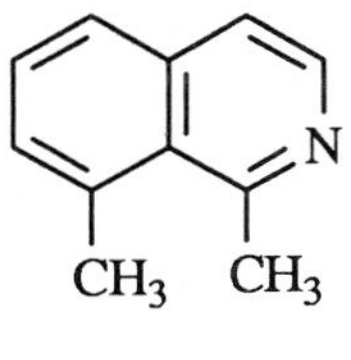

18.20 (a) 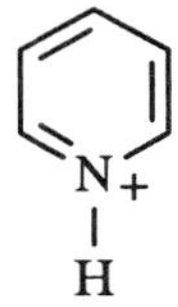 Br^-

(b) 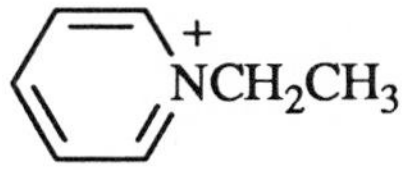 Br^-

(c) no reaction

18.21 Reaction (b) has the faster rate because the intermediate anion is stabilized by electron-withdrawal by the chlorine.

18.22 (a)

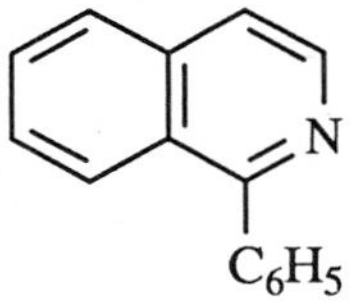

N

C_6H_5

(b)

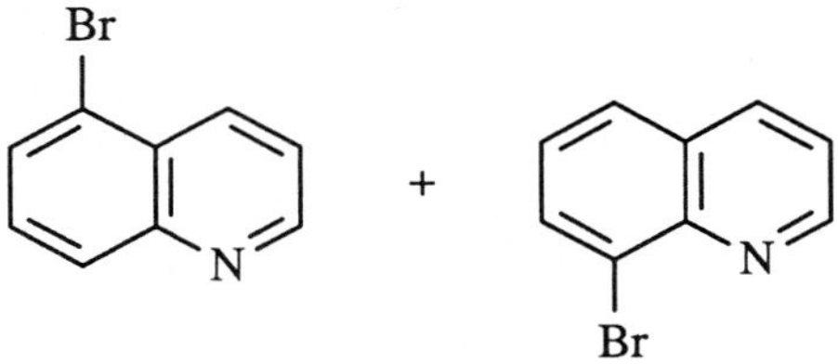

Br

N

+

N

Br

(c)

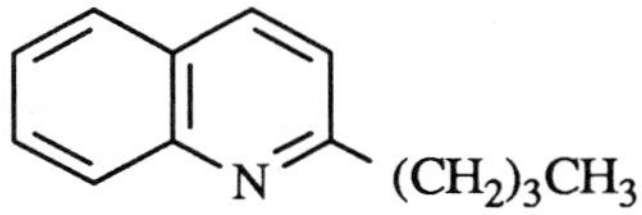

N

$(CH_2)_3CH_3$

18.23 (a)

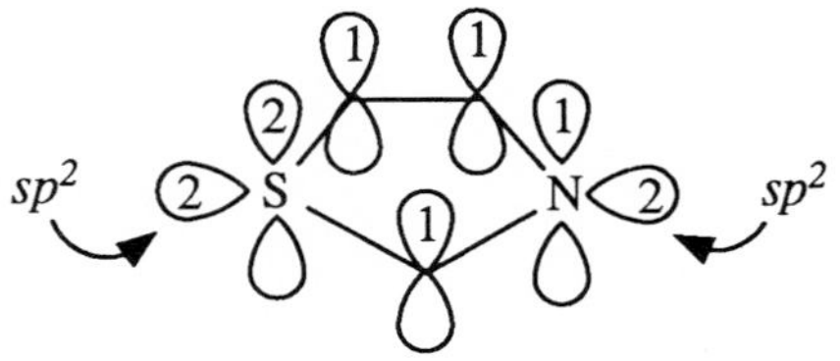

sp^2 S N *sp*^{2}

aromatic

(b)

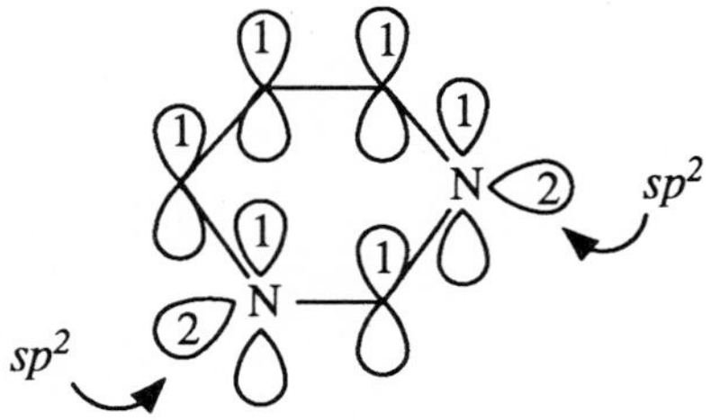

N N *sp*^{2} *sp*^{2}

aromatic

(c)

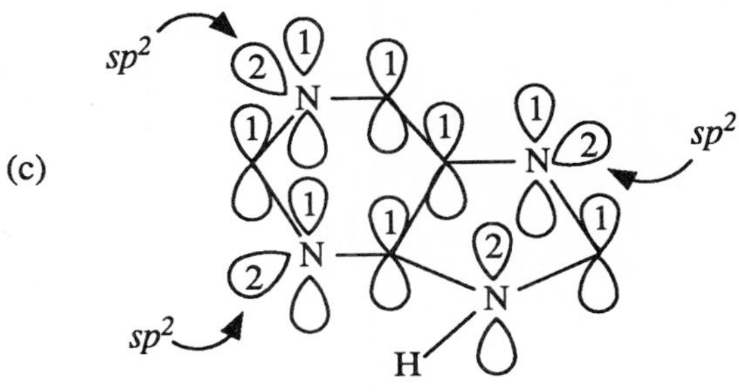

aromatic

(d)

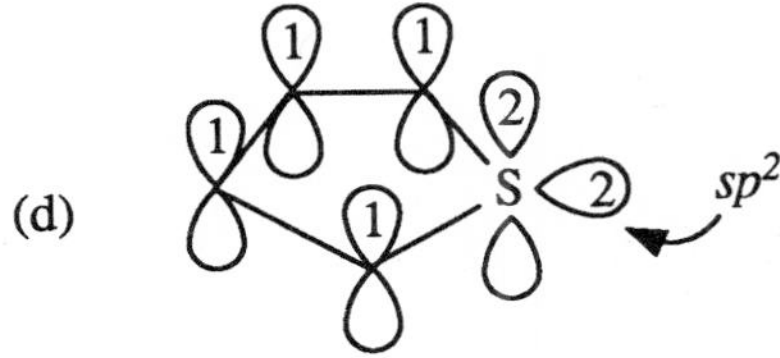

aromatic

(e)

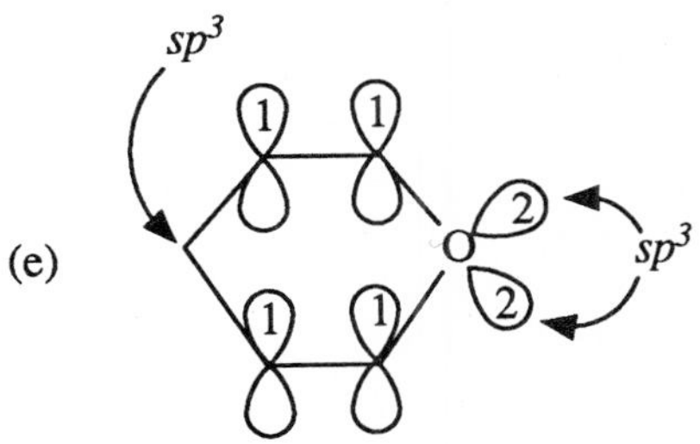

not aromatic

18.24 (a) N CH_3 (b) N H CH_2CH_3

(c) N

18.25 (a) 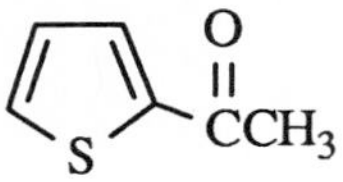 2-acetylthiophene [structure: thiophene–$\overset{O}{\overset{||}{C}}CH_3$]

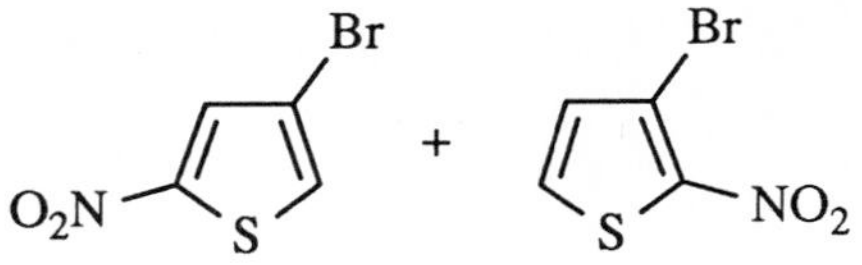

(b) [structures: O_2N-substituted 3-bromothiophene (5-nitro) + 3-bromo-2-nitrothiophene (NO_2)]

(c) [structure: 5-bromothiophene-3-carboxylic acid, Br and CO_2H]

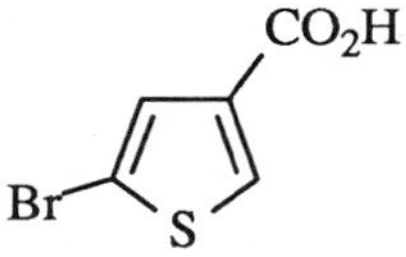

In (c), we would not expect substitution at the 2-position because the transition state leading to its intermediate is destabilized compared to that for 5-substitution.

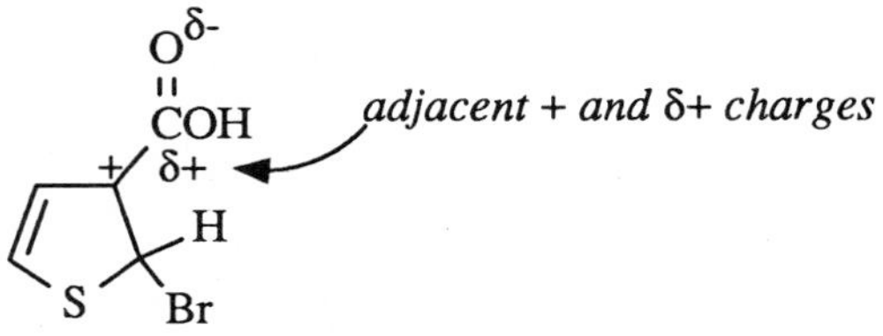

adjacent + and δ+ charges

18.26 The $-CO_2H$ group is electron-withdrawing and deactivates the ring toward electrophilic substitution.

18.27 The nitrogen in thiazole has a pair of unshared electrons, while the nitrogen in pyrrole does not. In thiazole, the sulfur atom, rather than the nitrogen, provides two electrons for the aromatic pi cloud.

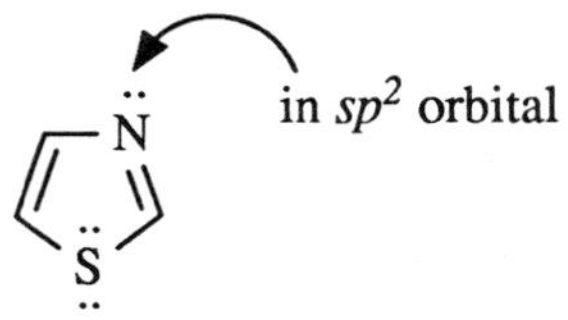

in sp^2 orbital

18.28 The negative charge is delocalized:

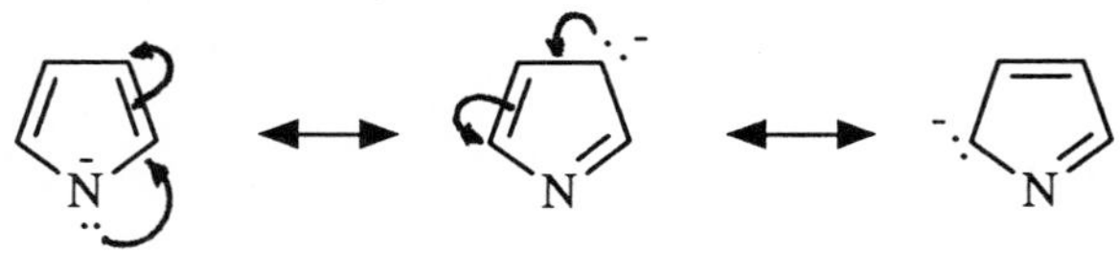

18.29 (b) < (c) < (a) < (d)

(b) is less reactive than (c) because the active positions (2 and 5) are blocked.

18.30

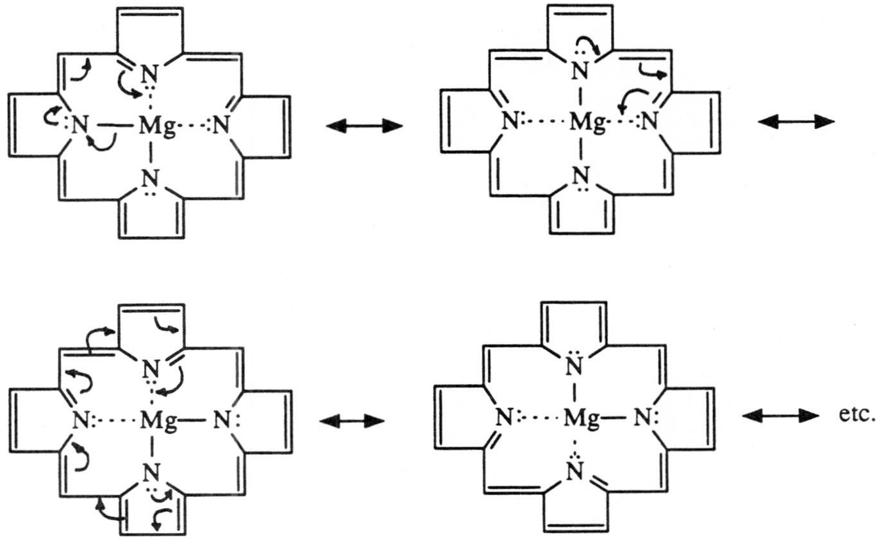

N — Mg ··· N ⟷ N ··· Mg ··· N ⟷

N ··· Mg — N ⟷ N ··· Mg — N ⟷ etc.

18.31 (a)

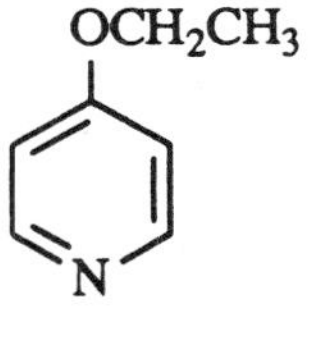

OCH_2CH_3

N

(b)

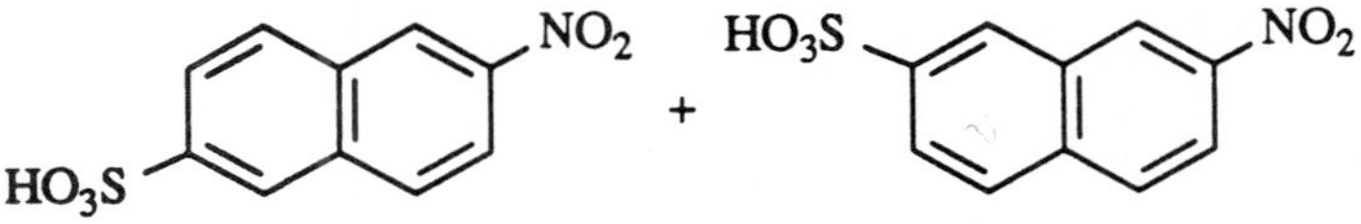

NO_2 + HO_3S NO_2

HO_3S

(c) H_3C S NO_2

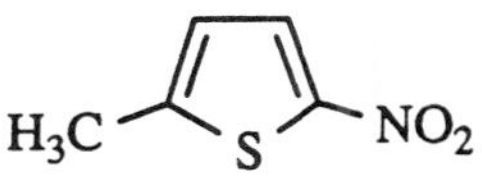

(d) CH_3

SO_3H

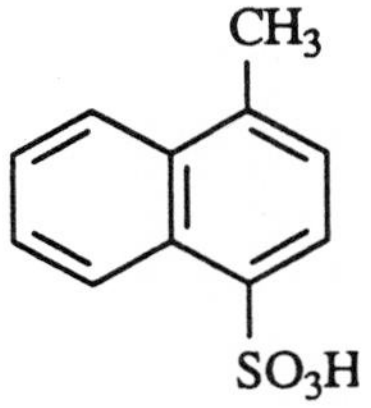

(e) 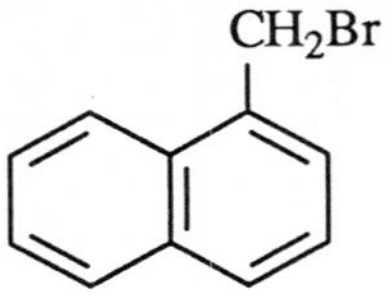 [1-(bromomethyl)naphthalene: naphthalene ring with CH_2Br]

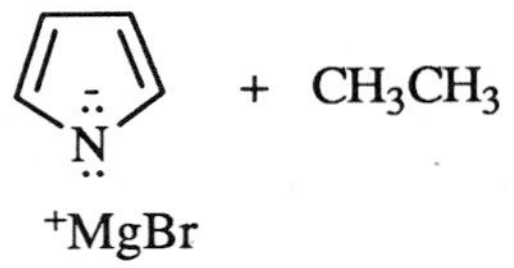

(f) 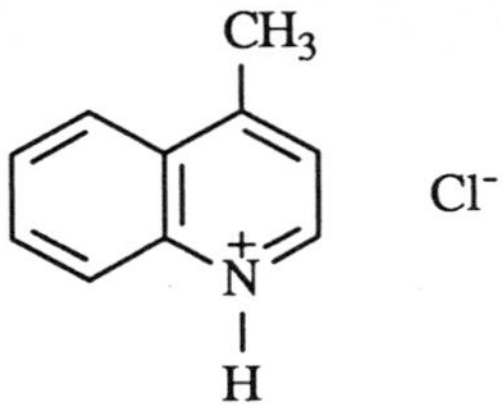 [4-methylquinolinium: quinoline ring with CH_3, $\overset{+}{N}$–H] Cl^-

(g) [pyrrole anion $\ddot{N}$ with ^+MgBr] + CH_3CH_3

(h) $\xrightarrow[\text{-CH}_3\text{OH}]{\text{(1) Wolff-Kishner}}$ [furan ring with CH_3] ^-O_2C … O … $(CH_2)_4CH_3$

$\xrightarrow{\text{(2) H}_2\text{O, H}^+}$ [furan ring with CH_3] HO_2C … O … $(CH_2)_4CH_3$

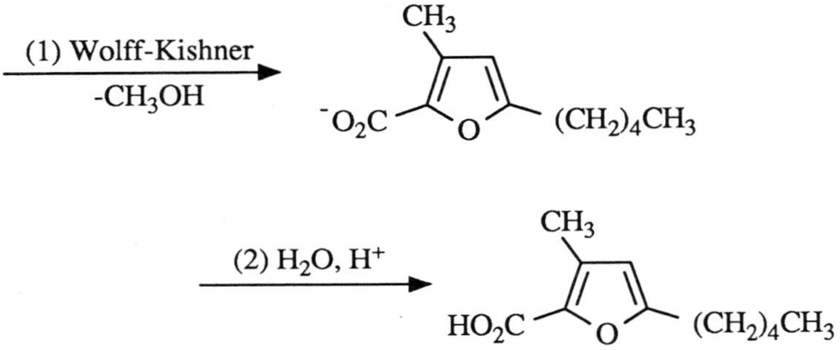

18.32 (a) 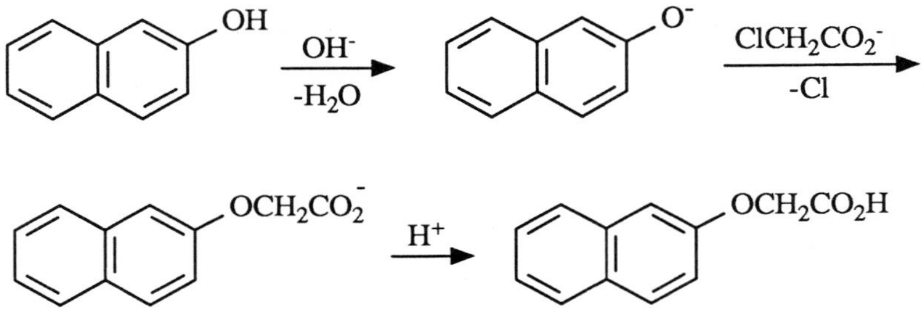 [naphthalene with OH] $\xrightarrow[\text{-H}_2\text{O}]{\text{OH}^-}$ [naphthalene with O^-] $\xrightarrow[\text{-Cl}]{\text{ClCH}_2\text{CO}_2^-}$

[naphthalene with $OCH_2CO_2^-$] $\xrightarrow{\text{H}^+}$ [naphthalene with OCH_2CO_2H]

2-naphthoxyacetic acid

(b) 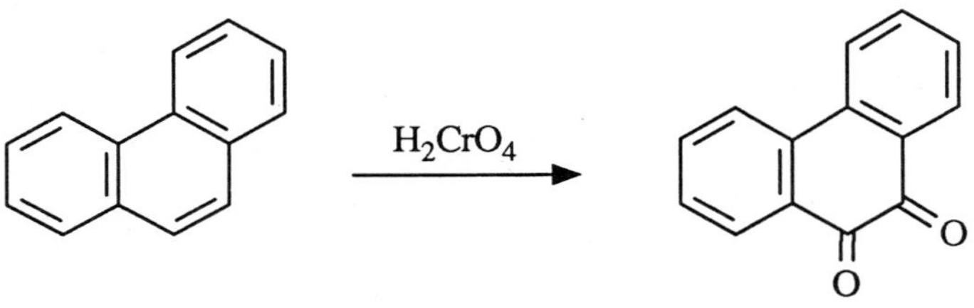 [phenanthrene] $\xrightarrow{\text{H}_2\text{CrO}_4}$ [phenanthrenequinone: O, O]

$\xrightarrow[(2)\ H_2O,\ H^+]{(1)\ LiAlH_4}$

HO OH

or $\xrightarrow{C_6H_5CO_3H}$ O $\xrightarrow{H_2O,\ H^+}$ *trans*-diol

(c) $\xrightarrow[H_2SO_4]{HNO_3}$ NO_2 $\xrightarrow[(2)\ OH^-]{(1)\ Fe,\ HCl}$

NH_2 $\xrightarrow{(CH_3\overset{O}{\overset{\|}{C}})_2O}$ $NH\overset{\|}{\underset{O}{C}}CH_3$ $\xrightarrow{H_2SO_4}$

SO_3H $NH\overset{\|}{\underset{O}{C}}CH_3$ $\xrightarrow[heat]{H_2O,\ H^+}$ SO_3H NH_3^+

adjust pH → (4-amino-1-naphthalenesulfonic acid: SO_3H / NH_2), or (SO_3^- / NH_3^+)

In order to put two substituents on the same ring, we must introduce an *o,p*-director first. Although the $-NH_2$ group is an *o,p*-director, it is protonated in strongly acidic solution and becomes an *m*-director. To circumvent this problem, an amide protecting group is used.

(d) naphthalene $\xrightarrow[H_2SO_4]{HNO_3}$ 1-nitronaphthalene (NO_2) $\xrightarrow[(2)\ OH^-]{(1)\ Fe,\ HCl}$

1-naphthylamine (NH_2) $\xrightarrow[0°]{NaNO_2,\ HCl}$ ($^+N_2$) $\xrightarrow{\text{furan (O)}}$

Furan is an activated ring and, like phenol, can be coupled with a diazonium salt. The 2, or alpha, position of furan is the most reactive position toward electrophilic substitution.

(e) $C_6H_5NH_2 \xrightarrow[S_N2]{BrCH_2CH_2CH_2CH_2Br} C_6H_5\overset{+}{N}H_2(CH_2)_4Br \xrightarrow[-H_2O]{OH^-}$

$C_6H_5\ddot{N}H$ CH_2—Br ($CH_2CH_2CH_2$) ⟶ $\xrightarrow{OH^-}$ product

(f) :ÖH, CH_2CH_2Br (cyclohexane) $\xrightarrow[\text{strong base}]{\text{NaH or other}}$:Ö:⁻, CH_2CH_2—Br $\xrightarrow{-Br^-}$ product

(g)

CH_2CO_2H

$SOCl_2$

CH_2CCl (C=O)

$AlCl_3$

O

$NaBH_4$

OH

H^+, heat

$-H_2O$

(h)

CH_3

NBS

CH_2Br

(1) NaCN

(2) H_2O, H^+, heat

CH_2CO_2H

(1) $SOCl_2$

(2) $AlCl_3$

O

$NaBH_4$

OH

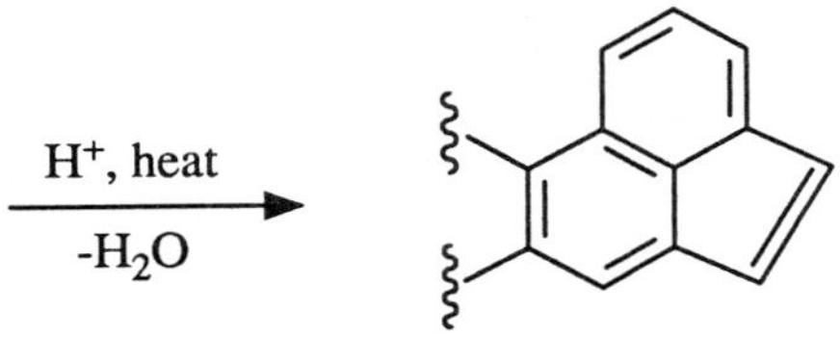
H+, heat
-H2O

(i)

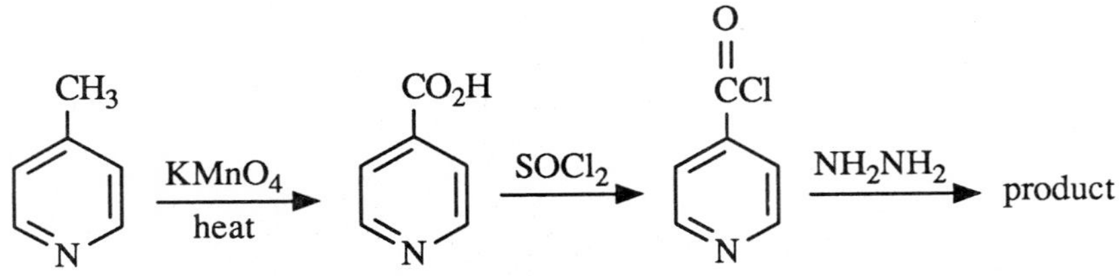
CH3
KMnO4
heat
CO2H
SOCl2
O
CCl
NH2NH2
product

(j) The starting material is at a higher oxidation state than is the product. Therefore, the synthesis must have a reduction as one step.

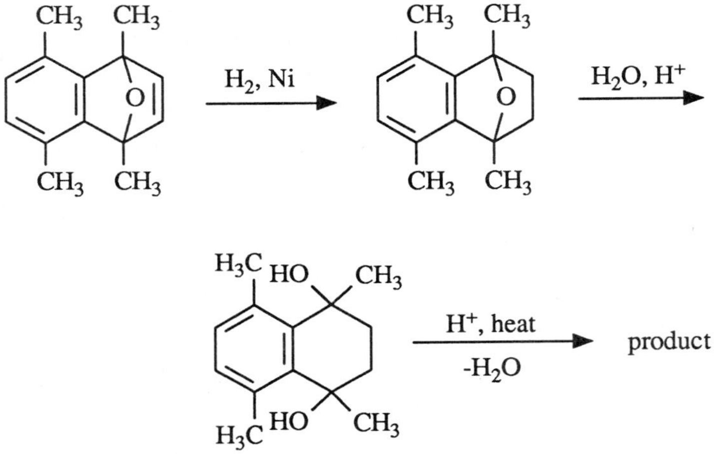
CH3 CH3
CH3 CH3
H2, Ni
H2O, H+
H3C HO CH3
H+, heat
-H2O
product

(k)

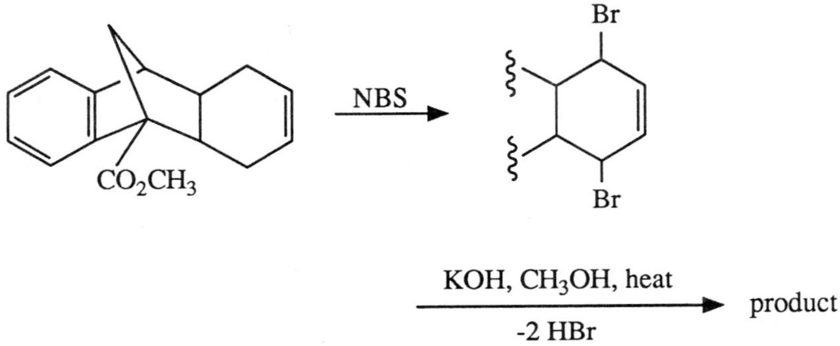
CO2CH3
NBS
Br
Br
KOH, CH3OH, heat
-2 HBr
product

In (k), the tertiary benzylic carbon does not react with NBS. Resonance stabilization of this benzylic radical cannot occur because the bond angles at this carbon atom cannot be deformed.

(l) Br $\xrightarrow{\text{Mg, ether}}$ MgBr

(1) CH_2-CH_2 (O bridged) (2) H_2O, H^+ → CH_2CH_2OH $\xrightarrow{H_2CrO_4}$

CH_2CO_2H (1) $SOCl_2$ (2) $AlCl_3$ (3) $NaBH_4$ (4) H^+, heat, $-H_2O$ → product

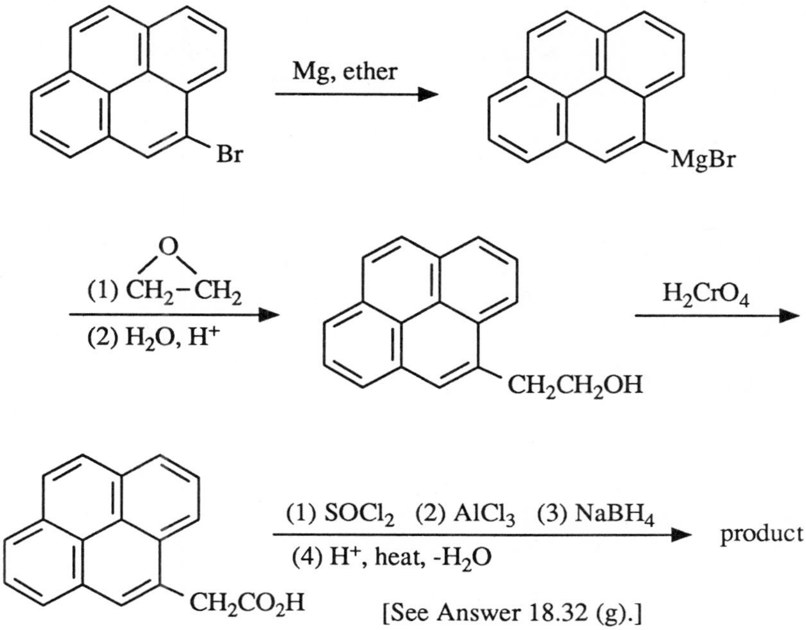

[See Answer 18.32 (g).]

18.33 The 2- and 4-hydroxypyridines undergo tautomerization to lactams that have a high degree of resonance stabilization. For example,

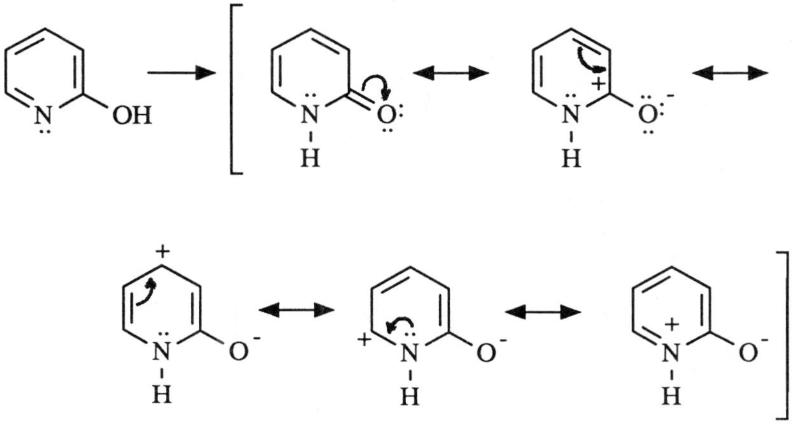

N OH → [N H O ↔ N H + O⁻ ↔

+ N H O⁻ ↔ + N H O⁻ ↔ + N H O⁻]

major contributor (aromatic)

The 3-hydroxypyridine does not form such a stabilized lactam; therefore, the enol form (which can act like a phenol) is favored.

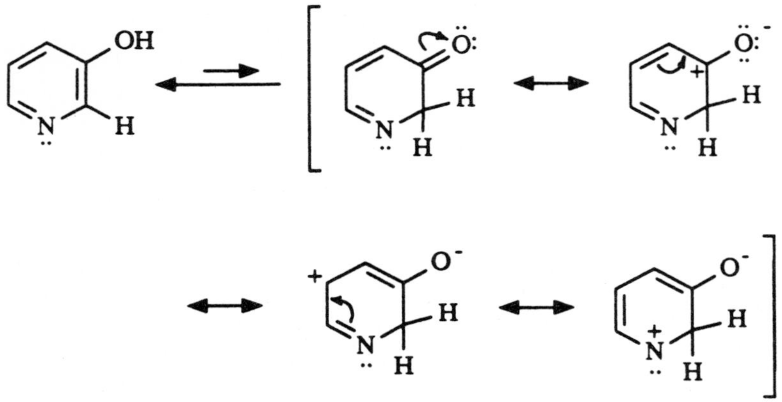

minor contributor (N has incomplete octet)

18.34 $\xrightarrow{H^+}$ etc.

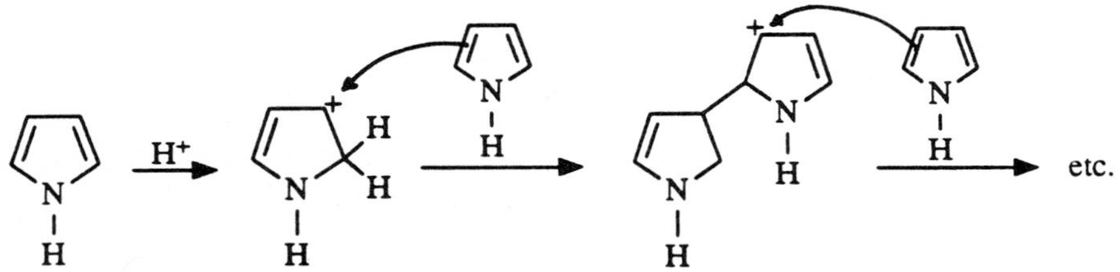

resonance stabilized

Other polymerization paths are possible.

18.35 Substitution in the 3-position is favored because (1) the nitrogen ring is electron-rich, and (2) the intermediate has three resonance structures with the aromatic benzene ring intact (compared to two such structures for 2-substitution).

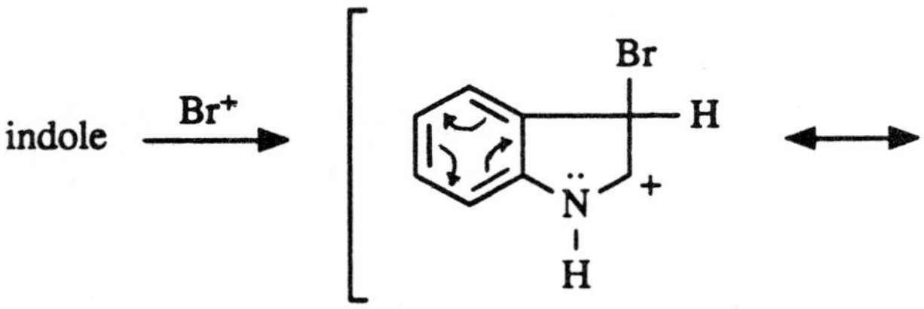

indole $\xrightarrow{Br^+}$

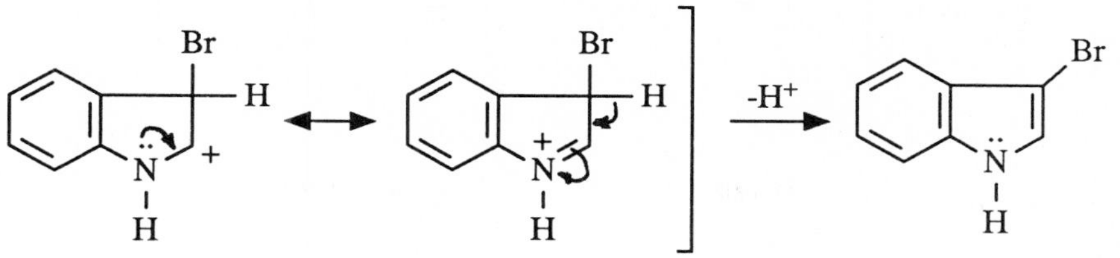

major contributors

18.36 Pyrimidine is less reactive than pyridine toward electrophilic substitution at a ring carbon, but more reactive toward nucleophilic attack because the two nitrogens of pyrimidine decrease the electron density of the ring carbons to a greater extent.

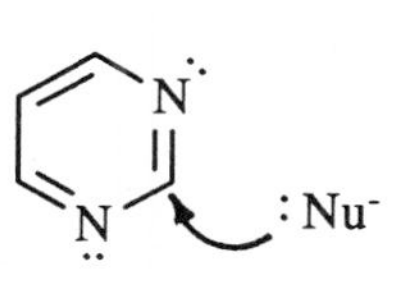

increased reactivity 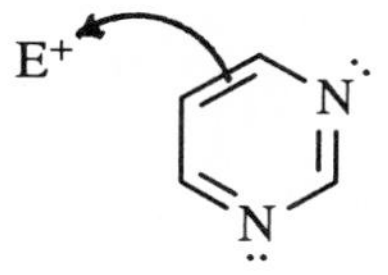 *decreased reactivity*

18.37 Imidazole and the imidazolium ion each has six electrons in an aromatic pi cloud, as the *p*-orbital pictures show. (Note that the H^+ becomes bonded to the N by a pair of sp^2 electrons, not by any of the pi electrons that form the aromatic pi cloud.)

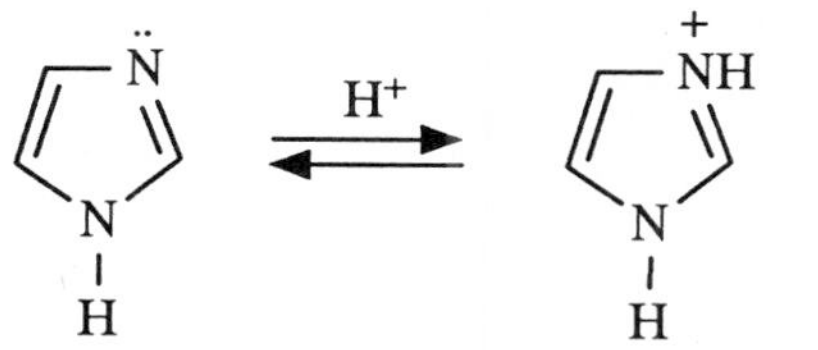

or

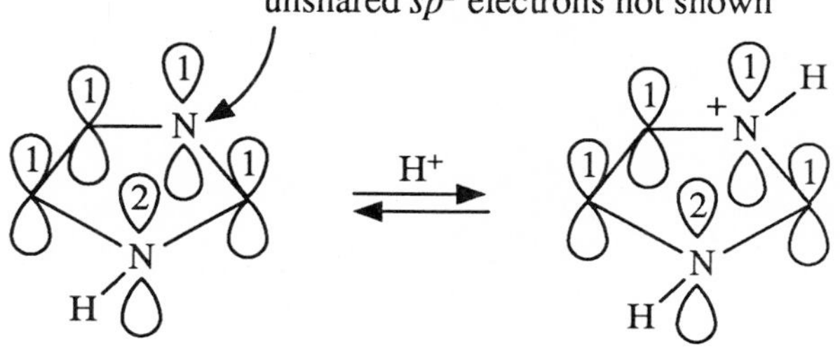

unshared sp^2 electrons not shown

18.38 Toluene (CH_3) + succinic anhydride $\xrightarrow{AlCl_3}$ (a) o-$CH_3C_6H_4C(=O)CH_2CH_2CO_2H$ (+ *p*-isomer)

$\xrightarrow[HCl]{Zn(Hg)}$ (b) o-$CH_3C_6H_4CH_2CH_2CH_2CO_2H$ $\xrightarrow{HF}$ (c) [5-methyl-1-tetralone] $\xrightarrow[heat]{Pd}$

[(methyl-dihydronaphthalenone, CH_3, O) + H_2] ⟶ (CH_3, OH) $\xrightarrow{-H_2O}$ (d) (CH_3-naphthalene)

18.39 (1)

$$CH_2OH-CH\ddot{O}H-CH_2OH \underset{}{\overset{H^+}{\rightleftharpoons}} CH_2OH-CH\overset{+}{O}H_2-CH_2OH \xrightarrow{-H_2O} CH_2OH-CH^+-CH_2OH \xrightarrow{-H^+} CH_2OH-CH{=}CHOH$$

$$\rightleftharpoons CH_2OH-CH_2-CHO \overset{H^+}{\rightleftharpoons} CH_2{-}\overset{+}{O}H_2-(H{-}CH)-CHO \xrightarrow[-H^+]{-H_2O} CH_2{=}CH-CHO$$

(2) $C_6H_5\ddot{N}H_2$ + $CH_2{=}CH{-}CH{=}\ddot{O}{:}$ $\xrightarrow{\text{1,4-addition}}$

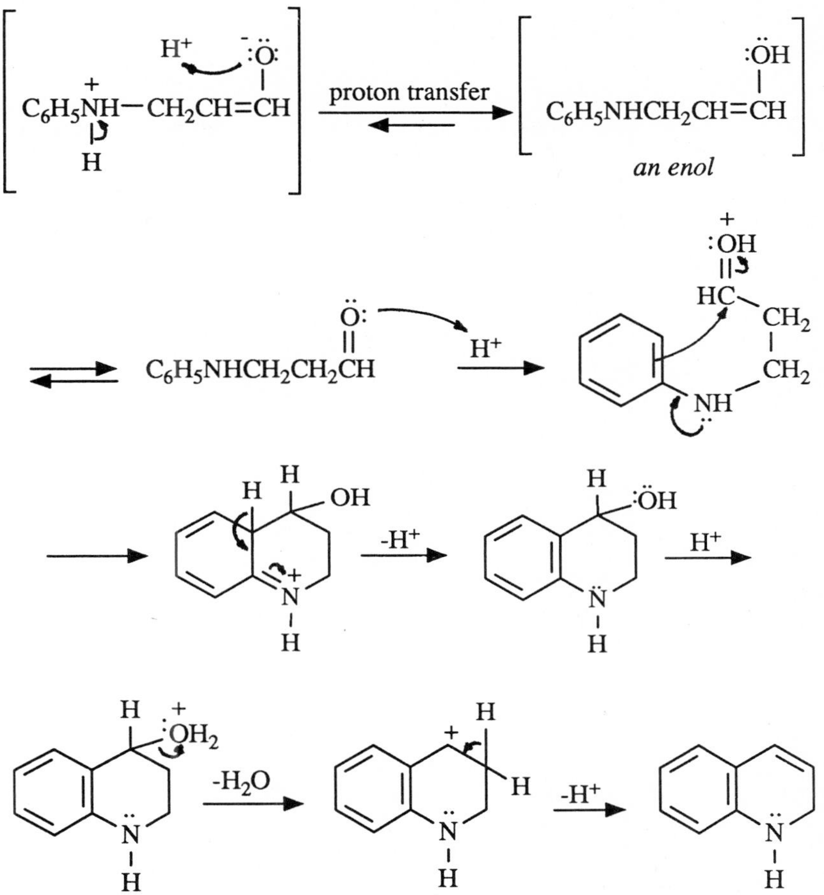

$[C_6H_5\overset{+}{N}H_2-CH_2CH=CH-\ddot{O}:^-]$ —proton transfer⇌ $[C_6H_5NHCH_2CH=CH-\ddot{O}H]$

an enol

⇌ $C_6H_5NHCH_2CH_2CH=\ddot{O}$ —H^+→ $HC(=\overset{+}{O}H)CH_2CH_2NH-C_6H_5$

→ (cyclized, $\overset{+}{N}$–H, H, OH) —$-H^+$→ (H, $\ddot{O}H$, $\ddot{N}$–H) —H^+→

(H, $\overset{+}{O}H_2$, $\ddot{N}$–H) —$-H_2O$→ (+, H, H, $\ddot{N}$–H) —$-H^+$→ ($\ddot{N}$–H)

18.40 $C_6H_5NH_2$ + (methyl vinyl ketone, O) → [(N, O)] → (CH_3, N–H)

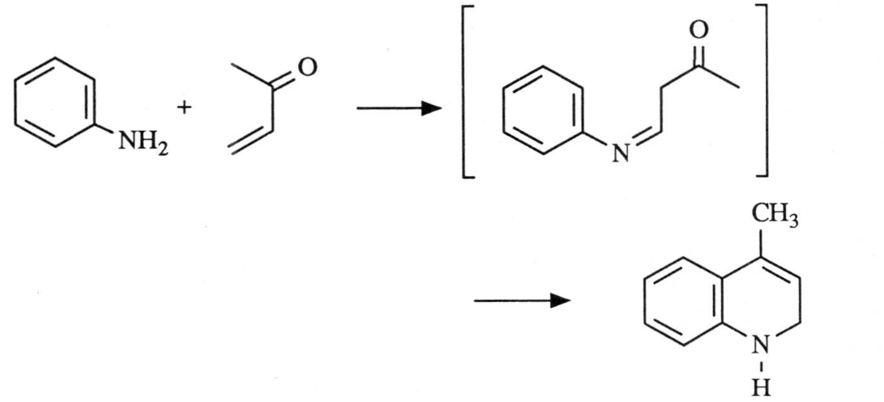

The mechanism in this series of reactions is similar to that in Problem 18.39.

18.41

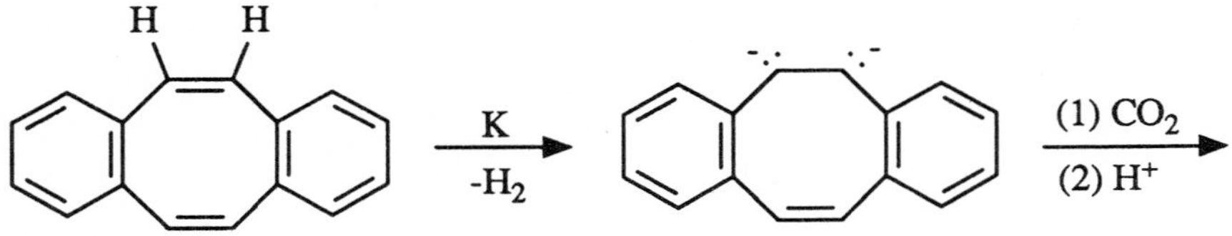

H H

K, $-H_2$ → A → (1) CO_2 (2) H^+ →

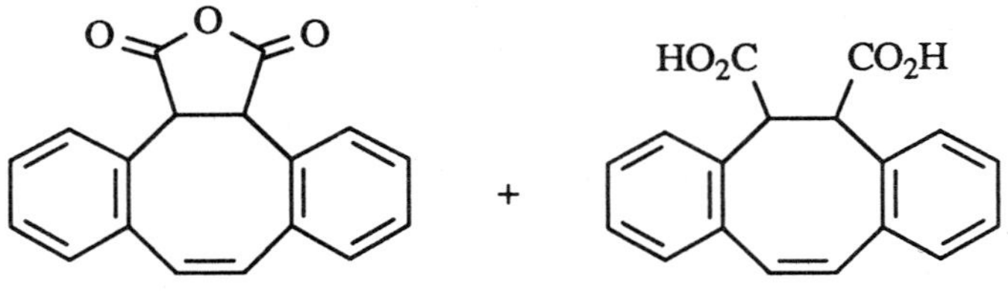

B + C

HO_2C CO_2H

nmr spectrum of B:

8 benzenoid protons:	δ7.2	(area 4)
2 vinylic protons:	δ6.98	(area 1)
2 benzylic protons:	δ4.64	(area 1)

18.42 A, CHO B, CH_2OH

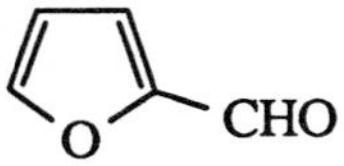

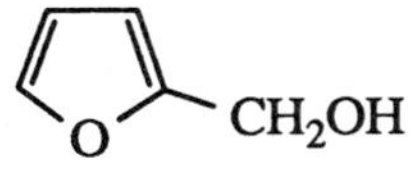

The 2-substituted ring is chosen over the 3-substituted ring because only one proton in B shows nmr absorption far downfield. (This proton is both aromatic and deshielded by O.)

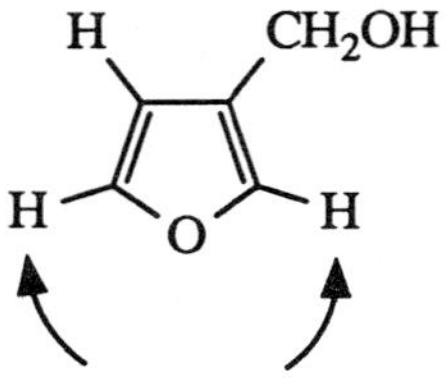

H CH_2OH

H O H

The nmr spectrum of the 3-substituted compound would show downfield aryl absorption for *two* protons.

Chapter 19

Natural Products: Studies in Organic Synthesis

19.9 (a) (b)

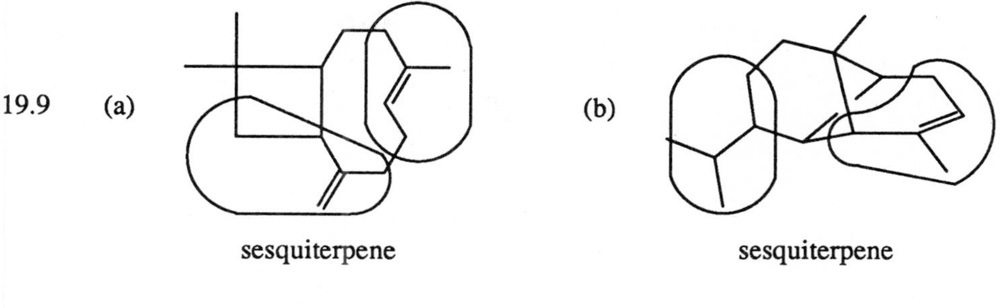

sesquiterpene sesquiterpene

(c)

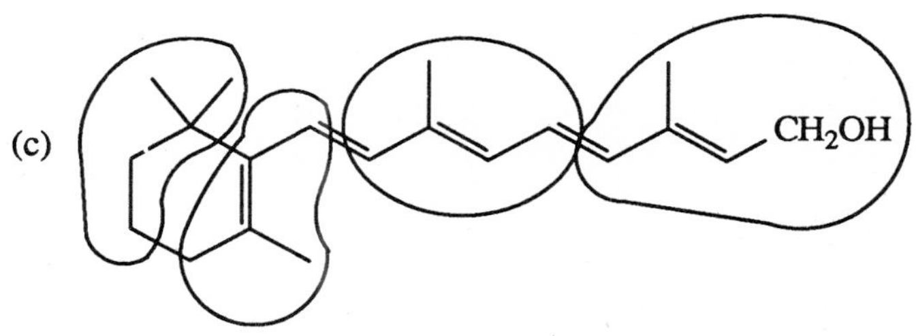

diterpene

19.10

1 2 3 4 5

The isoprene links are shown by dashed lines and numbers.

(a) six isoprene units: a triterpene

(b) (1) head-to-tail (2) head-to-tail (3) tail-to-tail

(4) head-to-tail (5) head-to-tail

(c) (6*E*,10*E*,14*E*,18*E*)-2,6,10,15,19,23-hexamethyl-2,6,10,14,18,22-tetracosa-hexaene

(d) CH_2OH + HOH_2C

[H], enzymes

link 3 in preceding formula

$CH_2—CH_2$

19.11 (a) $CH_3CH_2\overset{O}{\overset{\|}{C}}H + HOCH_2CH_2CH_2OH \xrightleftharpoons{H^+}$ product

(b) [2-methylene-1-ethyl-1-methylcyclohexane: ring with $=CH_2$, CH_2CH_3, CH_3] $\xrightarrow[\text{heat}]{CrO_3,\ H^+}$ product

(c) $CH_3CH_2C{\equiv}CH + LiNH_2 \longrightarrow$ product

19.12 (a) (*S*)-$CH_3\overset{OH}{\overset{|}{C}}HCH_2CO_2CH_2CH_3$ or [Fischer projection: CH_3 top, H left, OH right, $CH_2CO_2CH_2CH_3$ bottom]

or $CH_2CO_2C_2H_5$ — C — H_3C, H, OH

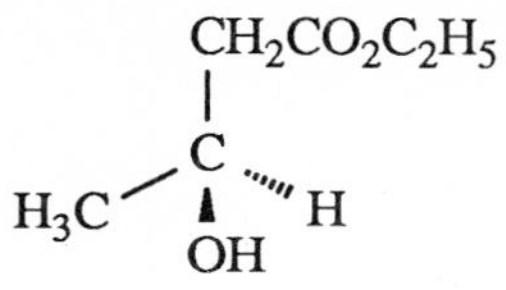

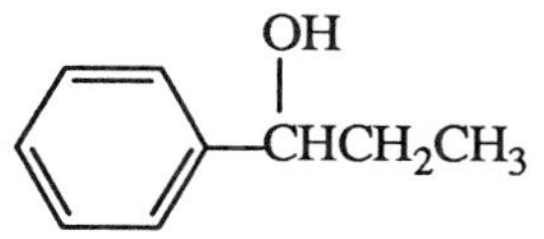

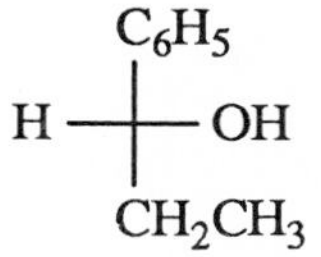

(b) (S)-C_6H_5-$CH(OH)CH_2CH_3$ or [Fischer projection: C_6H_5 top; H left; OH right; CH_2CH_3 bottom]

or [C_6H_5 top; CH_3CH_2, H, OH on C]

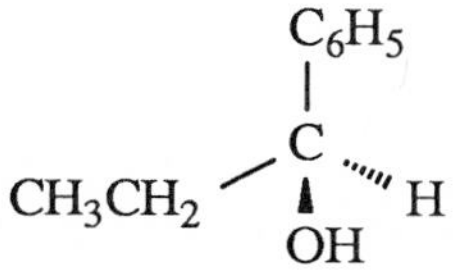

19.13 RAMP is (*R*)-(+)-1-amino-2-methoxymethylpyrrolidine

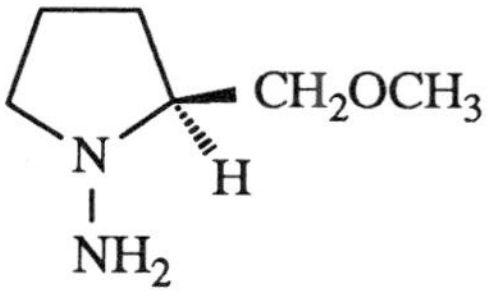

SAMP 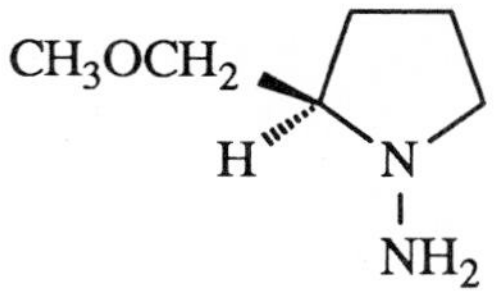 RAMP

19.14 (a) 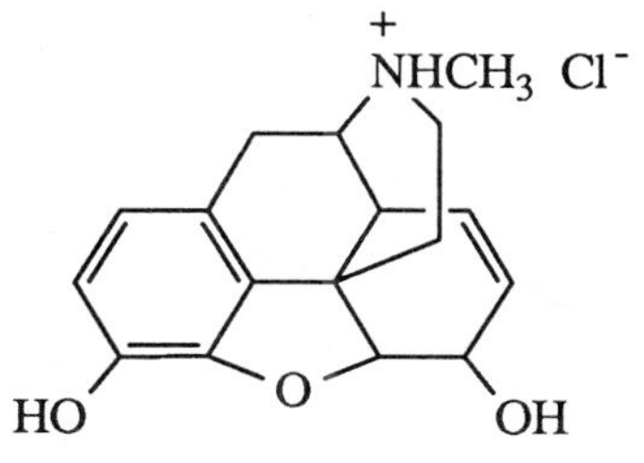 [structure: $\overset{+}{N}HCH_3$ Cl^-, HO, O, OH] (b) 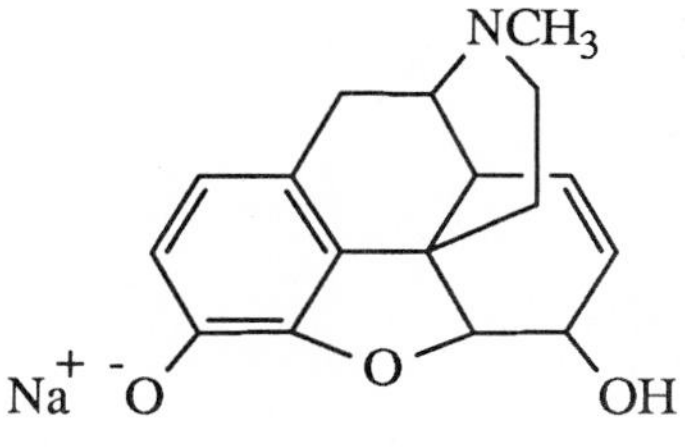 [structure: NCH_3, Na^+ ^-O, O, OH]

(c) no reaction (not a phenol)

(d) 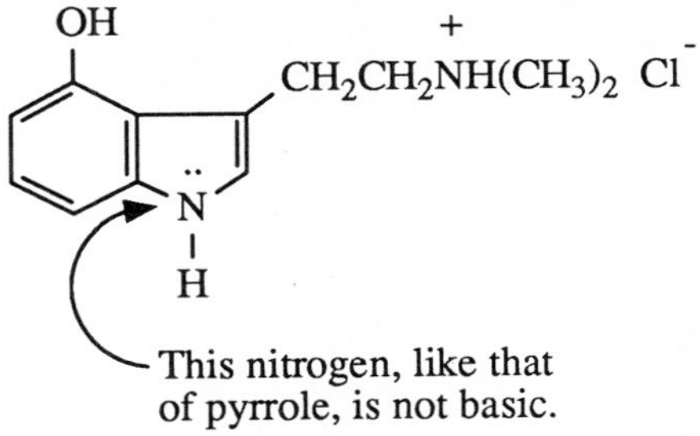 OH; $\overset{+}{CH_2CH_2NH(CH_3)_2}$ Cl^-; N–H

This nitrogen, like that of pyrrole, is not basic.

19.15 (a) 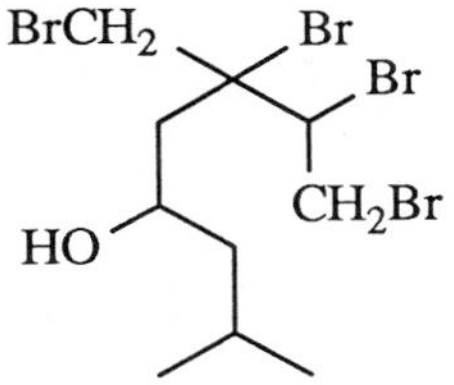 $BrCH_2$, Br, Br, CH_2Br, HO

(b) 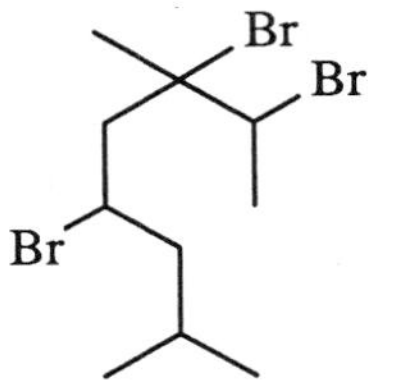 Br, Br, Br

(c) 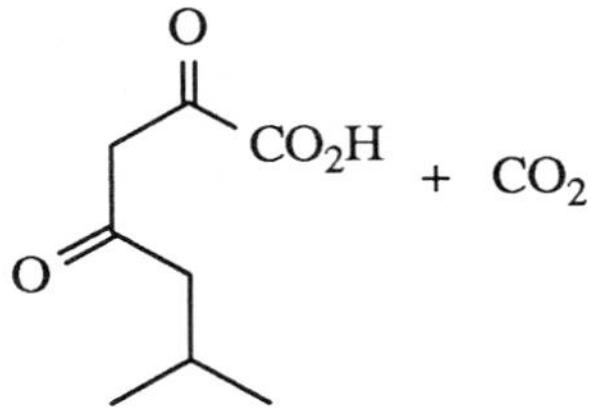 O, CO_2H, O + CO_2

19.16 (a) 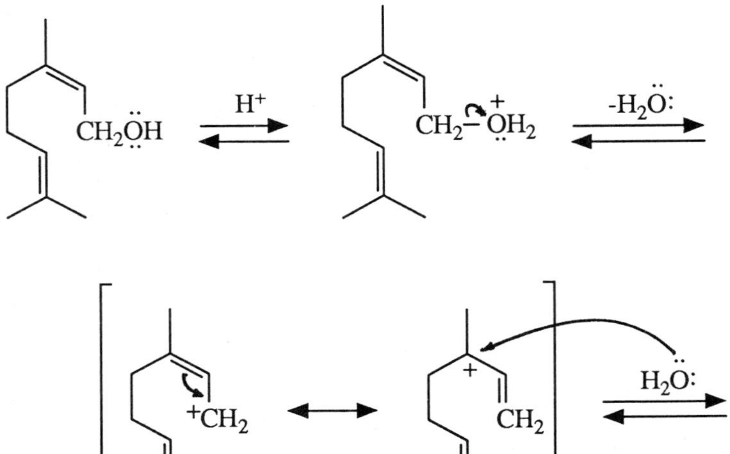 $CH_2\ddot{O}H$ $\xrightarrow{H^+}$ CH_2–$\overset{+}{O}H_2$ $\xrightarrow{-H_2\ddot{O}:}$

[$^+CH_2$ ⟷ + CH_2] $\xrightarrow{H_2\ddot{O}:}$

an allylic cation

$\ddot{O}H_2^+$ $\xrightarrow{-H^+}$ $\ddot{O}H$

(b) $CH_2\ddot{O}H$ $\xrightarrow{H^+}$ $CH_2-\overset{+}{O}H_2$ $\xrightarrow{-H_2\ddot{O}:}$

CH_2^+ ⇌ ⇌ $H_2\ddot{O}:$

an allylic cation
(one resonance structure)

OH_2^+ $\xrightarrow{-H^+}$ $\ddot{O}H$

(c) O $\xrightarrow{H^+}$ O ⇌

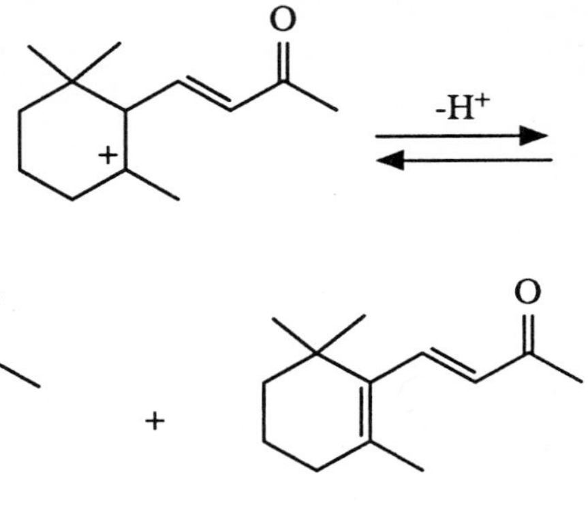

$\xrightarrow{-H^+}$ (equilibrium)

+

19.17 (*Z*)-$CH_2{=}CH(CH_2)_3CH{=}CH(CH_2)_3CH_2OH$ $\xrightarrow{\text{PCC}}$

(*Z*)-$CH_2{=}CH(CH_2)_3CH{=}CH(CH_2)_3CHO$

A

$\xrightarrow[\text{(2) } H_2O,\ H^+]{\text{(1) } CH_3(CH_2)_9MgBr}$ (*Z*)-$CH_2{=}CH(CH_2)_3CH{=}CH(CH_2)_3CH(OH)(CH_2)_9CH_3$

B

$\xrightarrow{\text{PCC}}$ $CH_2{=}CH(CH_2)_3(H)C{=}C(H)(CH_2)_3C(=O)(CH_2)_9CH_3$

the pheromone

19.18 citronellal (skeletal structure with CHO)

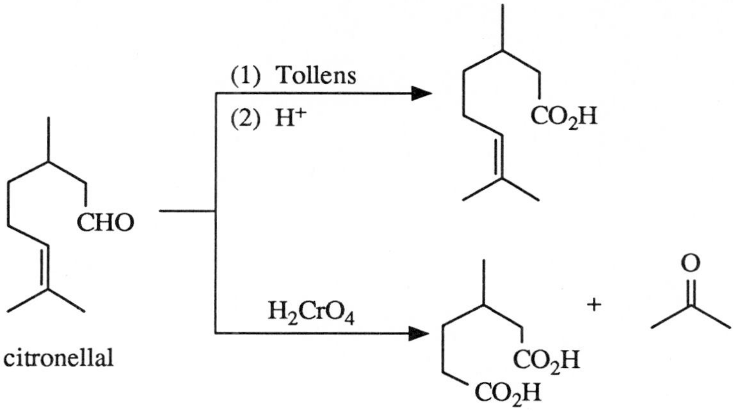

(1) Tollens; (2) H^+ → (skeletal structure with CO_2H)

H_2CrO_4 → (skeletal diacid with CO_2H, CO_2H) + acetone (O)

(skeletal structure with CHO)

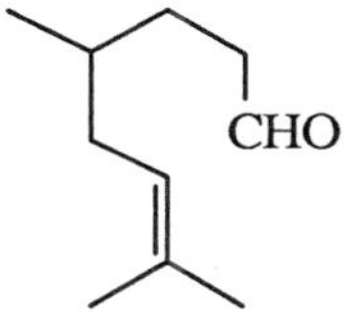

would be an incorrect answer because it is not a terpenoid.

19.19

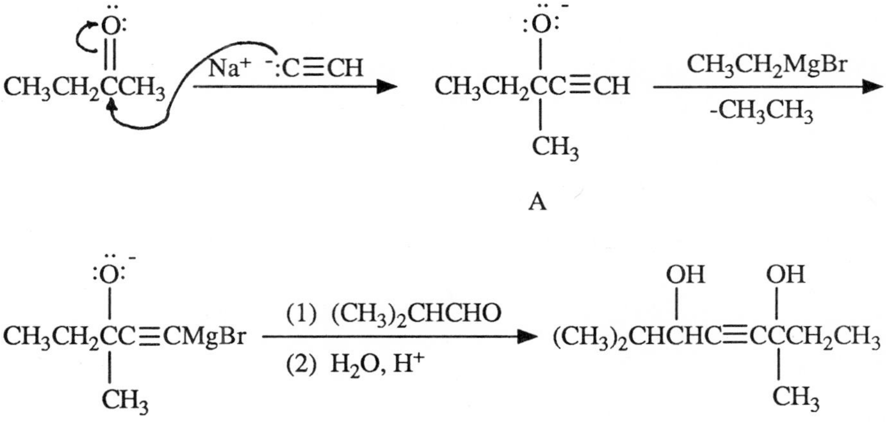

$$CH_3CH_2\overset{\overset{\ddot{O}:}{\|}}{C}CH_3 \xrightarrow{Na^+ \ ^-:C{\equiv}CH} CH_3CH_2\underset{\underset{CH_3}{|}}{\overset{\overset{:\ddot{O}:^-}{|}}{C}}C{\equiv}CH \xrightarrow[-CH_3CH_3]{CH_3CH_2MgBr}$$

A

$$CH_3CH_2\underset{\underset{CH_3}{|}}{\overset{\overset{:\ddot{O}:^-}{|}}{C}}C{\equiv}CMgBr \xrightarrow[(2)\ H_2O,\ H^+]{(1)\ (CH_3)_2CHCHO} (CH_3)_2CH\overset{\overset{OH}{|}}{C}HC{\equiv}C\underset{\underset{CH_3}{|}}{\overset{\overset{OH}{|}}{C}}CH_2CH_3$$

B C

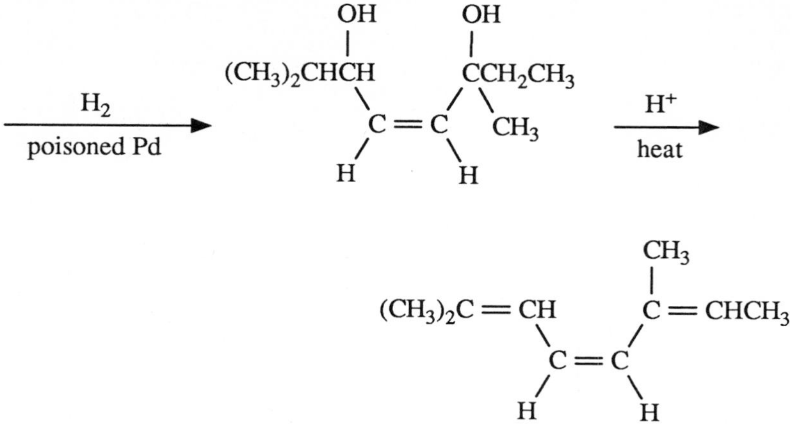

$\xrightarrow[\text{poisoned Pd}]{H_2}$ (CH$_3$)$_2$CHCH(OH)–CH=CH–C(OH)(CH$_3$)CH$_2$CH$_3$ $\xrightarrow[\text{heat}]{H^+}$

(CH$_3$)$_2$C=CH–CH=CH–C(CH$_3$)=CHCH$_3$

alloocimene

19.20 (CH$_3$)$_2$COH, H, CH$_3$ $\xrightarrow{Br_2}$:ÖH, (CH$_3$)$_2$C, Br, CH$_3$, H, H, Br

$\xrightarrow{NaOC_2H_5}$:Ö:$^-$, (CH$_3$)$_2$C, Br, CH$_3$, H, H, Br ⇌ (CH$_3$)$_2$C–Ö:$^-$, H, Br, Br, CH$_3$

$\xrightarrow{-Br^-}$ (CH$_3$)$_2$C, O, Br, CH$_3$, H, H $\xrightarrow[-HBr]{NaOC_2H_5}$

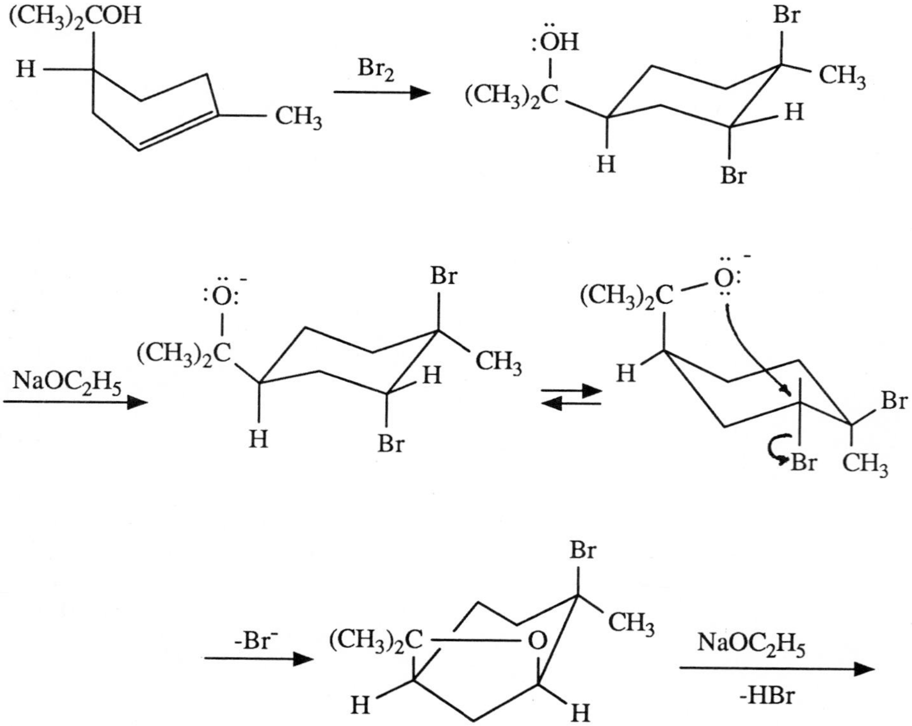

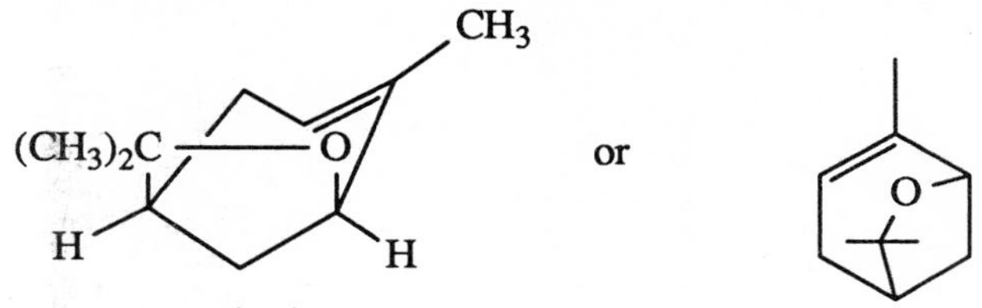

CH_3

$(CH_3)_2C$ O H H

pinol

or

O

19.21 $HC(=CH_2)C(CH_3)=CH_2$ + $CH_2=CHCCH_3$ (O) $\xrightarrow[\text{(Diels-Alder)}]{\text{heat}}$

H_3C (cyclohexenyl) CCH_3 (O) + H_3C (cyclohexenyl) CCH_3 (O)

(1) CH_3MgI

(2) H_2O, H^+

HO

$C(CH_3)_2$

H_3C

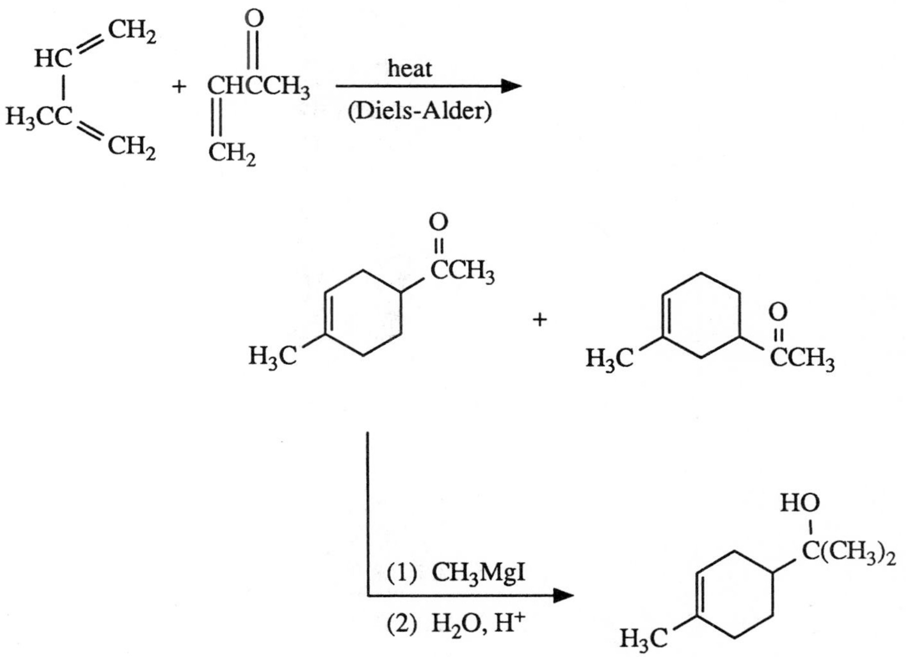

α-terpineol

Note that the Diels-Alder reaction would lead to a mixture of isomers, which must be separated before the Grignard reaction. You might have devised a synthetic scheme using a Dieckmann ring closure; however, such a route would require many more steps.

Chapter 20

Pericyclic Reactions

20.12 (a) cycloaddition (b) sigmatropic

(c) electrocyclic (d) electrocyclic

(e) electrocyclic

20.13 (a)

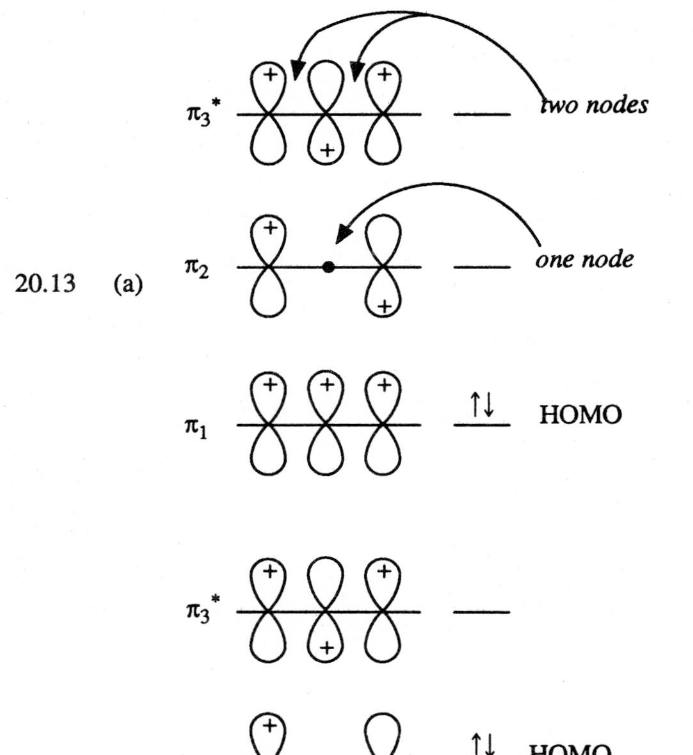

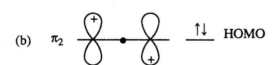

In (a) and (b), three carbon atoms are in the sp^2-hybrid state. The structure in (a) contains two pi electrons, while the structure in (b) contains four pi electrons.

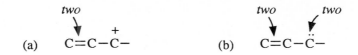

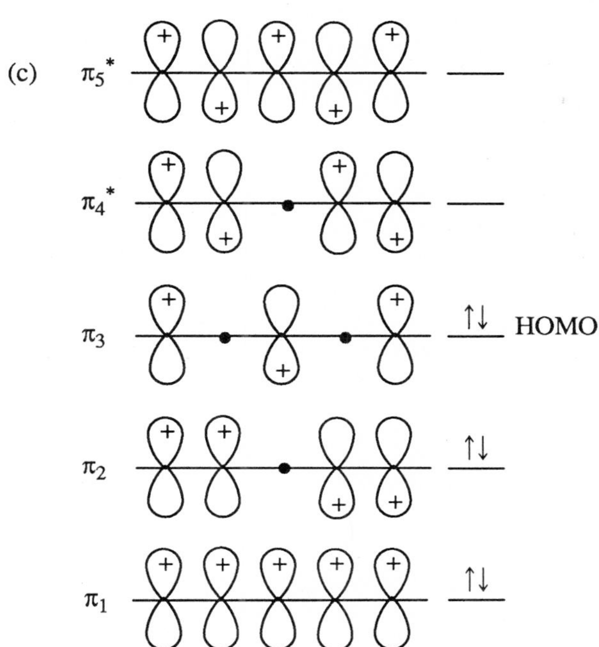

The anion in (c) contains five sp^2-hybridized carbon atoms; therefore, five *p* orbitals result. The anion has six pi electrons

two *two* *two*

−C=C−C=C−C̈−

20.14 (a) [2 + 2] (b) [4 + 2] (c) [8 + 2]

In each example, the reactants contain more pi electrons than are involved in the cyclization. The nonparticipating pi electrons are not used in the classification system.

20.15 (b) and (c). For a quick rule of thumb, remember that a reaction involving $4n$ pi electrons requires light while a reaction involving $4n + 2$ pi electrons requires heat.

20.16 The reaction is a [2 + 2] cycloaddition. Therefore, the skeleton of the product is as shown in the following equation:

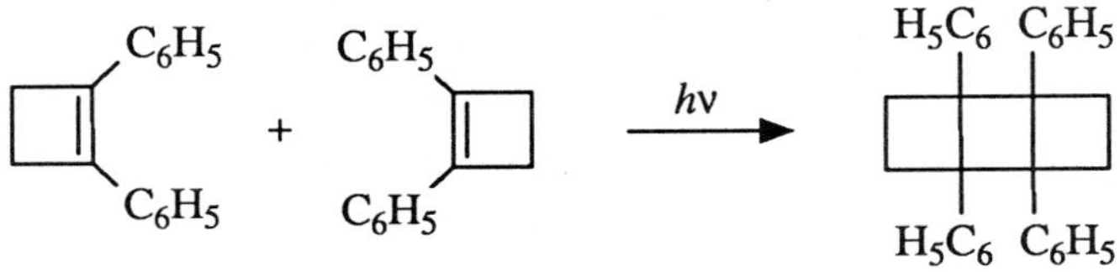

$$\text{1,2-diphenylcyclobutene} + \text{1,2-diphenylcyclobutene} \xrightarrow{h\nu} \text{tetraphenyl tricyclic product}$$

The stereochemistry can be determined by envisioning the end-to-end overlap of the *p*-orbital components:

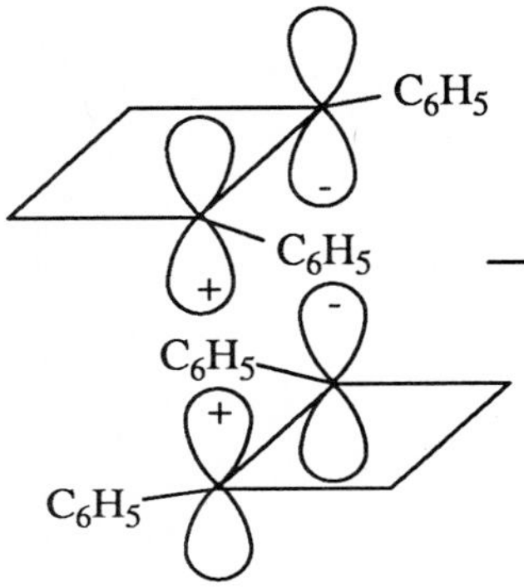

Orbitals of the same phase overlap.

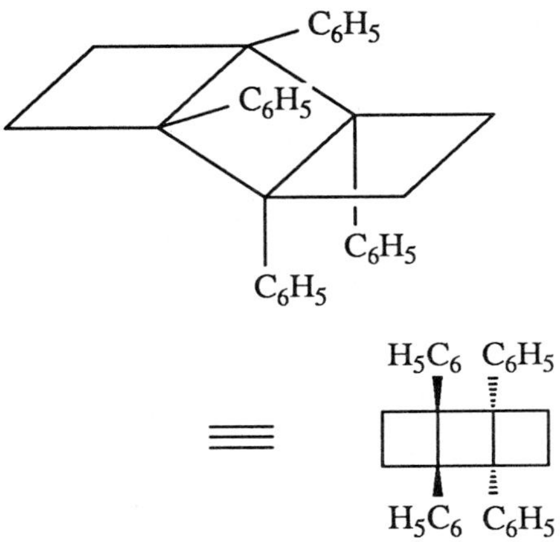

20.17 (a) [4 + 2] cycloaddition: C_6H_5-substituted diene + $C_2H_5O_2C-C{\equiv}C-CO_2C_2H_5 \xrightarrow{\text{heat}}$

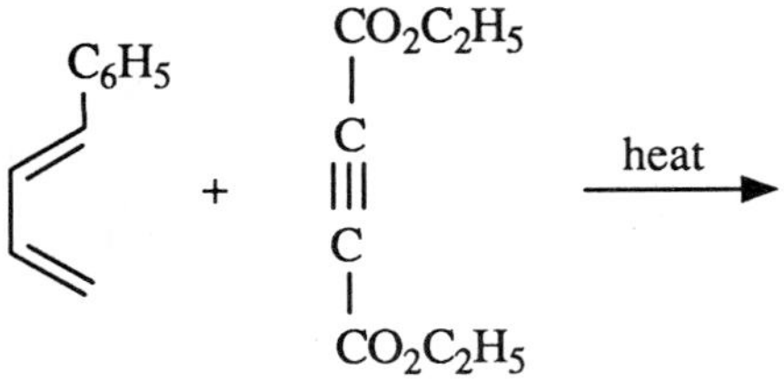

(b) [2 + 2] cycloaddition: cyclopentenone + $CH_3-C{\equiv}C-CH_3 \xrightarrow{h\nu}$

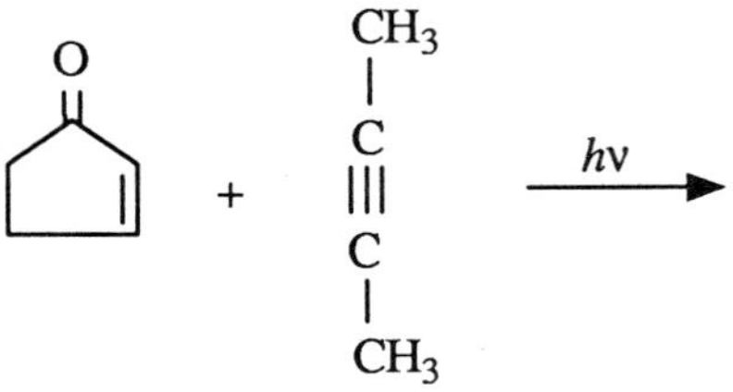

(c) [4 + 2] cycloaddition: CH_3O_2C (diene) + (maleic anhydride) $\xrightarrow{\text{heat}}$

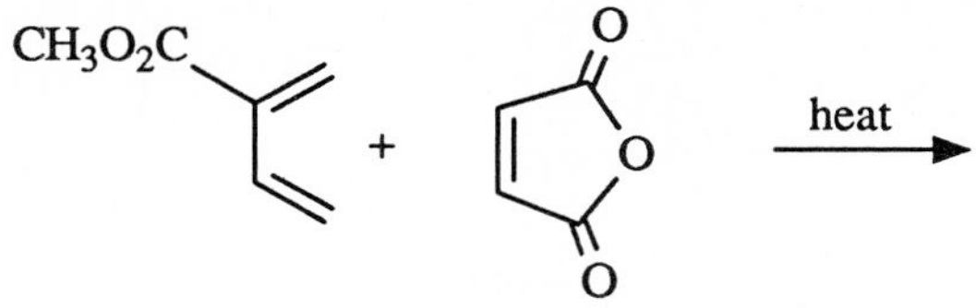

(d) [4 + 2] cycloaddition:

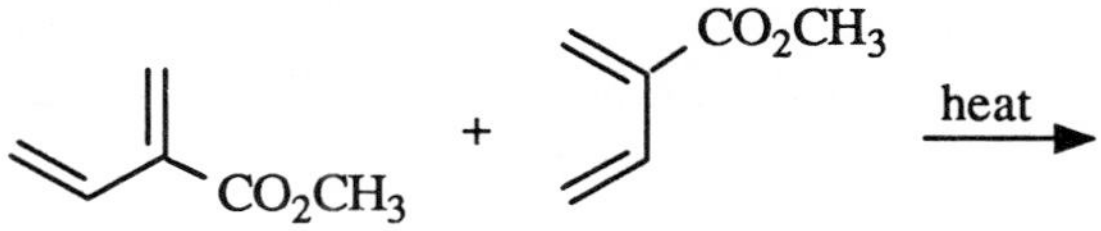

(diene) CO_2CH_3 + (diene) CO_2CH_3 $\xrightarrow{\text{heat}}$

In (d), the following reaction would probably also occur, lowering the yield and making it necessary to separate the product from the by-product.

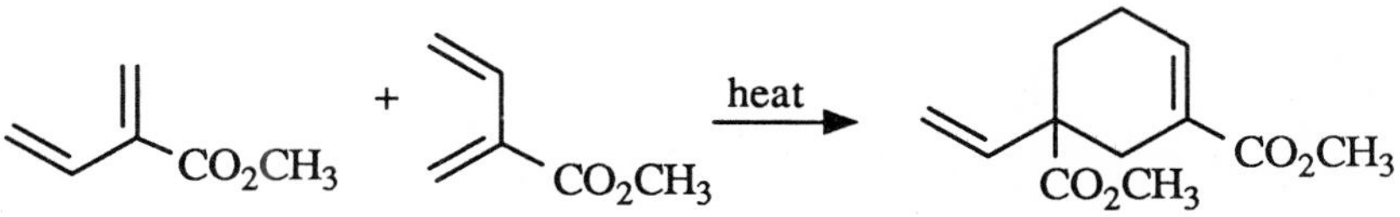

CO_2CH_3 + CO_2CH_3 $\xrightarrow{\text{heat}}$ CO_2CH_3 CO_2CH_3

When a four-membered ring is the desired product, a [2 + 2] cycloaddition (light-promoted) should be considered. When a six-membered ring is desired, then a [4 + 2] cycloaddition (heat-promoted; Diels-Alder reaction) should be considered. Note that rings containing an odd number of ring atoms cannot be formed in cycloaddition reactions.

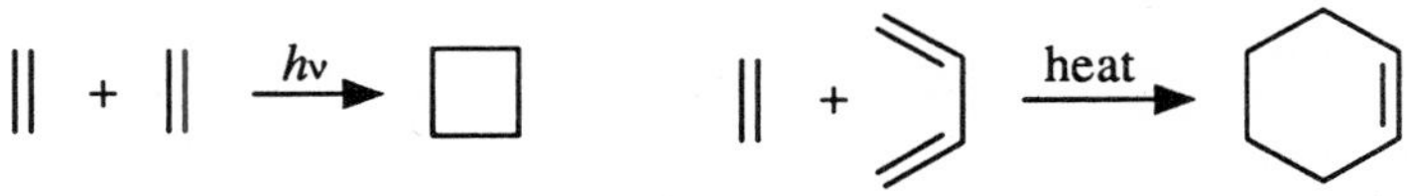

|| + || $\xrightarrow{h\nu}$ □ || + (diene) $\xrightarrow{\text{heat}}$ (cyclohexene)

In a Diels-Alder reaction, the ring double bond of the cyclohexene product must arise from the diene reactant.

from the diene reactant

CH_3O_2C (product) $\xleftarrow{\text{heat}}$ CH_3O_2C (diene) + (maleic anhydride)

from the dienophile reactant

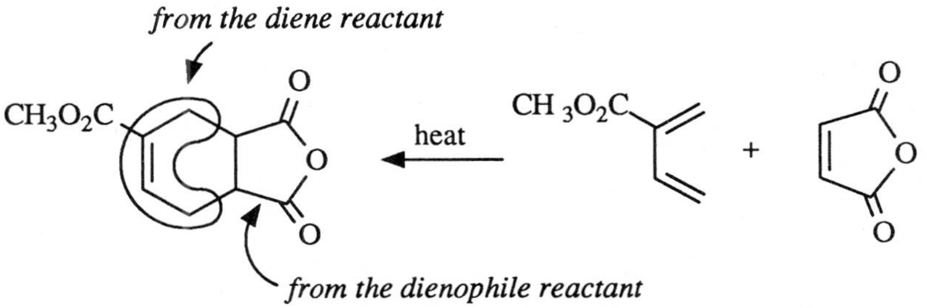

20.18 The mode of rotation can be determined with the aid of Table 20.1 in the text.

(a) 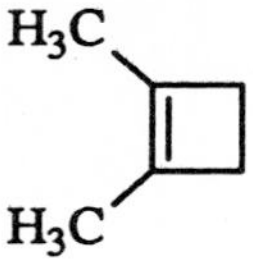 H_3C, H_3C

$4n$ (heat): conrotatory

(b) 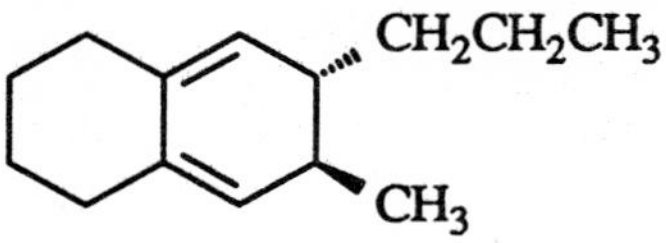 $CH_2CH_2CH_3$, CH_3

$(4n + 2)$ ($h\nu$): conrotatory

(c) 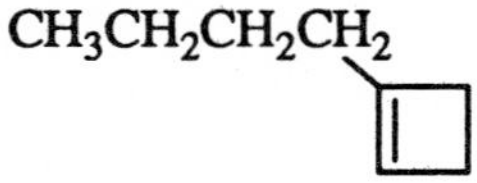 $CH_3CH_2CH_2CH_2$

$4n$ ($h\nu$): disrotatory

(d)

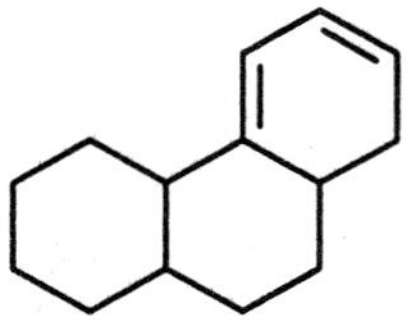

$(4n + 2)$ (heat): disrotatory

20.19 (a)

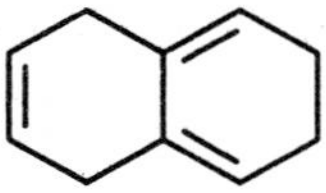

(b) 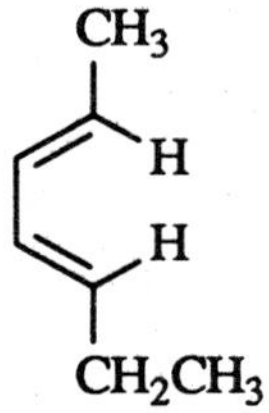 CH_3, H, H, CH_2CH_3 and H, CH_3, CH_2CH_3, H

(2*E*,4*E*) (2*Z*,4*Z*)

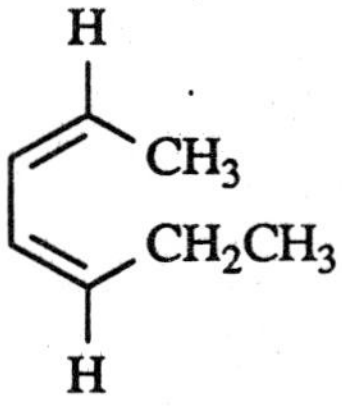

by conrotatory motion

(c) 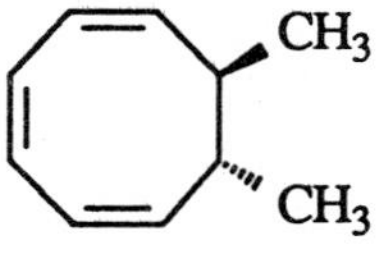 CH_3, CH_3

trans methyls
by conrotatory motion

(d) 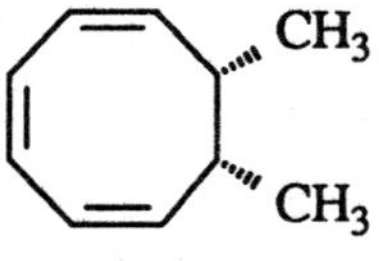 CH_3, CH_3

cis methyls
by disrotatory motion

(e) 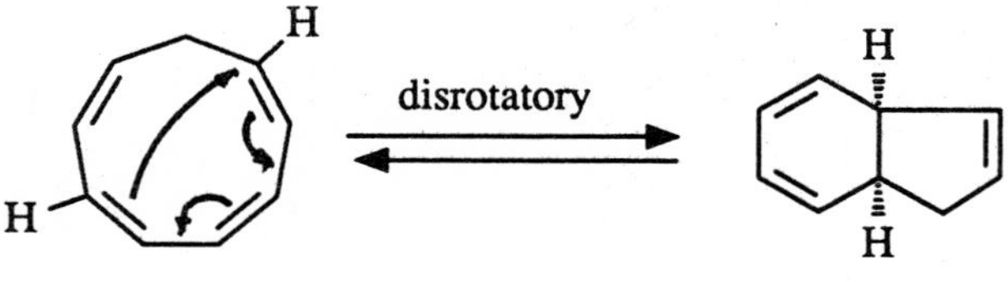 H, H — disrotatory → H, H

$(4n + 2)$

cis ring juncture

and

H H conrotatory H H

$4n$ *trans* ring juncture

Because electrocyclic reactions are reversible, the first product shown would predominate because it is of lower energy (less-strained ring system).

20.20 (a) H H $h\nu$ disrotatory H H

$4n$ and (E,Z)

(b) H H $h\nu$ disrotatory H H

$4n$ and (Z,Z)

20.21 (a) CH_3 CH_3 heat conrotatory CH_3 H H CH_3 $h\nu$ disrotatory CH_3 CH_3

$4n$, *trans* (E,E) *cis*

The steps could be reversed, using $h\nu$ first; the result is the same.

(b) 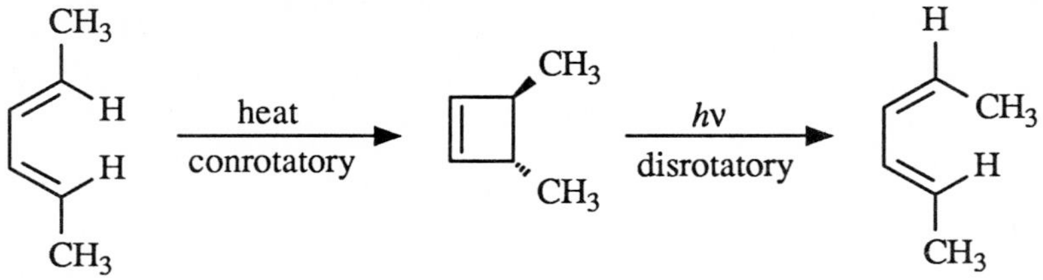

4*n*, (*E*,*E*) *trans* (*E*,*Z*)

Again, the steps could be reversed.

20.22 In each case, number the migrating group and the alkenyl chain starting at their point of original attachment. Then number the product using the same numbering system.

(a) 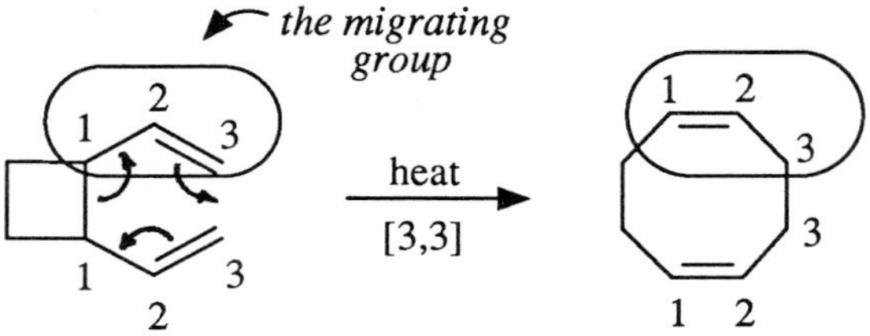

(b) 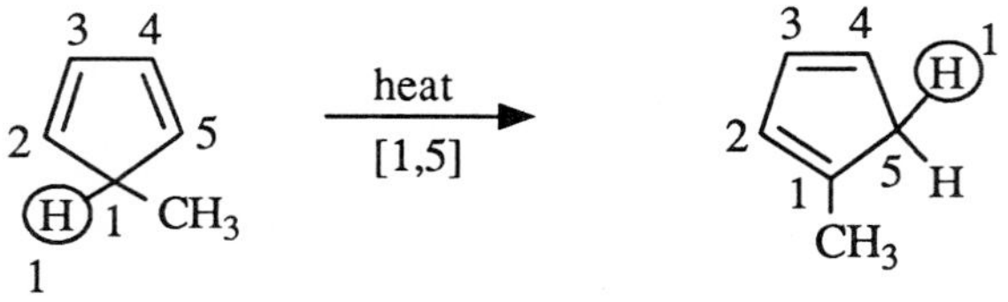

(c) 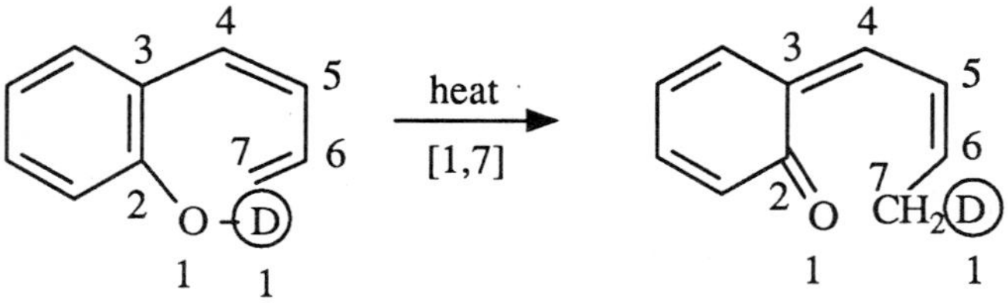

In (a), both the migrating group and the alkenyl chain have three atoms. In (b) and (c), the migrating group has only one atom (H and D, respectively); therefore, the numbering of the migrating group stops with the number 1.

20.23 Inspection of the structure shows that either a [1,3] or a [1,5] sigmatropic rearrangement of deuterium to another position is possible. [1,3]-Sigmatropic rearrangements are rare but [1,5]-sigmatropic rearrangements are common. Therefore, one would predict that the reaction will proceed by a [1,5]-shift.

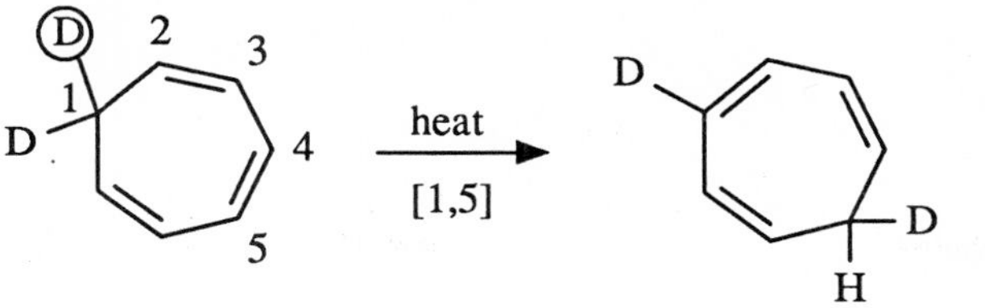

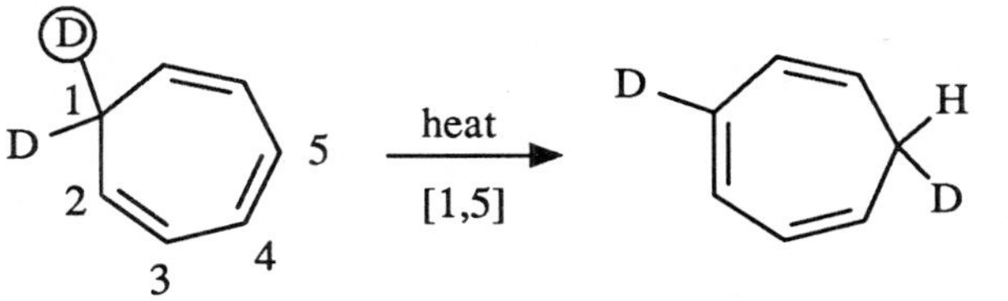

Because of symmetry in the product, either rearrangement shown results in the same compound. (A [1,7] shift is also possible.)

20.24

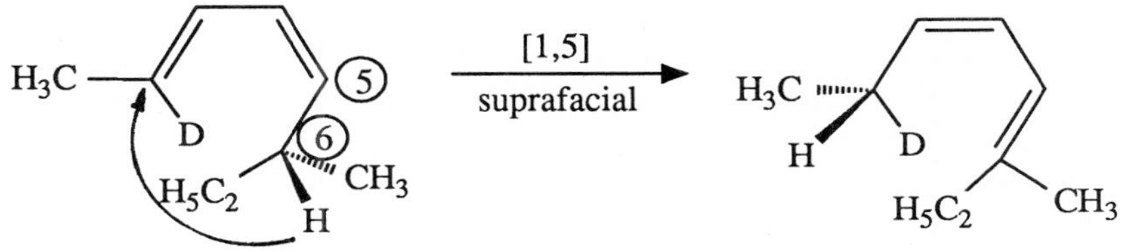

H attacks from "top"

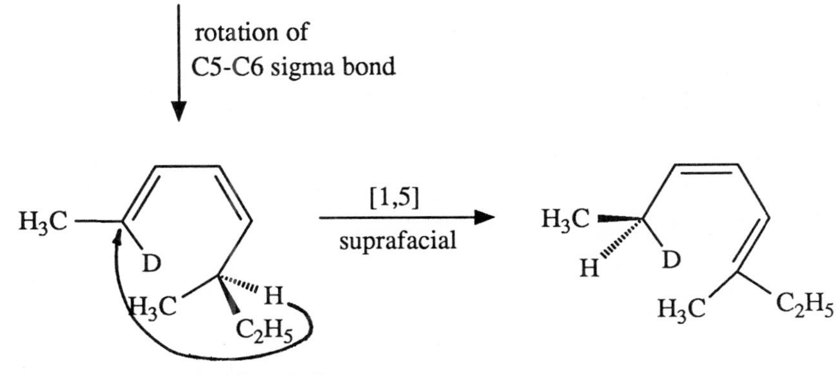

H attacks from the "bottom"

The stereochemistry is easier to see with models.

20.25 (a) 4*n* electrocyclic reaction, disrotatory, *h*ν. The reaction must take place with disrotatory motion because conrotatory motion would place one of the H's inside the ring and the other outside.

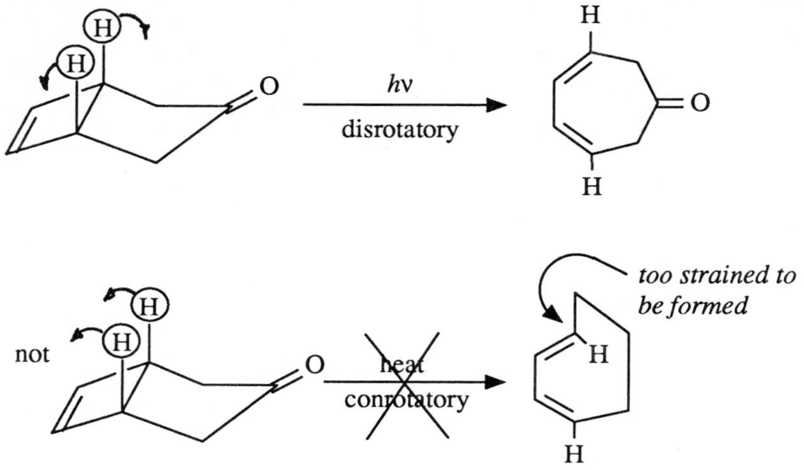

(b) [2 + 2] cycloaddition, $h\nu$

(c) [3,3] sigmatropic rearrangement, heat. The rearrangement is easier to see if the structures are redrawn:

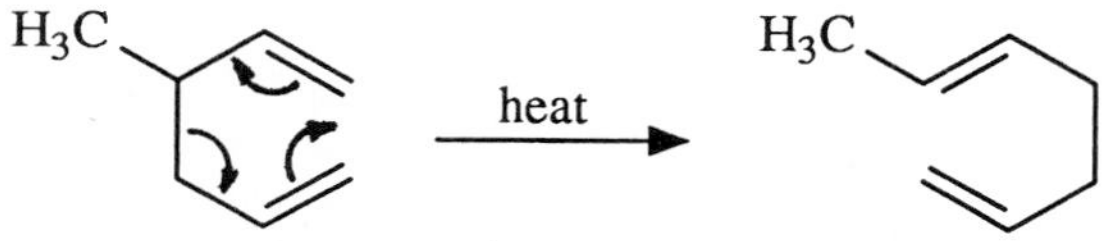

20.26 The thermal reaction is a [4 + 2] cycloaddition, or Diels-Alder reaction.

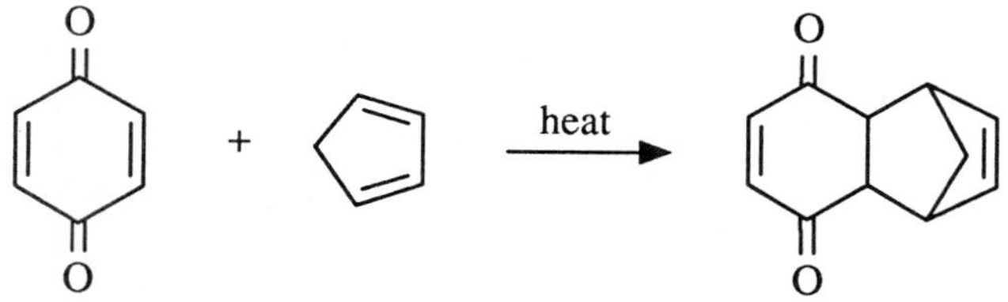

Diels-Alder product

The second reaction is an internal photo-induced [2 + 2] cycloaddition. In order to see how this second reaction takes place, redraw the Diels-Alder product to show the proximity of the double bonds.

$h\nu$

the isomer

20.27 (a) H_3C + C_6H_5 —heat [4 + 2]→ H_3C … C_6H_5 + H_3C … C_6H_5

(b) OCH_3 H H OCH_3 + CO_2CH_3 C ≡ C CO_2CH_3 —heat [4 + 2]→ OCH_3 CO_2CH_3 CO_2CH_3 OCH_3

(c) H H_2C C CH_2 O: OCH_3 —heat [3,3] sigmatropic rearrangement→ H H_2C C CH_2 H O: OCH_3

—tautomerization→ $CH_2CH{=}CH_2$ OH OCH_3

The last step in (c) is labeled "tautomerization" instead of "[1,3] sigmatropic rearrangement" because tautomerization is a stepwise ionic reaction and not a concerted pericyclic reaction.

(d) $C_6H_5CH{=}CHC_6H_5$ + $H_3CCH{=}CHCO_2CH_3$ $\xrightarrow[\text{[2 + 2]}]{h\nu}$

C_6H_5 CO_2CH_3 H H H H C_6H_5 CH_3 + C_6H_5 H H H_3C H CO_2CH_3 C_6H_5 H

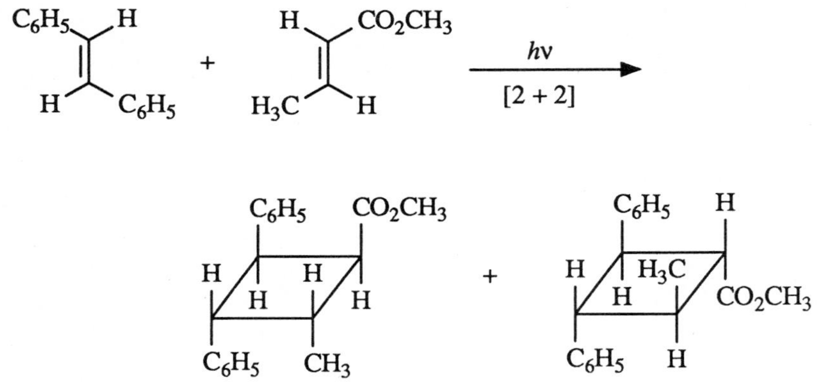

To determine the stereochemistry in (d), view the molecules as they approach each other before reaction occurs. For example,

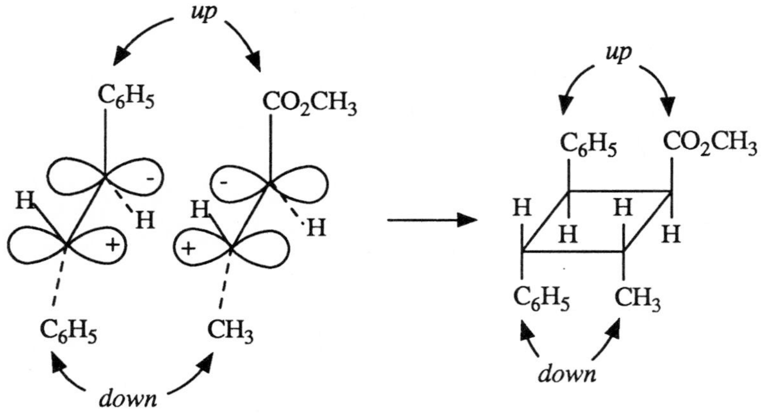

up C_6H_5 CO_2CH_3 H H - + H H + - C_6H_5 CH_3 down → up C_6H_5 CO_2CH_3 H H H H C_6H_5 CH_3 down

(e) H H $\xrightarrow[\text{[4n + 2] photochemical conrotatory}]{h\nu}$

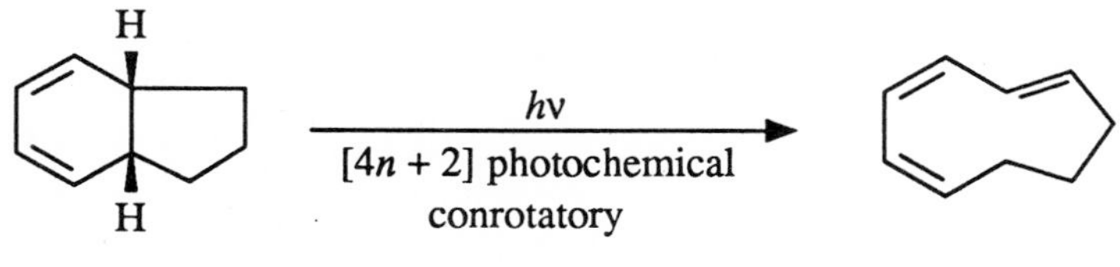

This reaction is very similar to the reverse reaction of the cyclization of 1,3,5-cyclohexatriene.

$h\nu$

To determine the direction of the movement of the groups in electrocyclic reactions, approach the reaction in steps (even though the actual reaction is concerted). Do not worry about the initial appearance of the product because you can redraw the structure to be more stereochemically correct.

H
H

conrotatory motion of H's

trans
cis
H
cis
H
cis
trans
cis
cis

H
H

apparent motion of electrons

(f)

heat
disrotatory

H
H

Compare this reaction with that in part (e). To return to a *cis* ring juncture would require light and conrotatory motion. Therefore, the ring juncture H atoms must be *trans* when heat is used.

20.28

H_3C O + CH_3 O heat [4 + 2] H_3C O CH_3 O

or H_3C O + O CH_3 heat [4 + 2] → O CH_3 H_3C O

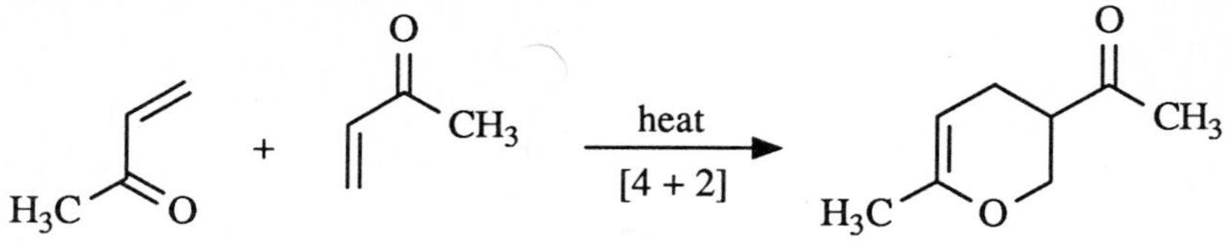

20.29 (a) 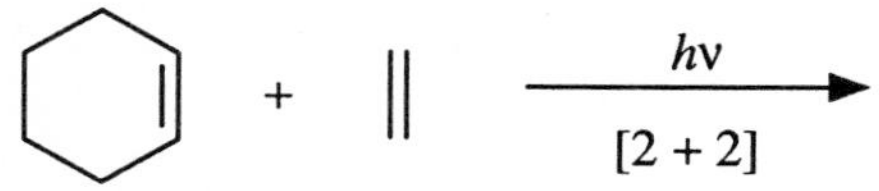 + hν [2 + 2] →

or H H hν disrotatory →

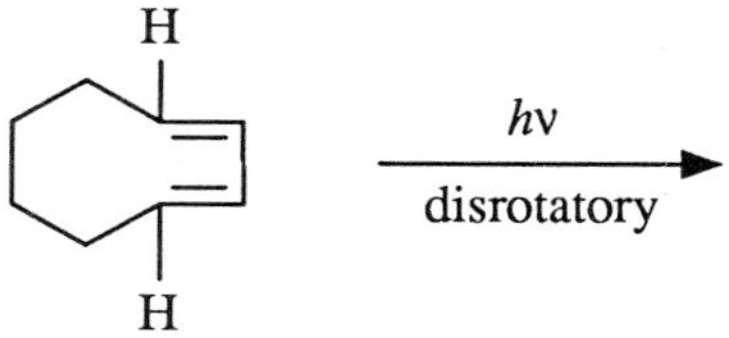

(b) 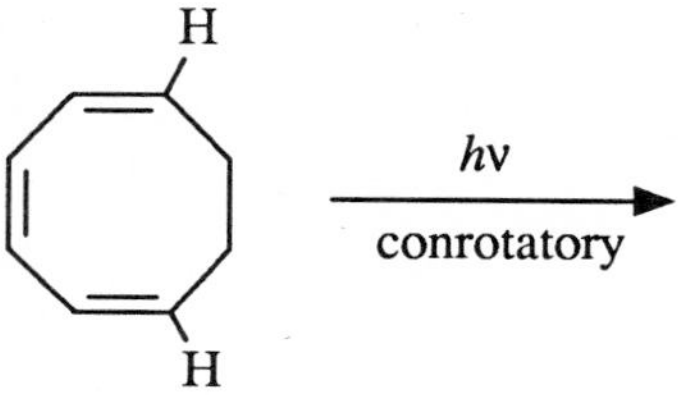 H H hν conrotatory →

(4n + 2)

(c) 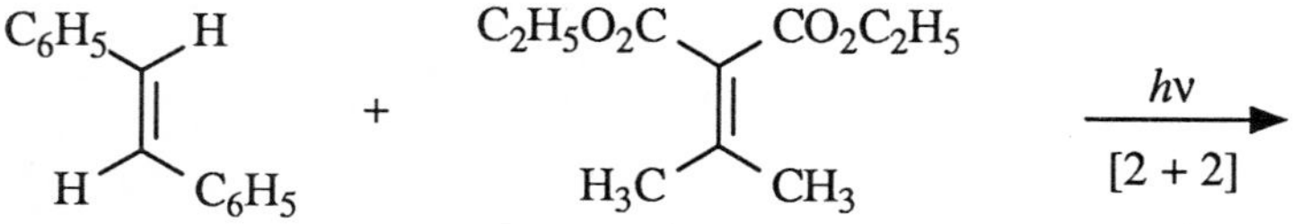 C_6H_5 H H C_6H_5 + $C_2H_5O_2C$ $CO_2C_2H_5$ H_3C CH_3 hν [2 + 2] →

(d) Note the cyclohexene ring.

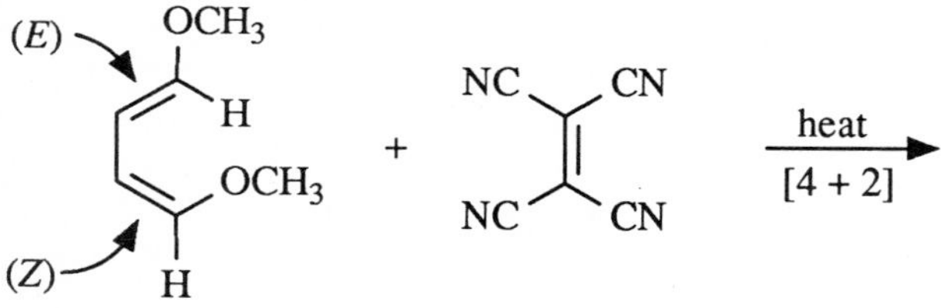

(*E*) OCH_3 H OCH_3 (*Z*) H + NC CN NC CN heat [4 + 2] →

(e) 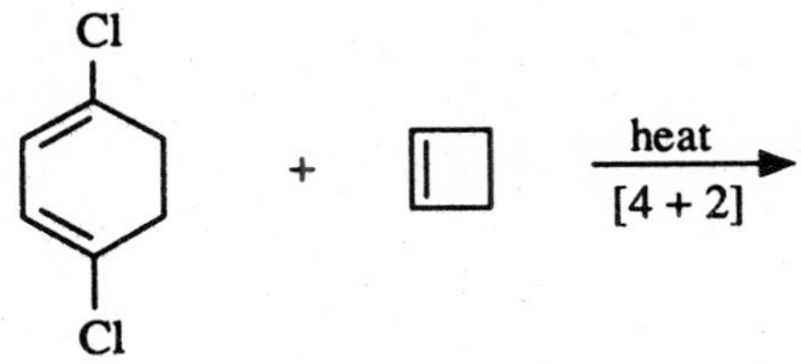 [1,4-dichloro-1,3-cyclohexadiene] + [cyclobutene] $\xrightarrow[\text{[4 + 2]}]{\text{heat}}$

This route yields the desired product with the desired stereochemistry.

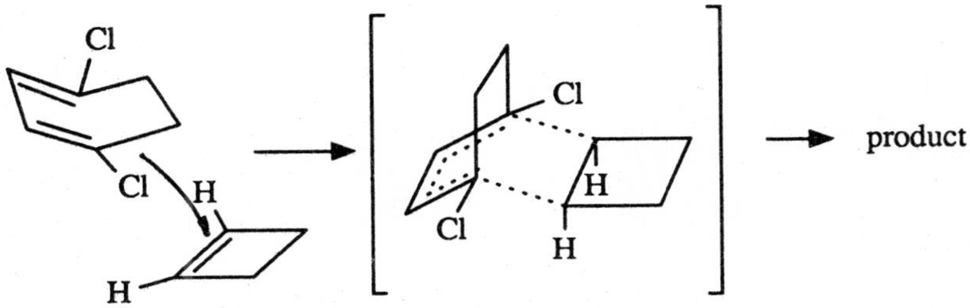

[diene Cl, Cl + cyclobutene H, H] → [transition state: Cl, Cl, H, H] → product

You may have chosen to use a different diene and dienophile, but this second route would be a poor choice because a mixture of products would probably result. Also, cycloaddition reactions with ethylene often give poor yields.

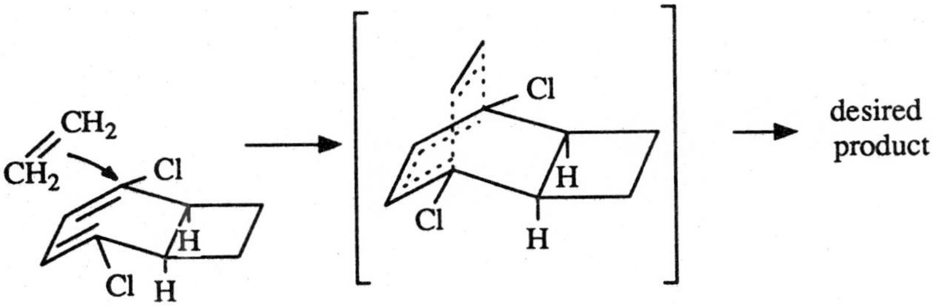

[CH_2=CH_2 + diene Cl, Cl, H, H] → [transition state: Cl, Cl, H, H] → desired product

or

[diene Cl, Cl, H, H + CH_2=CH_2] → [transition state: Cl, Cl, H, H]

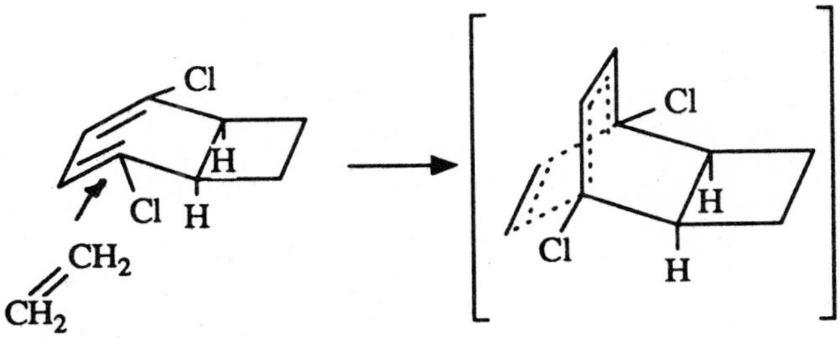

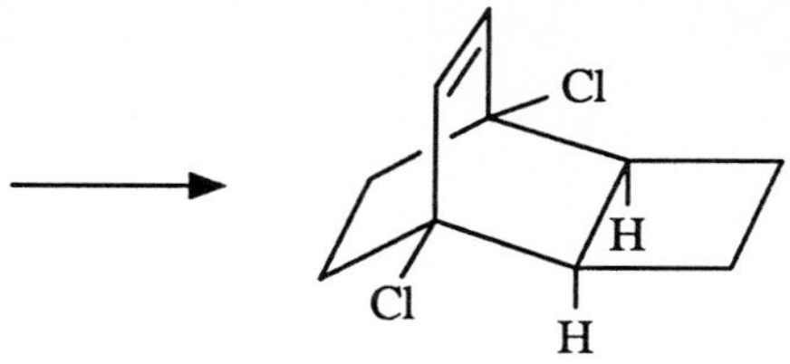

Cl

Cl

H

H

wrong stereochemistry

(f) 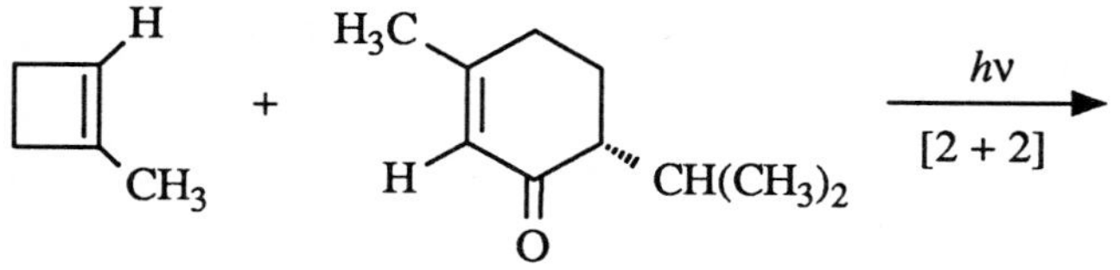 [cyclobutene with H and CH_3] + [H_3C, H, O, $CH(CH_3)_2$ enone] $\xrightarrow[\text{[2 + 2]}]{h\nu}$

In (f), the stereochemistry is dictated by the steric hindrance of the isopropyl group.

attack trans to the isopropyl group

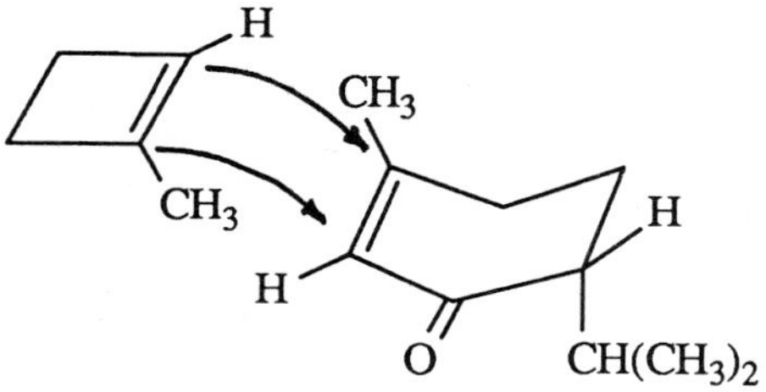

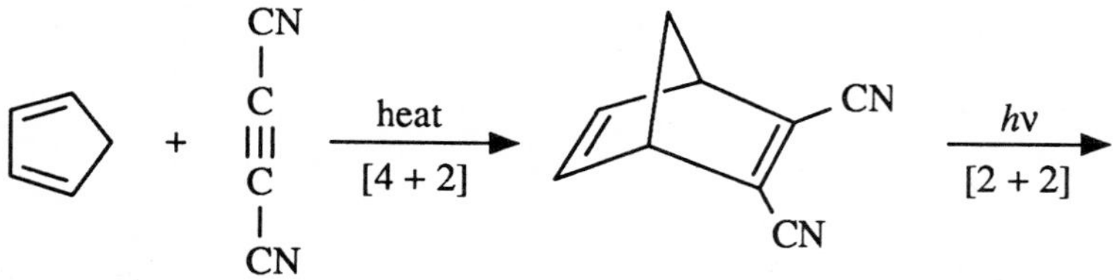

H

CH_3

CH_3

H

H

O

$CH(CH_3)_2$

(g) [cyclopentadiene] + $NC-C\equiv C-CN$ $\xrightarrow[\text{[4 + 2]}]{\text{heat}}$ [CN, CN bicyclic adduct] $\xrightarrow[\text{[2 + 2]}]{h\nu}$

(h) [NC, CN, NC, CN alkene] + [NCO_2CH_3 azepine] $\xrightarrow[\text{[4 + 2]}]{\text{heat}}$

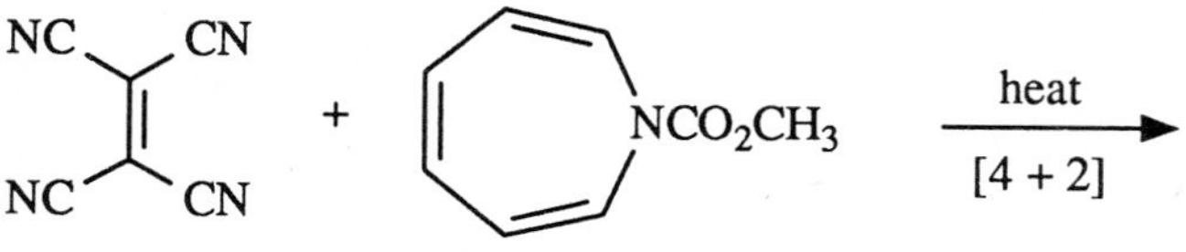

20.30 (a) *Step 1:*

CH_3O_2C ... electrocyclic ring opening, 4*n*, conrotatory → CH_3O_2C ... H H H H

not involved in first step

(A ten-membered ring can accommodate a *trans* double bond with H "inside" the ring.)

Step 2: Redraw the structure.

CH_3O_2C, CO_2CH_3, H, H, H, H → ring closure, (4*n* + 2), disrotatory → CH_3O_2C, CH_3O_2C, H, H

(b) CH_3, O, $^{14}CH_2$, CH, CH_2, CH_3 → [3,3] → CH_3, O, CH_3, $H_2{}^{14}C$, CH, CH_2 → [3,3] →

CH_3, O:, H, CH_3, $H_2{}^{14}C$, CH, CH_2 → $-H^+$ → CH_3, $\ddot{O}:^-$, CH_3, $CH_2{=}CH^{14}CH_2$ → H^+ →

(c) 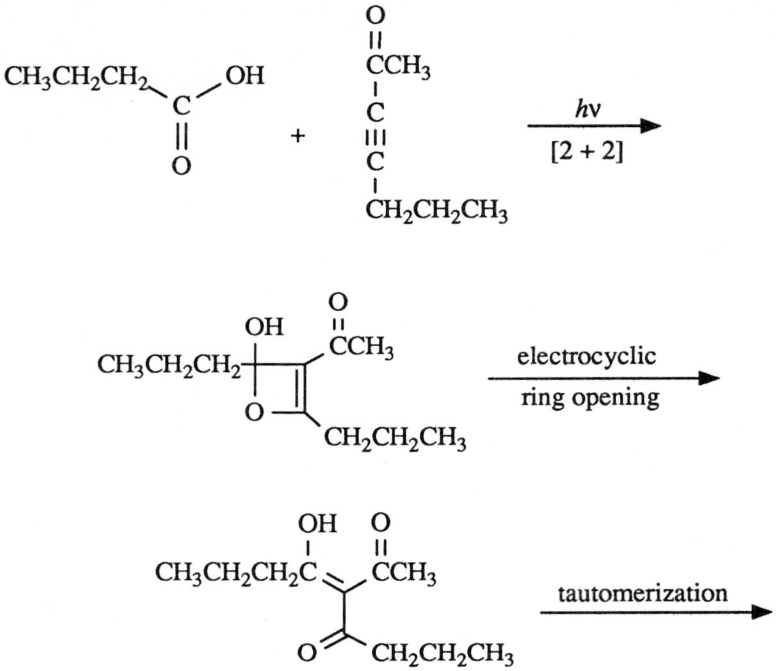
hν
[2 + 2]
electrocyclic
ring opening
tautomerization

20.31 (a) 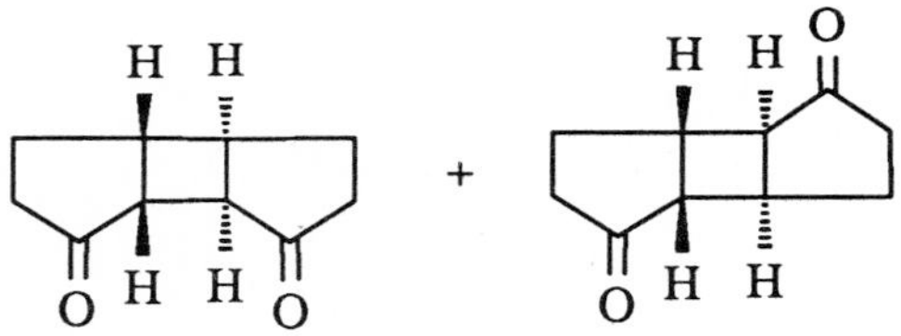 [2 + 2]

(b) 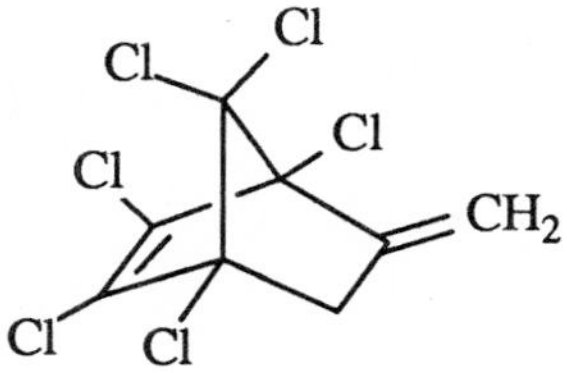 [4 + 2]

(c) 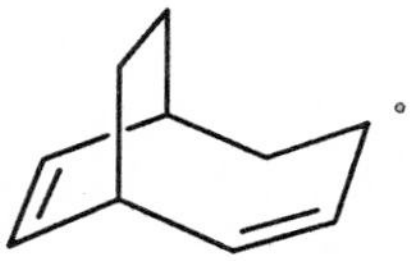 [4 + 2]

(d) $\xrightarrow{[4+2]}$ CO_2CH_3 CO_2CH_3 $\xrightarrow{[4+2]}$

CO_2CH_3 CH_3O_2C or CO_2CH_3 CO_2CH_3

Chapter 21

Spectroscopy II: Ultraviolet Spectra, Color and Vision, Mass Spectra

21.14 In the uv spectra, the tetraene would show a λ_{max} at a longer wavelength than the triene.

21.15 (a) The two compounds cannot be distinguished with only uv. Both contain the same type of π system (an ester).

(b) The carbonyl group in conjugation with the double bond exhibits different absorption from that of the conjugated diene.

(c) The ketone with the conjugated double bond absorbs at a longer wavelength than the ketone with the nonconjugated double bond.

(d) One structure is a conjugated triene, while the other is a conjugated diene; therefore, these compounds can also be distinguished by their uv spectra.

21.16 (a) $$\epsilon = \frac{A}{c\ l} = \frac{1.25}{(9.54 \times 10^{-5}\ M)(1.0\ \text{cm})} = 1.31 \times 10^{4}$$

(b) $$\epsilon = \frac{0.75}{(0.038)(1.0)} = 20$$

21.17 (a) $n \rightarrow \pi^*$ and $\pi \rightarrow \pi^*$ transitions

(b) $\pi \rightarrow \pi^*$ transition at 219 nm

21.18 (a) $\pi \rightarrow \pi^*$ (b) $\pi \rightarrow \pi^*$ and $n \rightarrow \pi^*$

(c) $\pi \rightarrow \pi^*$ and $n \rightarrow \pi^*$ (d) $\pi \rightarrow \pi^*$ and $n \rightarrow \pi^*$

(All could exhibit $\sigma \rightarrow \pi^*$ transitions also.)

21.19 (a) < (c) < (b) < (d), the same order as that of increasing conjugation.

21.20 A conjugated diene, $CH_2{=}CHCH{=}CHCH_3$ or $CH_2{=}C(CH_3)CH{=}CH_2$

21.21 Because of van der Waals repulsions, *cis*-stilbene cannot be planar; therefore, the degree of conjugation is less and the wavelength of uv absorption is shorter.

21.22

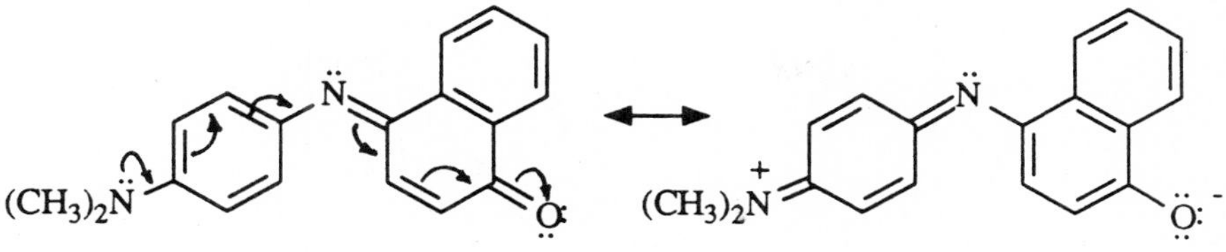

$(CH_3)_2\ddot{N}$ ... $\ddot{N}$ ... $\ddot{O}$ ⟷ $(CH_3)_2\overset{+}{N}$ ... $\ddot{N}$ ... $\ddot{O}:^-$

plus benzenoid resonance structures

21.23 At pH 6, the structure is as shown in the problem. At pH 9:

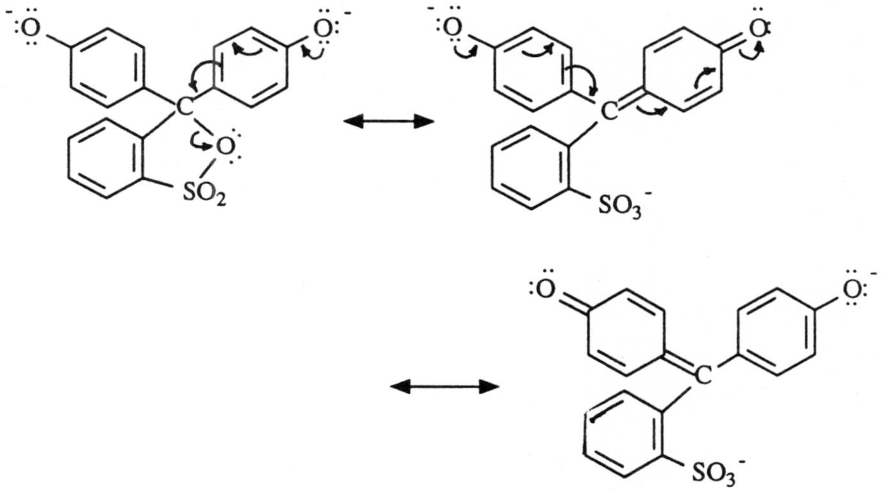

$^-:\ddot{O}$ ... $\ddot{O}:^-$ ... C ... O ... SO_2 ⟷ $^-:\ddot{O}$ ... $\ddot{O}$ ... C ... SO_3^-

⟷ $:\ddot{O}$ ... $\ddot{O}:^-$ ... C ... SO_3^-

21.24 (a) Norrish type I

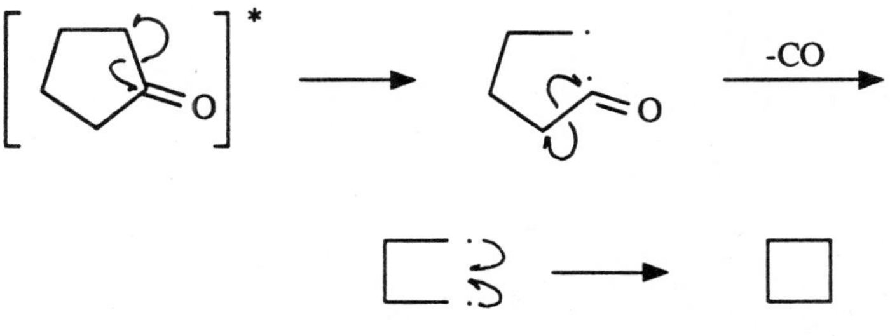

[cyclopentanone]* ⟶ O ⟶ (-CO)

⟶

(b) Norrish type II

$$\left[C_6H_5-\overset{\overset{O}{\parallel}}{C}\underset{CH_2}{\diagdown}\,CH_2-CH_2-H \right]^* \longrightarrow C_6H_5-\overset{OH}{\underset{\cdot}{C}}-CH_2-CH_2-\cdot CH_2$$

$$\longrightarrow \text{(cyclobutanol: } C_6H_5C(OH)-CH_2-CH_2-CH_2 \text{ ring)}$$

$$\xrightarrow{-\,CH_2{=}CH_2} C_6H_5\overset{OH}{\overset{|}{C}}{=}CH_2 \longrightarrow C_6H_5\overset{O}{\overset{\parallel}{C}}CH_3$$

(c) Norrish type I

$$\left[CH_3\overset{O}{\overset{\parallel}{C}}-CH_3 \right]^* \longrightarrow CH_3\overset{O}{\overset{\parallel}{C}}\cdot \;+\; \cdot CH_3$$

$$2\ CH_3\overset{O}{\overset{\parallel}{C}}\cdot \longrightarrow CH_3\overset{O}{\overset{\parallel}{C}}-\overset{O}{\overset{\parallel}{C}}CH_3$$

$$2\ \cdot CH_3 \longrightarrow CH_3CH_3$$

21.25 (a) a $4n$ photo-induced electrocyclic reaction with disrotatory motion ($n = 1$):

$$\text{cycloheptadiene (H, H)} \xrightarrow{h\nu} \text{bicyclo[3.2.0]hept-6-ene (cis H, H)}$$

(b) a photodimerization reaction

2 (2-cyclohexenone) $\xrightarrow{h\nu}$ (head-to-head cyclobutane dimer, O O, H H / H H) + (head-to-tail cyclobutane dimer, O, H H / H H, O)

(c) a photoisomerization reaction

H_5C_6–N=N–C_6H_5 (trans) $\xrightarrow{h\nu}$ H_5C_6–N=N–C_6H_5 (cis)

21.26 (a) $CH_3CH(H)-CH_2\overset{+}{\ddot{O}}H$ $\longrightarrow$ $CH_3\dot{C}H-CH_2(H-\overset{+}{\ddot{O}}H)$ $\xrightarrow{-H_2O}$

$CH_3\dot{C}H\overset{+}{C}H_2$ or $[CH_3CHCH_2]^{+\cdot}$

(b) benzene $\xrightarrow{-e^-}$ (cyclohexadienyl radical cation, H, + —H) or $[\ldots]^{+\cdot}$ (H, —H)

(c) cyclohexyl–$\overset{+}{\ddot{O}}H$ (H) $\xrightarrow{-H\cdot}$ cyclohexylidene=$\overset{+}{\ddot{O}}H$ or $[\text{cyclohexylidene}=OH]^{+}$

(d) $(CH_3)_2CH-\overset{+}{\dot{C}l}:$ $\xrightarrow{-:\dot{C}l:}$ $(CH_3)_2\overset{+}{C}H$ or $[(CH_3)_2CH]^{+}$

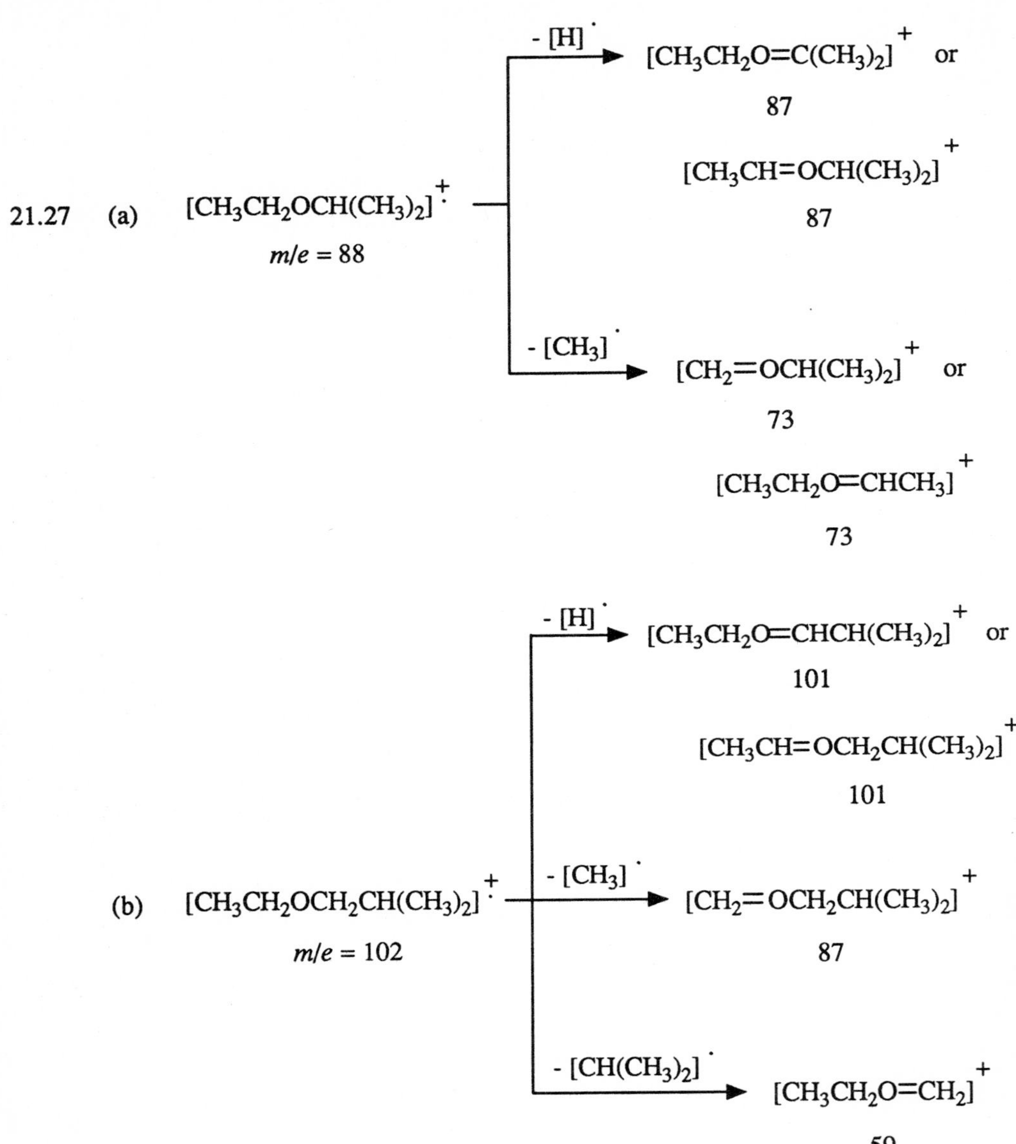
21.27 (a) $[CH_3CH_2OCH(CH_3)_2]^{+\cdot}$
$m/e = 88$
$- [H]^{\cdot}$
$[CH_3CH_2O{=}C(CH_3)_2]^{+}$ or
87
$[CH_3CH{=}OCH(CH_3)_2]^{+}$
87
$- [CH_3]^{\cdot}$
$[CH_2{=}OCH(CH_3)_2]^{+}$ or
73
$[CH_3CH_2O{=}CHCH_3]^{+}$
73
(b) $[CH_3CH_2OCH_2CH(CH_3)_2]^{+\cdot}$
$m/e = 102$
$- [H]^{\cdot}$
$[CH_3CH_2O{=}CHCH(CH_3)_2]^{+}$ or
101
$[CH_3CH{=}OCH_2CH(CH_3)_2]^{+}$
101
$- [CH_3]^{\cdot}$
$[CH_2{=}OCH_2CH(CH_3)_2]^{+}$
87
$- [CH(CH_3)_2]^{\cdot}$
$[CH_3CH_2O{=}CH_2]^{+}$
59

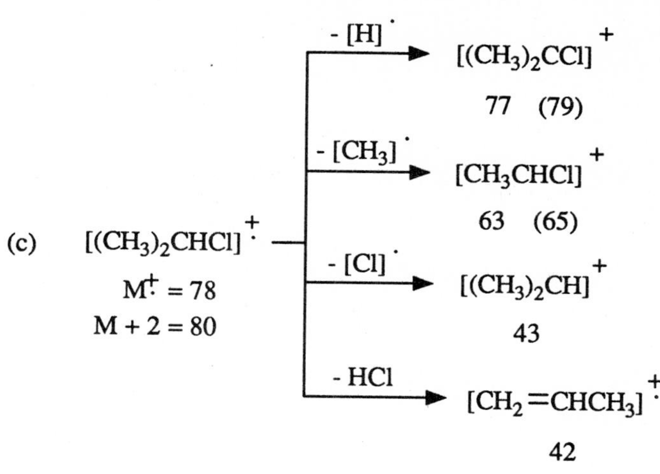

(c)
[(CH3)2CHCl]+·
M+· = 78
M + 2 = 80
- [H]·
[(CH3)2CCl]+
77 (79)
- [CH3]·
[CH3CHCl]+
63 (65)
- [Cl]·
[(CH3)2CH]+
43
- HCl
[CH2=CHCH3]+·
42

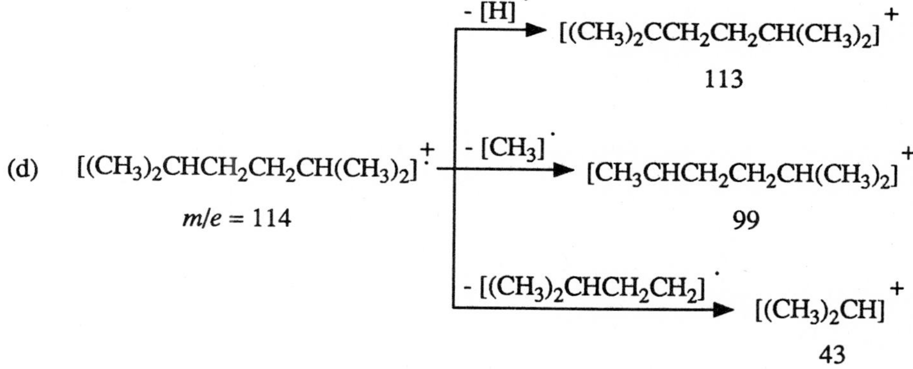

(d)
[(CH3)2CHCH2CH2CH(CH3)2]+·
m/e = 114
- [H]·
[(CH3)2CCH2CH2CH(CH3)2]+
113
- [CH3]·
[CH3CHCH2CH2CH(CH3)2]+
99
- [(CH3)2CHCH2CH2]·
[(CH3)2CH]+
43

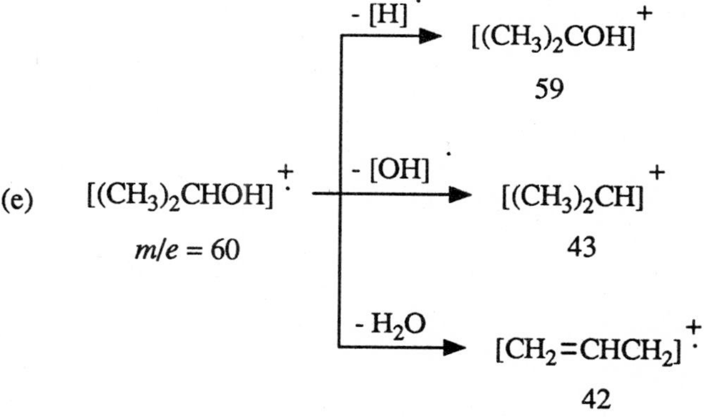

(e)
[(CH3)2CHOH]+·
m/e = 60
- [H]·
[(CH3)2COH]+
59
- [OH]·
[(CH3)2CH]+
43
- H2O
[CH2=CHCH2]+·
42

(f) $\left[\text{C}_5\text{H}_9\text{–CH(H)–CH}_2\text{–CH}_2\text{–CH=O} \right]^{+\cdot}$ $\xrightarrow{-\ \text{C}_5\text{H}_9\text{–CH=CH}_2}$ $\left[\text{CH}_2\text{=CH–OH} \right]^{+\cdot}$

$m/e = 140$ → 44

21.28 (a) $[CH_3CH_2CH_2CH_3]^{+\cdot}$ (58) ⟶ $[CH_3CH_2CH_2CH_2]^{+}$ (57) ⟶

$[CH_3CH_2CH_2]^{+}$ (43) ⟶ $[CH_3CH_2]^{+}$ (29) ⟶ $[CH_3]^{+}$ (15)

(b) $[C_6H_5C(=O)NH_2]^{+\cdot}$ (121) ⟶ $[C_6H_5C{=}O]^{+}$ (105) ⟶ $[C_6H_5]^{+}$ (77)

(c) $[CH_3CH_2CH_2Br]^{+\cdot}$ ⟶ $[CH_3CH_2CH_2]^{+}$ (43) ⟶

$M^{+\cdot} = 122$

$M + 2 = 124$

$[CH_3CH_2]^{+}$ (29) ⟶ $[CH_3]^{+}$ (15)

(d) $[C_6H_5CH_2OCH_3]^{+\cdot}$ ⟶ $[C_6H_5CHOCH_3]^+$

$m/e = 122$ ⟶ 121

⟶ $[C_6H_5]^+$ 77

⟶ $[C_6H_5CH_2]^+$ 91

It has been shown that the benzyl cation undergoes rearrangement to the tropylium ion:

$$C_6H_5\overset{+}{C}H_2 \longrightarrow C_7H_7^+ \text{ (cycloheptatrienyl cation, } \overset{+}{C}\text{–H) or (ring with } + \text{)}$$

(e) $\left[(CH_3)_2CHCH_2CH_2\overset{O}{\overset{\|}{C}}CH_3\right]^{+\cdot}$ ⟶ $[(CH_3)_2CHCH_2CH_2]^+$

$m/e = 114$ (not observed) ⟶ 71

⟶ $[CH_3C{=}O]^+$ 43

⟶ $[CH_2{=}\overset{OH}{\overset{|}{C}}CH_3]^{+\cdot}$ or $[CH_2\overset{OH}{\overset{\|}{C}}CH_3]^{+\cdot}$

58 58

In (e), the following McLafferty rearrangement occurs:

$$\left[(CH_3)_2C(H)\text{–}CH_2\text{–}CH_2\text{–}C(=O)\text{–}CH_3\right]^{+\cdot} \longrightarrow (CH_3)_2C{=}CH_2 + \left[CH_2{=}C(OH)\text{–}CH_3\right]^{+\cdot}$$

21.29 $H_2N-C_6H_4-C_6H_4-NH_2 \xrightarrow[0°]{NaNO_2,\ HCl}$

$Cl^- \ {}^+N_2-C_6H_4-C_6H_4-N_2^+ \ Cl^-$ $\xrightarrow{2\ \text{(o-HOC}_6H_4CO_2H)}$

OH, 2, CO_2H

21.30 $C_6H_6 \xrightarrow[H_2SO_4]{HNO_3} C_6H_5NO_2 \xrightarrow[(2)\ OH^-]{(1)\ Fe,\ HCl} C_6H_5NH_2 \xrightarrow{(CH_3C(=O))_2O}$

$C_6H_5NHC(=O)CH_3 \xrightarrow{ClSO_3H} HO_3S-C_6H_4-NHC(=O)CH_3 \xrightarrow[(2)\ OH^-]{(1)\ H_2O,\ H^+}$

$HO_3S-C_6H_4-NH_2 \xrightarrow[0°]{NaNO_2,\ HCl} HO_3S-C_6H_4-N_2^+$

as a dipolar ion

$\longrightarrow HO_3S-C_6H_4-N=N-C_6H_4-N(CH_3)_2$

methyl orange

$C_6H_5NH_2 \xrightarrow[(2)\ OH^-]{(1)\ CH_3I} \xrightarrow[(2)\ OH^-]{(1)\ CH_3I} C_6H_5N(CH_3)_2$

21.31

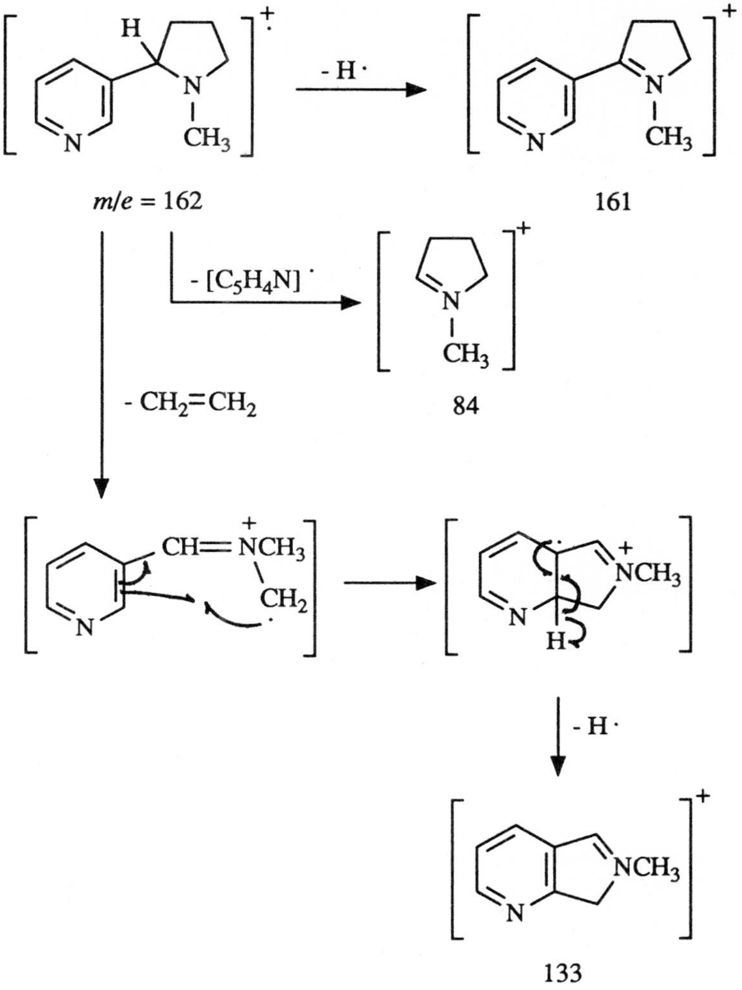

[Reaction scheme: molecular ion $m/e = 162$ loses H· to give ion 161; loses $[C_5H_4N]^{\cdot}$ to give ion 84; loses $CH_2{=}CH_2$ to give $[\text{pyridyl}-CH{=}\overset{+}{N}CH_3-CH_2^{\cdot}]$, which cyclizes and loses H· to give ion 133.]

$m/e = 162$ 161

- H·

- $[C_5H_4N]^{\cdot}$

84

- $CH_2{=}CH_2$

$CH{=}\overset{+}{N}CH_3$

CH_2

$\overset{+}{N}CH_3$

- H·

NCH_3

133

21.32 The phosphorescence would have a lower frequency (lower energy) than fluorescence. Energy is lost in intersystem crossing and, consequently, the triplet states are of lower energy than their corresponding singlet states.

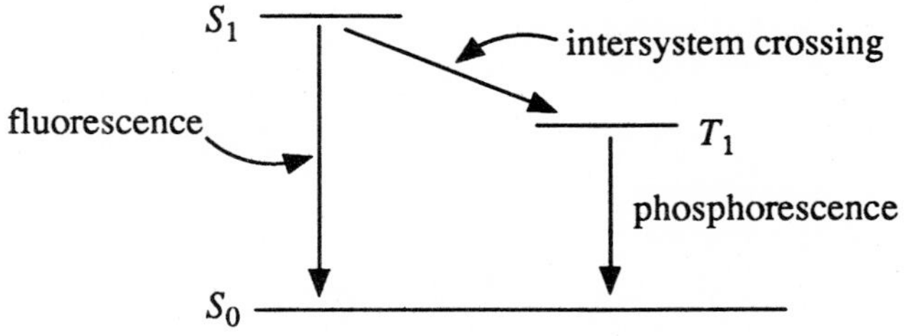

S_1

intersystem crossing

fluorescence

T_1

phosphorescence

S_0

21.33 The addition of each alkyl substituent to the conjugated system adds about 5 nm to the λ_{max}.

21.34 (a) 217 + 15 = 232 nm (b) 217 + 30 + 15 = 262 nm

21.35 (a) 229 nm (b) 234 nm (c) 224 + 30 - 5 = 249 nm

21.36 (b), (c), (d). All are methyl ketones and yield acetyl fragments (43), $[C_4H_9]^{+\cdot}$ (57), pentanoyl fragments (85), and a molecular ion (100).

21.37 $CH_3\overset{O}{\overset{\|}{C}}CH_3$

21.38 A, $ClCH_2CH_2CH_2OH$ B, $H_3C-C_6H_4-CN$ (para) C, C_6H_5I

21.39 From the mass spectrum, we determine that the formula weight is 156. The infrared spectrum shows -OH absorption. Since the compound is a terpenoid, it must contain ten carbons. The closest molecular formula is $C_{10}H_{20}O$ (MW = 156). A $C_{10}H_{20}O$ compound (unsaturation number = 1) can have one ring or one double bond. Because a downfield alkene absorption is observed in the proton nmr spectrum, the compound must be open chain and contain a double bond.

We determine the carbon skeleton using the fact that the compound is a terpenoid. A ten-carbon terpene or terpenoid contains two isoprene units joined together in a head-to-tail manner.

The position of the hydroxyl group is determined by the downfield absorption at about δ3.8. This absorption is a triplet and, therefore, consistent with a $-CH_2\underline{CH_2}OH$ group.

Possible structures are:

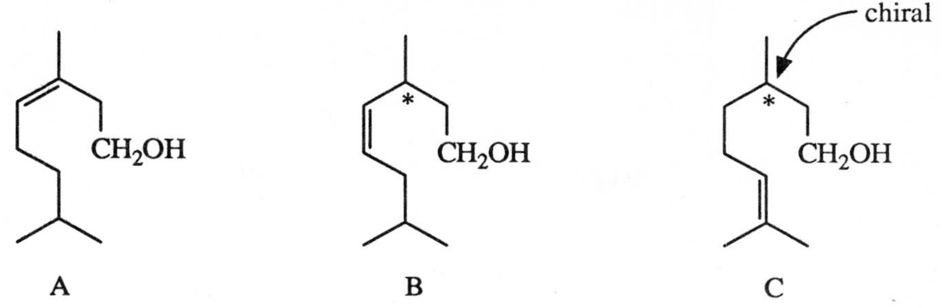

Of these possibilities, the structure is assigned to C. It is the only structure with both a chiral carbon and proton nmr alkene absorption to agree with a triplet in the alkene region (δ5-6).

$$—CH_2C\underline{H}{=}C(R)(R)$$

R, R: not protons

Chapter 22

Carbohydrates

22.20

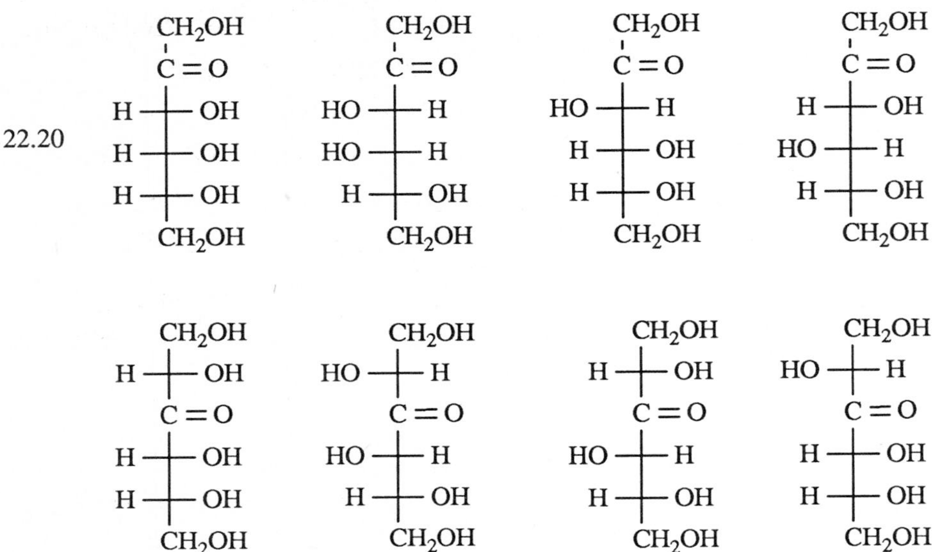

Each cross (—|—) represents a chiral carbon.

In Fischer projections, the keto group should be in position 2 or 3, toward the top of the projection. If the carbonyl group is placed in position 4 or 5, the formula will either be identical to one of the preceding formulas or be a member of the L series.

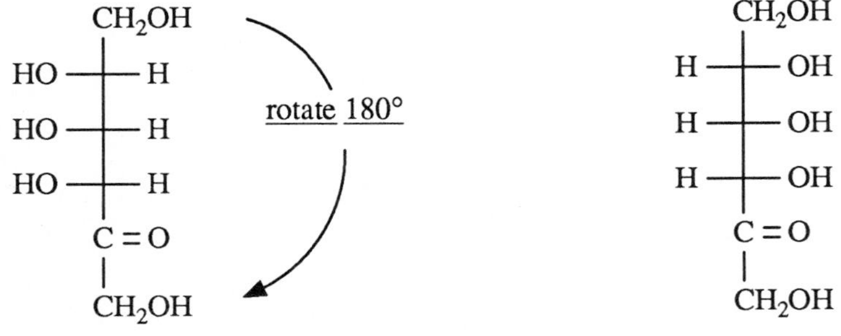

identical to the first formula shown

a member of the L series

22.21 (a) A four-carbon aldehyde monosaccharide is an aldotetrose.

(b) A three-carbon ketone monosaccharide is a ketotriose or triulose.

(c) A five-carbon aldehyde monosaccharide is an aldopentose.

22.22 (a) belongs to the D series.

(b) belongs to neither series because it contains no chiral carbons.

(c) belongs to the L series.

22.23 (a)

CHO; H—C—OH; CH_3 $\xrightarrow{HCN}$ ① CN; ② H—C—OH; ③ H—C—OH; ④ CH_3 + CN; HO—C—H; H—C—OH; CH_3

(2*S*,3*R*) (2*R*,3*R*)

(b) Yes. Although carbon 2 in the product is an equal mixture of (*R*) and (*S*), carbon 3 is (*R*) in each case. The compounds are diastereomers.

22.24

(*R*) → CHO; H—C—OH; CH_2OH $\xrightarrow{HCN}$ $\xrightarrow[H^+]{H_2O}$ $\xrightarrow[(2)\ H_2O,\ H^+]{(1)\ LiAlH_4}$

D-glyceraldehyde

CH_2OH; H—C—OH; H—C—OH; CH_2OH + ① CH_2OH; ② HO—C—H; ③ H—C—OH; ④ CH_2OH

meso (2*R*,3*R*)

The product mixture is a mixture of diastereomers and thus can be separated by physical means. Each isomer can then be compared with the unknown tetraol. The unknown tetraol is either *meso* (has no optical rotation); (2*R*,3*R*) (same optical rotation as the tetraol synthesized from D-glyceraldehyde); or (2*S*,3*S*) opposite rotation).

22.25 (a) (4) (b) (3) (c) (1) (d) (2)

An *-ose* ending refers to an aldehyde or hemiacetal (or ketone or hemiketal), while an *-oside* ending refers to a glycoside (an acetal or ketal).

22.26 all D

22.27 (1) The formula in Problem 22.25 (1) is the β-diastereomer. Its anomer would be α.

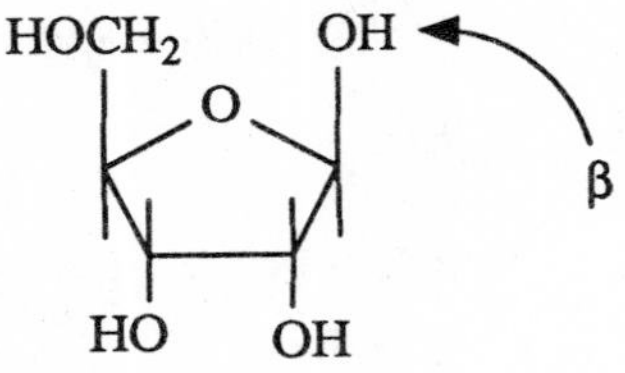

β

formula in 22.25 (1)

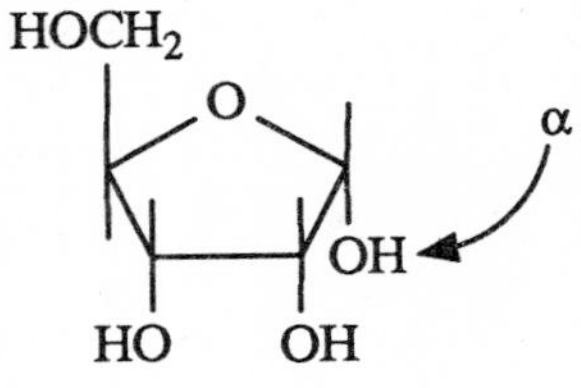

α

the α-anomer

(4)

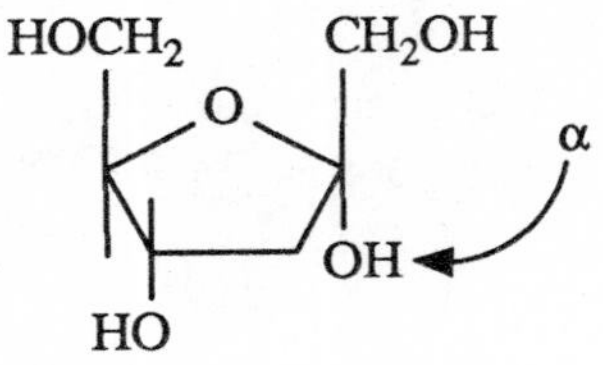

α

formula in 22.25 (4)

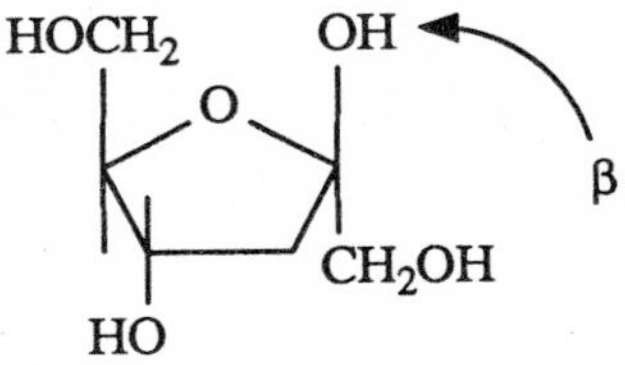

β

the β-anomer

22.28 (a)

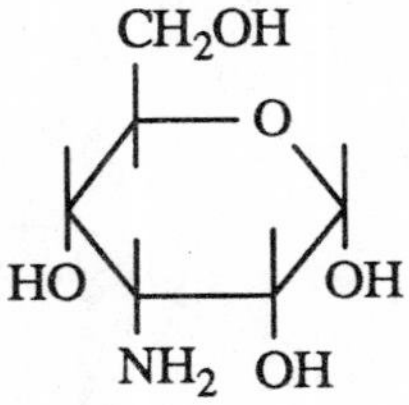

(b)

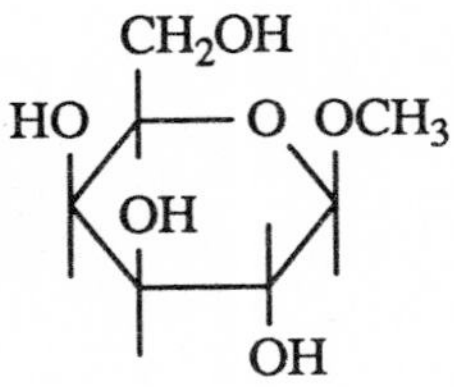

(c)

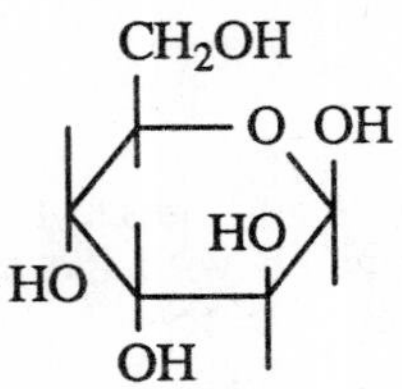

(d)

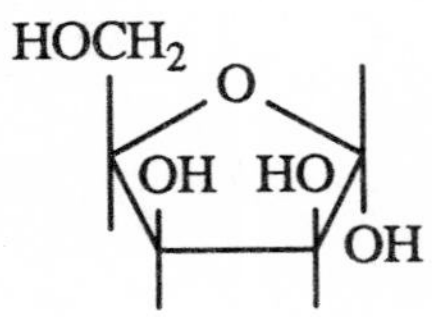

The Haworth formulas of (c) and (d) are easily determined. Compare the Fischer projections to that of D-glucose and position the hydroxyl groups accordingly.

22.29 (a)

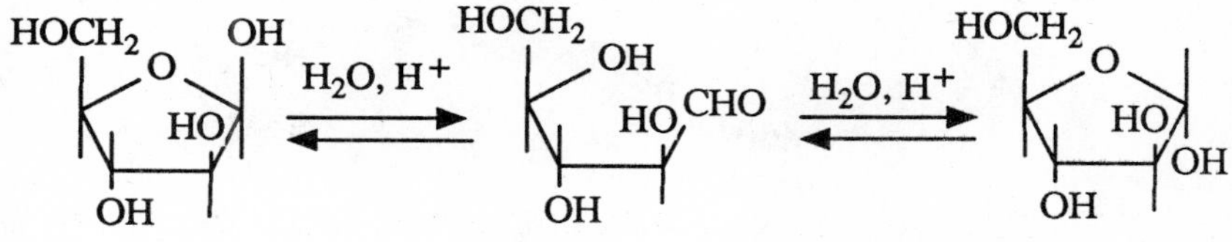

HOCH$_2$ O OH HO OH $\xrightarrow[]{H_2O,\ H^+}$ (reversible) HOCH$_2$ OH HO CHO OH $\xrightarrow[]{H_2O,\ H^+}$ (reversible) HOCH$_2$ O HO OH OH

β α

(b)

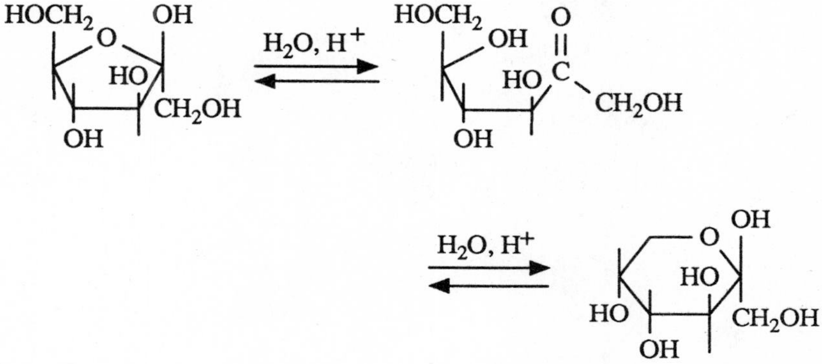

HOCH$_2$ O OH HO CH$_2$OH OH $\xrightarrow[]{H_2O,\ H^+}$ (reversible) HOCH$_2$ OH O HO C CH$_2$OH OH

$\xrightarrow[]{H_2O,\ H^+}$ (reversible) OH O HO HO CH$_2$OH OH

(c)

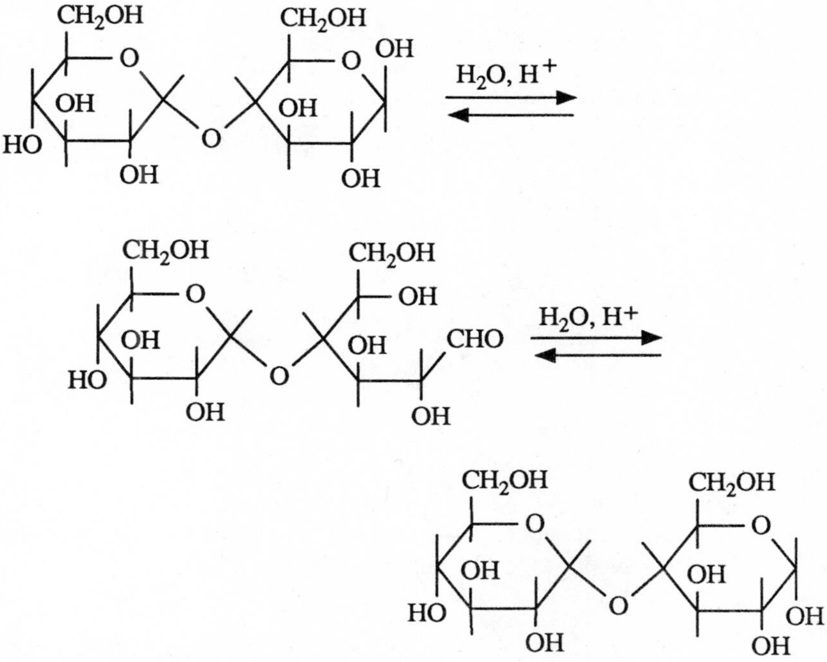

CH$_2$OH O OH HO OH O CH$_2$OH O OH OH OH $\xrightarrow[]{H_2O,\ H^+}$ (reversible)

CH$_2$OH O OH HO OH O CH$_2$OH OH OH CHO OH $\xrightarrow[]{H_2O,\ H^+}$ (reversible)

CH$_2$OH O OH HO OH O CH$_2$OH O OH OH OH

22.30 (a) 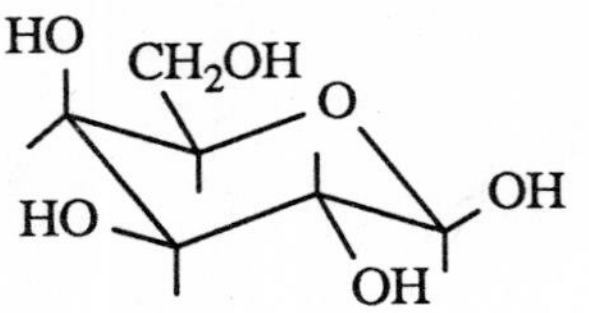 (b)

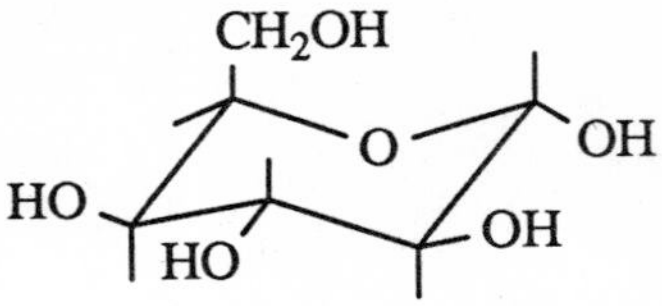

In each case, the most stable conformation is the one in which the greatest number of groups are equatorial.

22.31 (a) 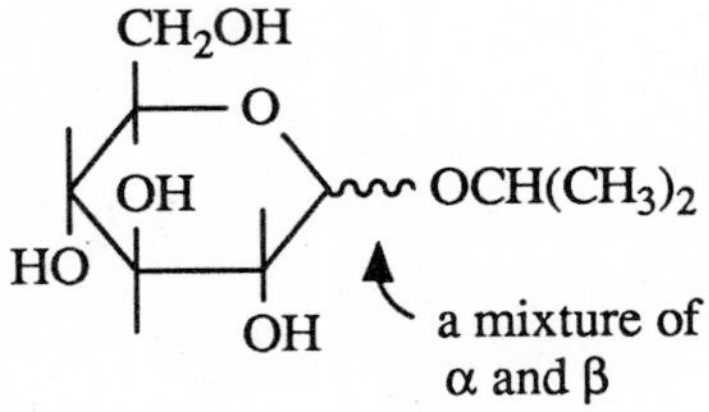 $\sim OCH(CH_3)_2$ — a mixture of α and β (b) 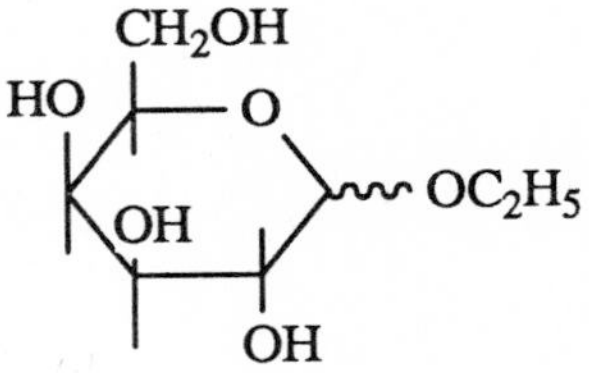 $\sim OC_2H_5$

(c) 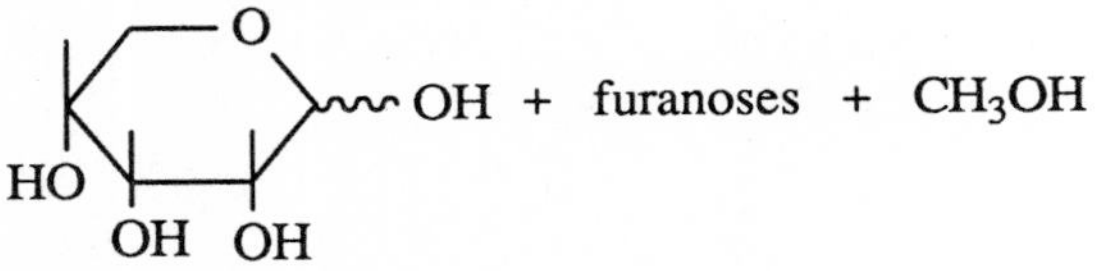 $\sim OH$ + furanoses + CH_3OH

22.32 Tollens reagent contains $Ag(NH_3)_2^+$ and OH^- ions. In base, D-fructose can be converted to D-glucose and D-mannose by way of an enediol intermediate. (These two sugars differ in configuration only at carbon 2.) Oxidation of the aldehyde groups in these sugars therefore yields aldonic acids of both sugars.

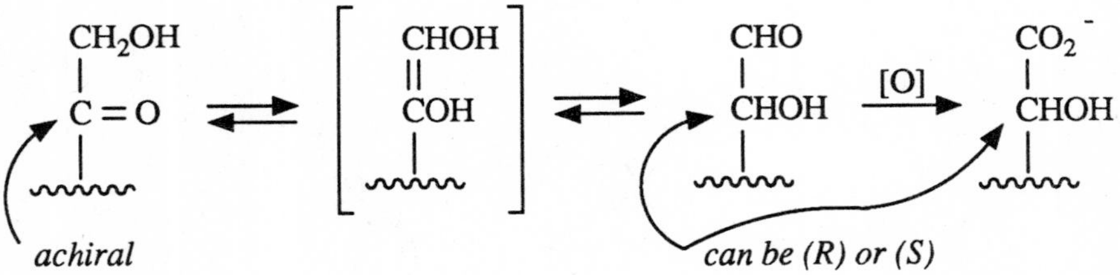

$$CH_2OH{-}C{=}O \rightleftharpoons [CHOH{=}COH] \rightleftharpoons CHO{-}CHOH \xrightarrow{[O]} CO_2^-{-}CHOH$$

achiral *can be (R) or (S)*

22.33 (a), (c), (d), and (f), because none of these compounds contains a carbonyl or hemiacetal group; all are glycosides, which are not in equilibrium with the carbonyl compounds or the anomers. (b) and (e) contain hemiacetal groups at carbon 1 and thus can undergo mutarotation.

22.34 (a), (c), (d), and (f), the same answers as for 22.33, because a carbonyl or hemiacetal group (not an acetal or glycoside group) is required for a sugar to be reducing sugar.

22.35 (a)

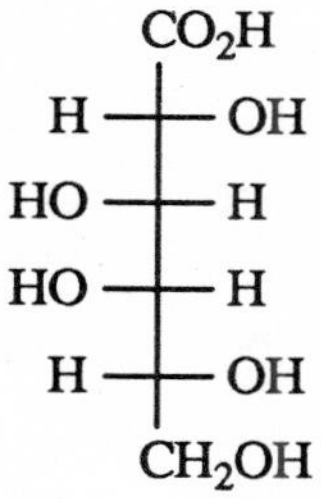

$$\begin{array}{c} CO_2H \\ H-C-OH \\ HO-C-H \\ HO-C-H \\ H-C-OH \\ CH_2OH \end{array}$$

D-galactonic acid

or (lactone, Fischer projection) or (lactone, ring structure)

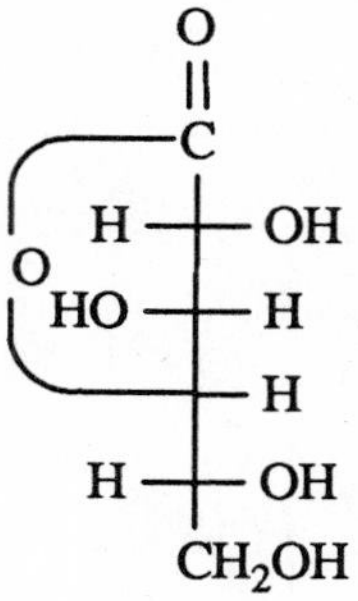

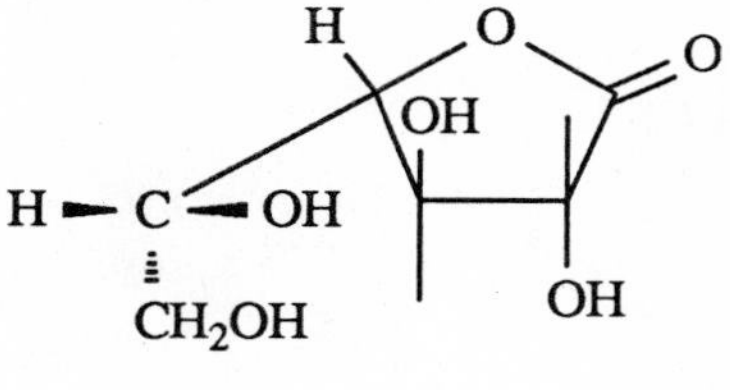

a lactone (Use models.)

(b)

$$\begin{array}{c} CO_2H \\ H-C-OH \\ HO-C-H \\ HO-C-H \\ H-C-OH \\ CO_2H \end{array}$$

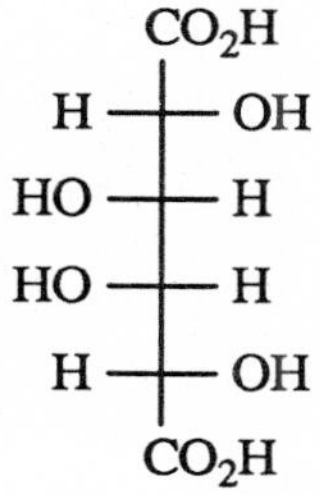

D-galactaric acid or lactones

(c)

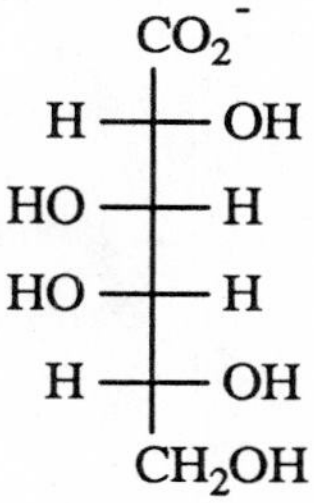

$$\begin{array}{c} CO_2^- \\ H-C-OH \\ HO-C-H \\ HO-C-H \\ H-C-OH \\ CH_2OH \end{array}$$

D-galactonate ion

(d)

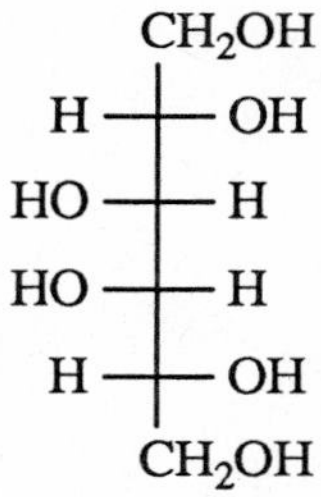

$$\begin{array}{c} CH_2OH \\ H-C-OH \\ HO-C-H \\ HO-C-H \\ H-C-OH \\ CH_2OH \end{array}$$

galactitol

22.36

$$\begin{array}{c} CHO \\ HO-C-H \\ HO-C-H \\ H-C-OH \\ H-C-OH \\ CO_2H \end{array}$$

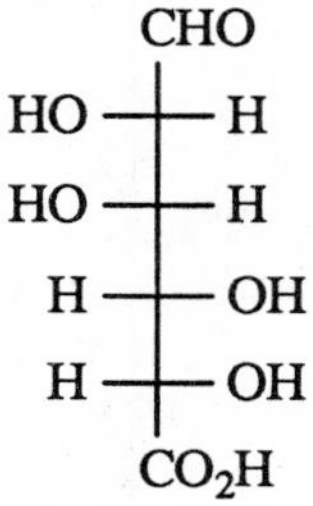

which would exist as a lactone: (lactone, Fischer projection) or (lactone, ring structure)

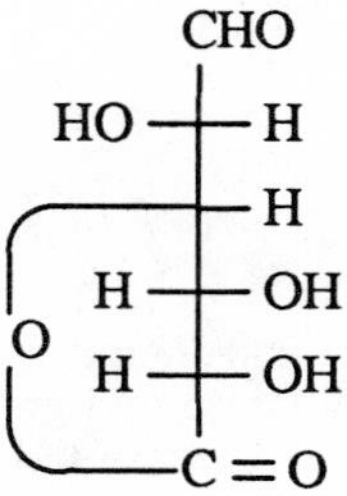

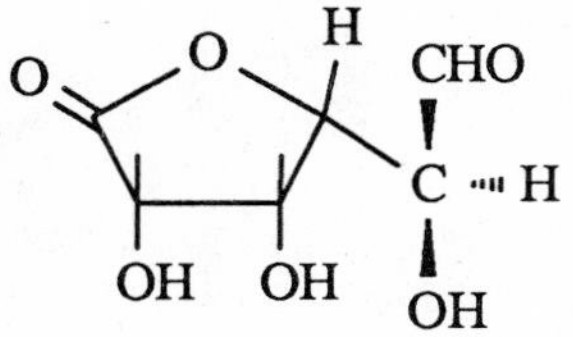

22.37 (a)

$$\begin{array}{c} CHO \\ H-\!\!\!-\!\!\!\!\!\!\!\!|\!\!-\!\!\!-OH \\ H-\!\!\!-\!\!\!\!\!\!\!\!|\!\!-\!\!\!-OH \\ CH_2OH \end{array} \xrightarrow{\text{hot } HNO_3} \begin{array}{c} CO_2H \\ H-\!\!\!-\!\!\!\!\!\!\!\!|\!\!-\!\!\!-OH \\ H-\!\!\!-\!\!\!\!\!\!\!\!|\!\!-\!\!\!-OH \\ CO_2H \end{array}$$

D-erythrose (the D-tetrose) → *meso*-tartaric acid

(b)

$$\begin{array}{c} CO_2^- \\ H-\!\!\!-\!\!\!\!\!\!\!\!|\!\!-\!\!\!-OH \\ H-\!\!\!-\!\!\!\!\!\!\!\!|\!\!-\!\!\!-OH \\ H-\!\!\!-\!\!\!\!\!\!\!\!|\!\!-\!\!\!-OH \\ CH_2OH \end{array}$$

D-ribonic acid

(c)

$$\begin{array}{c} CHO \\ H-\!\!\!-\!\!\!\!\!\!\!\!|\!\!-\!\!\!-OH \\ H-\!\!\!-\!\!\!\!\!\!\!\!|\!\!-\!\!\!-OH \\ H-\!\!\!-\!\!\!\!\!\!\!\!|\!\!-\!\!\!-OH \\ H-\!\!\!-\!\!\!\!\!\!\!\!|\!\!-\!\!\!-OH \\ CH_2OH \end{array} \text{ or } \begin{array}{c} CHO \\ H-\!\!\!-\!\!\!\!\!\!\!\!|\!\!-\!\!\!-OH \\ HO-\!\!\!-\!\!\!\!\!\!\!\!|\!\!-\!\!\!-H \\ HO-\!\!\!-\!\!\!\!\!\!\!\!|\!\!-\!\!\!-H \\ H-\!\!\!-\!\!\!\!\!\!\!\!|\!\!-\!\!\!-OH \\ CH_2OH \end{array} \xrightarrow[\text{(2) } H_2O,\ H^+]{\text{(1) } NaBH_4}$$

D-allose or D-galactose

$$\begin{array}{c} CH_2OH \\ H-\!\!\!-\!\!\!\!\!\!\!\!|\!\!-\!\!\!-OH \\ H-\!\!\!-\!\!\!\!\!\!\!\!|\!\!-\!\!\!-OH \\ H-\!\!\!-\!\!\!\!\!\!\!\!|\!\!-\!\!\!-OH \\ H-\!\!\!-\!\!\!\!\!\!\!\!|\!\!-\!\!\!-OH \\ CH_2OH \end{array} \text{ or } \begin{array}{c} CH_2OH \\ H-\!\!\!-\!\!\!\!\!\!\!\!|\!\!-\!\!\!-OH \\ HO-\!\!\!-\!\!\!\!\!\!\!\!|\!\!-\!\!\!-H \\ HO-\!\!\!-\!\!\!\!\!\!\!\!|\!\!-\!\!\!-H \\ H-\!\!\!-\!\!\!\!\!\!\!\!|\!\!-\!\!\!-OH \\ CH_2OH \end{array}$$

D-allitol or D-galactitol

Both are *meso* alditols.

22.38 Both five-membered and six-membered lactones can be formed. Because D-glucaric acid contains two carboxyl groups, a total of four simple lactones could be formed. (Dilactones are also possible.)

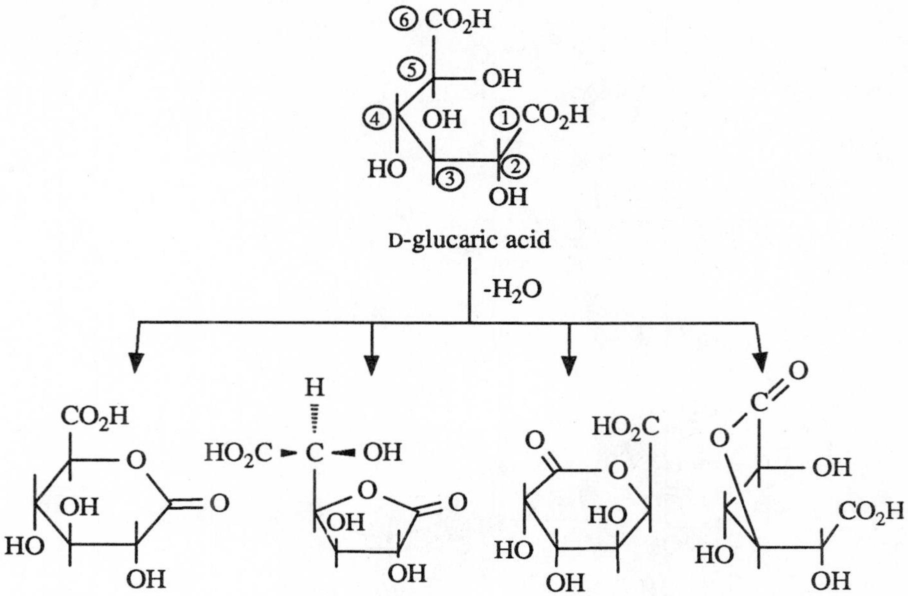

⑥ CO_2H

⑤ OH

④ OH ① CO_2H

HO ③ ② OH

D-glucaric acid

$-H_2O$

carboxyl 1, hydroxyl 5

carboxyl 1, hydroxyl 4

carboxyl 6, hydroxyl 2

carboxyl 6, hydroxyl 3

22.39 (a) $CH_3C(=O)-CH(OH)CH_2CH_3$ — to $-CO_2H$; to -CHO

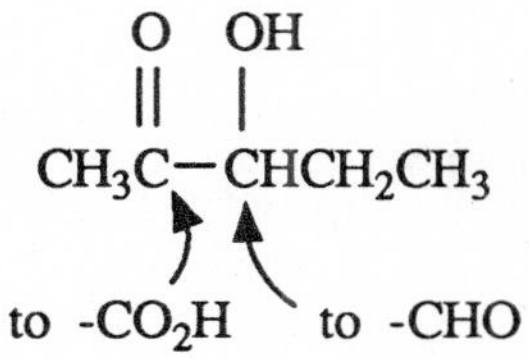

(b) 1-methylcyclopentane-1,2-diol — CH_3, OH: to ketone; OH: to -CHO

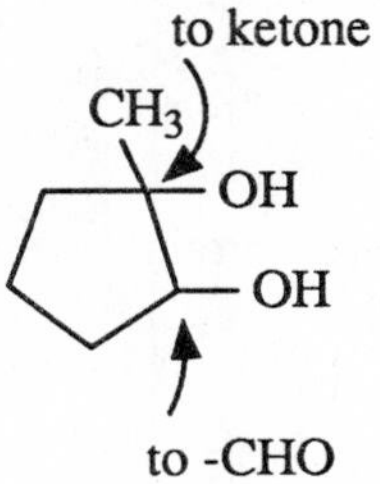

(c) $HOCH_2-C(OH)(CH_3)_2$ — to HCHO; to ketone

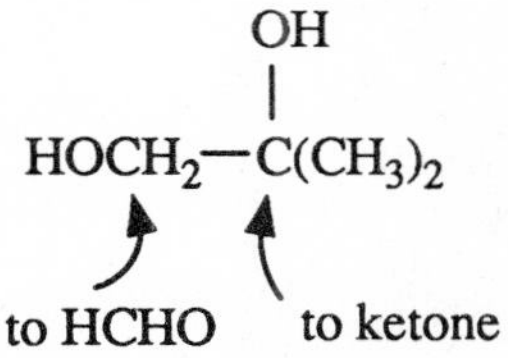

To solve this type of problem, rewrite the formulas of the products so that the carbonyl groups are next to each other. These carbonyl carbons would be bonded together in the reactant. Using (b) as an example:

$CH_3C(=O)CH_2CH_2CH_2C(=O)H \xleftarrow{HIO_4}$ 1-methylcyclopentane-1,2-diol (CH_3, OH, OH, H)

These carbon atoms are bonded together in the reactant.

22.40

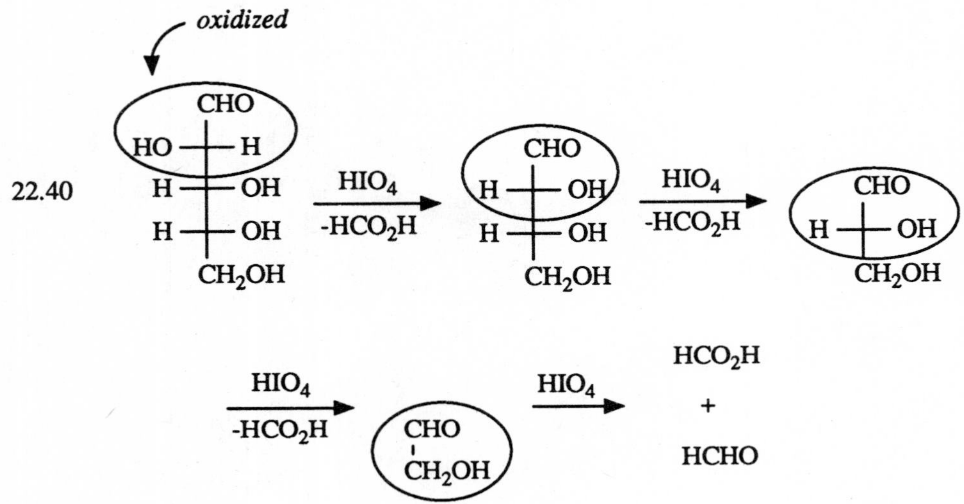

The net products would be 4 HCO_2H + 1 HCHO

22.41 (a) furanose ($HOCH_2$, OH, HO, OH) + excess $C_2H_5OSO_2OC_2H_5$ + excess OH^-

$\longrightarrow$ furanoside ($C_2H_5OCH_2$, OC_2H_5, C_2H_5O, OC_2H_5) + SO_4^{2-} + H_2O

(b) [furanose structure: $HOCH_2$, OH, HO, OH] + 4 $(CH_3\overset{O}{\overset{||}{C}})_2O$ ⟶

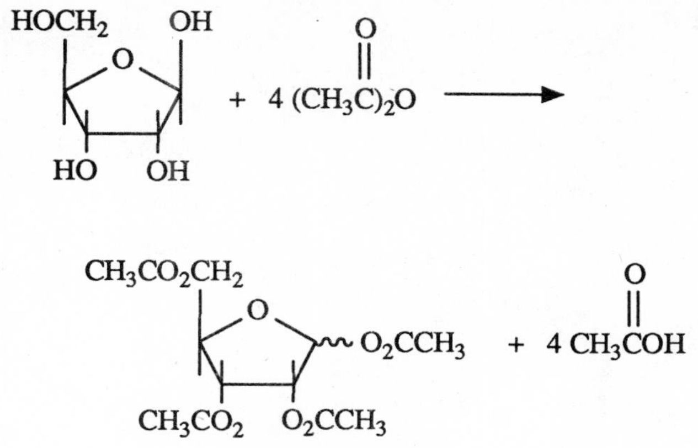

[furanose structure: $CH_3CO_2CH_2$, O_2CCH_3, CH_3CO_2, O_2CCH_3] + 4 $CH_3\overset{O}{\overset{||}{C}}OH$

22.42 [pyranose structure: CH_2OH, HO, O, OCH_3, HO, OH] $\xrightarrow{(1)}$ [pyranose structure: CH_2OCH_3, CH_3O, O, OCH_3, CH_3O, OCH_3] $\xrightarrow{(2)}$

[Fischer projection: CHO; H—OCH_3; H—OCH_3; CH_3O—H; H—OH; CH_2OCH_3] $\xrightarrow{(3)}$ [Fischer projection: CO_2H; H—OCH_3; H—OCH_3; CH_3O—H; CO_2H]

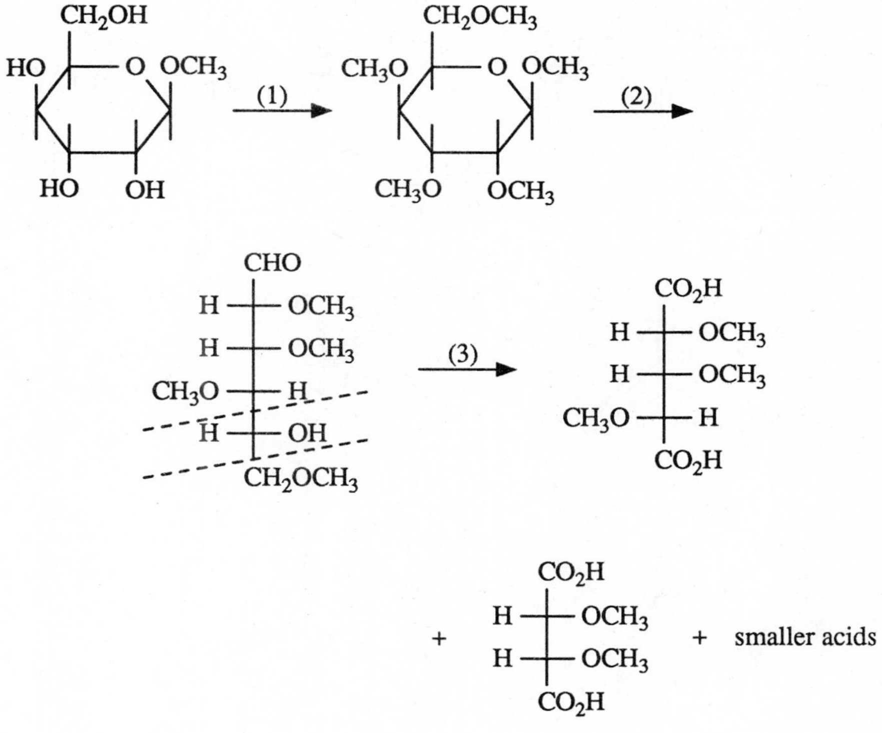

+ [Fischer projection: CO_2H; H—OCH_3; H—OCH_3; CO_2H] + smaller acids

22.43 [pyranose chair structure: OH, CH_2OH, O, HO, HO, OH] + $C_6H_5\overset{O}{\overset{||}{C}}H$ $\overset{H^+}{\rightleftharpoons}$

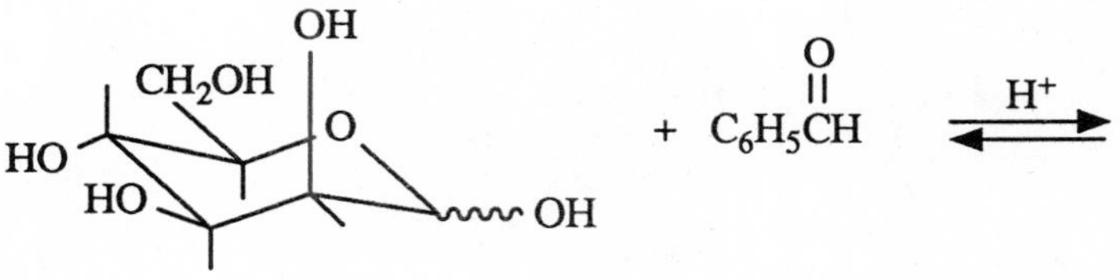

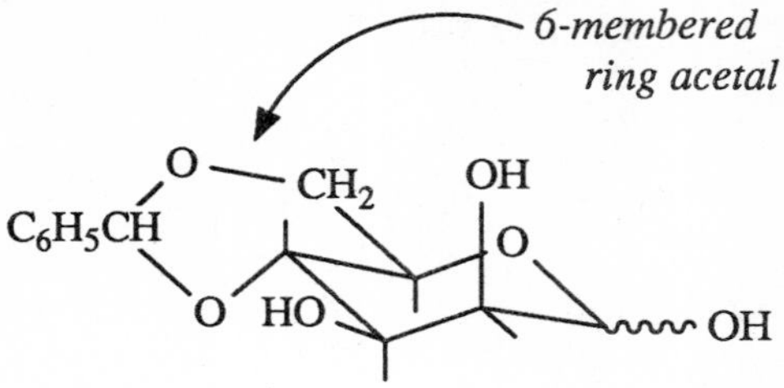

6-membered ring acetal

C_6H_5CH O — CH_2 OH O O HO OH

22.44 (a)

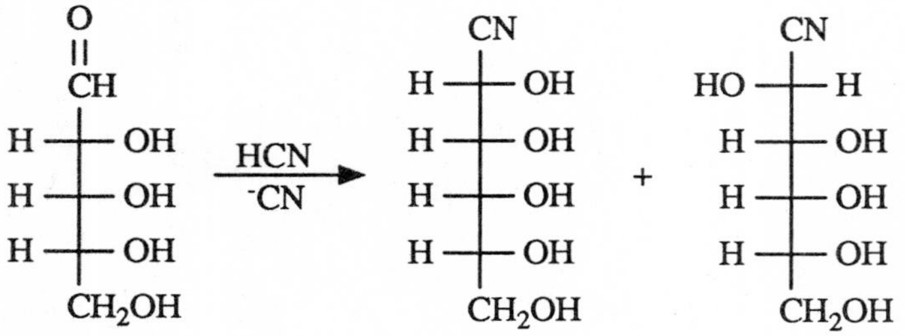

D-ribose: CHO; H–OH; H–OH; H–OH; CH_2OH

$\xrightarrow[^{-}CN]{HCN}$

CN; H–OH; H–OH; H–OH; H–OH; CH_2OH + CN; HO–H; H–OH; H–OH; H–OH; CH_2OH

$\xrightarrow[(2)\ H_2O,\ H^+]{(1)\ \text{separate}}$ CO_2H; H–OH; H–OH; H–OH; H–OH; CH_2OH $\xrightarrow[CO_2]{Na(Hg)}$ CHO; H–OH; H–OH; H–OH; H–OH; CH_2OH

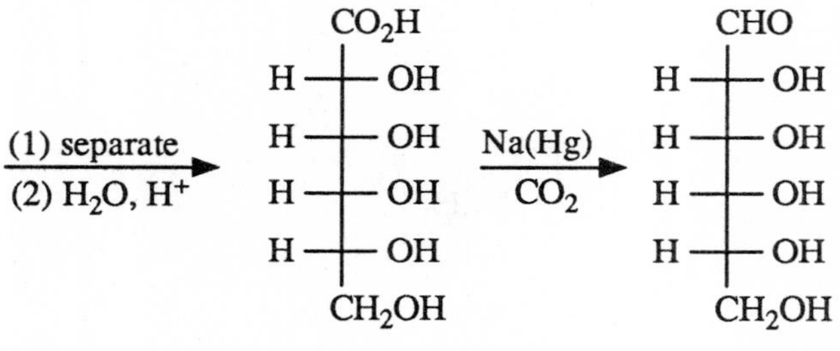

D-allose

(b) Oxidation of D-allose with hot nitric acid would yield a *meso* aldaric acid, while oxidation of D-altrose would yield an optically active acid. (Reduction to an alditol could similarly be used.)

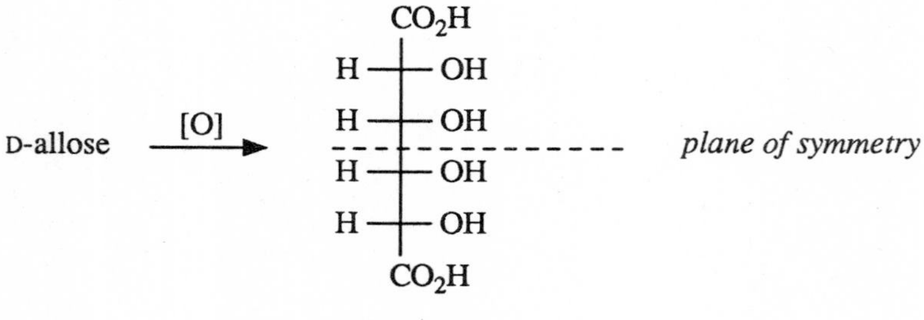

D-allose $\xrightarrow{[O]}$ CO_2H; H–OH; H–OH; H–OH; H–OH; CO_2H — *plane of symmetry*

meso

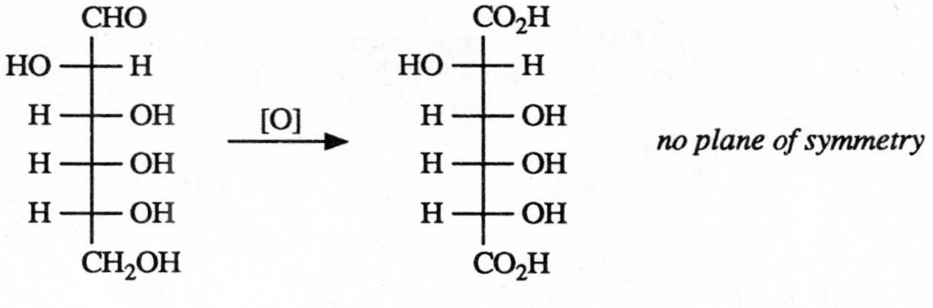

D-altrose → [O] → no plane of symmetry

22.45 excess CH_3-OSO_3CH_3, NaOH → ; H_2O, H^+, heat →

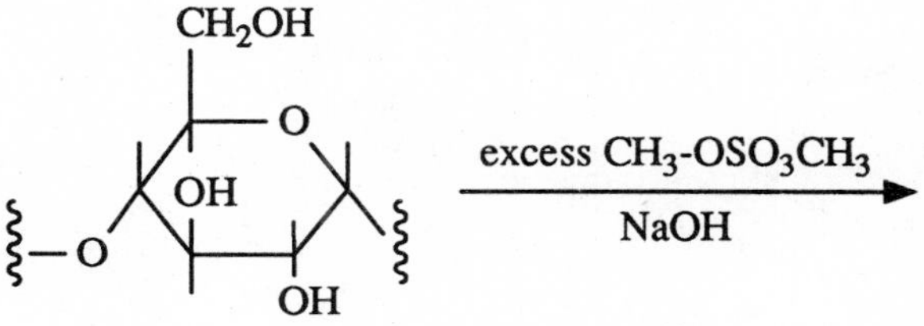

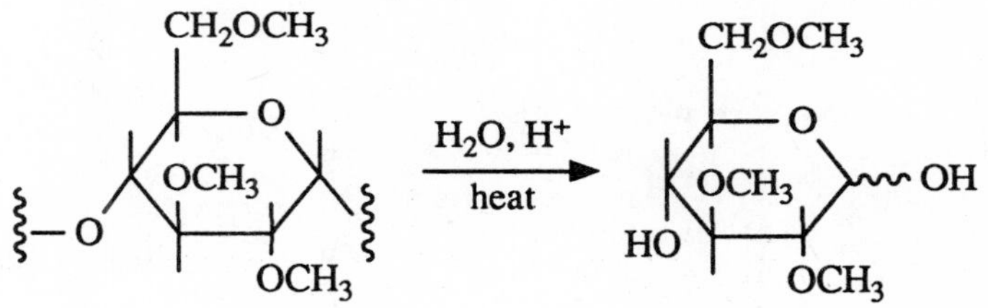

22.46 eleven: α,α-1,1′; α,β-1,1′; β,β-1,1′; α-1,2′; β-1,2′; α-1,3′; β-1,3′; α-1,4′; β-1,4′; α-1,6′; β-1,6′. Two examples follow:

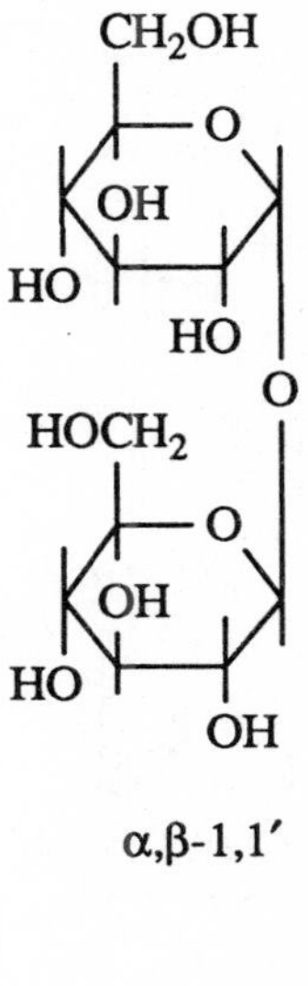

α,β-1,1′

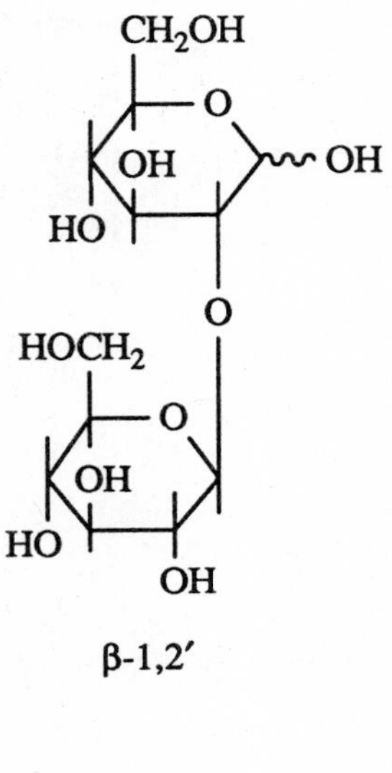

β-1,2′

22.47 α,α′-, α,β′-, or β,β′-1,1′-D-glucopyranoside. (Because trehalose is nonreducing, we know that carbon 1 of the first unit is joined to carbon 1 of the second unit.)

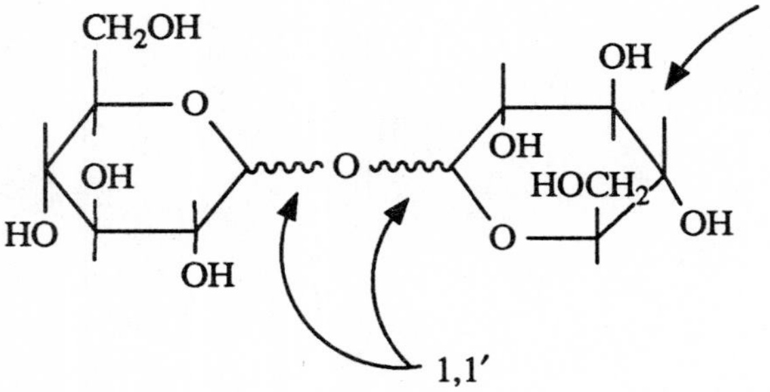

a D-glucopyranose ring rotated 180° from the conventional formula to show the 1,1′ link

22.48 α and β maltose and

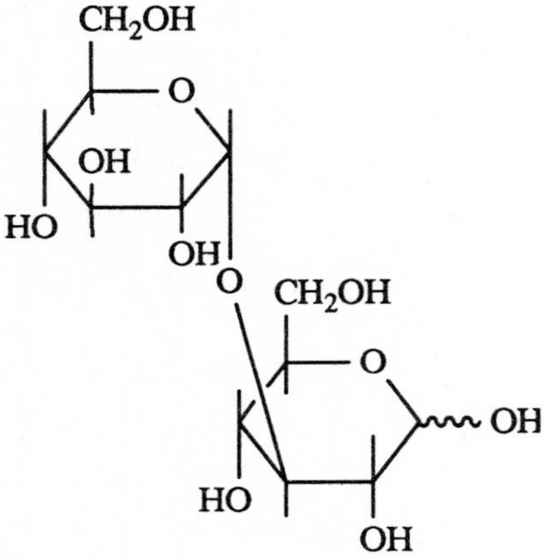

22.49 (a)

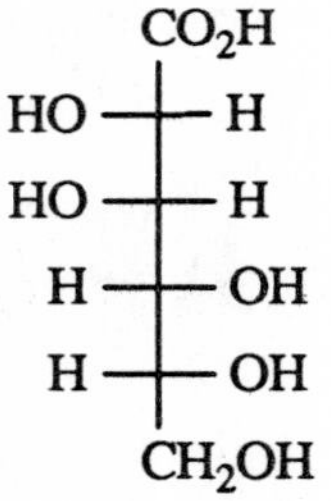

or lactone

(b)

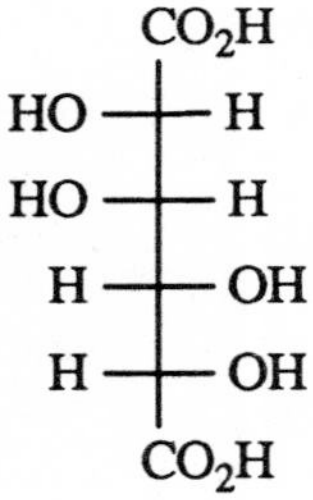

or lactones

(c)

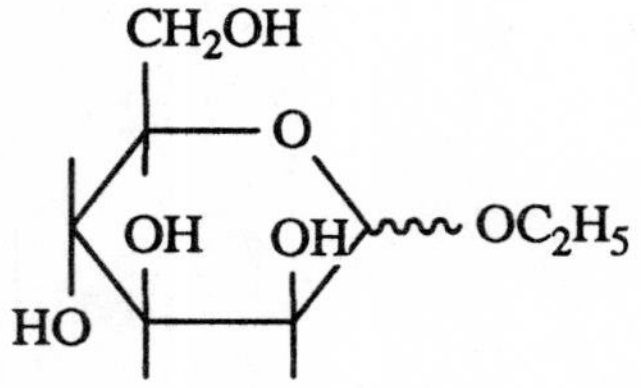

(d)

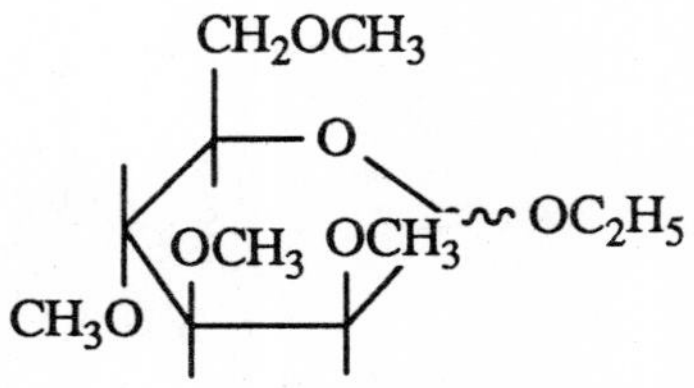

(e) same as (d)

(f)

CH_2OAc

O

OAc OAc ~OAc + CH_3CO_2H

AcO

Ac = $CH_3CO—$

(g) same as (f)

(h)

CH_2OH

HO — H

HO — H

H — OH

H — OH

CH_2OH

(i)

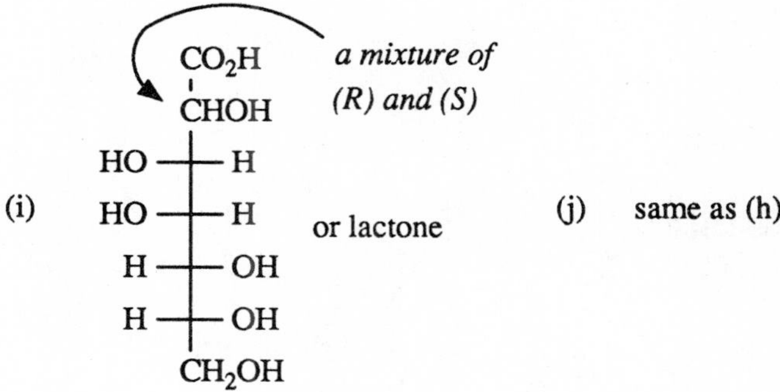

(j) same as (h)

(k) no reaction unless heat and pressure are used, in which case the product would be the same as in (h)

(l)

CH_2OH

O

OH OH ~OH

HO

22.50 (a) maltose, positive Tollens test; sucrose, negative

(b) Treat with HNO_3, heat; D-xylose → *meso*-diacid and D-lyxose → optically active diacid.

(c) mannose, positive Tollens test; mannitol, negative

(d) Treat with $NaHCO_3$; the acid would liberate CO_2 while the aldose would not.

(e) Treat with $NaBH_4$; glucose would yield an optically active alditol while galactose would yield a *meso* alditol. (Oxidation to aldaric acids could be used similarly.)

(f) glucose, positive Tollens test; the glucoside, negative

22.51 The fact that two stereoisomers (α and β) of glucose exist; mutarotation; the fact that glycoside formation requires only an equimolar quantity of an alcohol; the fact that totally methylated glucose contains five methoxyl groups, not four.

22.52

CH_2OH, O, OH, OH, HO, OH $\xrightarrow[H^+]{C_6H_5CHO}$ O — CH_2, C_6H_5CH, O, OH, OH, O, OH

$\xrightarrow[NaOH]{(CH_3O)_2SO_2}$ O — CH_2, C_6H_5CH, O, OCH_3, OCH_3, O, OCH_3

all OH groups methylated

$\xrightarrow{H_2O,\ H^+}$ CH_2OH, O, OCH_3, OH, HO, OCH_3

both acetal groups hydrolyzed

22.53

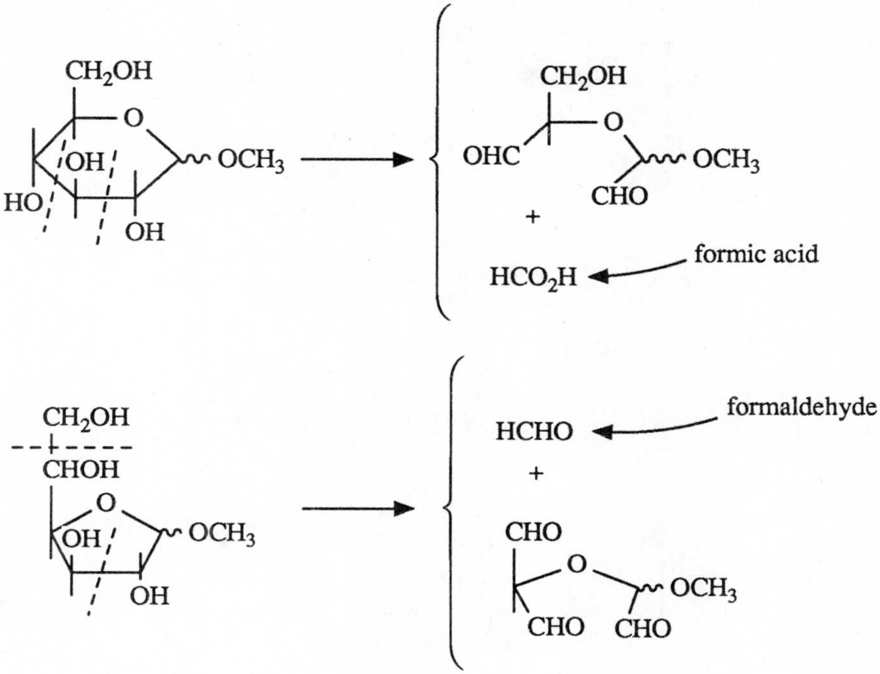

CH_2OH / O / OH / OCH_3 / HO / OH → CH_2OH / O / OHC / OCH_3 / CHO + HCO_2H ← formic acid

CH_2OH / CHOH / O / OH / OCH_3 / OH → HCHO ← formaldehyde + CHO / O / OCH_3 / CHO / CHO

22.54 A,

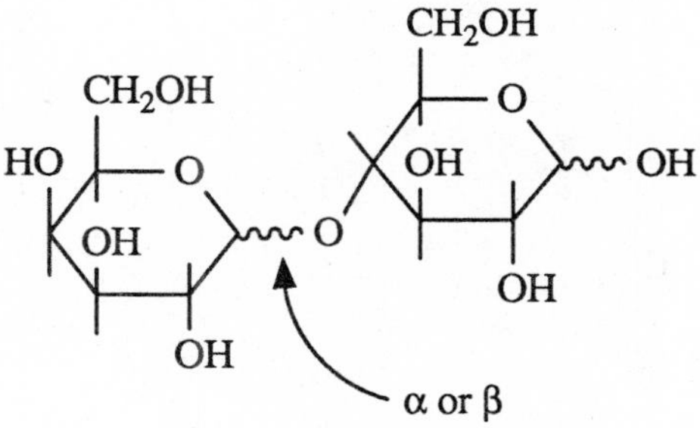

CH_2OH / HO / O / OH / OH / O / CH_2OH / O / OH / OH / OH

α or β

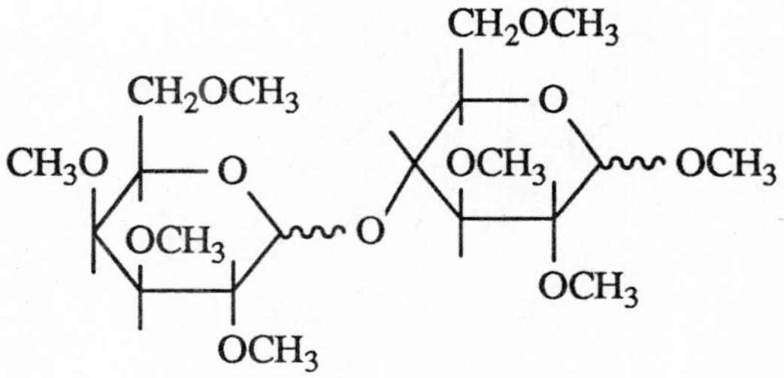

B,

CH_2OCH_3 / CH_3O / O / OCH_3 / OCH_3 / O / CH_2OCH_3 / O / OCH_3 / OCH_3 / OCH_3

C, CH_2OH HO O OH OH O CH_2OH OH OH CO_2H OH

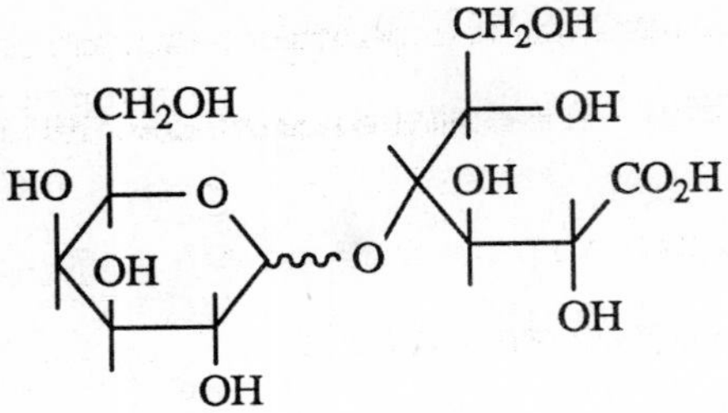

22.55 raffinose:

CH_2OH HO O OH HO OCH_2 ← α-D-galactopyranose

O OH HO HO ← α-D-glucopyranose

O $HOCH_2$ O HO CH_2OH OH ← β-D-fructofuranose

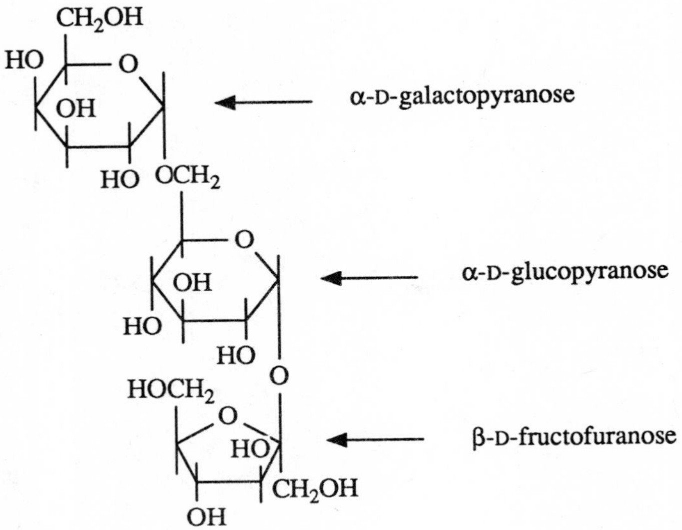

melibiose:

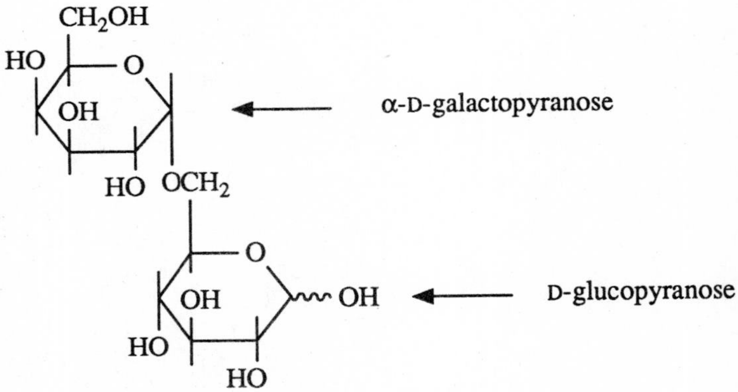

CH_2OH HO O OH HO OCH_2 ← α-D-galactopyranose

O OH OH HO HO ← D-glucopyranose

22.56 $C_{12}H_{22}O_{11}$ $\xrightarrow{H_2O}$ D-mannose + L-gulose

A

a reducing α-D-mannoside

α-D-mannose: CH_2OH O OH HO HO OH

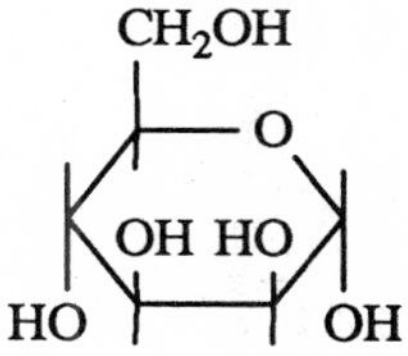

L-gulose: CHO HO—H HO—H H—OH HO—H CH_2OH or O CH_2OH OH OH HO HO

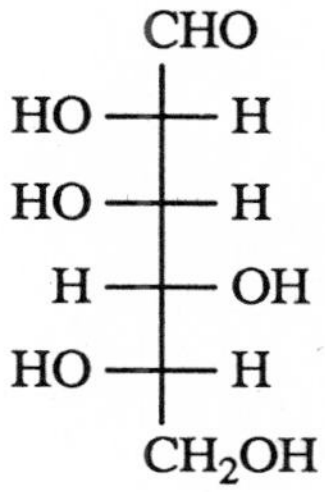

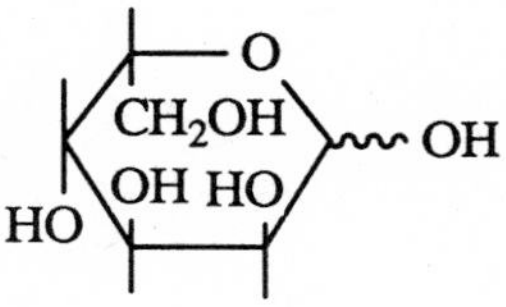

A $\xrightarrow[NaOH]{(CH_3O)_2SO_2}$ $\xrightarrow{H_2O, H^+}$

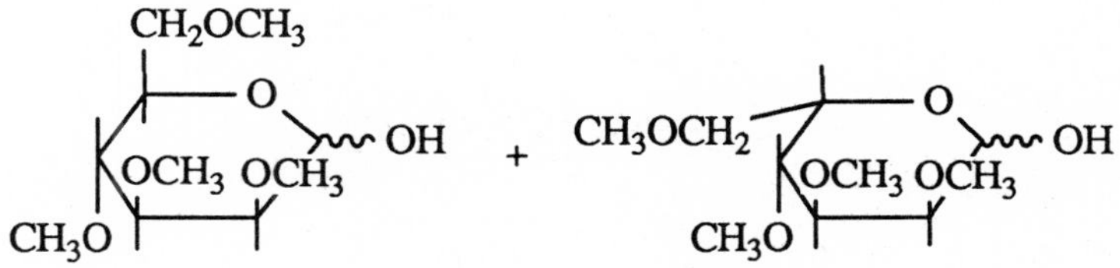

CH_2OCH_3 O OH OCH_3 OCH_3 CH_3O + CH_3OCH_2 O OH OCH_3 OCH_3 CH_3O

The methylation and hydrolysis show the position of attachment of L-gulose to α-D-mannose. The structure of compound A follows:

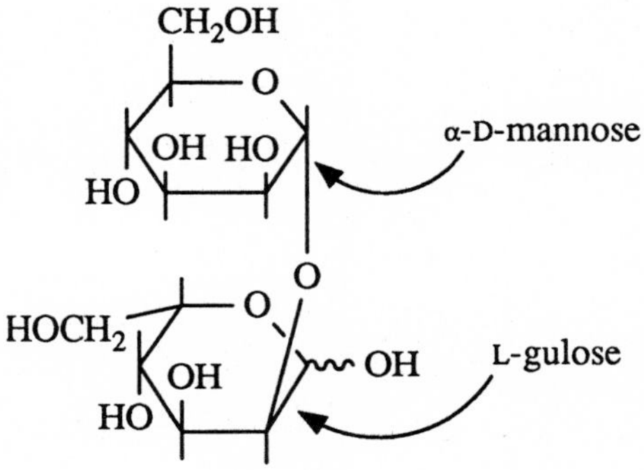

CH_2OH O OH HO HO α-D-mannose O $HOCH_2$ O OH OH HO L-gulose

22.57 linustatin $\xrightarrow[\beta\text{-D-glucosidase}]{H_2O}$ D-glucose + a D-glucoside

$C_{16}H_{27}NO_{11}$

a β-D-glucoside

From the formula and the chemical information, a partial structure can be drawn:

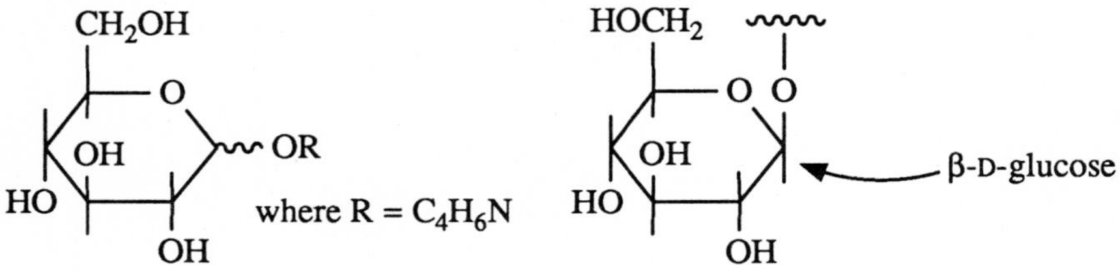

CH_2OH ... OR where $R = C_4H_6N$

$HOCH_2$ ... β-D-glucose

^{13}C nmr absorption:

two $-CH_3$

one $-\overset{|}{\underset{|}{C}}-$

one $-CH_2OH$

infrared absorption:

3400 cm^{-1} (2.94 μm): -OH

2240 cm^{-1} (4.46 μm): C≡C or C≡N

With only one $-CH_2OH$ group, the two glucose residues must be joined 1,6′. The ^{13}C nmr and infrared spectra show R to be:

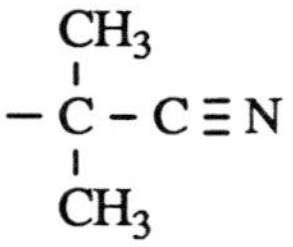

$$-\overset{CH_3}{\underset{CH_3}{C}}-C\equiv N$$

Therefore, the structure of linustatin must be as follows:

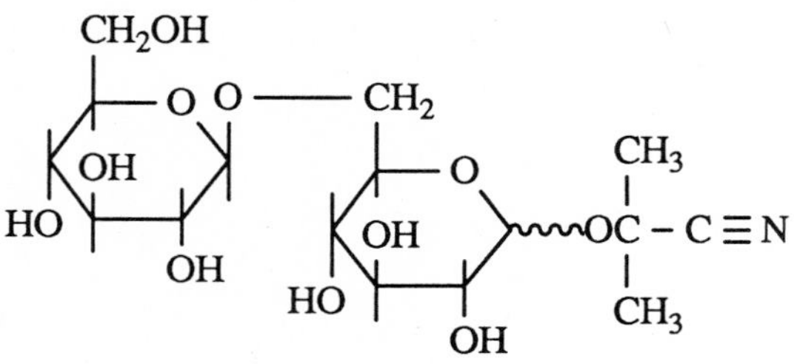

Chapter 23

Lipids

23.5

$$\begin{array}{l} CH_2O_2C(CH_2)_{12}CH_3 \\ | \\ CHO_2C(CH_2)_{12}CH_3 \\ | \\ CH_2O_2C(CH_2)_{12}CH_3 \end{array} \xrightarrow[-HOCH_2CH(OH)CH_2OH]{H_2O,\ H^+} 3\ CH_3(CH_2)_{12}CO_2H$$

trimyristin

myristic acid

$$CH_3(CH_2)_{11}Br \xrightarrow[-Br^-]{^-CH(CO_2C_2H_5)_2} CH_3(CH_2)_{11}CH(CO_2C_2H_5)_2 \xrightarrow[-CO_2]{H_2O,\ H^+,\ heat} CH_3(CH_2)_{12}CO_2H$$

23.6

$$\begin{array}{l} ③\ CH_2O_2C(CH_2)_{14}CH_3 \\ \quad | \\ ②\ CHO_2C(CH_2)_{14}CH_3 \\ \quad | \\ ①\ CH_2O_2C(CH_2)_7CH{=}CH(CH_2)_7CH_3 \end{array}$$

The oleic acid residue must be bonded to carbon 1 of glycerol; otherwise, carbon 2 would not be chiral.

$$\downarrow 3\ Na^+\ OH^-,\ H_2O$$

$$\begin{array}{l} CH_2OH \\ | \\ CHOH \\ | \\ CH_2OH \end{array} + 2\ CH_3(CH_2)_{14}CO_2^-\ Na^+ + CH_3(CH_2)_7CH{=}CH(CH_2)_7CO_2^-\ Na^+$$

$$\downarrow H_2O,\ H^+$$

$$2\ CH_3(CH_2)_{14}CO_2H + CH_3(CH_2)_7CH{=}CH(CH_2)_7CO_2H$$

23.7

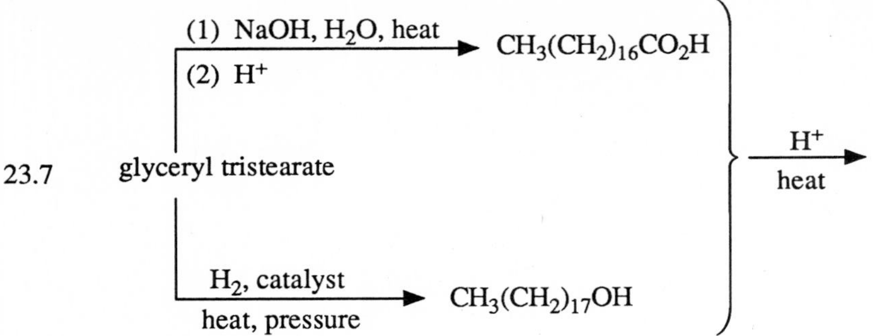

$$CH_3(CH_2)_{16}CO_2(CH_2)_{17}CH_3$$

The saponification would also yield glycerol (water soluble), which would have to be separated from the stearic acid. Glycerol would also have to be separated from the octadecanol.

23.8 (a) $CH_3(CH_2)_7CH{=}CH(CH_2)_7CO_2^-\ Na^+ \ + \ CH_3(CH_2)_{16}CO_2^-\ Na^+$

$$+ \ Na_2HPO_4 \ + \ HOCH_2CH_2\overset{+}{N}(CH_3)_3\ Cl^- \ + \ HOCH_2\underset{}{CH(OH)}CH_2OH$$

(b) $CH_3(CH_2)_7CH{=}CH(CH_2)_7CO_2^-\ Na^+ \ + \ CH_3(CH_2)_{16}CO_2^-\ Na^+$

$$+ \ Na_2HPO_4 \ + \ HOCH_2CH_2NH_2 \ + \ HOCH_2CH(OH)CH_2OH$$

The only difference between (a) and (b) is the amino alcohol; (a) yields a quaternary amine salt while (b) yields a 1° amine.

23.9 $CH_3(CH_2)_{22}CO_2H$ +

CH_2OH
HO O
OH OH
OH

The other product would be the following amino alcohol:

$$\textit{trans}\text{-}CH_3(CH_2)_{12}CH{=}CH\underset{}{\overset{OH}{\overset{|}{C}}}H\underset{\underset{CH_2OH}{|}}{C}HNH_2$$

23.10	(a) *trans*	(b) *trans*	(c) *trans*
	(d) *cis*	(e) *cis*	(f) *cis*

23.11 (a) (1) because the ring juncture is *trans*.

(b) (1) because the OH group is equatorial.

23.12 (a)

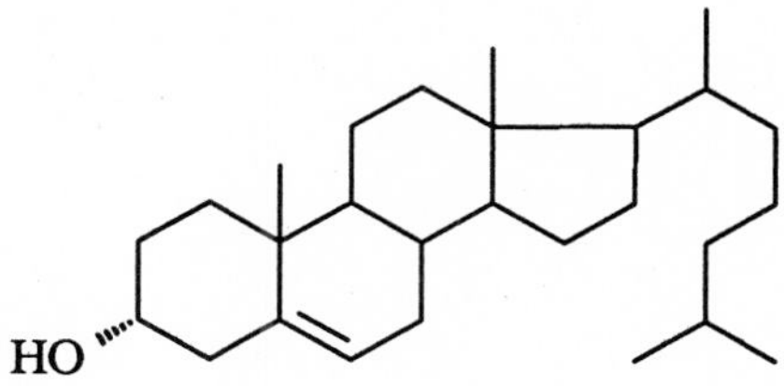

(b)

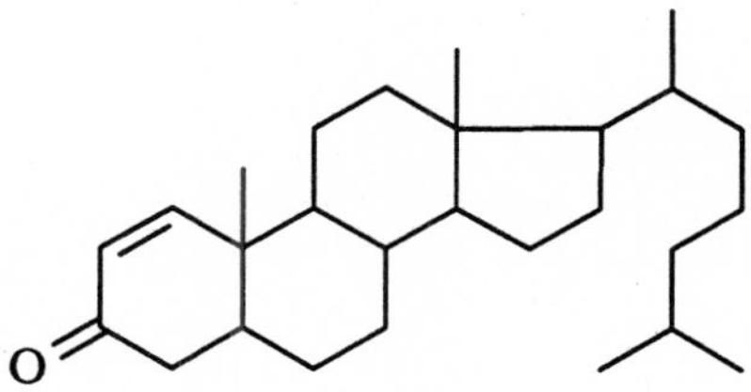

(c)

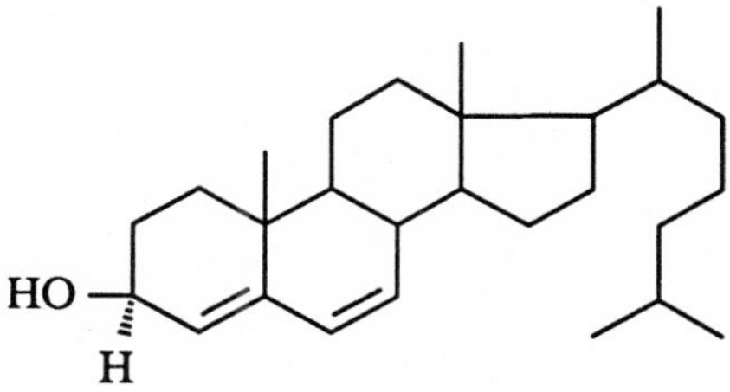

(d) 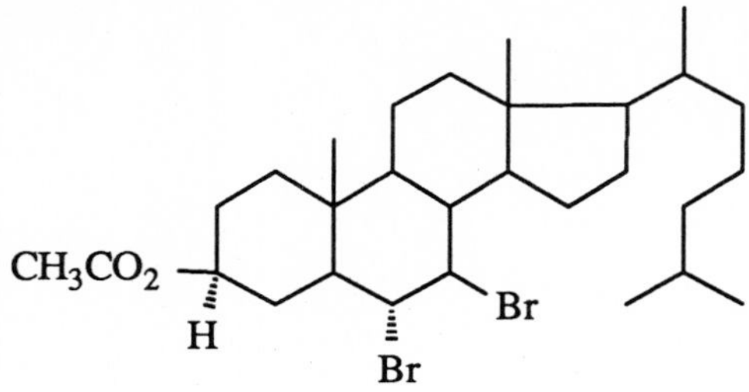 CH_3CO_2 ... H ... Br ... Br

23.13 CH_3 ... H ... CH_3 ... CH_3 ... $CH(CH_2)_2CO_2H$ ... H ... H ... OH ... H ... OH

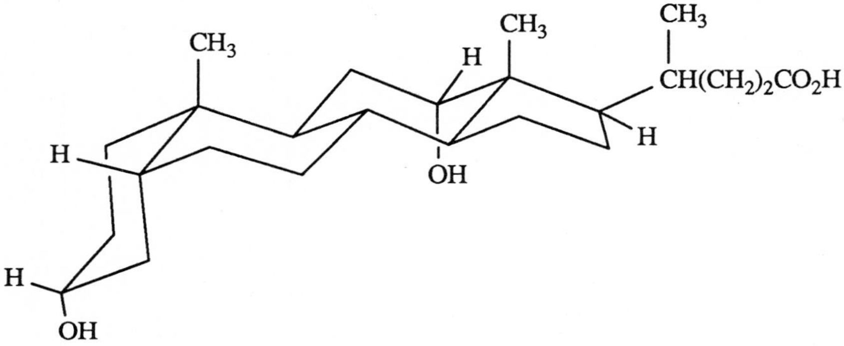

23.14 (a) $CH_3(CH_2)_5CH{=}CH(CH_2)_7CO_2H \xrightleftharpoons{\text{excess } C_2H_5OH,\ H^+}$

$CH_3(CH_2)_5CH{=}CH(CH_2)_7CO_2C_2H_5 + H_2O$

(b) ethyl palmitoleate from (a) $\xrightarrow{H_2,\ Ni}$ $CH_3(CH_2)_{14}CO_2C_2H_5$

(c) $\xrightarrow[\text{heat}]{H_2CrO_4}$ $CH_3(CH_2)_5CO_2H + HO_2C(CH_2)_7CO_2H$

nonanedioic acid

(d) ethyl palmitate from (b) $\xrightarrow[H_2O]{NaOH}$ $CH_3(CH_2)_{14}CO_2^- Na^+$

$\xrightarrow{H_2O,\ H^+}$ $CH_3(CH_2)_{14}CO_2H$ $\xrightarrow[(2)\ H_2O]{(1)\ Cl_2,\ PCl_3}$ $CH_3(CH_2)_{13}CH(Cl)CO_2H$

(e) $CH_3(CH_2)_5CO_2H$ from (c) $\xrightarrow{SOCl_2}$ $CH_3(CH_2)_5\overset{O}{\overset{\|}{C}}Cl$

$\xrightarrow[(2)\ H_2O,\ H^+]{(1)\ LiAlH[OC(CH_3)_3]_3}$ $CH_3(CH_2)_5CHO$

(f) ethyl palmitoleate from (a) $\xrightarrow[(2)\ H_2O,\ H^+]{(1)\ 2\ CH_3MgI}$

Remember that a 3° alcohol with two identical R groups can be synthesized by the Grignard reaction of an ester or an acid halide.

23.15 (a) $HOCH_2\underset{}{\overset{OH}{\overset{|}{C}}}HCH_2OH$ + $2\ CH_3(CH_2)_{16}CO_2^-\ Na^+$

+ $CH_3(CH_2)_5CH{=}CH(CH_2)_7CO_2^-\ Na^+$

(b) $HOCH_2\overset{OH}{\overset{|}{C}}HCH_2OH$ + $2\ CH_3(CH_2)_{16}CH_2OH$ + $CH_3(CH_2)_{14}CH_2OH$

(c)

$CH_2O_2C(CH_2)_7CHBrCHBr(CH_2)_5CH_3$
|
$CHO_2C(CH_2)_{16}CH_3$
|
$CH_2O_2C(CH_2)_{16}CH_3$

or

$CH_2O_2C(CH_2)_{16}CH_3$
|
$CHO_2C(CH_2)_7CHBrCHBr(CH_2)_5CH_3$
|
$CH_2O_2C(CH_2)_{16}CH_3$

23.16 (a) no reaction

(b) $CH_3(CH_2)_5CH \neq CH(CH_2)_7CO_2H \xrightarrow[(2)\ [O]]{(1)\ O_3}$

$CH_3(CH_2)_5CO_2H + HO_2C(CH_2)_7CO_2H$

(c) $CH_3(CH_2)_4CH \neq CHCH_2CH \neq CH(CH_2)_7CO_2H \xrightarrow[(2)\ [O]]{(1)\ O_3}$

$CH_3(CH_2)_4CO_2H + CH_2(CO_2H)_2 + HO_2C(CH_2)_7CO_2H$

(d) $CH_3CH_2CH{=}CHCH_2CH{=}CHCH_2CH{=}CH(CH_2)_7CO_2H \xrightarrow[(2)\ [O]]{(1)\ O_3}$

$CH_3CH_2CO_2H + 2\ CH_2(CO_2H)_2 + HO_2C(CH_2)_7CO_2H$

23.17 (a) Tripalmitolein decolorizes Br_2/CCl_4 solution; tripalmitin does not.

(b) Beeswax is hydrolyzed to a fatty acid and a water-insoluble monoalcohol; beef fat is hydrolyzed to a fatty acid and glycerol (water soluble). Also, beef fat will consume approximately twice as much NaOH per gram when saponified. Also, beef fat contains some unsaturated fatty acids and would thus react with bromine to decolorize it; beeswax will not decolorize bromine.

(c) Paraffin wax (a mixture of alkanes) does not undergo hydrolysis.

(d) Linoleic acid neutralizes dilute aqueous NaOH; linseed oil does not.

(e) Sodium palmitate precipitates with Ca^{2+}; sodium *p*-decylbenzenesulfonate does not.

(f) A vegetable oil is hydrolyzed to a fatty acid and glycerol; a motor oil (a mixture of hydrocarbons) does not undergo hydrolysis.

23.18 Estradiol is a phenol and can be extracted from the mixture by aqueous NaOH.

23.19 The OH at position 3 of cholic acid is equatorial, while the other two OH groups are axial. The esterification takes place at the 3-OH group because it is the least sterically hindered.

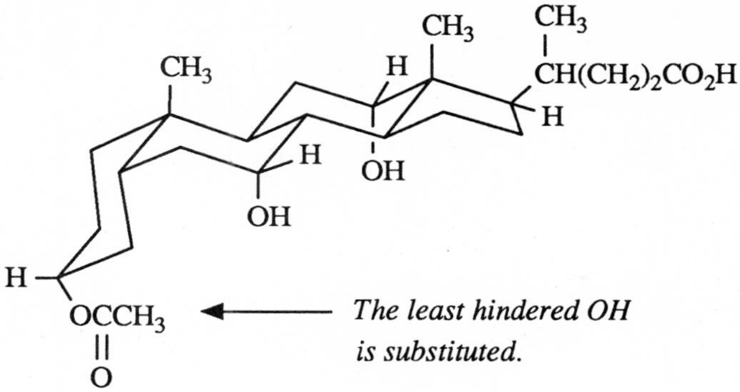

CH$_3$ CH$_3$ CH(CH$_2$)$_2$CO$_2$H H CH$_3$ H H OH OH H OCCH$_3$ (=O)

← *The least hindered OH is substituted.*

23.20 (a) none because the ^{14}C is lost as $^{14}CO_2$

(b) $^{14}CH_2OH$ + $^{14}CH_2OH$

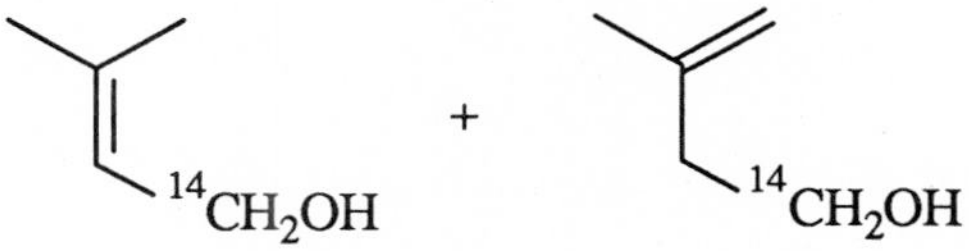

(c) $^{14}CH_3$ CH$_2$OH + $^{14}CH_2$ CH$_2$OH

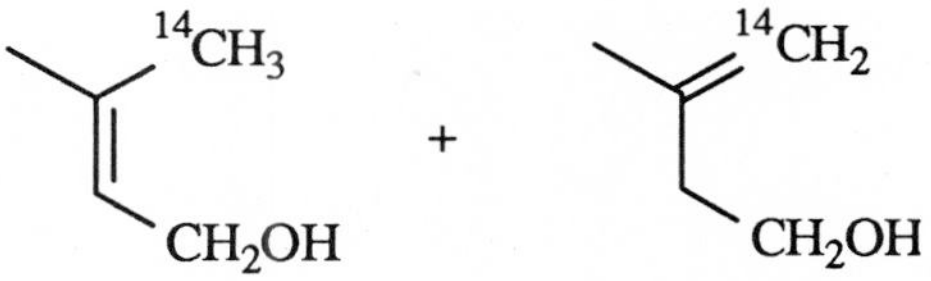

23.21 $CH_3C(OH)(CH_2{}^{14}CH_2OH)CH_2CO_2H$ —many steps→ $^{14}CH_2OH$ + $^{14}CH_2OH$

isopentenyl alcohols

→ $^{14}CH_2OH$ $^{14}CH_2$

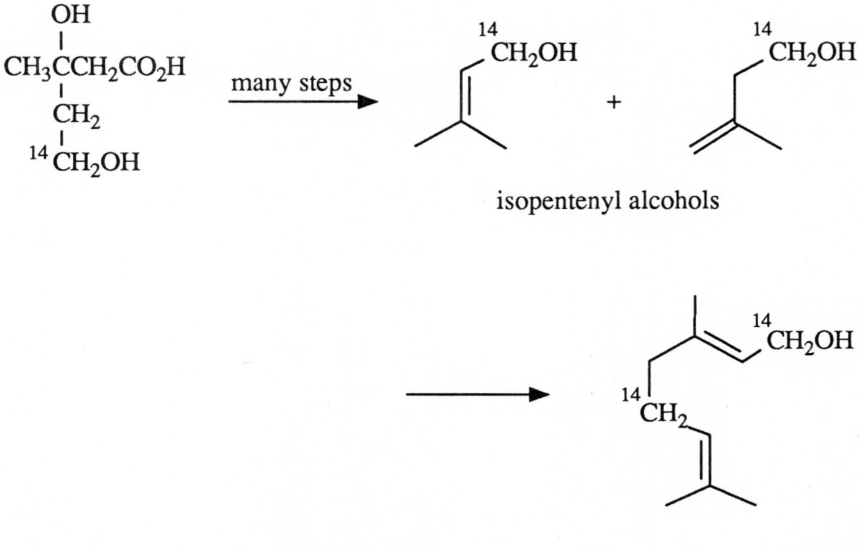

23.22 Let R represent the side chain, which is not affected in the reaction.

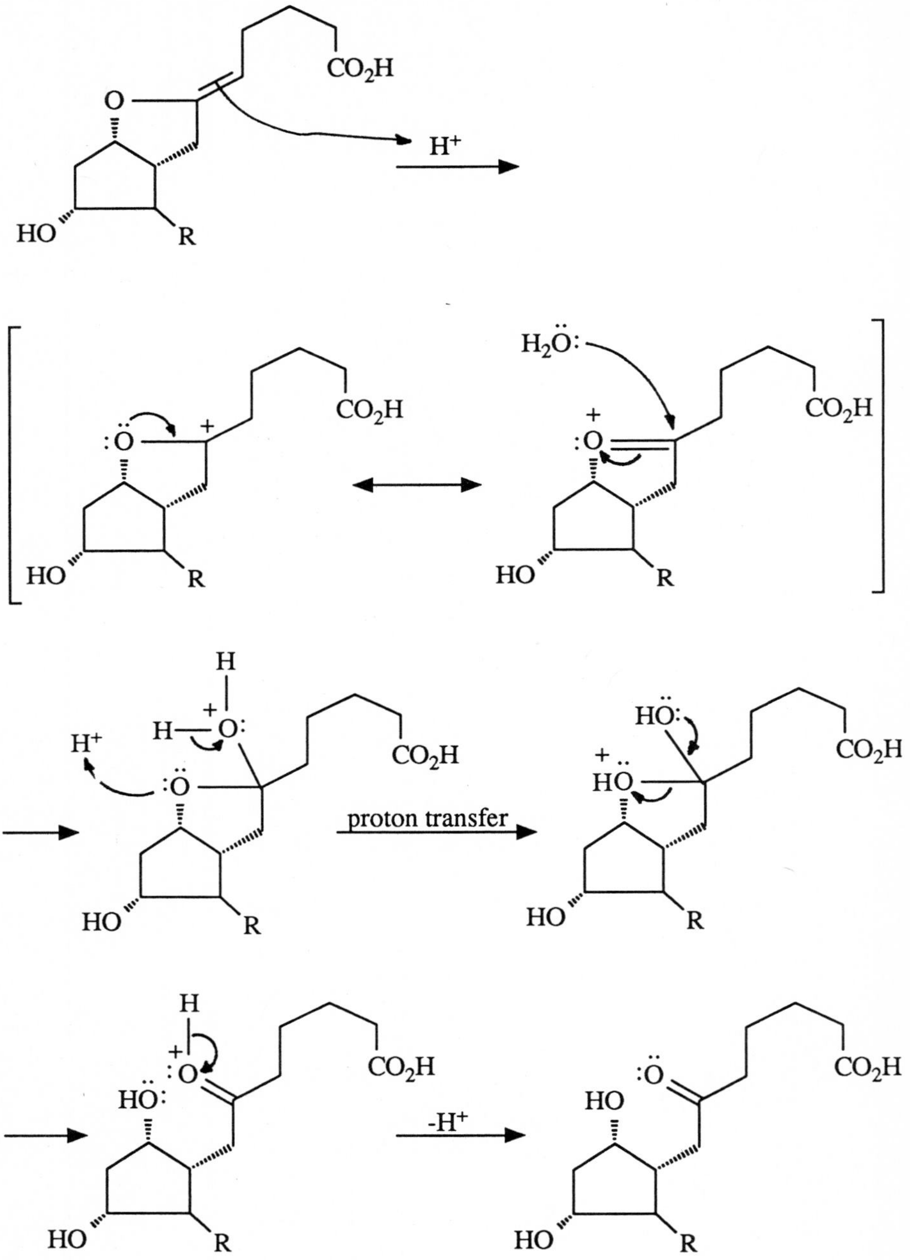

H^+

proton transfer

$-H^+$

Chapter 24

Amino Acids and Proteins

24.13 (a) neutral (b) acidic (c) neutral

(d) neutral (e) basic (f) neutral

24.14 isoleucine: $CH_3CH_2\overset{*}{C}H(CH_3){-}\overset{*}{C}H(NH_2)CO_2H$ threonine: $CH_3\overset{*}{C}H(OH){-}\overset{*}{C}H(NH_2)CO_2H$

24.15

$$\begin{array}{l} SCH_2C(=O)NHCH(CH_2SH)CO_2H \\ | \\ SCH_2CH(NH_2)CO_2H \end{array}$$

$$HSCH_2C(=O)NHCH(CO_2H)CH_2S{-}SCH_2CH(NH_2)CO_2H$$

22.16 (a) $C_6H_5CH_2CHO \xrightarrow[HCN]{NH_3} C_6H_5CH_2CH(NH_2)CN \xrightarrow[(2)\ \text{neutralize}]{(1)\ H_2O,\ H^+}$

$C_6H_5CH_2CH(NH_2)CO_2H$ or $C_6H_5CH_2CH(\overset{+}{N}H_3)CO_2^-$

(b) $CH_3CH_2CH(CH_3)CHO \xrightarrow[HCN]{NH_3} CH_3CH_2CH(CH_3)CH(NH_2)CN \xrightarrow[(2)\ \text{neutralize}]{(1)\ H_2O,\ H^+}$

$$\underset{\substack{| \\ CH_3}}{CH_3CH_2\overset{\overset{NH_2}{|}}{CH}CHCO_2H} \quad \text{or} \quad \underset{\substack{| \\ CH_3}}{CH_3CH_2\overset{\overset{\overset{+}{NH_3}}{|}}{CH}CHCO_2^-}$$

(c) both racemic

24.17 $(CH_3)_2CHCH_2CO_2H \xrightarrow[PBr_3]{Br_2} (CH_3)_2CHCHBrCO_2H$

achiral *racemic*

$$\xrightarrow{\text{excess } NH_3} (CH_3)_2CH\overset{\overset{NH_2}{|}}{C}HCO_2^-\ NH_4^+ \xrightarrow{H^+}$$

racemic

$$\underbrace{(CH_3)_2CH\overset{\overset{NH_2}{|}}{C}HCO_2H \quad \text{or} \quad (CH_3)_2CH\overset{\overset{NH_3^+}{|}}{C}HCO_2^-}_{racemic}$$

24.18 $(R)\text{-}CH_3\overset{\overset{OH}{|}}{C}HCO_2H \xrightarrow[\text{or } H_2CrO_4]{\text{hot } KMnO_4} CH_3\overset{\overset{O}{||}}{C}CO_2H$

achiral

$$\xrightarrow{H_2,\ NH_3} \underbrace{CH_3\overset{\overset{NH_2}{|}}{C}HCO_2H \quad \text{or} \quad CH_3\overset{\overset{NH_3^+}{|}}{C}HCO_2^-}_{racemic}$$

24.19 (a) $CH_3\underset{\underset{NH_2}{|}}{C}HCO_2^-\ NH_4^+$ (b) $Cl^-\ \overset{+}{N}H_3CH_2CO_2H$ + (benzene ring with two adjacent CO_2H groups: phthalic acid)

(c) $C_6H_5CH_2CHCO_2H$ + 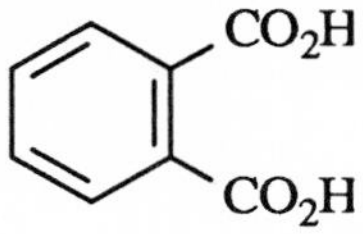

$\quad\quad |$

$\quad {}^{+}NH_3\ Cl^-$

(d) $(CH_3)_2CHCH_2CHCN$

$\quad\quad |$

$\quad NH_2$

(e) $C_6H_5CH_2CHCO_2CH_3$

$\quad\quad |$

$\quad {}^{+}NH_3\ Cl^-$

24.20 (a) CH_3CHCO_2H

$\quad |$

$NH_3^+\ Cl^-$

(b) $CH_3CHCO_2^-\ K^+$

$\quad |$

NH_2

(c) $CH_3CHCO_2CH_3$

$\quad |$

$NH_3^+\ HSO_4^-$

(d) CH_3CHCO_2H + CH_3CO_2H

$\quad |$

$NHCCH_3$

$\quad ||$

$\quad O$

24.21 (a) $H_3\overset{+}{N}CHCO_2^-$ + Cl^-

$\quad |$

$CH(CH_3)_2$

(b) $H_2NCHCO_2^-$ + Cl^-

$\quad |$

$(CH_2)_4NH_3^+$

(c) $H_3\overset{+}{N}CHCO_2^-$ + Na^+ + H_2O

$\quad |$

$CH(CH_3)_2$

(d) $H_3\overset{+}{N}CHCO_2^-$ + Na^+ + H_2O

$\quad |$

$(CH_2)_4NH_3^+$

In each case, look for the more acidic or more basic group in the reactant molecule.

24.22 (a) 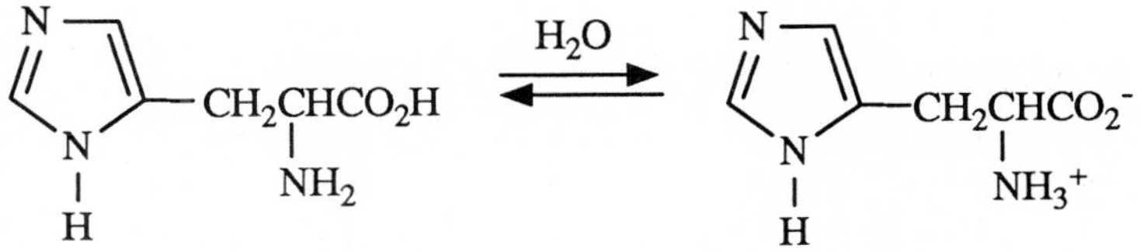

histidine

(b) 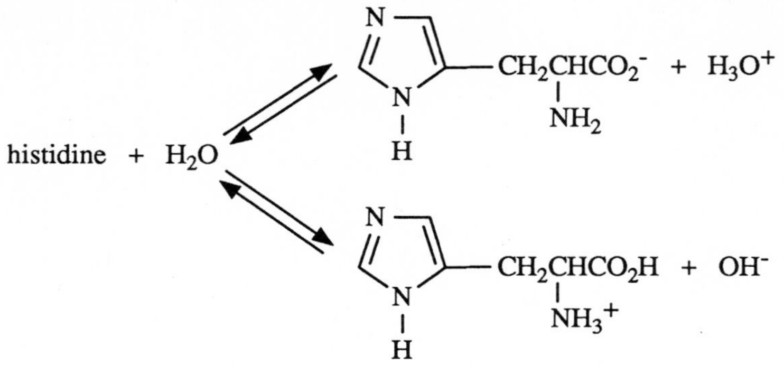 histidine + H_2O ⇌ (imidazole ring, N–H) $-CH_2CHCO_2^- + H_3O^+$ (with NH_2 on CH)

⇌ (imidazole ring, N–H) $-CH_2CHCO_2H + OH^-$ (with NH_3^+ on CH)

(c) 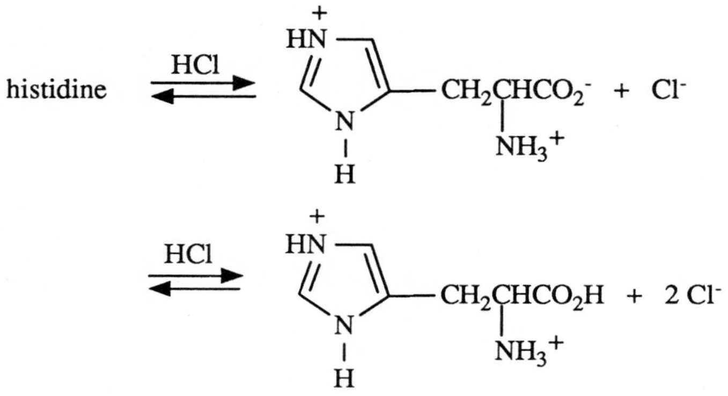 histidine $\xrightarrow{HCl}$ (imidazolium ring, $\overset{+}{H}N$, N–H) $-CH_2CHCO_2^- + Cl^-$ (with NH_3^+ on CH)

$\xrightarrow{HCl}$ (imidazolium ring, $\overset{+}{H}N$, N–H) $-CH_2CHCO_2H + 2\,Cl^-$ (with NH_3^+ on CH)

(d) 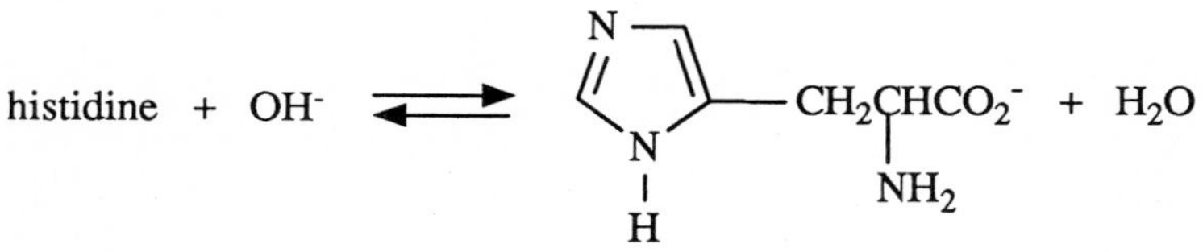 histidine + OH^- ⇌ (imidazole ring, N–H) $-CH_2CHCO_2^- + H_2O$ (with NH_2 on CH)

24.23 (a) An amino acid contains a carboxylate group ($-CO_2^-$) rather than a carboxyl group ($-CO_2H$).

(b) When the solution is acidified, the carboxyl group is generated.

24.24 (a) a neutral amino acid: (3) (b) an acidic amino acid: (4)

(c) a neutral amino acid: (2) (d) a basic amino acid: (1)

Note that cysteine (a) is slightly more acidic than proline (c). The reasons are that proline contains a more basic 2° amino group and that cysteine also contains an -SH group (weakly acidic).

24.25 (a) neutral, 6 (b) slightly basic, 8 (See Answer 24.49)

(c) acidic, 3 (d) neutral, 6 (e) basic, 10 (See Answer 24.49)

24.26 (a) [Ruhemann's purple structure: indane-1,3-dione =N– enolate indanone, with O, O, O^-, O labels] $+ H_2NC(=O)CH_2CH_2CH(=O) + CO_2 + H_2O$

$$\overset{O}{\overset{\|}{}}$$

(b) $H_2NC(=O)CH_2CH_2CH(NHC(=O)CH_3)CO_2H + CH_3CO_2H$

(c) $CH_3C(=O)O–C_6H_4–CH_2CH(NHC(=O)CH_3)CO_2H + 2\ CH_3CO_2H$

(d) $CH_3C(=O)OCH_2CH(NHC(=O)CH_3)CO_2H + 2\ CH_3CO_2H$

(e) [same structure as (a), with O, O, O^-, O labels and =N–] $+ CH_3CH(=O) + CO_2 + H_2O$

24.27 $(CH_3)_2CHCH_2CH(NH_2)C(=O)-NHCH(CH_2CH_2SCH_3)C(=O)-NHCH_2CO_2H$

(a) arrows point to the two amide (peptide) bonds; (b) $(CH_3)_2CHCH_2CH(NH_2)$ portion; (c) $NHCH_2CO_2H$ portion

(d) tripeptide (e) neutral

24.28 (a) $H_2NCH_2C(=O)-NHCH_2CO_2H$ or, more correctly, $H_3\overset{+}{N}CH_2C(=O)-NHCH_2CO_2^-$

(b) $H_2NCH(CH_3)C(=O)\text{-}NHCH(CH_2CH(CH_3)_2)C(=O)\text{-}NHCH(CH_2CH_2SCH_3)CO_2H$ or

$H_3\overset{+}{N}CH(CH_3)C(=O)\text{-}NHCH(CH_2CH(CH_3)_2)C(=O)\text{-}NHCH(CH_2CH_2SCH_3)CO_2^-$

24.29 (a) serylglycylleucine, ser-gly-leu

(b) prolylthreonylmethionine, pro-thr-met

24.30 (a) $H_3\overset{+}{N}CH_2C(=O)\text{-}NHCH((CH_2)_4NH_3^+)CO_2H$ (b) $H_2NCH_2C(=O)\text{-}NHCH(CH_2CH_2CO_2^-)CO_2^-$

(c) $H_2NCH_2\overset{O}{\overset{\|}{C}}\text{-}NH\underset{CH_2-C_6H_4-O^-}{\underset{|}{C}}HCO_2^-$

24.31 The protein contains either an unsubstituted amide group, $-\underset{O}{\underset{\|}{C}}NH_2$, or arginine.

$$R\overset{O}{\overset{\|}{C}}NH_2 \xrightarrow{H_2O,\ H^+} R\overset{O}{\overset{\|}{C}}OH + NH_4^+$$

$$NH_4^+ \xrightarrow{OH^-} NH_3 + H_2O$$

$$RNH\overset{NH}{\overset{\|}{C}}NH_2 \xrightarrow{H_2O,\ H^+} R\overset{+}{N}H_3 + 2\ NH_4^+ + H_2O + CO_2$$

$$2\ NH_4^+ \xrightarrow{2\ OH^-} 2\ NH_3 + 2\ H_2O$$

24.32 (a) 3-phenyl-2-thioxoimidazolidin-4-one (C_6H_5N, ring C=S, NH, CH_2, C=O) + $H_3\overset{+}{N}\underset{CH_3}{\underset{|}{C}}HCO_2H$

(b) 5-methyl-3-phenyl-2-thioxoimidazolidin-4-one (C_6H_5N, ring C=S, NH, $CHCH_3$, C=O) + $H_3\overset{+}{N}CH_2CO_2H$

(c) [thiohydantoin: C_6H_5N, C=S, NH, C=O, CH_2OH] + $H_3\overset{+}{N}CH(CH_2C_6H_5)C(=O)-NHCH(CH_2CH_2SCH_3)CO_2H$ ⟶

$H_3\overset{+}{N}CH(CH_2C_6H_5)CO_2H$ + $H_3\overset{+}{N}CH(CH_2CH_2SCH_3)CO_2H$

24.33 (a) $O_2N-C_6H_3(NO_2)-NHCH(CH(CH_3)_2)CO_2H$

(b) $O_2N-C_6H_3(NO_2)-NHCH(CH_3)C(=O)-NHCH(CH(CH_3)_2)CO_2H$ ⟶

$O_2N-C_6H_3(NO_2)-NHCH(CH_3)CO_2H$ + $H_3\overset{+}{N}CH(CH(CH_3)_2)CO_2H$

(c) $H_2N\overset{O}{\overset{\|}{C}}CH_2CH_2\underset{NH-C_6H_3(NO_2)_2}{CH}\overset{O}{\overset{\|}{C}}\text{-}NHCH_2CO_2H \longrightarrow$

$NH_4^+ + HO_2CCH_2CH_2\underset{NH-C_6H_3(NO_2)_2}{CH}CO_2H + H_3\overset{+}{N}CH_2CO_2H$

Trypsin catalyzes hydrolysis here.

24.34 tyr-leu-asp-ser-arg-/- $\xrightarrow[\text{-tyr}]{\text{Edman}}$ leu-asp-ser-arg

$\xrightarrow[\text{-leu}]{\text{Edman}}$ asp-ser-arg $\xrightarrow[\text{-asp}]{\text{Edman}}$ ser-arg

24.35 Trypsin catalyzes hydrolysis of the amide bond at the carboxyl group of lysine or arginine.

(a) lys-/-asp-gly-ala-ala-glu-ser-gly $\xrightarrow{\text{trypsin}}$

lys + asp-gly-ala-ala-glu-ser-gly

(b) tyr-cys-lys-/-ala-arg-/-arg-/-gly $\xrightarrow{\text{trypsin}}$

try-cys-lys + ala-arg + arg + gly

(c) ala-ala-his-arg ⁄ glu-lys ⁄ phe-ile-gly-glu-gly-glu $\xrightarrow{\text{trypsin}}$

ala-ala-his-arg + glu-lys + phe-ile-gly-glu-gly-glu

24.36 Sanger reagent reacts with *N*-terminal amino acids:

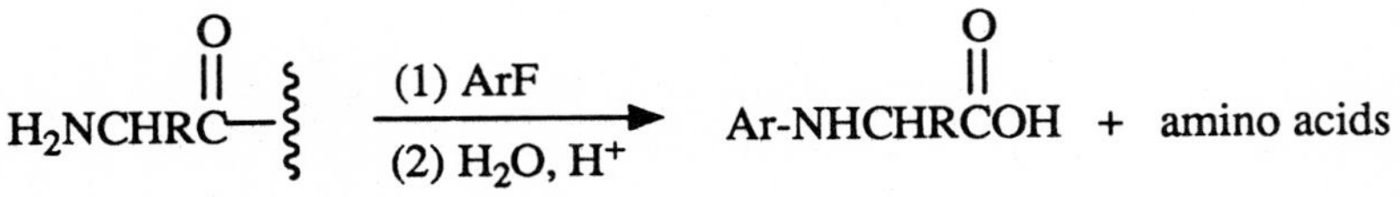

$$H_2NCHRC(=O)\text{—}\!\!\sim \xrightarrow[(2)\ H_2O,\ H^+]{(1)\ ArF} \text{Ar-NHCHRC(=O)OH} + \text{amino acids}$$

Thus, the following products would be obtained:

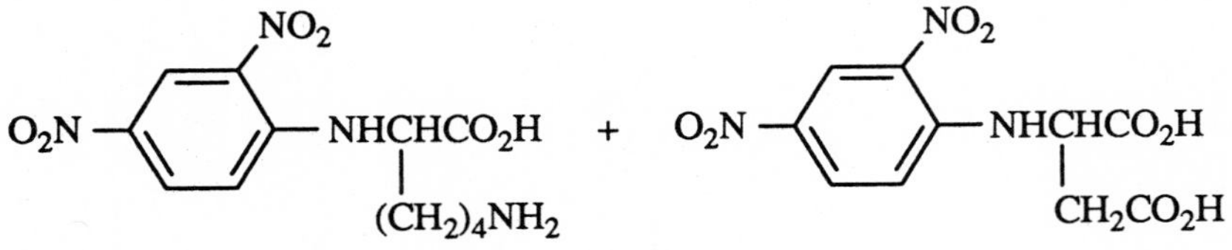

O_2N—(2-NO_2-C_6H_3)—$NHCHCO_2H$ with side chain $(CH_2)_4NH_2$ + O_2N—(2-NO_2-C_6H_3)—$NHCHCO_2H$ with side chain CH_2CO_2H

+ 2 gly + 2 ala + glu + ser

The Sanger reagent could also react with the side-chain amino group in lys. (See Problem 24.50)

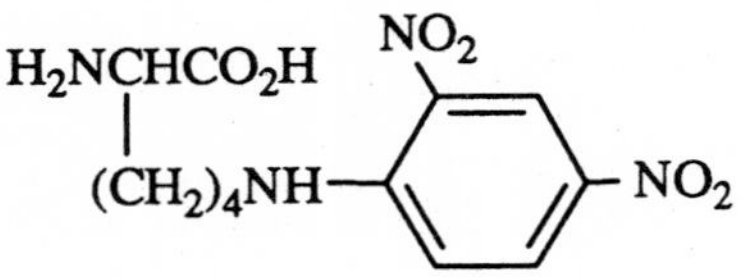

H_2NCHCO_2H with side chain $(CH_2)_4NH$—(2-NO_2-4-NO_2-C_6H_3)

or

O_2N—(2-NO_2-C_6H_3)—$NHCHCO_2H$ with side chain $(CH_2)_4NH$—(2-NO_2-4-NO_2-C_6H_3)

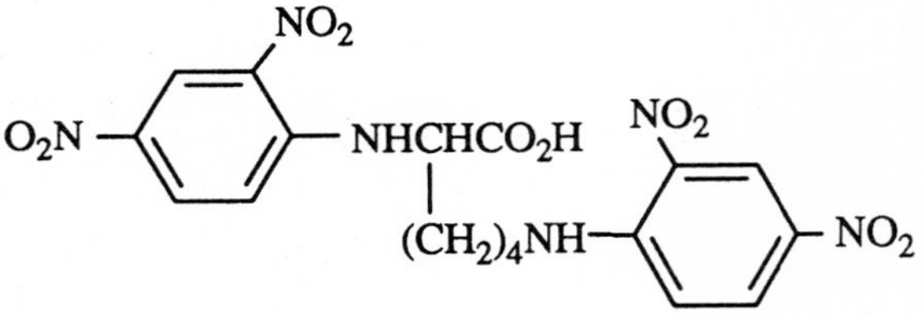

24.37 (a) and (b), $H_2NCH_2CO_2H$ + H_2NCHCO_2H (with CH_3 side chain)

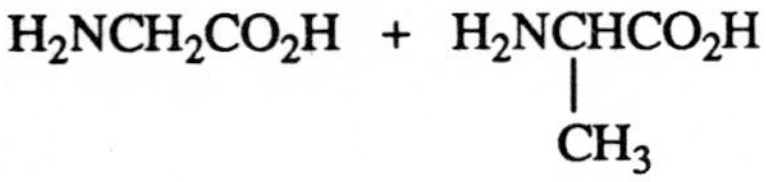

(c) 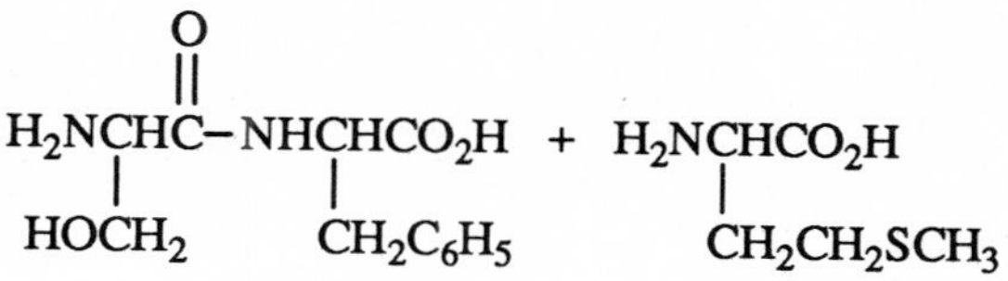 $H_2NCH(CH_2OH)C(=O)-NHCH(CH_2C_6H_5)CO_2H + H_2NCH(CH_2CH_2SCH_3)CO_2H$

24.38 arg-pro-pro-gly-phe-ser-pro-phe-arg

24.39 $H_2NCH(CH(CH_3)_2)CO_2H \xrightarrow[(a)]{SOCl_2} Cl^- \; H_3\overset{+}{N}CH(CH(CH_3)_2)COCl \xrightarrow[(b)]{H_2NCH(CH_3)CO_2H}$

$$H_2NCH((CH_3)_2CH)C(=O)-NHCH(CH(CH_3)_2)CO_2H + H_2NCH((CH_3)_2CH)C(=O)-NHCH(CH_3)CO_2H$$

val-val   val-ala

$$+ H_2NCH(CH_3)C(=O)-NHCH(CH(CH_3)_2)CO_2H + H_2NCH(CH_3)C(=O)-NHCH(CH_3)CO_2H$$

ala-val   ala-ala

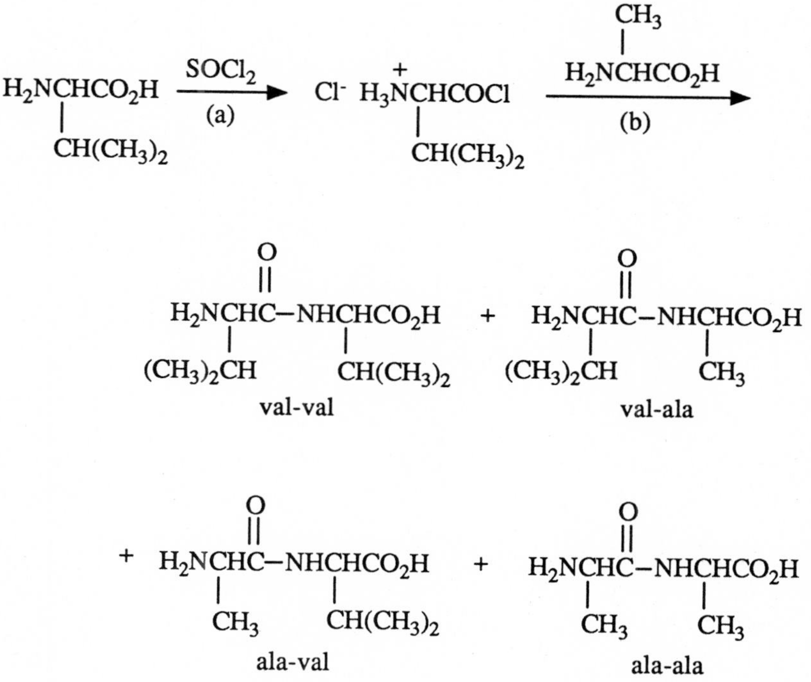

In (b), the acid halide of val can undergo an exchange reaction with the new acid (ala).

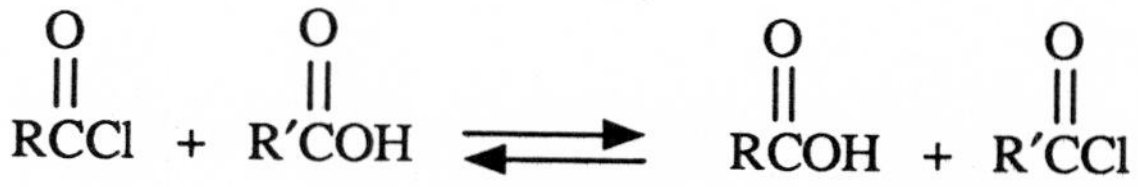

$$RC(=O)Cl + R'C(=O)OH \rightleftharpoons RC(=O)OH + R'C(=O)Cl$$

24.40 (a)

$$\underset{\text{ala}}{CH_3CH(NH_2)CO_2H} \xrightarrow{C_6H_5CH_2O_2CCl} CH_3CH(NHCO_2CH_2C_6H_5)CO_2H$$

$$\xrightarrow{ClCO_2C_2H_5} CH_3CH(NHCO_2CH_2C_6H_5)CO_2CO_2C_2H_5 \xrightarrow[\text{or ester}]{H_2NCH_2CO_2H}$$

$$CH_3CH(NHCO_2CH_2C_6H_5)C(=O)NHCH_2CO_2H \xrightarrow{H_2,\ Pd} \underset{\text{ala-gly}}{CH_3CH(NH_2)C(=O)NHCH_2CO_2H}$$

(b)

$$\underset{\text{phe}}{C_6H_5CH_2CH(NH_2)CO_2H} \xrightarrow[\substack{(3)\ (CH_3)_2CHCH(NH_2)CO_2H \\ (4)\ H_2,\ Pd}]{\substack{(1)\ C_6H_5CH_2O_2CCl \\ (2)\ ClCO_2C_2H_5}} \underset{\text{phe-val}}{C_6H_5CH_2CH(NH_2)C(=O)NHCH(CH(CH_3)_2)CO_2H}$$

(c)

$$\text{ala-gly} \xrightarrow[\substack{(3)\ \text{phe-val} \\ (4)\ H_2,\ Pd}]{\substack{(1)\ C_6H_5CH_2O_2CCl \\ (2)\ ClCO_2C_2H_5}} \text{ala-gly-phe-val}$$

Except for the amino acids, the series of reagents used in (b) and (c) are the same as those in (a).

24.41 (a), (d), (e)

24.42 (a) The peptide bond is an amide C-N bond formed from the alpha amino group and the carboxyl group of the amino acids. This bond is used to form the backbone of a protein molecule.

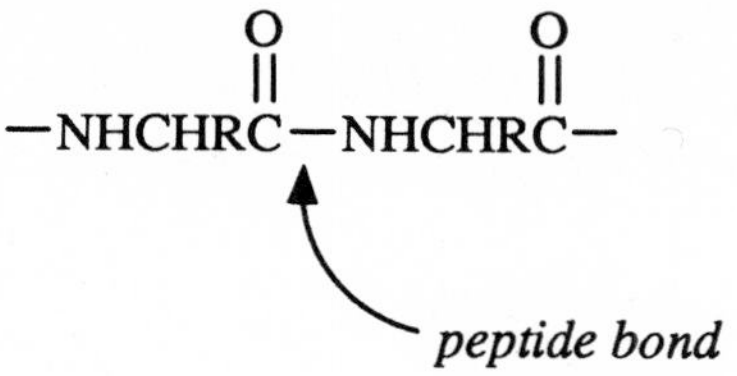

$$\begin{array}{c} \quad\;\; O \qquad\qquad\quad\;\; O \\ \quad\;\; \| \qquad\qquad\quad\;\; \| \\ -NHCHRC-NHCHRC- \end{array}$$

↑ *peptide bond*

(b) The disulfide bond, RS-SR, is a cross-linking bond that forms links between two different peptide or protein chains.

(c) and (d) The hydrogen bonds and ionic salt-bridge bonds hold a protein molecule together in its unique conformation. They are also used to form attractions between different protein molecules as in silk fibroin or in hemoglobin.

24.43 (a) The acetic acid in vinegar forms hydrogen bonds with protein groups, disrupting the hydrogen bonds that hold the protein molecules together.

(b) Aqueous sucrose can disrupt the hydrogen bonding and tenderize the meat to some extent, but not to the degree that the strongly hydrogen bonding acetic acid can.

24.44 (a) 4 (b) 3, 5, 6 (c) 6 (d) 2 (e) 1 (f) 7

24.45 (b) because it is most similar structurally and electronically.

24.46 $HO_2CCH_2CH_2CH(^+NH_3\,Cl^-)CO_2H$ + $H_3\overset{+}{N}CHCO_2H$ (with CH_2–imidazolium ring: NH, $H\overset{+}{N}$) + $H_2\overset{+}{N}-CHCO_2H$ (pyrrolidine ring) + $NH_4^+\,Cl^-$

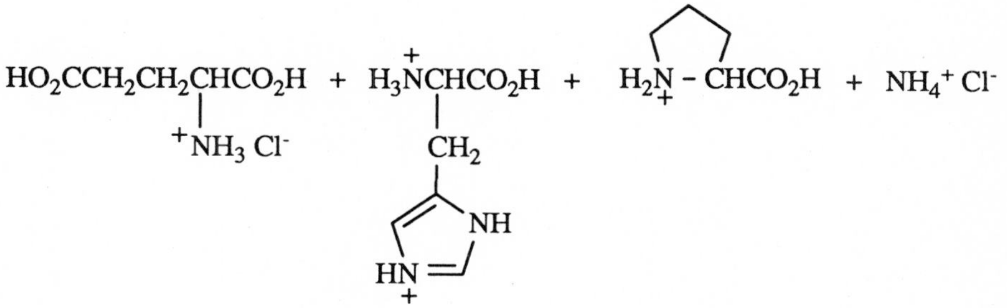

24.47 (a) $(HO_2C)_2CHCH_2CH(NH_2)CO_2H \xrightarrow[\text{heat}]{H^+} HO_2CCH_2CH_2CH(^+NH_3)CO_2H + CO_2$

A β-dicarboxylic acid can undergo decarboxylation when heated.

(b) $(HO_2C)_2CHCH_2CH(NH_2)CO_2H + 3\,OH^- \longrightarrow (^-O_2C)_2CHCH_2CH(NH_2)CO_2^- + 3\,H_2O$

We would expect the three carboxyl groups to form three stable carboxylate groups in base.

24.48 The peptide is cyclic. For example,

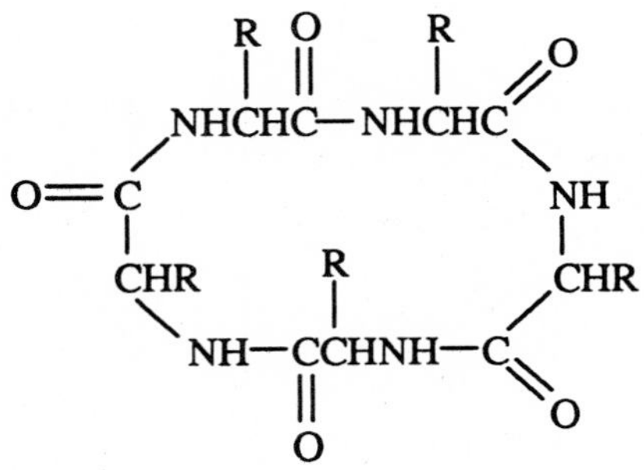

cyclo(–NHCHRC(=O)–NHCHRC(=O)–NHCHRC(=O)–NHCHRC(=O)–)

24.49 (a) Lysine contains a side-chain amino group as well as an alpha amino group. (This side-chain group is more basic than the alpha group.) By contrast, histidine contains (besides the alpha amino group) two sp^2 ring nitrogens. One of these ring nitrogens is not basic (*cf.* pyrrole, Section 18.9). The other ring nitrogen, being sp^2, is less basic than lysine's side-chain amino group is. Histidine thus has an isoelectric point closer to those of neutral amino acids.

(b) The side-chain of arginine is more basic than that of lysine because of resonance stabilization of the cation.

lysine: $RCH_2\ddot{N}H_2 + H_2O \rightleftharpoons RCH_2\overset{+}{N}H_3 + OH^-$

arginine: $R\ddot{N}HC(=\ddot{N}H)\ddot{N}H_2 + H_2O \rightleftharpoons OH^-$

$+ \left[R\ddot{N}HC(=\overset{+}{N}H_2)\ddot{N}H_2 \longleftrightarrow R\ddot{N}H-C(:NH_2)=\overset{+}{N}H_2 \longleftrightarrow R\overset{+}{N}H=C(:NH_2)\ddot{N}H_2 \right]$

24.50 (a) 1-Fluoro-2,4-dinitrobenzene reacts with amino groups. The peptide ala-lys-ala-gly has two free amino groups--one as part of the *N*-terminal amino acid and one on the side-chain of lys. Therefore, the two dinitrophenylamino acids are the following:

2,4-$(O_2N)_2C_6H_3$–$NHCH(CH_3)CO_2H$

from ala

2,4-$(O_2N)_2C_6H_3$–$NH(CH_2)_4CH(NH_2)CO_2H$

from lys

(b) No, the product from lysine would not interfere with the structure determination. Only the amino acid with the dinitrophenyl group bonded to the *alpha* amino group is the *N*-terminal amino acid.

24.51 (a) Because the reaction is slow for acidic and basic D-amino acids, but fast for neutral D-amino acids, we conclude that electronic effects are more important in determining the rate.

(b) For the same reason, we predict that the binding site is nonpolar.

(c) D-Cysteine, a neutral amino acid, undergoes oxidation faster than D-arginine, a basic amino acid.

24.52 val-[glu-leu]-lys-phe-tyr-asp-ala-gly or val-[leu-glu]-lys-phe-tyr-asp-ala-gly

The two possibilities could be differentiated in a number of ways. One of the simplest is *N*-terminal analysis (two cycles) to see if leu or glu is the second residue.

24.53 $H_2NCH(CO_2H)CH_2CH_2CH_2NH$ ⁄ $C(=NH)NH_2$ $\xrightarrow[\text{enzyme}]{H_2O}$ $H_2NC(=O)NH_2$ + $H_2NCH(CO_2H)CH_2CH_2CH_2NH_2$

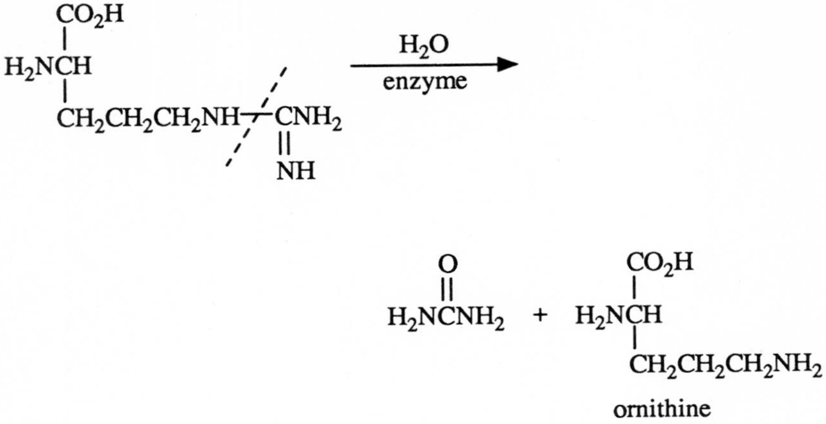

ornithine

24.54 ~~~ $CH_2CH_2CH_2\ddot{N}H_2$ + $\overset{\delta-}{H_2N}-\overset{\delta+}{C}\equiv\overset{\delta-}{N}:$ $\longrightarrow$

~~~ $CH_2CH_2CH_2\overset{+}{N}H(H)$–$H_2NC=\ddot{N}:^-$ ($\rightarrow H^+$) $\longrightarrow$ ~~~ $CH_2CH_2CH_2\ddot{N}H$–$H_2NC=\ddot{N}H$

24.55 (a) glu-cyS-gly

(b) $HO_2CCH(NH_2)CH_2CH_2C(=O)\text{-}NHCH(CH_2SH)C(=O)\text{-}NHCH_2CO_2H$
~~~

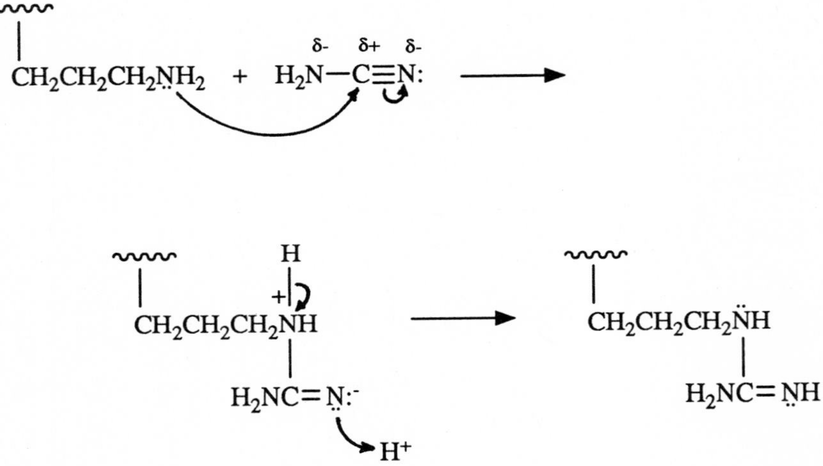

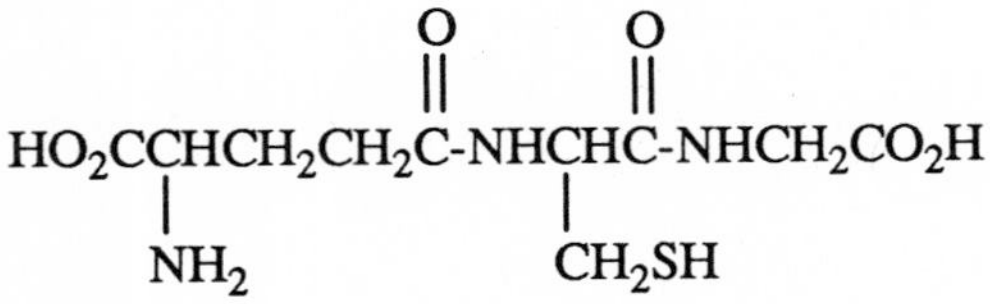

Chapter 25

Nucleic Acids

25.5 (a) [structure: furanose ring with $HOCH_2$ and OH at top, $^{2-}O_3PO$ at bottom]

HOCH$_2$ OH

O

$^{2-}O_3PO$

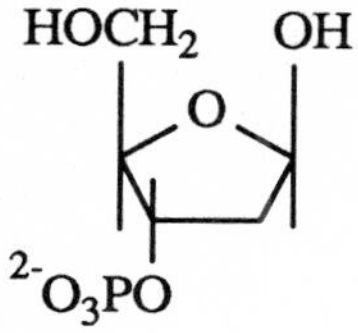

(b) [structure]

$^{2-}O_3POCH_2$ OH

O

OH

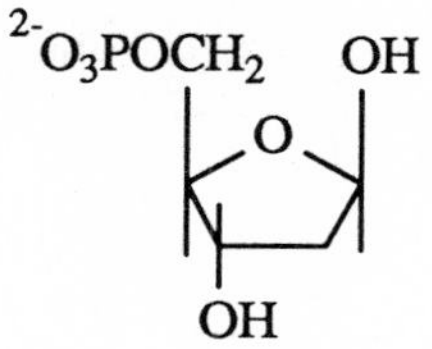

25.6

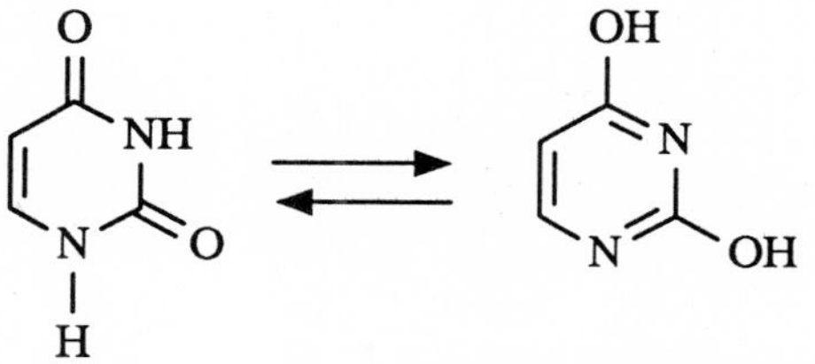

O

NH

N O

H

uracil

OH

N

N OH

tautomer of uracil

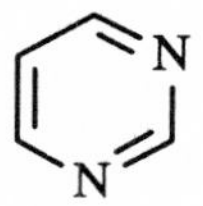

N

N

pyrimidine

25.7 (a)

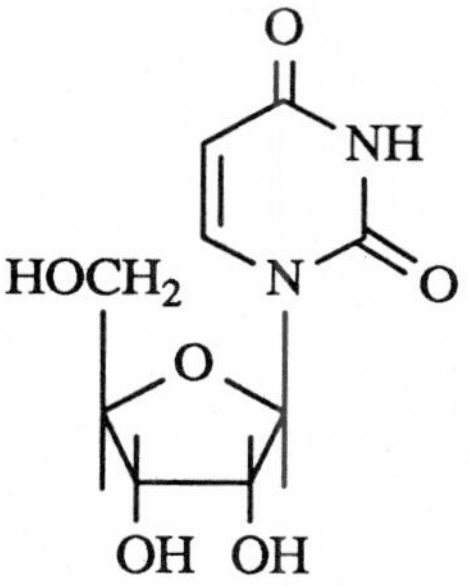

O

NH

HOCH$_2$ N O

O

OH OH

(b)

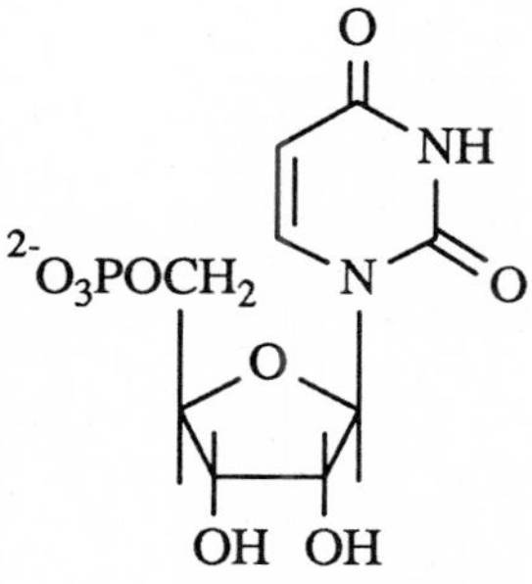

O

NH

$^{2-}O_3POCH_2$ N O

O

OH OH

a 5′-nucleotide

O

NH

HOCH$_2$ N O

O

$^{2-}O_3PO$ OH

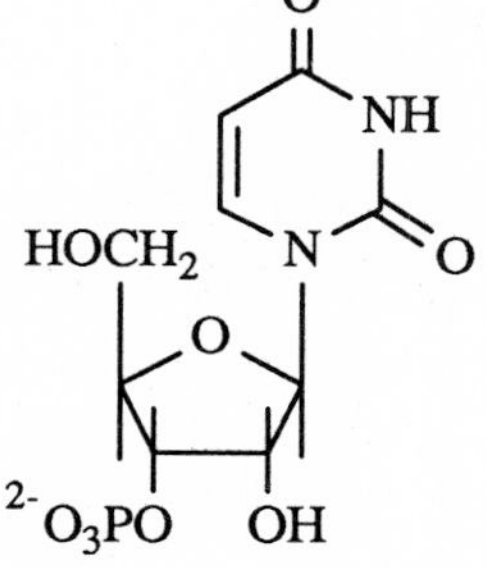

a 3′-nucleotide

25.8

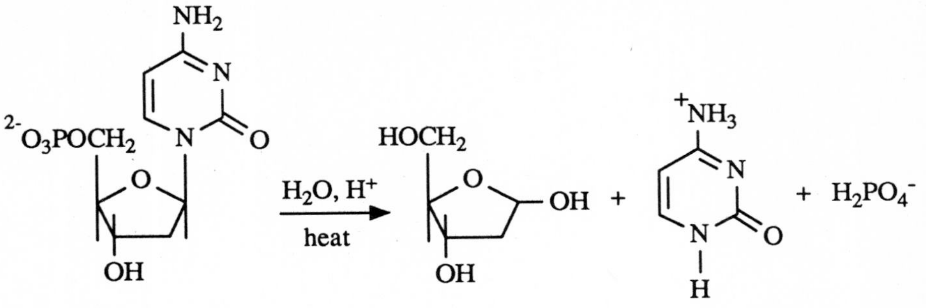

$$^{2-}O_3POCH_2\text{-(2-deoxycytidine)} \xrightarrow[\text{heat}]{H_2O,\ H^+} HOCH_2\text{-(2-deoxyribose)-OH} + \text{cytosine}\ (^+NH_3) + H_2PO_4^-$$

2-deoxycytidine 5′-phosphate — 2-deoxyribose — cytosine

25.9

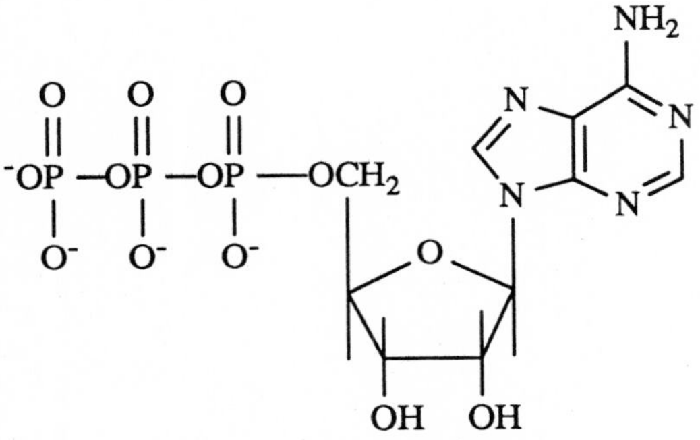

$^-$OP—OP—OP—OCH$_2$ (each P with =O and O$^-$), adenine (NH_2) on ribose with OH OH

adenosine 5′-triphosphate

25.10 Guanine and cytosine can form three hydrogen bonds while adenine and cytosine can form only one.

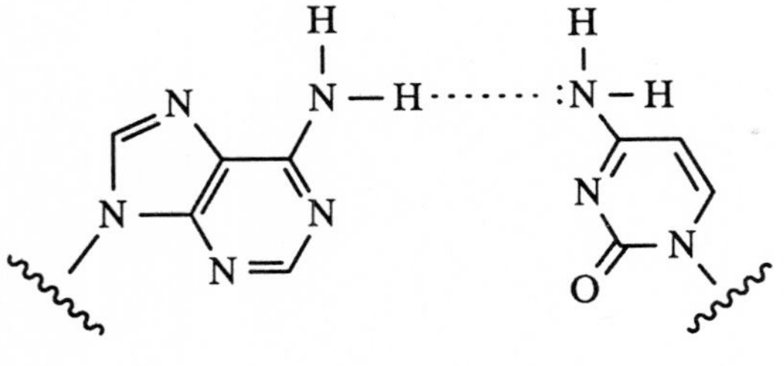

N—H······ :N—H

adenine — cytosine

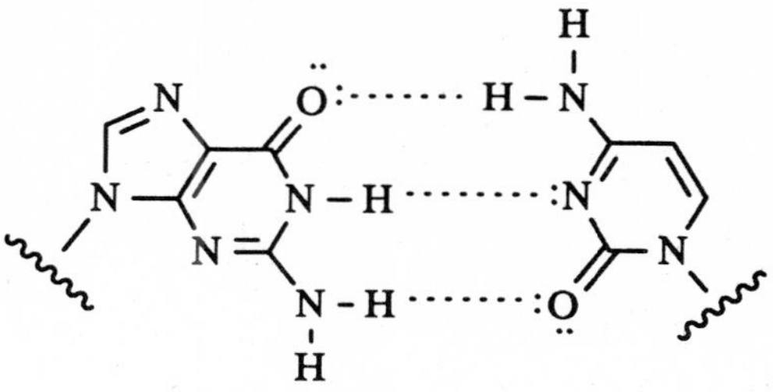

guanine cytosine

25.11 (a)

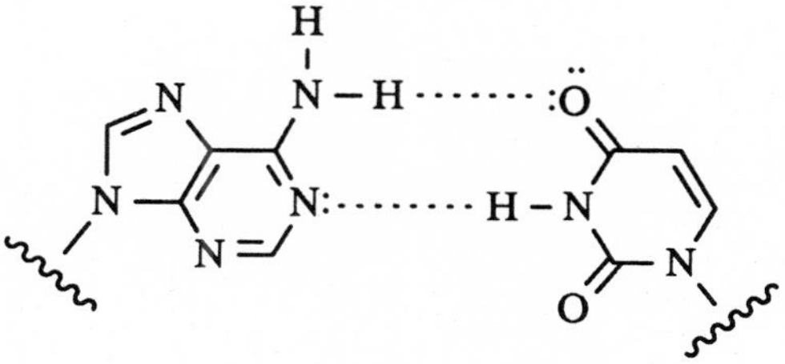

adenine uracil

(b)

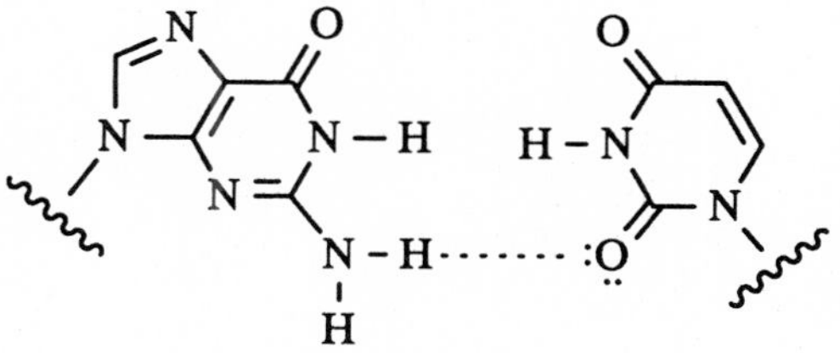

Guanine and uracil are unsuitable for forming more than one hydrogen bond.

(c)

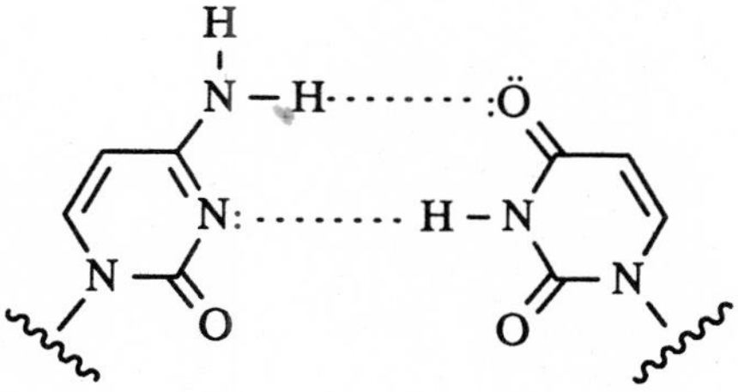

cytosine uracil

25.12 Only uracil and adenine for two reasons: (1) Natural base pairing occurs only between pyrimidine and purine bases. (2) Of the two potential pyrimidine-purine base pairs in the problem, only uracil and adenine can form two hydrogen bonds.

25.13 The guanosine-cytosine base pair is joined by three (rather than two) hydrogen bonds. More energy as heat is needed to unravel (denature) the two chains of a DNA if they contain a preponderance of these hydrogen-bonded base pairs.

25.14 tRNA < mRNA << DNA

25.15 *replication:* the creation of two daughter DNA helices from a single helix

transcription: the synthesis of mRNA using a small segment of DNA as the template

translation: the synthesis of protein using mRNA as the template

25.16 (a) 5′-A T T C G G T A T-3′ DNA
3′-T A A G C C A T A-5′ complementary strand of DNA

(b) 3′-U A A G C C A U A-5′ complementary RNA strand
(In RNA, uracil replaces thymine.)

25.17 (a) TAA, TAG, TAT (b) UAA, UAG, UAU

25.18 (a) (b)

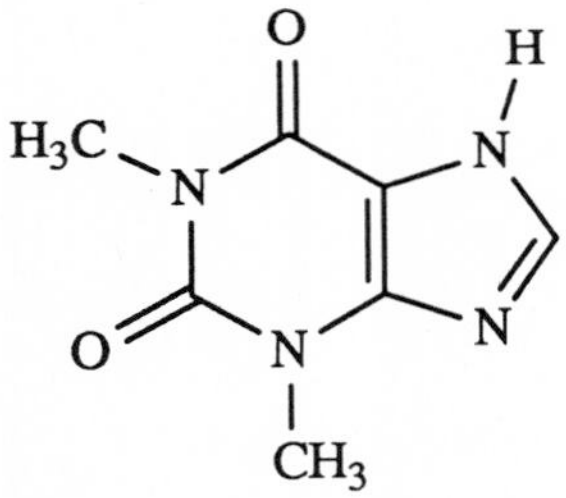

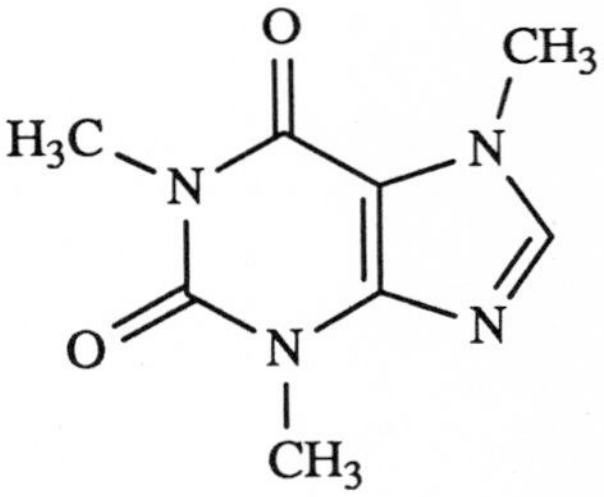

25.19 (a) phe-tyr-ser (b) phe-tyr-ser (same sequence)

25.20 To signal valine instead of glutamic acid, the central base in the codon in RNA must be changed from A to U. Therefore, in DNA, the central base corresponding to this codon must be adenine instead of the correct thymine. Instead of containing the correct sequence (CTT or CTC), the defective DNA contains the incorrect CAT or CAC.

25.21 Codons for glu are GAA and GAG. Codons for arg are CGU, CGC, CGA, CGG, AGA, AGG. The possible combinations for glu-arg are:

glu	arg	glu	arg
GAA	CGU	GAG	CGU
GAA	CGC	GAG	CGC
GAA	CGA	GAG	CGA
GAA	CGG	GAG	CGG
GAA	AGA	GAG	AGA
GAA	AGG	GAG	AGG